Environmental Division 2017

Core Programming Area at the 2017 AIChE Annual Meeting

Minneapolis, Minnesota, USA
29 October - 3 November 2017

ISBN: 978-1-5108-5798-8

Printed from e-media with permission by:

Curran Associates, Inc.
57 Morehouse Lane
Red Hook, NY 12571

Some format issues inherent in the e-media version may also appear in this print version.

Printed by Curran Associates, Inc. (2018)

For permission requests, please contact AIChE
at the address below.

AIChE
120 Wall Street, FL 23
New York, NY 10005-4020

Phone: (800) 242-4363
Fax: (203) 775-5177

www.aiche.org

Additional copies of this publication are available from:

Curran Associates, Inc.
57 Morehouse Lane
Red Hook, NY 12571 USA
Phone: 845-758-0400
Fax: 845-758-2633
Email: curran@proceedings.com
Web: www.proceedings.com

TABLE OF CONTENTS

Some abstracts in this book refer to images that were not made available by the publisher

491441 Synthesis of Graphene Based Iron-Copper Bimetllic Nanocomposite and Its Environmental Applications

Abstract:

Magnetic bimetallic Fe-Cu nanoparticles (NPs) supported on the reduced graphene oxide (rGO-FeCu) were synthesized successfully, and characterized by transmission electron microscopy (TEM) and X-ray diffraction (XRD). At the same concentration and incubation time, rGO-FeCu showed much higher antibacterial activity compared to the rGO-supported iron. To better understand the antibacterial effect of rGO-FeCu, antibacterial activities of NPs with various compositions towards *Escherichia coli* were analyzed. Results show that the antibacterial activity is time and concentration dependent. rGOFeCu with 5% copper loading exhibits excellent removal efficiency of TCA as compared to others. This rGO-FeCu represents a highly potential material for antibacterial activity and remediation of other contaminants in wastewater.

490876 Chemisorption of Elemental Mercury Onto $CuCl_2$-Impregnated Activated Carbon Sorbent

Abstract:

Elemental mercury (Hg(0)) is an environmentally toxic heavy metal that is released during the combustion of coal in power plants. It is imperative that Hg(0) be removed due to its toxicity and bioaccumulation. Since oxidized mercury (Hg(2+)) is highly soluble in water and can be easily adsorbed, it is normal practice to remove mercury by oxidizing the elemental form. In our studies, $CuCl_2$-impregnated activated carbon (AC) sorbent has shown great potential to oxidize Hg(0) without losing any chlorine species in a typical flue gas temperature window while AC sorbent adsorbs the oxidized form. The $CuCl_2$/AC sorbent with three different $CuCl_2$ loadings of 0.2, 0.6, and 1%(wt) was prepared using wetness incipient method. The breakthrough curves for Hg(0) and Hg(2+) were determined as a function of $CuCl_2$ loading. A rigorous mathematical model was developed to describe the oxidation of Hg(0) with $CuCl_2$ and the subsequent adsorption of Hg(2+) based on our previous detailed kinetic studies of the oxidation and adsorption. In this presentation, the kinetic model results for Hg(0) oxidation reaction followed by subsequent Hg(2+) adsorption will be compared with experimental results for the prediction of Hg(0) chemisorption onto $CuCl_2$/AC sorbent.

502782 CO_2 Driven Chemical Looping Gasification on Doubly Doped Ceria

Abstract:

Climate change is one of the most talked about issues in today's world. Scientists and researchers have been putting their best efforts to find an effective method to capture CO2 emitted by fossil fuels as a viable solution to this important matter. Chemical Lopping Combustion (CLC) is an advanced combustion system that facilitates the efficient capture of CO_2.[1] However, CLC does not propose a solution for CO2 utilization. Replacing air with CO2 in oxidation reaction, which is called "Chemical Looping Dry Combustion (CLDR)" has emerged as an interesting way to sequester the CO2. Reducing CO2 to CO in the oxidation reactor produces a useful and versatile gas, which can be the feed gas into many chemical processes. [2][3]

We propose to use a doubly doped ceria system that would have high oxygen capacity as well as high oxygen uptake and release rates in oxidizing and reducing atmospheres respectively. To aid in the increase in oxygen capacity as well as the oxygen release rates, we have doped ceria with zirconia. Additionally, to enhance the O_2 uptake, we have doped the zirconia doped ceria with iron while the same was doped with copper to facilitate the oxygen release in the fuel reactor. It should be noted that the level of doping allows the solids to maintain the cubic flurorite structure of Ce_{O2} The results from these studies at different doping levels were then compared and an optimal oxygen carrier was designed, synthesized and evaluated. The results of these CLDR studies and the characterization of the materials will be presented.

References:

[1] Hossain, Mohammad M., and Hugo I. de Lasa. "Chemical-looping combustion (CLC) for inherent CO2 separations—a review." *Chemical Engineering Science* 63.18 (2008): 4433-4451.

[2] Song, C. S. Global challenges and strategies for control, conversion and utilization of CO2 for sustainable development involving energy, catalysis, adsorption and chemical processing. *Catal. Today* 2006, *115*, 2-32. 17.

[3] Teuner, S. Make CO from CO2. *Hydrocarb. Process.* 1985, *64*, 106-107.

499291 Mercury Capture from Coal-Fired Power Plant Flue Gas Using Zeolite Based Sorbent Polymer Composite (SPC) Materials

Abstract:

The recently implemented Mercury and Air Toxics Standard (MATS) regulation requires that coal-fired power plants in the US achieve very low mercury emissions. Similar regulations are about to be implemented in the EU and several Asian countries. To address this issue several technologies are developed and implemented (i.e. activated carbon injection, use of chemical additives etc.). W.L. Gore & Associates developed an innovative technology, called the GOREâ„¢ Mercury Control System (GMCS) [1], for capturing gas phase mercury from industrial flue gas. The GMCS is a unique fixed sorbent system that also provides SO_2 removal as a co-benefit. It is based on discrete stackable modules comprised of Sorbent Polymer Composite (SPC) materials with an open channel geometry. The SPC effectively chemisorbs both elemental and oxidized mercury from the gas phase. At the same time, part of the SO_2 present in the flue gas is converted to sulfuric acid, which is expelled from the material due to the hydrophobic nature of the SPC material. Corrosion-resistant materials of construction are utilized due to the sulfuric acid generated in the process.

In this talk, the mercury capture technology based on the GMCS will initially be presented and its main features and full scale performance will be highlighted. The second part of the talk will place emphasis on our recent efforts exploring the use of modified zeolites as sorbents in the SPC material. Some zeolites do not catalyze SO_2 conversion to sulfuric acid, which could lead to advantages in certain applications for the GMCS. The evaluation of zeolite powders and zeolite based SPC composites in the lab will be presented and discussed.

References

[1] https://www.gore.com/products/gore-mercury-control-systems

CO_2 Capture by Amine-functionalized Metal-Organic Frameworks (MOFs), Graphene Oxide (GO), and MOF/GO Composites

Jeewan Pokhrel[1], Nidhika Bhoria[1], Theodore Tsoufis[2], George Romanos[2], Georgios N. Karanikolos[1*]

[1]Department of Chemical Engineering, The Petroleum Institute, Khalifa University of Science & Technology, Abu Dhabi, UAE

[2]Institute of Nanoscience and Nanotechnology, Demokritos National Research Centre, Athens, Greece

KHALIFA UNIVERSITY OF SCIENCE AND TECHNOLOGY جامعة خليفة للعلوم والتكنولوجيا

جامعة خليفة KHALIFA UNIVERSITY

Background & Objective

The Problem

- CO_2 emissions are being increased due to higher energy consumption, and continued reliance on fossil fuels.
- Global warming consequences. Environmental regulations apply stricter CO_2 permissible levels.
- Current technologies need help... Energy demanding regeneration causes itself a significant carbon footprint

Objective

- Energy-efficient CO_2 capture: Investigation on strategies to combine with (or even replace) current thermally driven capture processes.

Promising option

- High-resolution separation and capture based on preferential adsorption and/or molecular sieving: Adsorbents and Membranes

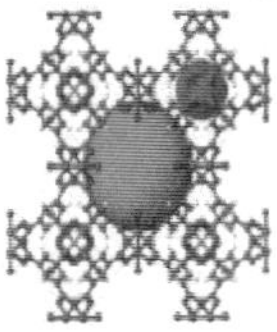

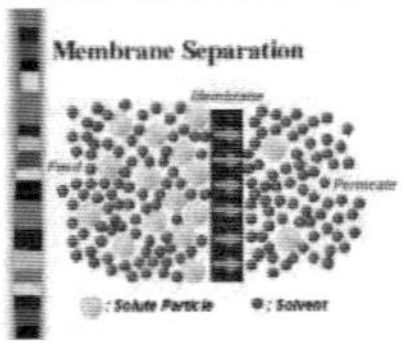

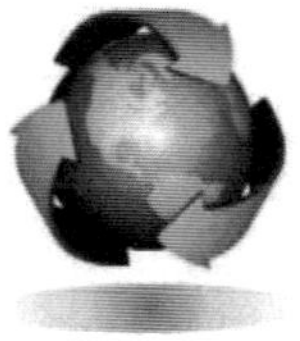

2

CCSU: CO_2 as Enhanced Oil Recovery Medium

- Oil Production: Key role of **EOR**
- Major EOR medium: **CO_2**

➢ ***High extraction cost***

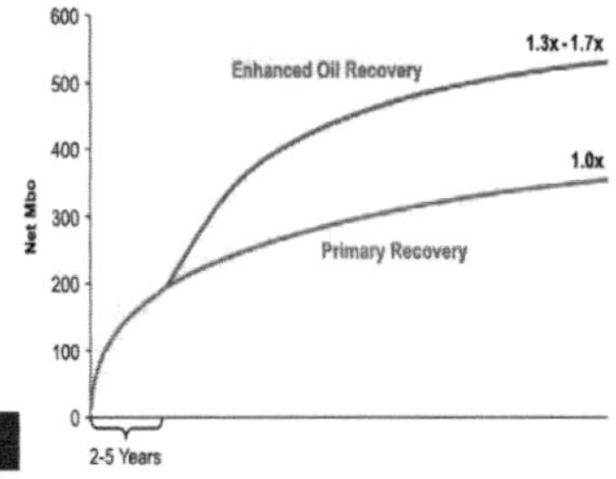

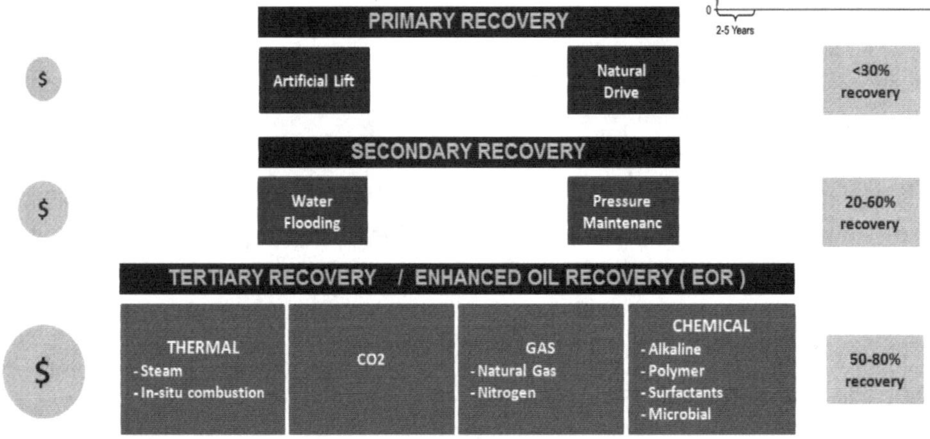

3

Adsorption-based CO_2 Capture

	Absorption	Adsorption
Definition	Absorption is the process in which a gas is dissolved by a liquid or a solid (absorbent, e.g. liquid aqueous amine solutions)	Adsorption is the process in which molecules, atoms, or ions from a substance (it could be gas, liquid or dissolved solid) adhere to a surface of an adsorbent (typically porous)
Phenomenon	It is a bulk phenomenon	It is a surface phenomenon
Heat exchange	Endothermic process	Exothermic process
Rate of sorption	It occurs at a uniform rate (typically diffusion-limited)	It is typically rapid and steadily reaches equilibrium
Energy	High energy consumption for sorption and regeneration of absorbent	Low energy consumption for sorption and regeneration of adsorbent

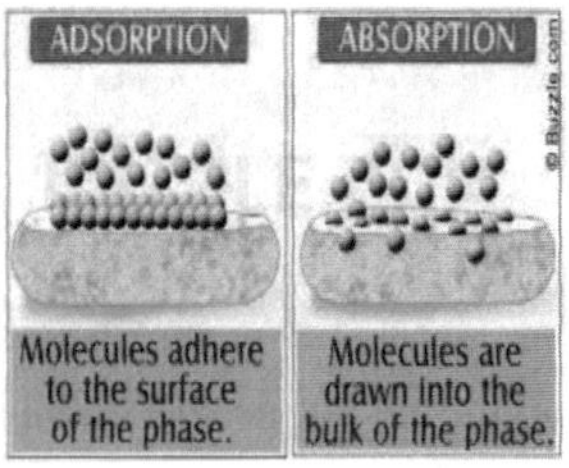

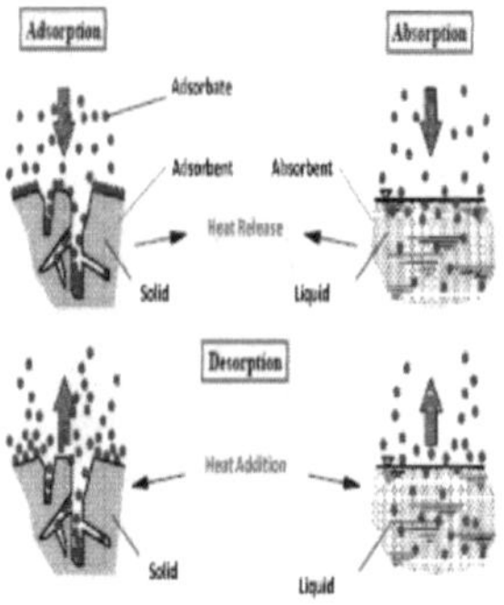

4

Porous Adsorbents

- They exhibit desired characteristics for CO_2 adsorption:
  - Porosity, high surface area, structural diversity.
  - Reversible uptake and release.
  - Low energy consumption in sorption/desorption cycles.
  - Fast kinetics.
- Different adsorbents promising for CO_2 capture:

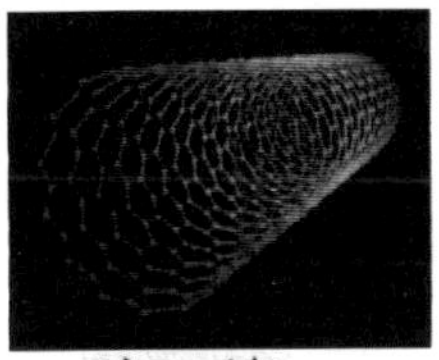
carbon nanotube

zeolite

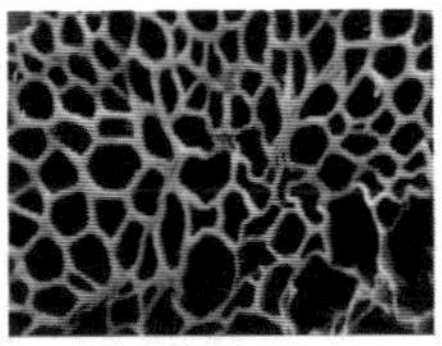
activated carbon

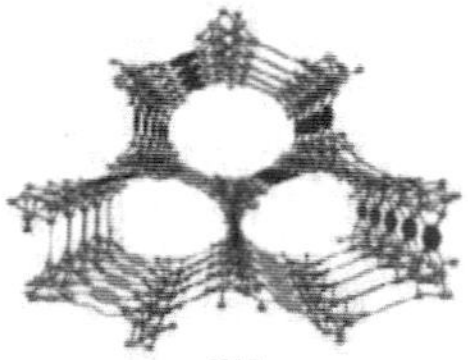
MOF

- For large-scale application, optimal adsorbents should exhibit:
  - High CO_2 capacity, selectivity, chemical and thermal stability, fast kinetics, sustainable performance for many cycles, low energy consumption, low manufacturing cost...Research is needed

5

Focus of current research:
Amine-grafted Graphene, MOFs, and Composites

Graphene Oxide Synthesis:

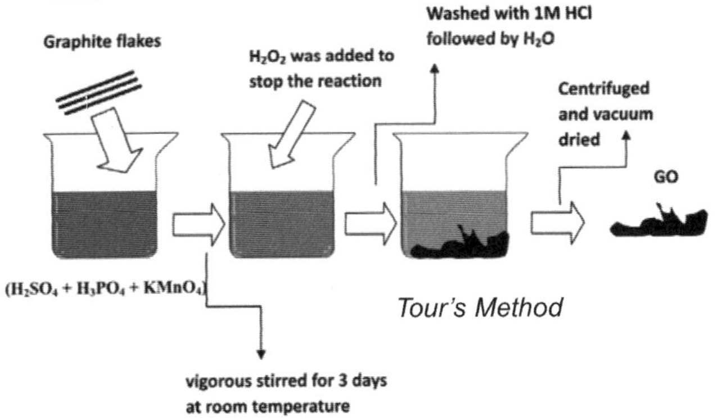

Tour's Method

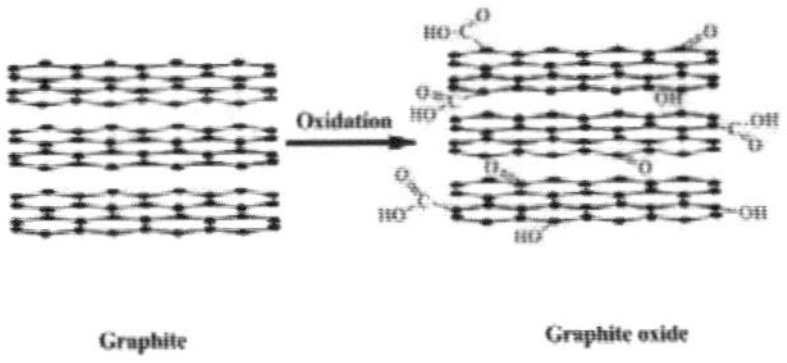

Amine Grafting:

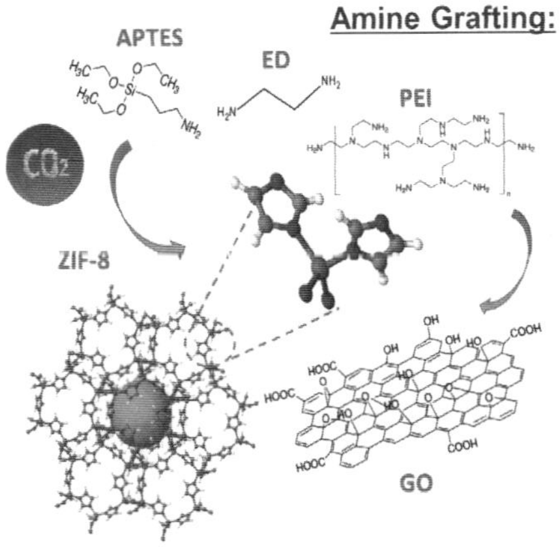

MOF Growth:

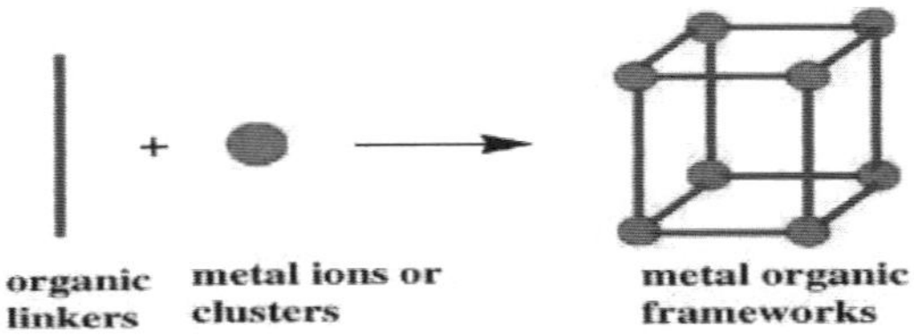

6

Functionalized Graphene Oxide (*f*-GO)

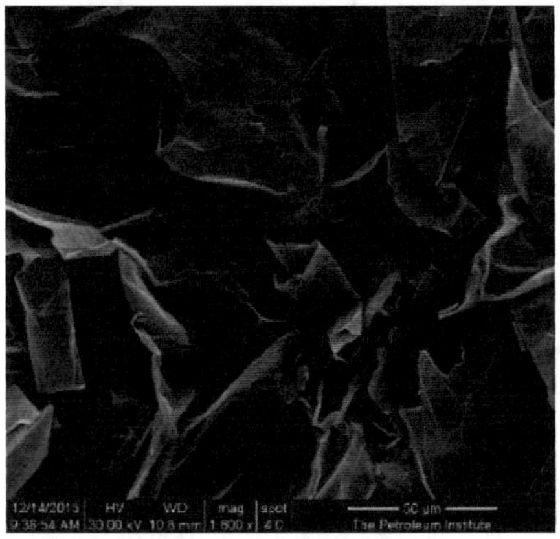

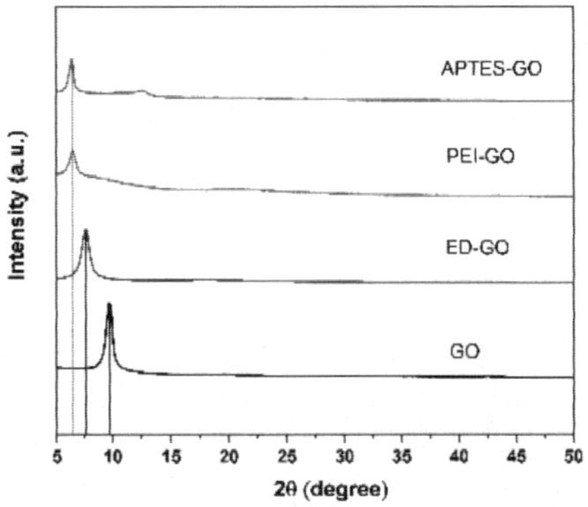

- GO: a well-defined peak at 2θ of 10 degrees.
- For functionalized samples, this peak was shifted to lower angles indicating increase in the interlayer spacing due to the chains of the intercalated amine molecules.

7

Elemental Analysis

Material	% N	% C	% H	O %
GO	1.12	44.46	2.50	51.92
PEI-GO	14.31	58.19	5.99	21.50
ED-GO	5.35	54.63	3.47	36.52
APTES-GO	5.38	42.59	3.90	48.12

- GO exhibits high oxygen content due to large number of carboxyl and epoxy groups present on the surface.
- After amine functionalization, the oxygen content decreases and the nitrogen content increases as the –COOH is converted to –CONH groups.
- PEI-GO exhibited the highest increase in nitrogen content compared to APTES-GO and ED-GO.

8

FTIR Analysis

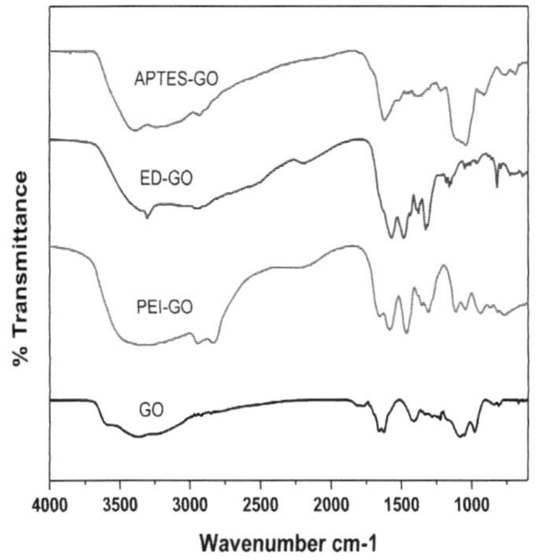

- GO:
OH stretching vibrations at 3430 cm^{-1}, C=O stretching vibration in the range of 1720-1740 cm^{-1}, C=C from unoxidized sp^2 bonds in the range of 1590-1620 cm^{-1} and C-O vibrations at 1250 cm^{-1}

- Amine- functionalized GO:
Additional peaks corresponding to vNH at 3294 cm^{-1}, NH bending at 1560 cm^{-1} and C=O stretching (amides) at 1650 cm^{-1}

9

CO_2 Adsorption on *f*-GO

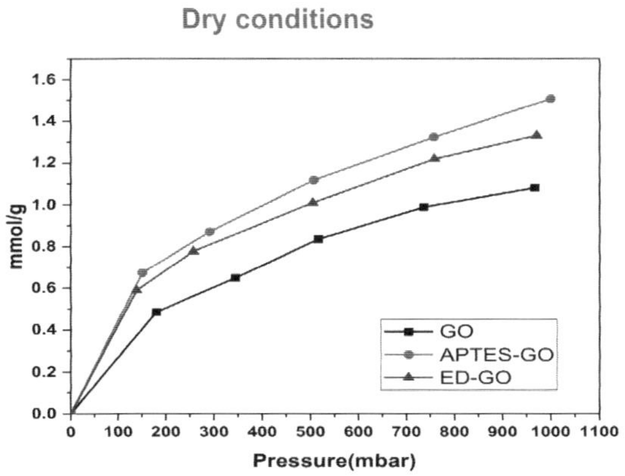

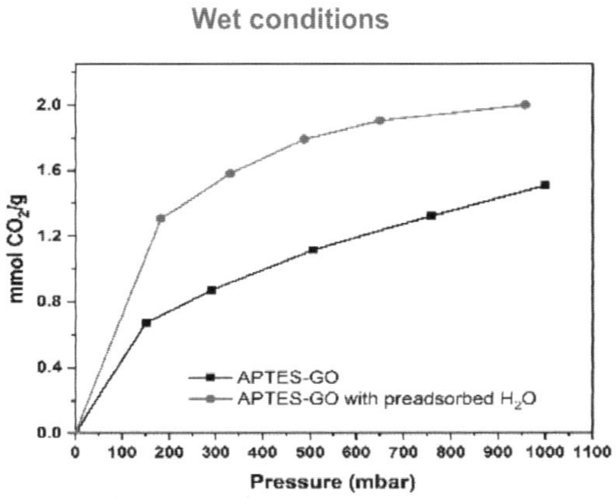

- APTES-GO > ED-GO > pure GO, under dry conditions.
- APTES-GO with pre-adsorbed water exhibits enhanced capacity. This is attributed to the formation of bicarbonates under humidity conditions that are able to engage additional CO_2.

10

Functionalized MOFs (*f*-MOFs)

ZIF-8 precursors: $Zn(NO_3)_2 6H_2O$ and 2-methylimidazole

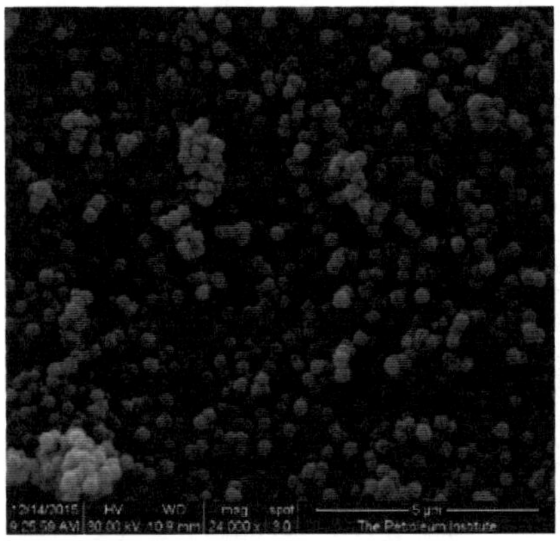

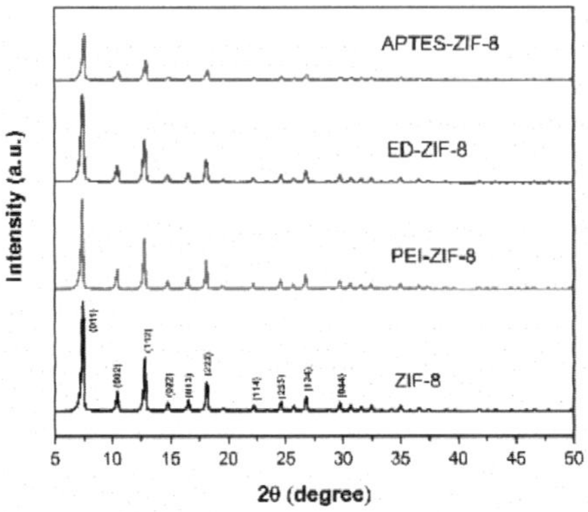

- Well-crystalline and monodisperse particles.
- Crystallinity is preserved after functionalization using three amines, namely, APTES, PEI, and ED.

11

CO_2 Adsorption on *f*-ZIF-8

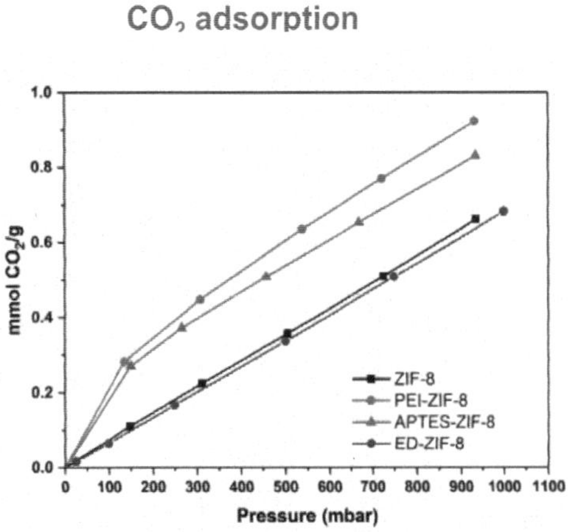

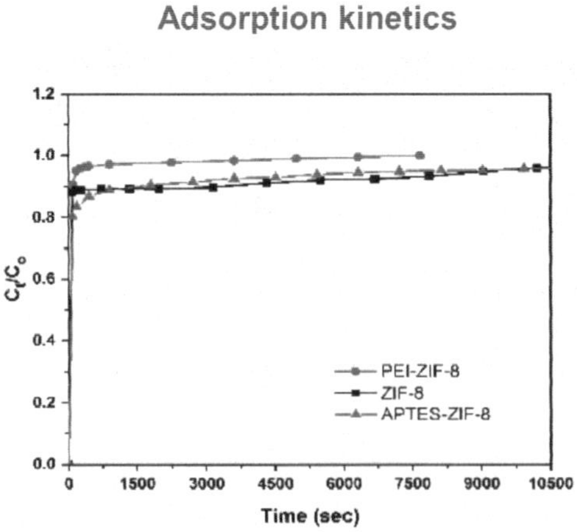

- PEI functionalization results in the highest capacity: 50% enhancement compared to unmodified ZIF-8.
- Fast kinetics: Maximum capacity reached in ~5 min.

12

MOF-GO composites

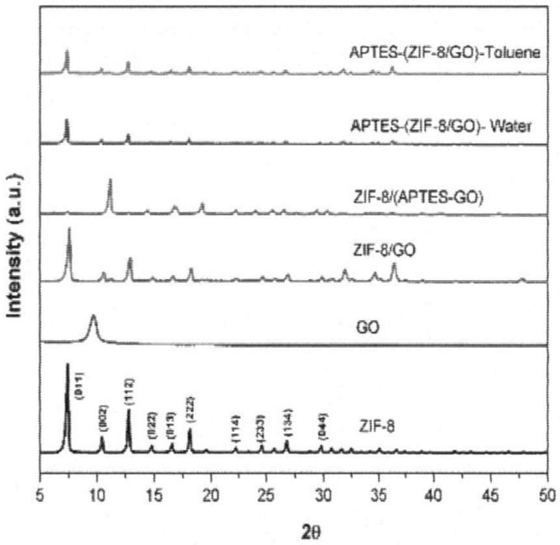

- "Sandwich"-type adsorbent consisting of graphene flakes intercalated *in situ* by ZIF-8 crystallites.
- MOF crystallinity is preserved. GO peak does not exists in the composites indicating complete exfoliation of the graphene sheets due to the intercalation of the MOF crystallites.

13

CO_2 Adsorption on ZIF-8/GO

CO_2 adsorption

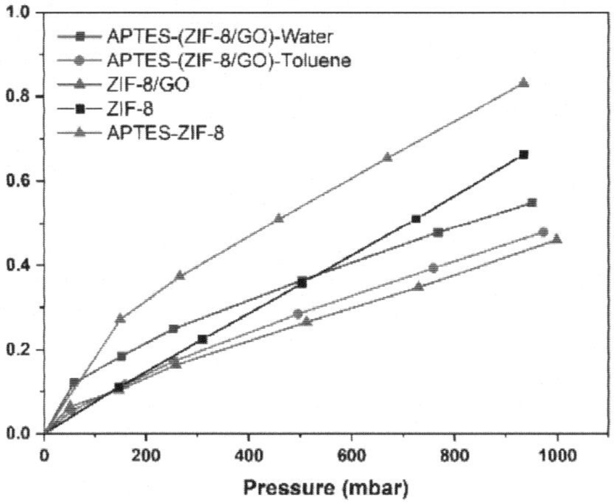

Adsorption kinetics

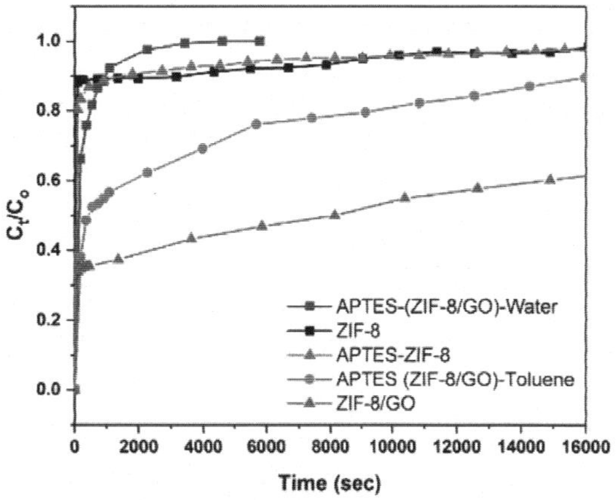

- APTES-(ZIF-8/GO) exhibits higher capacity than ZIF-8 at low pressures
- APTES-ZIF-8 still best adsorbent

14

MOF-GO composites

HKUST-1

- Solvothermal growth
- Copper nitrate hexahydrate and 1,3,5-benzene tri-carboxylic acid in DMF,ethanol and water
- Mix, stir and sonication
- Heat at 85 °C for 20 h under reflux
- Wash with dichloromethane
- Dry the crystals at 170 °C for 28 h under vacuum
- **Blue** HKUST-1 crystals

Cu based nodes

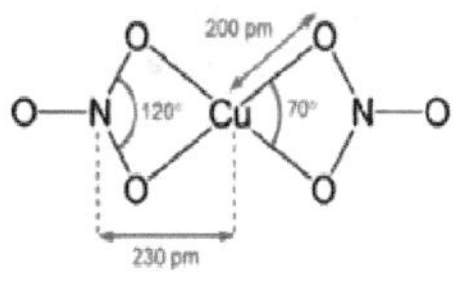

Organic linker

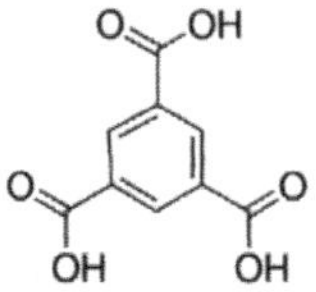

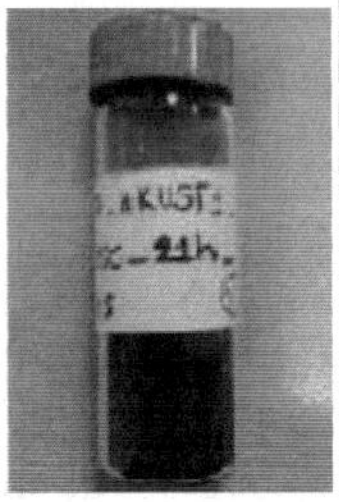

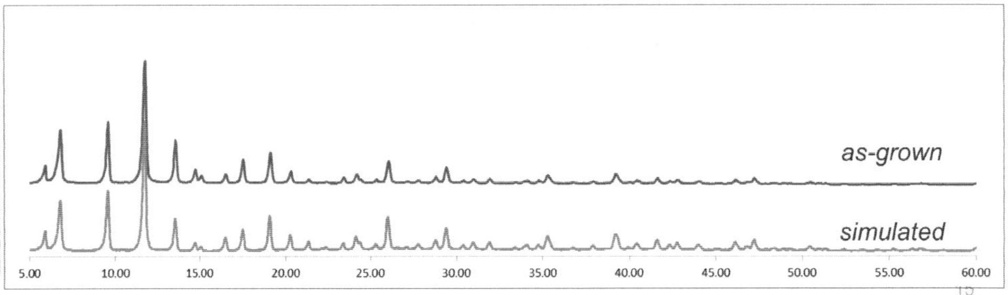

15

HKUST-1/GO Composites

Replace the common copper nitrate hexahydrate metal precursor with copper nitrate hemi-pentahydrate, and apply post-growth treatment with ammonium chloride.

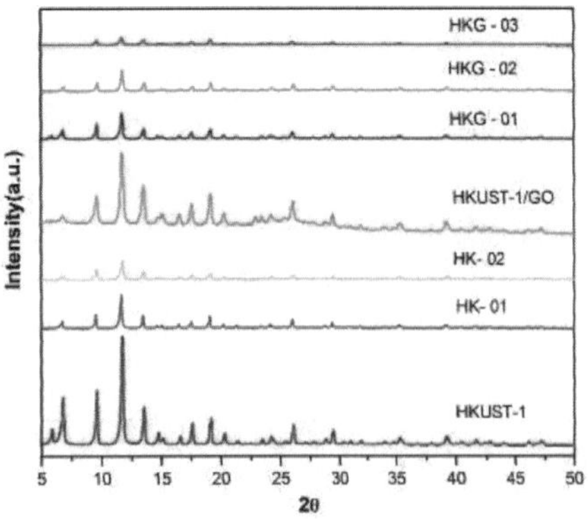

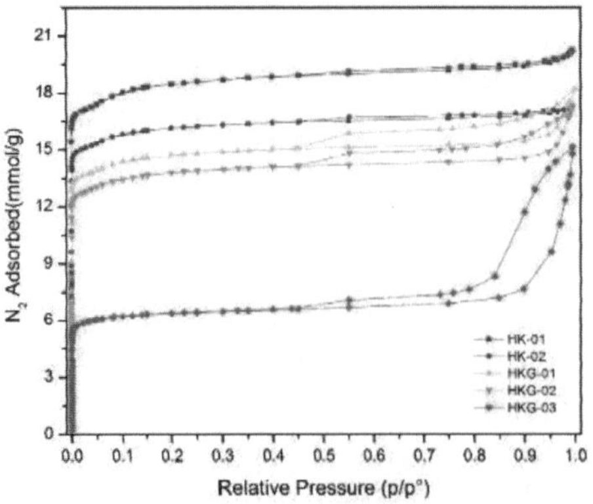

- BET surface area of HKUST-1 reaches 1380 m^2/g, HKUST-1/GO sample with 1% of GO is 1096 m^2/g, with 2% of GO is 1027 m^2/g and with 3% of GO is 474.7 m^2/g
- Hysteresis in HKUST-1/GO samples indicative of mesopores formed in GO/MOF interstitial spaces

16

CO_2 adsorption and cyclability of HKUST-1 and HKUST-1/GO

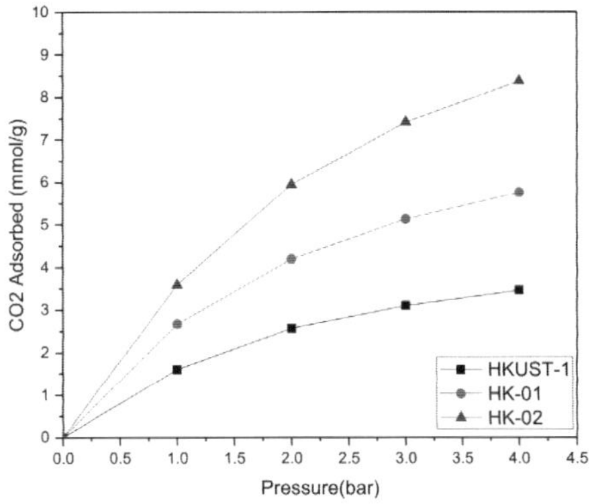

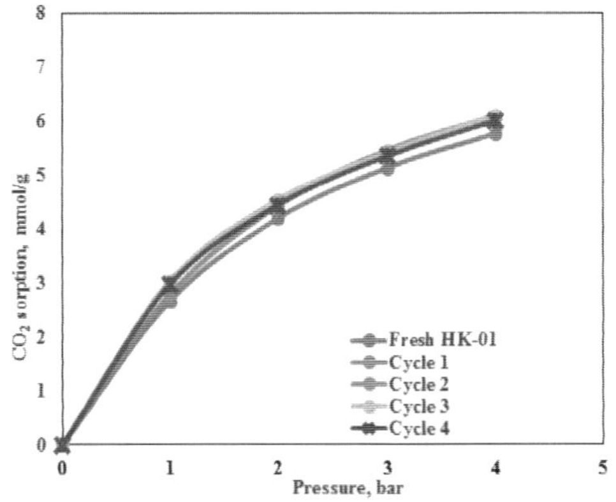

- Higher capacity of improved HKUST-1 (~3.7 mmol/g at 1 bar and 8.5 mmol/g at 4 bar)
- Capacity unchanged upon adsorption/desorption cycles

17

GO Reduction

- The increase in capacity of the amine-functionalized adsorbents under humid conditions is attributed to the formation of bicarbonates that are able to engage additional CO_2.
- In the amine-functionalized solid adsorbents, water is beneficial for the kinetics, stability, and energy demand during regeneration as urea-type linkage is avoided. Bond strength is weaker than in carbamates.
- However: The amount of co-adsorbed water should be kept at minimum required (~25% of the adsorbed CO_2), otherwise a lot of energy must be spent to evaporate the excess water.
- Solution: Apply reduction to the *f*-GO to get rid off un-grafted carboxylic groups

18

Post reduction with Sodium Borohydride - Effect on H_2O adsorption

Grafting of PEI and DAB on the GO is achieved via chemical bonds between the primary amine groups and the epoxy groups of GO. Therefore the carboxylic functional groups are useless and only add to the hydrophilicity of the adsorbent.

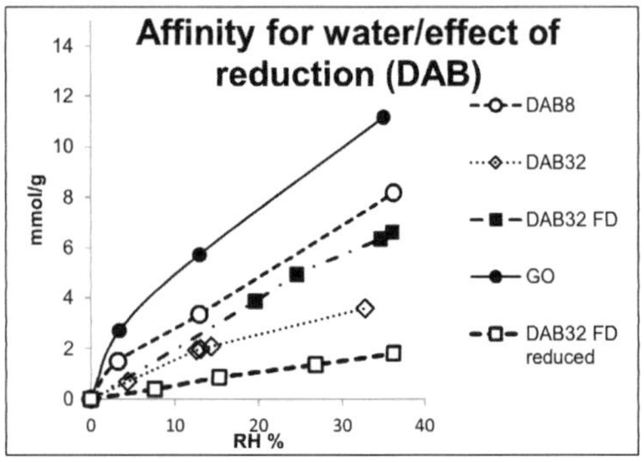

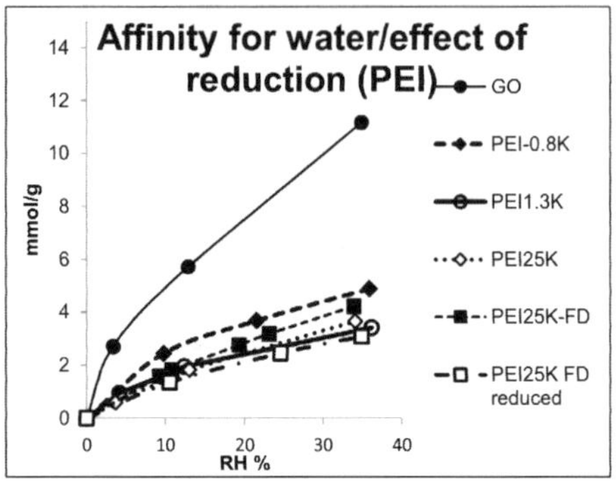

- Significant decrease in hydrophilicity upon reduction, particularly in the DAB32 FD and PEI25K FD
- H_2O/CO_2 molar ratio: from 3.1 to 0.78 for DAB 32 and from 1.14 to 0.88 for PEI25K

19

CO_2 adsorption on the reduced adsorbents

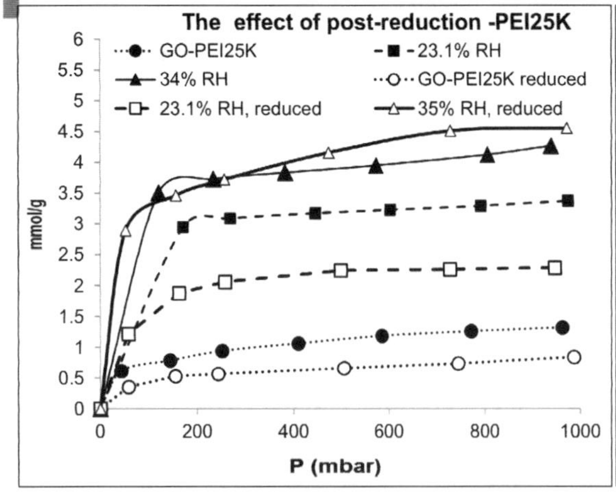

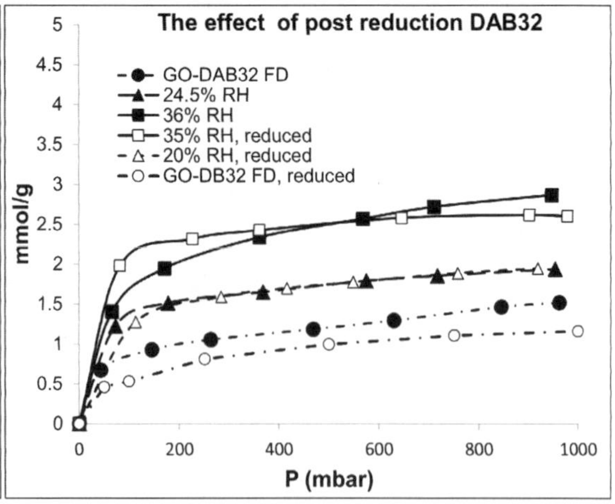

✓ Insignificant impact of reduction on the ability of the materials to adsorb CO_2

20

Conclusions

- Amine-functionalized GO, MOF and MOF/GO composite adsorbents were developed and studied for CO_2 adsorption under dry and wet conditions.
- Most amine-functionalized materials, such as *f*-GO, exhibited capacity enhancement in the presence of water.
- This is attributed to the formation of bicarbonates that are able to adsorb more CO_2, and is important as most CO_2 emission mixtures contain water.
- co-Adsorption of excess of water was controlled by reduction post-treatment of the functionalized materials to minimize available –COOH groups in the GO surface.
- This controlled treatment significantly decreased amount of water adsorbed without affecting CO_2 capacity.

21

THANK YOU

Funding: ADNOC R&D

22

500148 Photodegradation of Pharmaceuticals in Partially Nitritated Wastewater

Abstract:

Certain pharmaceutically active compounds may not be readily removed in anaerobic wastewater treatment, and the effluents from anaerobic wastewater treatment processes often contain high levels of ammonia. A polishing step is therefore necessary to remove nitrogen and trace organic contaminants, in particular for reuse considerations. For nitrogen removal, an increasingly popular, sustainable biotechnology is the anammox process, which which involves a prepatory step for partial conversion of ammonia to nitrite. Given that nitrite is an efficient sensitizer for production of hydroxyl radical, a UV irradiation step after partial nitrification might also be used to create a de-facto advanced oxidation process during disinfection to remove pharmaceuticals. The direct and indirect photolysis of select pharmaceutical compounds was investigated in systems containing nitrite, ammonia, and effluent collected from a bench-scale reactor using a 450 W medium-pressure mercury vapor lamp. Results indicate that the presence of nitrite enhances the removal of pharmaceutical compounds.

493283 Oxidation of High Molecular Weight Dye Using Ozone Isotopomers

Abstract:

Ozone has several isotopomers ranging in mass from $^{48}O_3$ and $^{54}O_3$. These isotopes can be formed in different proportions in the corona discharge process depending on the distribution of oxygen isotopes ^{16}O: ^{17}O: ^{18}O, in the feed gas. The symmetric ozone molecules $^{48}O_3$, $^{51}O_3$, and $^{54}O_3$ can have different reactivities than the asymmetric ozone molecules $^{49}O_3$, $^{50}O_3$, and $^{52}O_3$. An anomalous distribution of ozone isotopomers has been found in the upper atmosphere and has been investigated to some extent. Laboratory corona discharge studies have shown that there is an anomalous enhancement of certain ozone isotopes in the product gas when natural oxygen is used as feed gas for ozone generation. The goal of this research is to evaluate the effects of an isotopically enriched oxygen feed gas on the enhancement of oxidation of a high molecular weight dye. Experiments were conducted in stirred tank reactors and tubular porous electrode ozonators with ozone generated using different proportions of $^{18}O_2$ and/or $^{17}O_2$ along with $^{16}O_2$ to determine the impact of the ozone isotopomers on oxidation efficiency. Comparative results with ozone generated from natural oxygen isotopes will be provided.

500071 Removal of Algal Matter By Coagulation Both in Presence and Absence of Pre-Oxidation

Abstract:

Removal of algal matter by coagulation both in presence and absence of pre-oxidation

Ziming Zhao, Hassan Gomaa and Madhumita B. Rayï€[a]

Department of Chemical and Biochemical Engineering, University of Western Ontario, London, Ontario, N6A 5B9, Canada

< style="padding-left: 5pt; text-align: justify;">Keywords: Algal organic matter, coagulation, pre-oxidation

The eutrophication of surface water is a threat to the safety of drinking water supply worldwide. The presence of algae in intake water poses challenges for water treatment facilities in terms of increased coagulant dose due to release of algal organic matters (AOM), membrane fouling, reduced adsorption efficiency in activated carbon, and increased formation of harmful disinfection by-products (DBP). DBP formation increases after the filtration step due to the accumulation of algal matters on the filter and their subsequent release. AOM consists of both extracellular organic matter (EOM) and intracellular organic matter (IOM). Due to their hydrophilic nature, IOM is not easily removed in coagulation causing formation of downstream membrane foulants. There remains a critical lack of understanding on the type and concentration of IOM which cause adverse effects on water treatment performance, as well as DBP formation. *Our initial research* with four different algae indicates that AOM release (measured as dissolved organic carbon (DOC) and subsequent DBP formation significantly depend on the type of algae. For example, *P. Tricornutum* (diatom) (with cell density 2.4 ± 0.2 × 106 cells/mL) released AOM up to 32 mg/L as compared to only 5 mg/L of *C. Vulgaris* (Green algae) (with cell density 2.5 ± 0.1 × 107 cells/mL), with much higher DBP formation potential for diatom. Typical natural organic matter (NOM) in surface water is in the range of 5 mg/L highlighting the impact of AOM on treatment plants. The objective of present work is to determine the optimum coagulation schemes in presence and absence of pre-chlorination and pre-ozonation for algal matters (both AOM and IOM) produced by six different algae namely, *Chlorella vulgaris, Scenedesmus quadricauda, Merismopedia sp, Microcystis aeruginosa, Cyclotella meneghiniana, Phaeodactylum tricornutum.*

The effect of background water quality parameters such as turbidity, dissolved organic carbon (DOC), and alkalinity on the coagulation will be measured. Central composite design of experiments involving pH, dosage of coagulant, algal cell density, dosage of ozone/NaOCl, and concentration of background NOM are being used to determine the optimal coagulation conditions. Standard jar tests are conducted using a Phipps & Bird programmable jar tester, with rapid mixing for 2 min at 150 rpm, followed by mixing at 25

Figure 1. DOC and UV 254 absorbance removal performance of EOM and IOM for

Chlorella Vulgaris

The DOC and UV 254 absorbance removal rate for EOM and IOM is enhanced with the increase of alum dose as shown in Figure 1. Interestingly, DOC removal rate for EOM of*Chlorella Vulgaris* was less than that of IOM probably due to higher amount of hydrophilic compounds present in the EOM solution, which is difficult to remove by tradition coagulation. On the contrary, limited research in literature on this topic indicates that it is IOM which is typically more hydrophilic. However, the UV 254 absorbance removal rate for EOM was higher than that of IOM indicating the greater removal of aromatic and double-bond compounds in EOM than that in IOM. These results indicate the variability of coagulation performance of different species of algae indicating the need for comprehensive research on this topic, which is the objective of the present study.

495748 Ozone Microbubble and Its Use in Arsenic Removal from Waste Water

Abstract:

Microbubble-based methods, in recent times, have been widely used for purification of water and wastewater. Microbubbles have several physicochemical properties, which make them eminently suitable for wastewater treatment. Many practical applications of ozone, oxygen and air microbubbles, some of which are currently at various stages of commercialization, have been presented in literature. In addition, directions for future research of microbubble technology and their potential applications have been identified. Treatment of waste water by ozonation is gaining importance due to its high oxidizing power. Arsenic is one the most toxic elements found in ground water. The ground water is majorly polluted with arsenic with its As(III) and As(V) sources. It is known that As(V) can be more easily removed from water by adsorptive methods than As(III). In this work, oxidation of more toxic As(III) to less toxic As(V) was studied in a pilot-plant by using ozone microbubbles. The microbubbles are widely acceptable in dissolving ozone in water. The role of ozone microbubble in the oxidation of As(III) under acidic conditions is enunciated and kinetics for the advance oxidation is evaluated based on the experimental results. Adsorption of As(V) on zirconium impregnated activated carbon (Zrâ€'AC) and amorphous zirconium oxide (Am-Zr) has been studied in this work. The adsorption of As(V) on both of these adsorbents was fast and effective. The As(V) adsorption capacity of Am-Zr was larger than that of Zrâ€'AC.

487660 An Efficient Degradation of Perchloroethylene By Nano Zero Valent Iron (nZVI) and Nzvi Activated Sodium Percarbonate with and without Addition of Ethylenediaminetetraacetic Acid Disodium Salt Dehydrate

Abstract:

An efficient degradation of perchloroethylene by nano zero valent iron (nZVI) and nZVI activated sodium percarbonate with and without addition of ethylenediaminetetraacetic acid disodium salt dehydrate

Muhammad Danish[1,2], Shuguang Lu[1], Usman Farooq[1], Sadiq Hussain[2], Ayyaz Ahmad[3], Muhammad Naqvi[4], Mujtaba Ashraf [2], Abdul Sattar Qureshi[5],Xiaori Fu[1]

[1]State Environmental Protection Key Laboratory of Environmental Risk Assessment and Control on Chemical Process, East China University of Science and Technology, Shanghai 200237, China

[2]Department of Chemical Engineering, NFC-Institute of Engineering and Technology, Multan, Pakistan

[3]Department of Chemical Engineering, Muhammad Nawaz Sharif University of Engineering and Technology, Multan, Pakistan

[4]Department of Energy, Building and Environment, Mälardalen University, Västerås 72123, Sweden

[5]Institute of Biotechnology and Genetic Engineering, University of Sindh, Jamshoro 76080, Pakistan

Abstract

This study primarily focused on the degradation of perchloroethylene (PCE) by using nano zero valent iron (nZVI) without sodium percarbonate (SPC) in the absence and presence of ethylenediaminetetraacetic acid disodium salt dehydrate (EDTA-Na_2). The degradation of PCE followed pseudo-first-order kinetic behavior in the absence of EDTA-Na_2. The rate constants increased exponentially with the increasing molar ratios of SPC and nZVI. Main by-products found were formic acid and ethylene in both nZVI alone as well as in the nZVI-SPC system. The PCE degradation significantly enhanced over a shorter reaction time of 3 hours in the presence of EDTA-Na_2 by using nZVI system alone whereas, the PCE degradation ceased after 20 hours in long term experiment (120 hours) showing low degradation of PCE than that of no addition of EDTA-Na_2. However, the PCE degradation severely decreased in the nZVI-SPC system. The advantageous aspect of the EDTA-Na_2 can be attributed to the possible prevention of Fe^{+2} and Fe^{+3} precipitation that could minimize the nZVI surface passivation. On the contrary, rapid corrosion of nZVI by EDTA-Na_2 and larger generation of Fe^{+2} ions in a short span of time that can possibly consume the reactive oxygen species at a faster rate as compared to PCE could be the adverse characteristics of the EDTA-Na_2.

487543 Heterogeneous Elemental Mercury Oxidation Kinetics over RuO_2/TiO_2 Catalyst

Abstract:

We have developed a RuO_2/rutile TiO_2 catalyst, which is highly active for Hg(0) oxidation at typical SCR unit temperatures of 300-400 °C. In our previous study, the experimental evidence of dissociative HCl adsorption on the RuO_2/rutile TiO_2 catalyst surface under reaction conditions was obtained using an in-situ Diffuse Reflectance Infrared Fourier Transform Spectroscopy (DRIFTS). In this study, heterogeneous Hg(0) oxidation kinetics over the RuO_2/rutile TiO_2 catalyst was investigated using a fixed-bed system in a temperature range of 300 to 400 °C under a steady-state condition. A kinetic model was developed based on our in-situ DRIFTS study results suggesting an Eley-Rideal mechanism and dissociative HCl adsorption in order to determine the reaction rate constant, activation energy, adsorption equilibrium constants and enthalpy of adsorption of HCl, NH_3 and SO_2. The activation energy value determined for the reaction was a low end value for the heterogeneous Hg(0) oxidation reaction using HCl gas reported in the literature, indicating the effectiveness of the RuO_2/rutile TiO_2 catalyst. The equilibrium adsorption analysis was used to determine the intrinsic reaction rate expression for Hg(0) oxidation kinetics with HCl under the competitive adsorption of NH_3 and SO_2. The kinetic expression could reasonably predict the Hg(0) oxidation performance under the competitive adsorption of NH_3 and SO_2 gases. The intrinsic kinetic expression obtained for the RuO_2/rutile TiO_2 catalyst is expected to be applicable to the design of an Hg(0) catalyst that can be installed at the tail end of a typical SCR unit.

487775 Fundamental Chemistry, Kinetics and Mass-Transfer Aspects of the Emerging Sulfate-Radical Based Environmental Technologies for Pollution Control

Abstract:

Pollutants emitted from combustion sources are of great public concern due to their detrimental effects on human health and ecosystems. Although multipollutants' reductions from flue gas using different processes such as selective catalytic reduction, lime/limestone wet flue gas desulfurization and activated carbon, respectively for NO_x, SO_2 and Hg removal, are efficient purification techniques, these complex treatment processes require high construction and operational costs, and large installation space. It is desirable to develop cost-effective especially those capable of providing simultaneous multipollutants' control. Such technologies, which could also possibly be retrofitted to existing power plant are believed to be more cost-effective. Sulfate-radical ($SO_4^{\bullet-}$) based advanced oxidation processes (AOPs) have drawn increasing attention in research and successful applications in water treatment and soil remediation as well as in flue gas purification for multipollutant treatments. The persulfate (PS) or peroxydisulfate anion ($S_2O_8^{2-}$) is a strong oxidant, kinetically slow at ordinary conditions but activation by heat, ultraviolet (UV), ultrasound, microwave, transition metal ions, electrodes, nanoparticles or base (and alkaline pH) resulting in the generation of $SO_4^{\bullet-}$ and subsequent production of hydroxyl ($OH^{\bullet}$) radicals. As a cheap, naturally abundant and environmentally friendly material, iron (Fe(0), Fe(II), Fe(III)) activation has been preferentially used to generate. The activation by Fe(II) requires a relatively lower energy (14.8 kcal/mol) than the thermal energy (33.5 kcal/mol), so the PS/(Fe(II) reaction system has greater potential for destruction of target pollutants. This work evaluates the chemistry, reaction pathways, kinetics and removal efficiencies of NO and SO_2 from flue gas induced by the combined temperature and Fe^{2+} activation of aqueous PS. The work involves experimental studies and the development of a kinetic-mass transfer model utilizing a comprehensive reaction scheme for process evaluation. The mathematical model utilizes the Bodenstein or pseudo-steady-state approximation (PSSA) and equilibrium-state hypothesis (ESH) techniques, which have broad applications in modeling reaction kinetics in chemical, biochemical, and environmental reacting systems. The model equations were solved numerically using the fourth order Runge-Kutta method in Matlab. The model results, which appeared to fit the experimental results remarkably well, were used to predict species concentrations, NO and SO_2 mass transfer coefficients, new kinetic data and other process parameters at different temperatures for the NO_x-S(IV)-$S_2O_8^{2-}$-Fe^{2+}system.

References:

1) Liu, Y., Adewuyi, Y.G. A Review on Removal of Elemental Mercury from Flue Gas Using Advanced Oxidation Process: Chemistry and Process. Chem. Eng. Research & Design. 2016, 112, 199-250.

2) Adewuyi, Y.G.; Khan, M.A.; Removal of Nitric Oxide (NO) by Combined Aqueous Persulfate and Ferrous-EDTA Systems: Effects of Persulfate and EDTA Concentrations, temperature, pH and SO_2. Chem. Eng. J. 2016, 304, 793-807.

3) Adewuyi, Y.G.; Khan, M.A. Nitric Oxide Removal by Combined Aqueous Persulfate and Ferrous-EDTA Reaction Systems. Chem. Eng. J. 2015, 281, 575-587.

4) Adewuyi, Y.G.; Khan, M.A., Sakyi, N.Y. Kinetics and Modeling of the Removal of Nitric Oxide by Aqueous Sodium Persulfate Simultaneously Activated by Temperature and Fe^{2+}. Ind. Eng. Chem. Res. 2014, 53, 828-839.

5) Adewuyi, Y.G.; Sakyi, N.Y. Simultaneous Absorption and Oxidation of Nitric Oxide and Sulfur Dioxide by Aqueous Sodium Persulfate Activated by Temperature. Ind. Eng. Chem. Res. 2013, 52, 11702-11711.

6) Adewuyi, Y.G.; Sakyi, N.Y. Removal of Nitric Oxide by Aqueous Sodium Persulfate Simultaneously Activated by Temperature and Fe^{2+} in a Lab-Scale Bubble Column Reactor. *Ind. Eng. Chem. Res.* 2013, 52, 14687-14697.

7.) Khan, N.E.; Adewuyi, Y.G. Absorption and Oxidation of Nitric Oxide (NO) by Aqueous Solutions of Sodium Persulfate in a Bubble Column Reactor. Ind. Eng. Chem. Res. 2010, 49 8749-8760.

AGC

AMOLEA

Minimum Ignition Energy Measurement for Explosive Disproportionation of New Refrigerant Trifluoroethylene

Tetsuo Otsuka[1], Hidekazu Okamoto[1], Katsuya Ueno[1], Masamichi Ippommatsu[2], Ritsu Dobashi[3]

[1] *Research Center, Asahi Glass Co., Ltd.*
[2] *Charmin UV Craftry Co., Ltd*
[3] *Department of Chemical System Engineering, The University of Tokyo*

Contents

AMOLEA

2

Contents

1. Background

2. Experiment
   - experimental condition
   - development of the ignition apparatus
   - statistical phenomena of trifluoroethylene disproportionation

3. Analysis
   - log-normal regression of experimental results

4. Conclusions

3

Needs to change hydrofluorocarbons to hydrofluoroolefins

Chlorofluorocarbons (CFCs)
Hydrochlorofluorocarbons (HCFCs)

Aggression on ozone layer

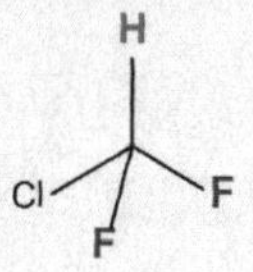

R-22

Hydrofluorocarbons (HFCs)

Global warming potential

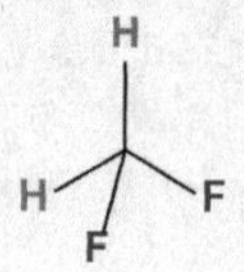

R-32

Hydrofluoroolefins (HFOs)

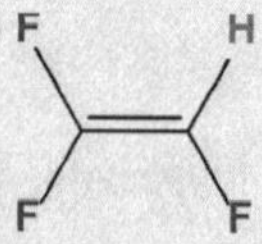

HFO-1123

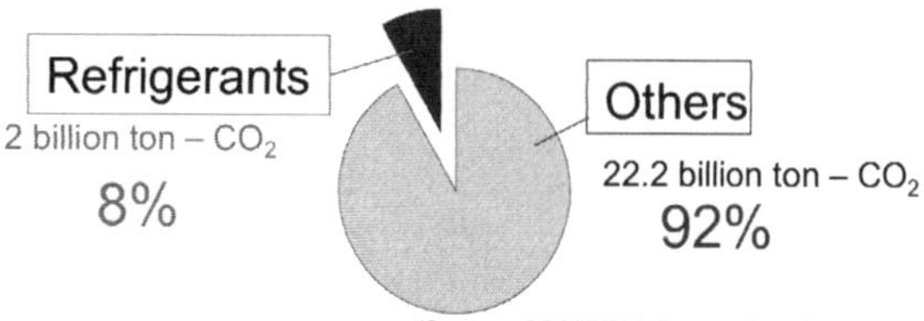

(Source : IPPC/ TEAP Report 2005, EDMC Data 2006)

Fig. Emission of global warming gas (worldwide)

Refrigerants with lower global warming potential (GWP) is globally required.

4

Basic properties of HFO-1123

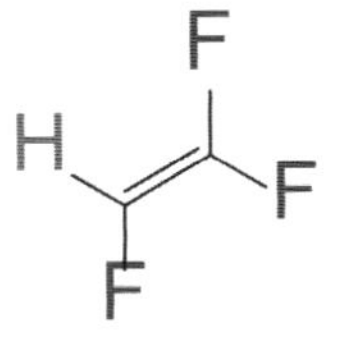

Trifluoroethylene
(HFO-1123)

Table. HFO-1123 and commercialized refrigerants for *air-conditioners*.

	HFO-1123	R-410A	HFC-32
Molecular weight	**82**	73	52
Boiling point [°C]	**-56**	-51	-51
Atmospheric lifetime	**1.6 day**	-	4.9 year
GWP(100year, CO_2 = 1)	0.3	2090	675

* R-410A : mixture of R-32 and HFC-125(pentafluoroethane)

Feature of HFO-1123

- Toxicity: LC_{50} >20,000 ppm
- Not or mildly –flammable
- Same performance as R-410A

HFO-1123 is very attractive as a refrigerant from the standpoint of GWP.

5

Undesirable feature of HFO-1123

HFO-1123 disproportionation reaction

$$CF_2{=}CHF \rightarrow 1.5C + 0.5CF_4 + HF + 250\ kJ/mol$$

(HFO-1123)

When the disproportionation occurred...

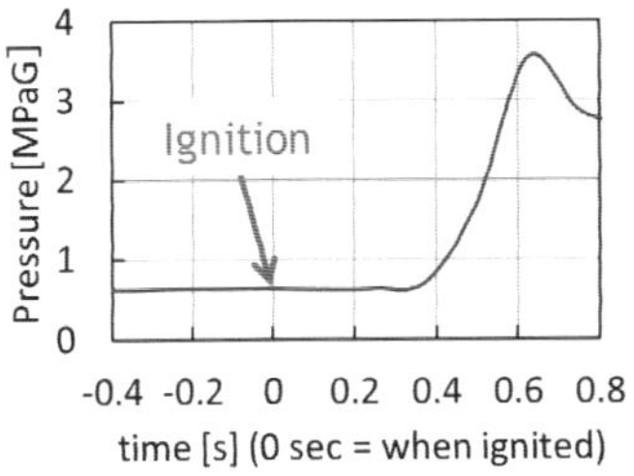

Fig. Pressure profile

Soot

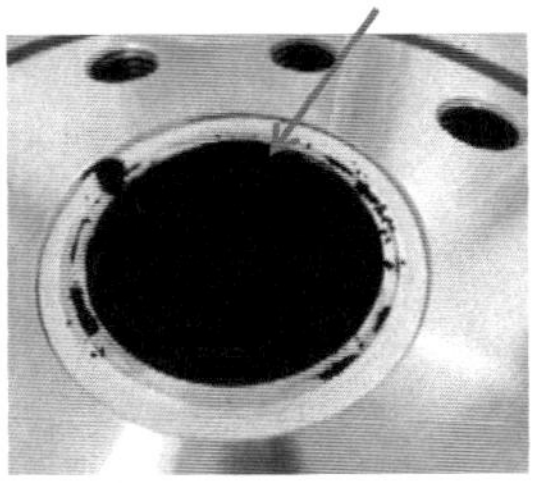

Fig. Test vessel

- Temperature
- Pressure

Rapidly increase

NOT allowed in air-conditioners

Prevention of the disproportionation is crucial for HFO-1123 application.

Measurement of minimum ignition energy is important to prevent HFO-1123 disproportionation.

6

Our hypothesis about HFO-1123 disproportionation

A example of statistical nature of explosion

*Moffett, S. P., et al. *Fall meeting of the Western States Section of the Combustion Institute, Paper 07F-42.* 2007

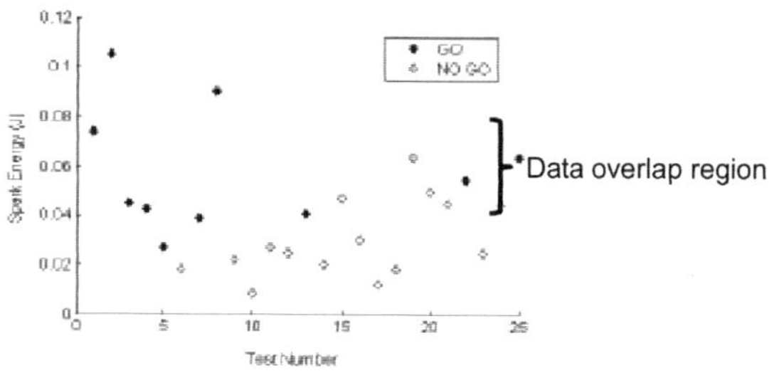

(a)
Jet A spark ignition data as a plot 25 tests versus spark energy(*)

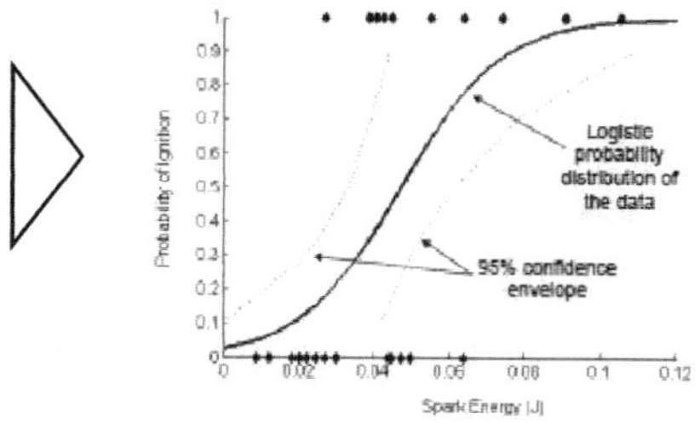

Logistic probability distribution and 95% confidence envelope for the Jet A spark ignition (*)

Hypothesis

HFO-1123 disproportionation would show statistical phenomena.

The results would be analyzed by statistical analysis.

7

Contents

8

Experiment condition

Table1 Experimental condition

HFO-1123 purity	99.5%(0.5% CTFE)
Temperature [K]	373
Pressure [MPa]	1
Ignition energy [J]	0-35

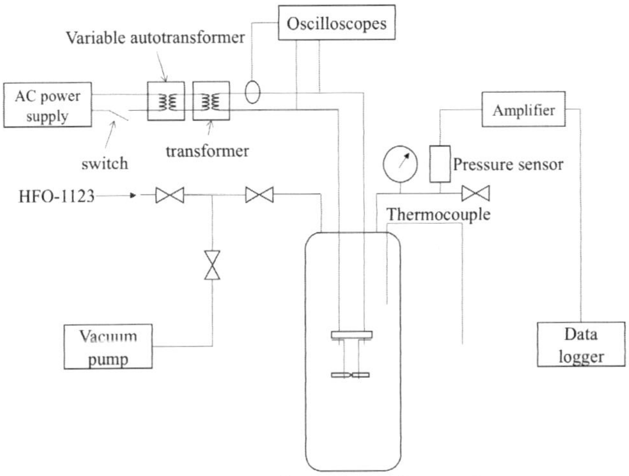

Fig. Schematic diagram of experimental apparatus

Fig. the picture of the autoclave

Table2 the dimension of the autoclave

Inner diameter [cm]	7.2
length [cm]	17.7
Volume [cm^3]	650
Ignition point from the ceiling the autoclave[cm]	10

9

Motivation to modify the ignition apparatus

"We want to evaluate minimum ignition energy under the same condition as the air conditioner "

◆The voltage to generate spark in conventional ignition apparatus is higher than 1 kV that is not generated in the air conditioners.

We newly developed the ignition apparatus that could generate spark using the lower voltage.

10

Developed ignition apparatus

(a) The initial situation of ignition apparatus

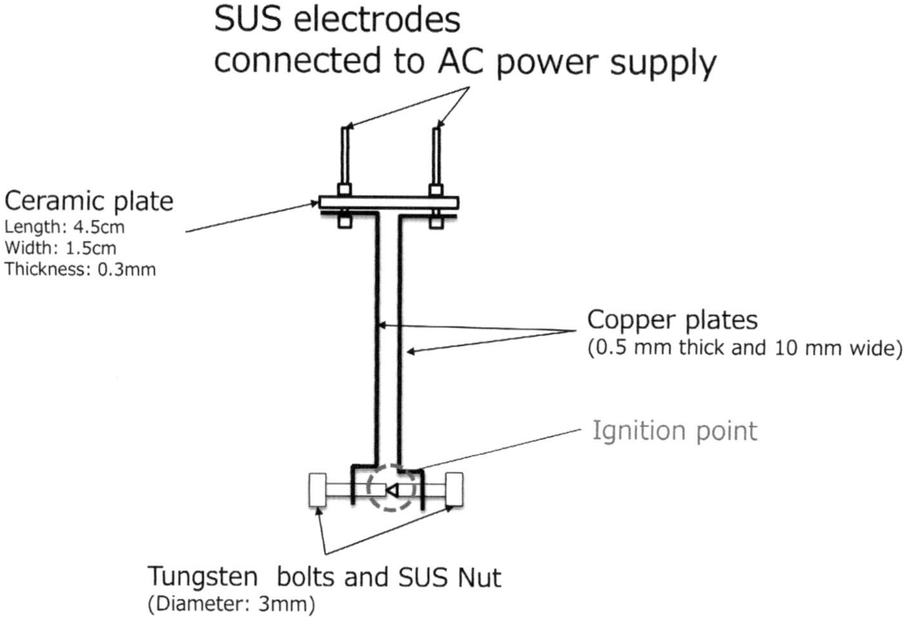

11

Movement of the ignition apparatus

(b) The situation at the start of current flow (c) The situation when spark is generated

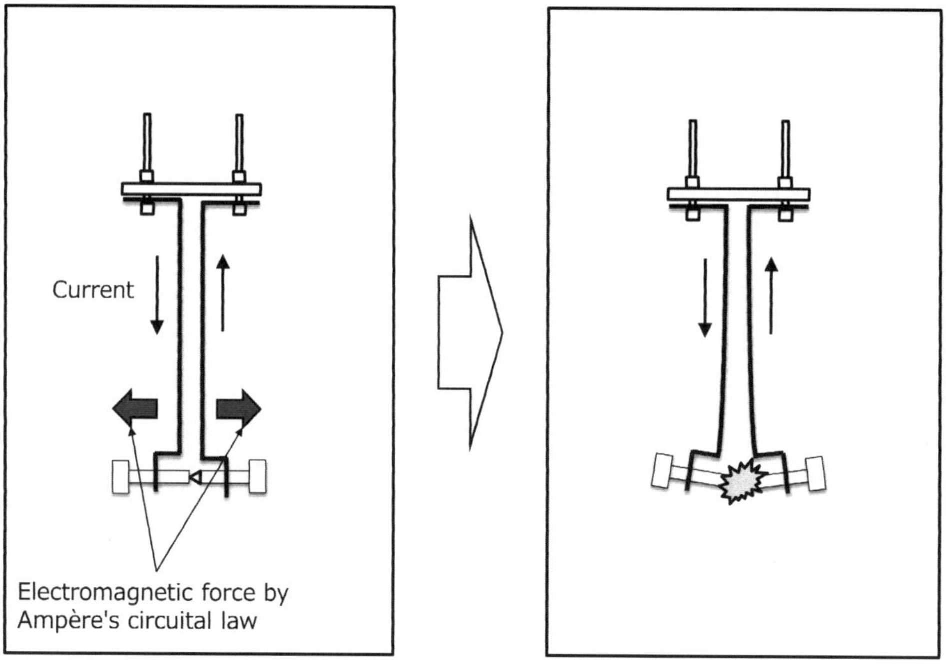

12

Statistical phenomena of HFO-1123 disproportionation

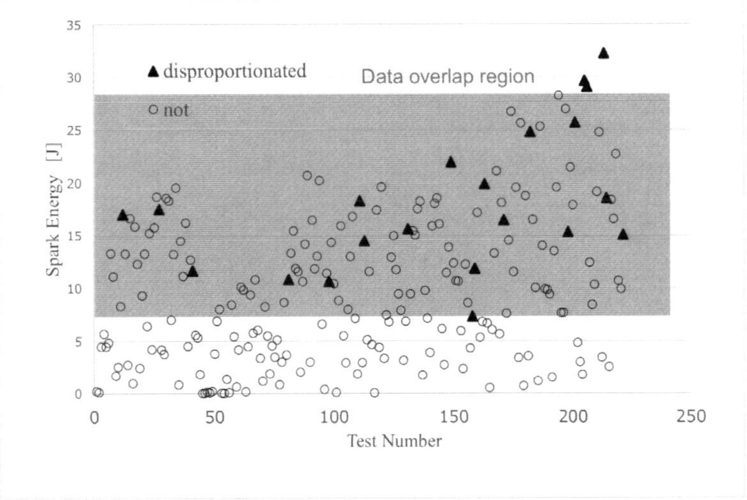

Fig. Plots of 242 tests versus spark energy at 1 MPa, 373 K

HFO-1123 disproportionation showed statistical phenomena.

13

Contents

14

Candidates of analytical methods

OAMOLEA

Logistic regression method

$$P(E) = \frac{1}{1 + exp(\beta_1 + \beta_2 E)}$$

⇒A large number of experimental data were required to calculate a precise probability on the order of 1‰ (*).

Normal distribution method

$$P(E) = \frac{1}{2}\left(1 + erf\left(\frac{E - \beta_1}{\beta_2\sqrt{2}}\right)\right)$$

Log-normal distribution method

⇒Log-normal distribution is the most suitable for especially low energies (**).

$$P(E) = \frac{1}{2}\left(1 + erf\left(\frac{ln(E) - \beta_1}{\beta_2\sqrt{2}}\right)\right)$$

$ß_1$ and $ß_2$ were estimated by maximizing the likelihood function.

Likelihood function: $L = \prod_i^n P(E_i)^{y_i}\left(1 - P(E_i)\right)^{1-y_i}$

* Wähner, A., Gramse, G., Langer, T., & Beyer, M. (2013). Determination of the minimum ignition energy on the basis of a statistical approach. *Journal of Loss Prevention in the Process Industries*, *26*(6), 1655-1660.

** Rudz, S., Chetehouna, K., Strozzi, C., & Gillard, P. (2014). Minimum ignition energy measurements for α-pinene/air mixtures. *Combustion Science and Technology*, *186*(10-11), 1597-1605.

15

Results of analysis

OAMOLEA

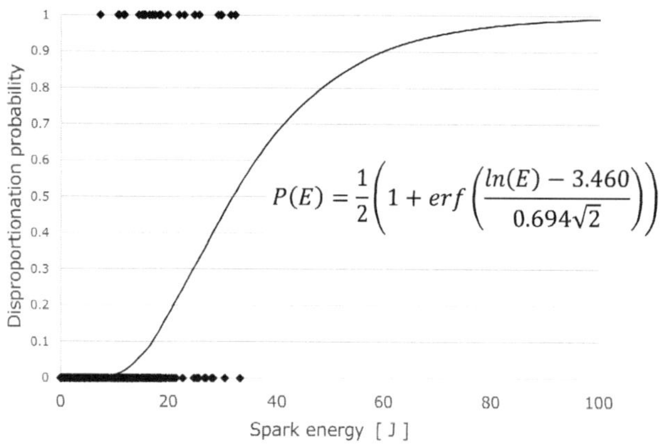

Fig. The ignition probability of HFO-1123 disproportionation at 1 MPa, 373 K

Table Calculated ignition probability of HFO-1123 disproportionation

Ignition probability	ignition energy
10^{-7}%	0.50J
1%	6.33J
5%	10.16J
10%	13.07J
50%	31.82J

Table Minimum ignition energy of hydrocarbons (*)

Methane	0.63mJ
Propane	0.46mJ
N-hexane	0.6mJ

◆ The ignition energy of HFO-1123 disproportionation at 1% probability is higher than hydrocarbons.

* Moorhouse, J., Williams, A., & Maddison, T. E. (1974). *Combustion and flame*, *23*(2), 203-213.

16

Contents

1. Background

2. Experiment
   - experimental condition
   - development of the ignition apparatus
   - statistical phenomena of HFO-1123 disproportionation

3. Analysis
   - log-normal regression of experimental results

4. Conclusions

17

Conclusions

1. The statistical nature of HFO-1123 disproportionation was observed.

2. The ignition probability of HFO-1123 disproportionation was evaluated under the same condition as air conditioners.

   · We newly developed the ignition apparatus using electromagnetic force.

   · The ignition energy of HFO-1123 disproportionation at 1% probability was 6.33J.

We can use HFO-1123 as a component of refrigerants more safely.

18

Thank you for your attention.

19

501472 Bioremediation of Petroleum-Contaminated Soil Using Three Simultaneously Operated Bioreactors

Abstract:

Nowadays, many of human activities such as heating, transportation and power production need oil-derivative fuels. There are many refineries that have been contaminated by petroleum hydrocarbons through various stages of exploration, production, transport and storage. Physiochemical and biological treatment methods have been known to remediate contaminations. For example, in biological remediation methods microorganismsis can be used to remediate contaminants. These methods are considered as a cost-effective and environmental friendly solution. In our research, two bacterial strains which had the highest total petroleum hydrocarbon (TPH) degradation efficiency were isolated from contaminated soil of a refinery. The strains were identified using biochemical tests and 16S rRNA analysis as *Pseudomonas aeruginosa* and *Bacillus Subtilis* . *Pseudomonas aeruginosa* and *Bacillus Subtilis* could respectively degrade 87% and 75% of total hydrocarbons in flasks containing 2% diesel and 98% water during 20 days of operation. Then four bioremediation modes in parallel-operated bioreactors were studied in order to find the best biological technique in remediation of petroleum contaminants. Natural attenuation, biostimulation, bioaugmentation, and combination of biostimulation and bioaugmentation were studied in the bioreactors. Total Petroleum Hydrocarbons (TPH) and microbial population as criteria to compare the efficiency of the methods were measured. Finally, first oreder kinetic model and Monod model were studied and kinetic parameters were determined in order to permit scale-up to field-scale operation of treatment processes.

502998 XPS Characterization of TiO_2 Photo-catalytic Adsorptive Desulfurization

Abstract:

Due to the stringent requirements about the maximum sulfur emissions have been put into effect by governments, ultra-deep desulfurization has been paid more and more attention recently. UV induced photo-assisted adsorptive desulfurization (PAD) technique has been proven to be capable to enhance the sulfur removal capacity of anatase TiO2 adsorbent. XPS characterization of TiO2 adsorbents before and after UV treatment can provide precise evidence to reveal the mechanism of the enhancement by UV irradiation. This work presents valuable evidence in support of the enhancement produced by UV irradiation is due to photo-catalytic reactions between adsorbed water and anatase TiO2. 2 types of anatase TiO2 adsorbents were applied to study the effect of UV irradiation and adsorbed H2O on the UV induced PAD. Thermogravimetric Analysis (TGA) results showed the relative amount of adsorbed H2O on each TiO2 adsorbent. XPS results illustrated that UV irradiation caused a photo-catalytic reaction between adsorbed H2O and anatase TiO2 to form additional hydroxyl groups, and the relative amount of adsorbed H2O on TiO2 is the key to determine how much additional hydroxyls can be produced. The photo-catalytic produced OH groups help improve the desulfurization capacity of TiO2 adsorbent.

Keywords: Desulfurization, Adsorption, Titanium dioxide, Photo-catalysis, UV, XPS, TGA

491786 Experimental Study of Hydrocarbon Trapping over Passive NO_x Adsorber

Abstract:

Experimental Study of Hydrocarbon Trapping over Passive NOx Adsorbers

Sam Malamis, Kyle Karinshak, Michael P. Harold*

Dept. of Chemical & Biomolecular Engineering, University of Houston, Houston, Texas

* corresponding author e-mail: mpharold@central.uh.edu

Introduction

Automotive catalysts including TWC, LNT, and SCR catalysts are all highly effective in reducing NO_x and HC emissions above 200°C [1]. Most engines, however, require a warm-up period during which the catalyst undergoes a "cold start" environment. During this period of up to 200s, large amounts of pollutants such as NOx, N_2O, CO and Hydrocarbons slip through the catalyst into the atmosphere [2]. To mitigate this occurrence, it is beneficial to develop a passive NOx adsorber, or PNA, that can store NOx and hydrocarbons at these low temperatures, and then to release them at higher temperatures to improve the effectiveness of the aftertreatment system [3].

Few comprehensive studies have been performed on this topic, and only recently has there been a movement toward understanding the chemistry behind these low temperature traps. To date, a wide variety of materials has been explored, including palladium on ceria-zirconia (Pd-CZO) [4], CeO_2-M_2O_3 with supported Pt or Pd [5], and Palladium on various zeolites including BEA, MFI, CHA [3]. While it has been reported that Pd-CZO is a potentially effective NO_x trap at low temperatures, it also strongly absorbs SO_2, which inhibits NO_x storage [3,4]. This study focuses on the use of zeolites as PNA materials in order to trap both NOx and HC during the cold start period, with the potential to act as a partial oxidation catalyst upstream of the SCR or LNT. The study mainly explores the effect of temperature and various feed gases on NOx and HC trapping.

Materials and methods

Experiments were performed using 2wt% Pd-BEA zeolite supported on a cordierite monolith. A feed gas was flowed over the catalyst in a vertical quartz-tube reactor, and the effluent species concentrations were measured using FT-IR. The catalyst was first pre-treated in 5% O_2 (bal. Ar) for 30 min at 550°C. To determine the amount of NOx stored on the catalyst during an experiment, a feed gas of 2% O_2 and 400ppm NO was flowed over the catalyst for 5 min at a fixed feed temperature, either 50°C, 80°C or 150°C. After 5 min., the NO was turned off and the reactor was maintained at the adsorption temperature until the concentration of NO dropped to less than 2ppm. To measure the species desorbing after adsorbing on the catalyst surface, a temperature programmed desorption (TPD) was performed at 20°C/min from the adsorption temperature to 550°C. The surface was then pretreated to re-oxidize the Pd to PdO to ensure proper behavior during subsequent experiments [6]. To measure the effect of hydrocarbons on catalyst performance, experiments similar to those described above were performed, using a feed with either 800ppm C_3H_6 and 2% O_2 or a mixture of 400ppm NO, 800ppm C_3H_6 and 2% O_2.

Results

It was found that using a feed of only NO and O_2, Pd-BEA stored between 0.4 and 1.0 mol NOx/gcat*10^4, but released it too early during the TPD (i.e. at a temperature <200°C) (Figure 1). It was also found that as the adsorption temperature increased, the amount of NOx stored and desorbed decreased (Figure 2). Experiments using only propylene in the feed indicate a very strong propensity for the catalyst to trap hydrocarbons at low temperatures (1.5-6.0 mol C_3H_6/gcat*10^4), more so than in trapping NOx (Figures 3). Adding C_3H_6 to the NO feed revealed a competitive adsorption effect of both NOx and C_3H_6, as the stored quantities of these species was significantly lower than the pure feeds, as shown in Figures 2 and 3. While the total amount of stored NOx decreased with the addition of C_3H_6, our studies show that the majority of NOx was released above 200°C, which is a promising result (Figure 1).

Figure 1. Percentage of NOx that desorbs at a catalyst temperature higher than 200°C for feeds with only NO and with added C_3H_6 on 2wt% Pd-BEA. Values are reported for each adsorption temperature. Note that the addition of propylene in the feed significantly delays NOx desorption. Feed: 400ppm NO with or without 800ppm C_3H_6, 2% O_2, bal. Ar.

Figure 2. NOx adsorption and desorption quantities at various adsorption temperatures and for feeds with or without C_3H_6. Values are reported as moles of NOx/g cat*10^4. Note that as the adsorption temperature increases, adsorption/desorption of NOx decreases. Feed: 400ppm NO with or without 800ppm C_3H_6, 2% O_2, bal. Ar.

Figure 3. C_3H_6 adsorption and desorption quantities at various adsorption temperatures and for feeds with or without NO. Adsorption values are reported as moles of C_3H_6/g cat*10^4, while desorption values are reported as moles of C/g cat*10^4. Note that as the adsorption temperature increases, adsorption/desorption of C_3H_6 decreases. Feed: 800ppm C_3H_6 with or without 400ppm NO, 2% O_2, bal. Ar.

References

1. Twigg M: **Progress and future challenges in controlling automotive exhaust gas emissions.** *Applied Catalyst B* 2007, **70**:2-15.

2. Wilenmann M, Favez J, Alvarez R: **Cold-start emissions of modern passenger cars at different low ambient temperatures and their evolution over vehicle legislation categories.** *Atmospheric environment* 2009, **43**:241-2429.

3. Chen HY *et al.*: **Low temperature NO storage of zeolite supported Pd or low temperature diesel engine emission control.** *Catal. Lett.* 2016.

4. Theis JR, Lambert C: **The effects of CO, C_2H_4, and H_2O on the NO_x storage performance of low temperature NO_x adsorbers for diesel applications.** *SAE International* 2017, **10.**

5. Jones S *et al.:* **CeO_2**
2016 **CLEERS Issue.**

498784 CuO/Co_3O_4: A Highly Active and Selective Low Temperature NO Decomposition Catalyst

Abstract:

Abstract:

The direct decomposition of NO is thermodynamically favorable within the temperatures encountered by an automotive catalytic converter; however, activity remains too low for practical application [1]. Platinum group metals (PGM), such as Pd, are active for NO decomposition at temperatures above 600 °C, and attempts to improve their activity below 600 °C have been unsuccessful due oxidation of the metal and subsequent deactivation [2]. Regarding non-PGM catalysts, Cu-ZSM5 is perhaps the most widely known material displaying low temperature (≤ 500 °C) direct NO decomposition activity, however it has propensity to produce an undesired side product: the greenhouse gas N_2O [3]. The spinel oxide Co_3O_4 does not exhibit direct NOx activity as high as Cu-ZSM5, but even at 350 °C it is able to decompose 50% of N_2O to N_2 [4,5]. To address the performance gap below 500 °C, we therefore propose combining the benefits of both CuOx and Co_3O_4 into a single NO decomposition catalyst. Supported copper oxide on cobalt oxide $CuOx/Co_3O_4$ was synthesized by incipient wetness impregnation. We hypothesized that this material can utilize the N_2O decomposition capabilities of Co_3O_4 to reduce the N_2O generated by NO decomposition over the CuOx species.

The $CuOx/Co_3O_4$ were synthesized with Cu surface densities ranging from 0.7 to 10.3 Cu/nm^2 and characterized by XRD, XPS, and TEM. The Cu surface density corresponding to the monolayer loading for $CuOx/Co_3O_4$ was found to be 6.7 Cu/nm^2, therefore, the catalysts set included sub-monolayer, monolayer, or 3D-microcrystalline CuOx species. Thus, this catalyst set formed a strong basis for identifying a direct NOx decomposition structure-activity relationship. The catalysts were evaluated for direct NOx decomposition activity using a packed bed reactor. The maximum activity for NO decomposition by $CuOx/Co_3O_4$ was found at the CuOx surface density corresponding to monolayer. There is a subsequent decrease in the activity at higher surface densities due to the presence of CuOx microcrystals. When compared to Cu-ZSM5, the specific activity from 300-400 °C is similar, however, the surface area normalized (areal) activity exceeds that of Cu-ZSM5 as the $CuOx/Co_3O_4$ has a specific surface area that is an order of magnitude lower. Additionally, at or below 400 °C, the N_2 selectivity is higher over $CuOx/Co_3O_4$ compared to Cu-ZSM5, as a result of decreased N_2O formation. The $CuOx/Co_3O_4$ catalysts maintain relatively good activity in the presence of O_2 and CO_2. Nitric oxide adsorption/desorption reveals an optimized adsorption strength over $CuOx/Co_3O_4$. Additional mechanistic insight from *in situ* FTIR will also be presented.

References:

1. H. Falsig, T. Bligaard, C.H. Christensen, J. K. Norskov, *Pure Appl. Chem.*, Vol. 79, No. 11, pp. 1895–1903, 2007.
2. G.K. Reddy, C. Ling, T. Peck, H. Jia, *RSC Adv.*, 2017, 7, 19645
3. B. Moden, P.D. Costa, D. K. Lee, E. Iglesia, *J. Phys. Chem. B* 2002**,** *106,* 9633-9641
4. P.W. Park, J.K. Kil, H.H. Kung, M. C. Kung, *Catal. Today*, Volume 42, Issues 1-2, June 1998, 51-60.
5. L. Xue, H. He, C. Liu, C. Zhang, and B. Zhang. *Environ. Sci. Technol.*, **2009**, 43 (3), 890-895.

488866 Effect of Different Structure Directing Agents on Deactivation of Cu/SAPO-34 during Low-Temperature NH_3-SCR

Abstract:

Cu/SAPO-34 and Cu/SSZ-13 having CHA structure have attracted significant attention because of their high activity and N_2 selectivity during SCR reaction as well as superior hydro carbon resistance. The Cu/SAPO-34 has shown better hydrothermal durability than Cu/SSZ-13 at high temperature [1]. However, we have earlier observed that Cu/SAPO-34 prepared by the morpholine route deteriorated under water exposure at low temperature, where the activity decreased from 87% conversion to 6% conversion after 9 h of low temperature experiment [2]. Nevertheless, there is still a lack of understanding of the deactivation mechanism of Cu/SAPO-34 during low temperature NH_3-SCR reaction.

In this study, Cu/SAPO-34 using three different templates, i.e., morpholine(MO), trimethylamine (TEA), and tetraethylammonium hydroxide (TEAOH), as structure directing agent were prepared by incipient wetness impregnation (IWI) method. Characterizations of those catalysts suggest that they have different physical/chemical properties as well as various species of Cu. For example, analysis of H_2-TPR profiles of Cu/SAPO-34 catalysts proposed that they contain different amount of Cu species depending on the structure directing agents. Those catalysts were systematically investigated for low-temperature NH_3-SCR system under water exposure in order to find the correlation of their chemical/physical structures and deactivation. It was observed that the activity of Cu/SAPO-34 declined when they were exposed to water at low temperature as observed in the previous study [2]. Interestingly, deactivation of Cu/SAPO-34 catalysts behave differently depending on the choice of template. Additionally we found that activity of Cu/SAPO-34 using MO as template after water exposure at low temperature declined regardless of a variety of regeneration methods including a dry atmosphere at 600 °C. To conclude, the choice of template over Cu/SAPO-34 catalysts significantly influence their activities and deactivation during low temperature NH_3-SCR.

[1] D.W. Fickel, E. Dâ€™Addio, J.A. Lauterbach, R.F. Lobo, Appl. Catal. B: Environ. 102 (2011) 441-448

[2] K. Leistner, L. Olsson, Appl. Catal. B: Environ. 165 (2015) 192-199

497426 A Perovskite-like Catalyst for the Simultaneous Removal of Soot and NO: Effect of the Synthesis Method

Abstract:

Due to the environmental problems associated with the use of diesel engines, and the strengthening of the regulations concerning pollutant emissions in the transport field. The study of the catalytic simultaneous removal of soot and NO at relatively low temperatures (within the range of diesel engine exhaust: 150-450°C) has taken a paramount importance in pollution control area.

Mixed oxides coming from perovskite-like materials, seems to be a promising catalyst for such purpose. In this way, the present work tested the $La_{0.7}Ag_{0.3}MnO_3$ perovskite-like material, prepared for different synthetic routes, for the simultaneous removal of soot and NO. Three different synthetic routes are compared: microwave-assisted hydrothermal (HMW), microwave-assisted (MW) and solid state (SS).

The catalytic activity of the prepared catalysts was tested in a temperature-controlled furnace inside which a soot-catalyst-sand fixed bed was placed. The mixture of carbonaceous material and catalysts was carried out with spatula (i.e. loose contact), keeping always a 1:9 weight ratio. The soot model used in this study consisted of a carbon black, FW200, from Orion Engineering. The soot-catalyst-sand mixture was exposed at a constant heating rate of 10°C/min up to 600°C, under a 2200 ppm NO/ 10% vol.O2/ He atmosphere. The evolution of CO_2, CO, NO, NOx, N_2,O_2 gases with temperature were followed by mass spectrometry and a KIGAZ 310 gas analyzer. The catalyst materials were also characterized by SEM, TEM, XRD, H_2-TPR and O_2-TPD.

The results show a good catalytic behavior of the materials tested for the oxidation of soot. Figure 1 and Table 1 present the temperature at which soot ignition begins (T_{ig}), and the temperatures at which the conversion of soot to CO_2 was 50%, 90% and 100%, named as T50, T90 and T100, respectively. When it is compared the catalytic activity results with the non-catalytic one; it was found an ignition temperature shift of 278°C when the HMW-synthesized catalyst was used, and a shift of 241°C for the MW-synthesized catalyst. Also, the maximum soot oxidation peak, (T50), for all the catalytic systems evaluated in this study was shifted to lower temperatures according to the following decreasing order: non-catalytic > SS > HMW > MW. Another important result regarding to catalytic soot oxidation, it is that no CO emission was found in the gas stream, indicating 100% selectivity toward CO_2. In contrast, for the non-catalytic one, the selectivity CO_2/CO ratio was near one.

Figure 1. CO_2 Yield as a function of soot oxidation temperature over $La_{0.7}Ag_{0.3}MnO_3$ perovskite-like material synthesized by different methodologies (MW, HMW, SS).

Table 1. Ignition temperature (Tig); temperatures at which the soot conversion was 50%, 90% and 100%, respectively; and CO_2 selectivity for catalytic and the non-catalytic test.

Test	Tig (°C)	T50 (°C)	T90 (°C)	T100 (°C)	S_{CO2} (%)
$NO+O_2$	428	634	676	728	N/A
HMW	150	409	465	577	100
MW	187	367	487	665	100
SS	170	439	502	679	100

Regarding NO removal, Figure 2 shows the NO evolution with temperature, for the experiments carried out with $La_{0.7}Ag_{0.3}MnO_3$ synthesized by methodologies. Figure 2 shows that during soot oxidation, the catalysts tested presented some NO adsorption/desorption events followed by a NO reduction toward N_2 around the soot oxidation peak. The catalyst synthesized through MW route shows a prominent desorption event between 210°C and 350°C compared with the other two materials. Nevertheless, its adsorption or reduction capacity was not as good as the one obtained from the HMW-synthesized catalyst.

Figure 2. NO evolution as a function of temperature for each of the synthesized catalysts tested.

Figure 3 shows the CO_2 and NO/NOx profiles as function of temperature when the HMW-synthesized catalyst was used. From figure, it is possible to see that this catalyst was able to remove 45% of the initial NO concentration as N2. Also, it is possible to see that the temperature for the NO/NOx peak reduction was found around 400 °C matching the temperature at which the evolution of CO_2 was maximum. This result shows the ability for the simultaneous removal of soot and NO of the $La_{0.7}Ag_{0.3}MnO_3$ perovskite-like material synthesized by the HMW method.

Figure 3. CO_2, NO/NOx and N_2 profiles as a function of temperature when the HMW-synthesized solid was used as catalyst.

The results presented above demonstrate that the selection of the synthetic route has a significant effect on the catalytic performance of a material. Also, it was evidenced that the perovskite-like material ($La_{0.7}Ag_{0.3}MnO_3$) synthesized by microwave-assisted hydrothermal (HMW) method has an excellent catalytic activity for the simultaneous removal of soot and NO by shifting the soot oxidation peak toward lower temperature and removing up to 45% of the initial NO as N2.

497655 Fast Cycling to Achieve High NO_x Conversion in Lean Exhaust: Role of Ceria

Abstract:

Fast Cycling to Achieve High NOx Conversion in Lean Exhaust: Role of Ceria

Zhiyu Zhou, Michael P. Harold*, Dan Luss**

Department of Chemical and Biomolecular Engineering, University of Houston, Houston, TX 77204

* *mharold@uh.edu*, ** *dluss@central.uh.edu*

The abatement of NOx (NO+NO_2) remains a challenge under the continuously more stringent emission standards. Commercialized deNOx systems, NOx storage and reduction (NSR) and selective catalytic reduction (SCR), function well from ~250 °C to ~400 °C, but have unsatisfactory performance outside that temperature range. Toyota researchers recently invented a new deNOx system, Di-Air (diesel NOx aftertreatment by adsorbed intermediate reductants), which involves fast injection of fuel into the exhaust feed to the NSR converter [1]. Previous studies show that ceria increases the deNOx performance of the Di-Air system in both low [2] and high [3] temperature regions. Data suggest that NOx conversion occurs via the reaction of an intermediate produced through the reaction with O_2 and NO [1, 2]. However, a recent study [3] suggests that catalysts containing ceria may involve an alternative pathway of NOx abatement via NO decomposition at high temperatures. We studied in a bench-scale flow reactor system the NO decomposition and reduction on ceria under different frequencies with CO as sole reductant.

In our study, a lean to rich feed time ratio 6 to 1 of was controlled to keep the same fuel penalty (fuel to oxidant ratio). The lean/rich cycling conditions included lean/rich times (in seconds) of 90/15, 60/10, 30/5 and 6/1. Different reaction conditions were applied to study NO conversion behavior under favorable and adverse conditions. 500 ppm NO was added to both rich phase and lean phase. 2.5% CO was only added to rich phase and O_2 with varied concentration was only added to lean phase. In some cases, CO_2 and H_2O were introduced continuously along with NO, O_2 and CO to simulate lean exhaust.

When CO is used as the sole reductant and no H_2O is introduced in the feed, NO is mainly converted to N_2 with only trace amounts of N_2O and NO_2 detected. Figure 1 shows the cycle-averaged NO conversion with four cycle times from 450 °C to 625 °C in two feed conditions. In the first case, when 1400 ppm O_2 is introduced in lean feed, Figure 1(a) shows that increase of lean/rich switching frequency enhances NO conversion. The optimized lean/rich switching frequency is always 6/1 above 500 °C. In the second case, when 2100ppm O_2 is introduced in the lean feed, Figure 1(b) shows that a different optimal lean/rich switching frequency exists for different temperatures. The optimal lean/rich switching frequency is 30/5 at intermediate temperatures (500 °C and 550 °C) and 6/1 at high temperatures (600 °C and 625 °C) respectively. Figure 1 shows that the optimal lean/rich switching frequency changes with reaction conditions (e.g., temperature and feed condition).

Figure 1. (a) NO conversion with different lean/rich switching frequency; (a) rich: 2.5%CO, 500ppm NO; lean: 500ppm NO, 1400ppm NO; (b) rich: 2.5%CO, 500ppm NO, lean: 500ppm NO, 2100ppm O_2

Figure 2 shows the cycle-averaged NO conversion with the same four cycle times at 600 °C with fixed ratio of lean and rich feeds. The short cycle limit represents NO conversion obtained with total mixed lean and rich feeds. The long cycle limit represents weight averaged NO conversion in lean phase and rich phase. The optimal lean/rich switching frequency of 30/5 exceeds those of both short and long cycle time limit. This shows that the lean/rich cycling operation is beneficial to NO conversion.

Figure 2. NO conversion at 600C, with a fixed switching frequency as 90s lean/15s rich; rich: 2.5%CO, 500ppm NO, lean: 500ppm NO, 3600ppm O_2

Figure 3 shows that NO conversion changes when CO_2 and H_2O are fed. By comparing the blue line (reference) with either red line (with H_2O only) or purple line (with CO_2 only), it is obvious that both H_2O and CO_2 inhibit NO decomposition on ceria but H_2O has a smaller detrimental effect than CO_2. The green line (with both CO_2 and H_2O) lies between the red line (with H_2O only) and the purple line (with CO_2 only), which indicates that H_2O can mitigate the detrimental effect of CO_2.

Figure 3. NO conversion at 550C, with a fixed switching frequency as 90s lean/15s rich; (rich: 2.5%CO, 500ppm NO, lean: 500ppm NO, varied O_2 concentration, CO_2 and H_2O in both lean and rich if introduced)

This study on NO decomposition on ceria during lean and rich cycling helps explain the beneficial function of ceria on NOx reduction at high temperatures, which also provide guidance for optimization of catalyst formulation and operation strategies.

Reference

1. Y. Bisaiji et al., *SAE Int. J. Fuels Lubr*. 5 (2012) 380-388
2. Y. Zheng et al., *Catal. Today*, 276 (2016), 192-201.
3. Y. Wang et al., *Top. Catal.*, 59 (2016), 854-860

495352 NO Reduction By CO over CeO_2 Supported Co_xO_y Catalysts

Abstract:

Nitrogen oxides (NO_x) are a main contributorâ€"in addition to hydrocarbons (HC), carbon monoxide (CO), sulfur oxides (SO_x), and particulate matter (PM)â€"to global air pollution [1]. Significant investigations have been performed for catalytic NO_x reduction resulting in technologies such as NH_3-SCR and NO_x storage and reduction [2]. In the last several decades, platinum group metals (PGMs) and zeolites have been extensively applied to NO_x abatement [3]. Although these catalysts exhibit a high level of catalytic activity, NO_x decomposition over PGMs and zeolites has some practical issues such as high cost and high light-off temperature [4]. To resolve these issues, transition metal oxide catalysts have been investigated for low temperature NO_x decomposition reaction by CO with varied O_2 concentrations, which is a primary reaction in three-way catalysis (TWC).

In this study, the series of CeO_2 supported Co_xO_y catalysts were prepared by incipient wetness impregnation method with different metal oxide loadings (0.5 wt%-30 wt%) for NO reduction by CO. Catalytic activity, catalyst physicochemical property, and molecular/electronic structure were investigated by FTIR, BET, Raman spectroscopy, X-ray diffraction (XRD), and X-ray absorption spectroscopy (XAS). No diffraction peaks due to crystalline CoO or Co_3O_4 were observed in XRD patterns under to 10 wt%, while nanocrystaline Co_3O_4 peaks at ~610cm^{-1} was observed in the Raman spectra with 6wt% sample. Both Raman and XRD spectroscopy results provided that 5~6wt% Co_xO_y /CeO_2 contains monolayer coverage. It was observed that both Co_xO_y coverage and oxidation state of Co are closely related to the NO reduction by CO. Monolayer and above monolayer coverage catalysts showed 80~90% CO and NO conversion at 300~400°C with high N_2 and CO_2 selectivity (>90%). Based on the time on stream results, it was also observed that the crystalline Co_3O_4 nanoparticles above monolayer coverage improve the stability of catalyst.

ACKNOWLEDGMENT: We gratefully acknowledge the financial support for this study from the Department of Materials Science & Chemical Engineering at Stony Brook University through start-up research funding.

[1]. K. Skalska, et al. Sci. Total Env., 408 (2010) 3976-3989, [2]. J.H. Kwak, et al., J. Catal., 287 (2012) 203-209, [3]. S. Roy, et al., Applied Energy, 86 (2009) 2283-2297, [4]. J. Liu, et al., Applied Catalysis B: Env., 84 (2008) 185-195

498251 Nb_2O_5 Promoted Base Metal Oxide Catalysts Used for Air Pollution Abatement in Stationary Plants

Abstract:

Catalytic oxidation is a very efficient way for the abatement of low concentrations of Volatile Organic Compounds (VOCs). The most active catalysts used for oxidation are supported noble metals. However the main drawback with noble metals is their high price. Our goal is to consider transition metal oxides as possible replacements for expensive precious metals for low-temperature stationary abatement applications. Current research in our group has found that niobium oxide (Nb_2O_5) is worthy of additional studies as a promoter in altering the oxidation states of base metal oxides. Consequently, bimetallic catalysts supported on Nb_2O_5 were synthesized to achieve the development of stable oxidation. Both Nb_2O_5 and Al_2O_3 carriers were pre-calcined at 800ËšC for 2hr. Then, the calcination temperature for metal-supported catalysts was experimentally determined using temperature programmed decomposition in air using thermogravimetric analysis (TGA). The C_3H_8 oxidation reactions were carried out in a fixed-bed quartz flow reactor at atmospheric pressure. The reaction conditions were as follows: C_3H_8: 0.1 vol. %, O_2: 18.9 vol. % and N_2 balance. Under ambient conditions, the gas hourly space velocity (GHSV) was 65 000 h^{-1} and catalysis volume is 0.1 ml. Effluent gases of were continuously analyzed by micro GC. Isothermal aging tests were conducted to assess the stability of the catalysts at 400 ËšC in the air. Results suggest a strong interaction between niobium oxide and the active components.

499320 N Doped TiO_2 nanoparticles Synthesized By One Step Liquid Flame Aerosol Method (LFSP): Understanding the Effect of Synthesis Parameters and Photocatalytic Degradation of VOCs

Abstract:

Environmental photocatalysis Nitrogen doped titanium dioxide is attracting a continuously increasing. Conventional wet synthesis methods need multiple processing steps, high energy input and long processing times, In contrast, flame synthesis, especially liquid flame spray pyrolysis (LFSP), is a versatile one-step process for synthesizing a variety of anion doped nanoparticles. Nitrogen-doped TiO

synthesis. A combination of various physico-chemical characterizations were performed. The photocatalytic activity and stability of the N-doped TiO_2 samples were evaluated through using the decomposition of phenol as model reaction under visible light irradiation. Characterization results show that the nitrogen dopants have a significant effect on the crystallite size and optical absorption of TiO_2. It is found that the N-doped TiO_2 catalysts have enhanced absorption in the visible light region, and exhibit higher activity for photocatalytic degradation of VOCs. We have produced The UV-vis spectroscopy results of N incorporated TiO_2 showed enhancement of light absorption in the visible range by selected composite materials in the UV range (250-400 nm). We have found the Ti/N ratio of 11% for the materials synthesized from our XPS data. To the best of our knowledge, this is the first report which demonstrates single-step, direct flame synthesis of N-doped TiO_2 and their application as a photocatalyst better performance. The method is extremely simple, low cost, energy efficient, very effective, and can be easily scaled up for large scale production. These interesting results will be discussed in the presentation.

508757 Lawrence K. Cecil Award Lecture: Environmental Aspects in the Utilization of Fossil Fuels

Abstract:

A major theme of my research has been investigating pollutant formation during the combustion of fossil fuels, namely solid fuels and liquids. This lecture will look at research and the resulting implications in environmental regulations, spanning from the early 1990's to today. A focus will be on more recent work dealing with global warming, namely carbon capture via chemical looping combustion and black carbon, or soot, formation and oxidation from engines.

Carbon dioxide is a major contributor to climate forcing with fossil energy combustion for being a component of these emissions. Carbon capture and storage is a potential method to the reduction of carbon dioxide to the environment. Oxy-fuel combustion, where a solid fuel is burned with oxygen diluted with recycled flue gas, is an effective method of carbon capture as the resulting flue gas is mainly carbon dioxide and water. This is in contrast to combustion in air, where the flue gas contains nitrogen as well. Chemical looping combustion has the potential to be a more energy effective method of oxygen generation as compared to air separation, which is energy intensive. Specifically, our group at the University of Utah has been studying chemical looping with oxygen uncoupling, where a copper oxide oxygen carrier is used in a fuel reactor (reducing copper oxide on an inert) and then "looped" to an air reactor (oxidizing the copper oxide). In the fuel reactor, the reduction of the cupric oxide to cuprous oxide results in oxygen release which is used to burn the solid fuel, in this case coal. The talk will focus on the role of chemical engineering fundamentals, i.e. mass and energy balances, mass transfer, and kinetics, in designing effective systems for solid fuel, in this case coal, combustion. Results from lab scale tests, ASPEN modeling, and computational fluid dynamics modeling (CFD) utilizing a commercial code, Barraucda-VR®, will be discussed.

Studies have suggested that the contribution of black carbon, or soot, to global warming could be secondary to that of carbon dioxide. In addition, the emission of black carbon or soot has important implications for ambient fine particulate concentrations, and hence, adverse health effects. An immediate remedy is to simply reduce the amount of soot emitted to the atmosphere. For industrialized countries, a major source of soot is from diesel engines. While reducing the formation of soot from engines is certainly an important aspect, our research has focused on the oxidation kinetics of soot. Results have suggested that soot nanostructure, which is a function of fuel type, is an important parameter in changes in the rates of soot oxidation. The development of effective soot oxidation rate equations for a particular fuel is critical in the ability to predict emissions from engines.

510405 Environmental Division Early Career Award Lecture: Environmental Life Cycle Assessment and Optimization of Process and Energy Systems

Abstract:

In the post COP21 era, environmental sustainability issues are becoming more and more prevalent in systems analysis, design and operations. This presentation will introduce the basic concepts and methods for environmental life cycle analysis and sustainability optimization for process and energy systems. Applications to perovskite photovoltaics, shale gas and biofuels will be presented.

503670 Digital Alchemy for Assembly Engineeringc

Abstract:

Back in the day, Stuart Churchill was a pioneer in the use of computers for modeling chemical and thermal processes. Since that time, the role of computing in chemical engineering has become indispensable. In this talk, we present new computational methods we call "digital alchemy" that allow for predictive modeling of thermodynamic assembly, and permit the design of building blocks optimized for assembly into target structures with desired properties. In this approach, standard thermodynamic ensembles are generalized to include alchemical variables and their thermodynamic conjugates, allowing for rapid, real-time computational investigation of large, complex design spaces.

503671 Improved Algebraic, Numerical, and Graphical Representations in Fluid Mechanics

Abstract:

The concept of equivalents is applied to fully developed flows through straight channels, and two applications are introduced: the equivalents that produce equal rates of flow through channels of different cross-section, and the diminishment that occurs between the regimes of laminar and turbulent flow. The validity of the eddy diffusivity and the invalidity of the mixing-length as measures of the shear stress due to turbulence have perhaps never before been demonstrated so clearly. The possible choices of primary variables and the consequences are also examined. The objective of these explorations is an improved understanding of the basic fluid mechanics by teachers, students, and industrial practitioners.

503672 The Scaling of Turbulence Near the Wall and the Churchill Turbulent Flux Correlation: Insights with Lagrangian Simulations

Abstract:

The evaluation of mean velocity and mean temperature in wall-bounded turbulent flows has been based on the "classical" theory of turbulence and the law of the wall. The viscous wall parameters, namely the friction velocity, the friction temperature, and the kinematic viscosity of the fluid have been used to scale not only the mean values of velocity, but also other mean turbulent quantities close to the wall, including the Reynolds stresses and the turbulent velocity and temperature fluctuations. Stuart Churchill, as he questioned the validity of the Reynolds analogy between momentum and heat or mass transfer in his works [1], he realized that extending the scaling of momentum to the scaling of scalar quantities in turbulence was not a sound approach. He proposed a scaling of turbulence that depends on the turbulent momentum flux and the turbulent heat flux. In his AIChE Institute Lecture (in Miami, 1998), Churchill presented a summary of his work on this topic, and closed form solutions for the mean velocity and temperature [2]. The turbulent Prandtl number, the most significant parameter for the modeling of turbulent transport not only in theory but also in modern CFD commercial software packages, was also interpreted physically based on these ideas [3]. We have used direct numerical simulation results and results from Lagrangian simulations [4] for the transport of passive scalars in turbulent channel flows to explore Churchill's ideas on turbulent flow scaling and to develop an updated theory for the scaling of heat transport in turbulent flows [5-7]. His passion to remove empiricisms and approximations from the development of predictive correlations in turbulent transport has been an inspiration. In this talk, we will revisit the Churchill scaling in view of our simulation results, and will discuss the counterintuitive finding that wall turbulence can be used to separate rather than mix substances under special circumstances [8].

1. Churchill, S. W., 1996, "A Critique of Predictive and Correlative Models for Turbulent Flow and Convection," Ind. Eng. Chem. Res., 35, pp. 3122–3140.
2. Churchill, S.W., 2000, "Progress in the Thermal Sciences: AICHE Institute Lecture," AIChE Journal, 46(9), pp. 1704-1722.
3. Churchill, S. W., 2002, "A reinterpretation of the Turbulent Prandtl Number," Ind. Eng. Chem. Res., 41, pp. 6393–6401.
4. Papavassiliou, D.V, 2002, "Turbulent transport from continuous sources at the wall of a channel," Int. J. Heat Mass Transfer, 45(17), pp. 3571-3583.
5. Le, P.M., and Papavassiliou, D.V., 2008, "On the scaling of heat transfer using thermal flux gradients for fully developed turbulent channel and Couette flows," Int. Commun. Heat Mass Transfer, 35(4), pp. 404-412.
6. Srinivasan, C., and Papavassiliou, D. V., 2011, "Prediction of Turbulent Prandtl Number in Wall Flows With Lagrangian Simulations," Ind. Eng. Chem. Res., 50(15), pp. 8881–8891.
7. Srinivasan, C., and Papavassiliou, D.V., 2013, "Heat transfer scaling for wall bounded turbulent flows," Applied Mechanics Review 65(3), Art. 031002.
8. Nguyen, Q., Srinivasan, C., and Papavassiliou, D.V., 2015, "Flow induced separation in wall turbulence", Phys Rev E, 91, Art. 033019.

503674 Flow Boiling Using a Piranha Pin Fin Heat Sink

Abstract:

Flow Boiling using a Piranha Pin Fin Heat Sink

Corey Woodcock, Xiangfei Yu, Yoav Peles[1] and Joel Plawsky

Rensselaer Polytechnic Institute; [1]University of Central Florida

The *Piranha Pin Fin* (*PPF*) microdevice is a heat exchanger engineered to dissipate extreme thermal loads using either dielectric fluids such as HFE7000 or water. Fabrication of the device is CMOS compatible and the first generations of the device have been fashioned from silicon substrates. The PPF operates on the principle that heat transfer would be more efficient if heated fluids could be removed nearly as fast as they are created. This process would keep the driving force as high as possible. Thus the heat exchanger is built as two layers; a primary substrate that dissipates the bulk of the heat and shunts the heated fluids to a booster section that continues to heat the fluid and extends the total surface area of the device. Three generations of devices were fabricated. The first two generations were single-layer systems designed to help understand the operation of the primary layer and optimize the shape and location of the pin fins. The third generation was the first, two-layer system to be fabricated and tested.

Figure 1 Representative PPF array

Correlations for the heat transfer coefficient, h_{tp}, based on the Boiling Number and the Weber Number in the PPF design was developed and validated with Gen II. The discrepancy of heat transfer coefficient between predicted and measured values was within 20%. The ratio of two-phase to single-phase pressure drop was correlated with the Boiling Number (Bo) and the Jakob Number (Ja). The discrepancy of the two-phase multiplier between prediction and experimental measurements were within 12%.

The third generation *PPF* heat sinks were the first series to provide both a primary substrate populated with *PPF*s, as well as a booster substrate populated with solid micro pin fins. The substrates are micro-aligned and fusion-bonded to provide a full three dimensional (3D) architecture with bulk silicon properties. Experimental results from the Gen III *PPF* microdevices provide a proof of concept of the true *PPF* heat sink concept with fluid venting into multiple substrates. At low flowrates (G = 670 [kg/s/m^2]) the Gen III *PPF* heat sink has been shown to dissipate heat loads approaching 1.1 [kW/cm^2] at an exit quality of 95%, while at higher flowrates (G = 3,300 [kg/s/m^2]) the thermal limit of the heat sink remains unknown due to exceeding the available power supply at >1.4 [kW/cm^2].

503673 Transport Problems in the Spirit of Stuart Churchill for Teaching and Research at the University of Michigan

Abstract:

Stuart Churchillâ€™s solutions to important heat transport problems have greatly influence both pedagogy and research. The correlation of Churchill and Chu for the natural convection heat transfer to or from a vertical plate, over an astonishing range of Rayleigh numbers is based partly on a similarity solution of a type common in heat, mass, and momentum transfer for problems in chemical engineering. One such problem is asphaltene or wax transport and deposition onto the wall of an oil pipeline, which can be solved using similar methods, as will be discussed. In addition, such problems, and the wealth of their applications make for excellent motivations and examples in the teaching of graduate or undergraduate transport, as will be discussed in the talk.

509054 Solve this! Fundamental Approach to Problem Solving in Industrial Processes I (Invited Talks)

Abstract:

Solving complex plant problems is an everyday exercise that affects all projects and budgets, and it is very rarely introduced to students at any level. In fact, by the time young engineers are few years into working in industry, most have lost the engineering skills that they developed in school. We hope to encourage future engineers to retain those skills and learn the importance of plant problem-solving.

492950 Meeting Challenges in a Land Constrained Solar Economy: Co-Production of Energy and Food Using Photovoltaic Systems over Farmland

Abstract:

The main purpose of this presentation is to demonstrate that contrary to the popular belief that ample land area is available to install photovoltaic modules, in an economy mainly supported by solar energy, there will be severe land availability constraints in many regions of the world—including many states in the USA. This will inevitably lead to competition for land between food and energy [1]. *We therefore propose several innovative PV systems that will allow for the co-production of food and energy from the same unit area of farmland thereby allowing local energy needs to be met by local supply of photons.* This allows us to overcome feedstock (land) and transport (long-distance transmission) limitations for electrical generation that would otherwise be difficult to overcome in a solar economy.

A recent study by MacKay showed that the energy consumption per unit area in densely populated countries such as the UK, Germany, Republic of Korea, and Japan is approaching the energy generation per unit area of photovoltaic farms [2]. This suggests that a transition to a solar powered economy in these nations would require deployment of solar on a large fraction of the nation's total land area, which will directly compete with land requirements for other needs (e.g. agriculture). Many other countries, including the United States are also trending toward this scenario [2]. A recent study by National Renewable Energy Laboratories confirms the large land area requirements of solar farms [3].

Using the available data on current U.S. energy consumption and land availability [4]-[5], we have estimated that as many as 31 states may be unable to meet their needs without importing energy long distances from other states or offshore. This scenario arises because much of the open land in the U.S. is devoted to cropland (21.5%), pasture land (32.3%), and forest land (30.4%). Only 3.6% of the land falls in the category of miscellaneous/other use land [5].

As it should remain a priority to preserve our native forest and wildlife areas, one of the greatest opportunities for increasing the availability of land for solar energy generation is over farmland. These land areas are often easily accessed and surround population centers. However, if PV is to be successful deployed over farmland, novel systems will need to be designed that do not negatively impact crop production. As most photosynthetic activity requires only the visible portion of the spectrum (400-750 nm), these wavelengths can be allowed to past through to plants while the IR portion of the spectrum is harvested as electricity. It is also clear that plants can go relatively extended periods of time without solar irradiance (night) and survive extended periods of reduced lighting (cloudy days) without significant impact to crop production so it may be possible to harvest a portion of the visible portion as well.

Figure 1 shows several possible configurations for harvesting solar energy over farmland. Fig. 1 (a)-(b) are systems that could operate optimally using current industrially-proven solar absorbers (e.g. silicon). Fig. 1 (c)-(f) shows systems for which light splitting would result in even higher performance by optimizing the band gap of the absorber for the portion of the spectrum it is absorbing. Fig. 1(a) is the simplest case in which south-facing PV modules are altered such that every other cell is left open or replaced with a light diffuser. In this system crops never experience shading for very long as the shadow will be constantly moving underneath the panel. Fig. 1(b) is the case in which single- or dual-axis tracking panels are used. Again, the shadow will move throughout the day. In the event that crops cannot tolerate extended periods of shade, these panels can also be patterned and be made twice the size to decrease the time intervals of shading, though the total time each plant would spend in the shade over the course of the day would be the same.

Fig. 1(c)-(f) are systems based on industrially available dichroic mirrors that split the spectrum [6]-[7]. These offer the advantage of concentrating the light for higher efficiencies. As a downside, diffuse portions of the spectrum may not be harvested efficiently. In Fig. 1(c), a tracking parabolic trough system is used in which the IR portion of the spectrum is reflected to a PV cell of appropriate band gap while the visible light passes through. As an option, a thin bifacial PV cells may be placed at the back (or a pattern thereof) to harvest a portion of the visible light and light scattered back from the crops. In Fig. 1(d), the system is non-tracking for mechanical simplicity and directs much of the IR spectrum to bifacial PV cells of appropriate band gap. During non-growing seasons, PV panels could be inserted in place of the transparent support. Fig. 1(e) depicts a tracking, long-pass system in which the visible light is concentrated and collected in fiber optic cables and then redistributed to plants underneath (and potentially other uses such as lighting. Fig. 1(f) depicts a non-tracking, long-pass system in which visible light is again collected in fiber optic cables and then redistributed to plants underneath, while the IR passes through to be collected on PV of appropriate band gap underneath. Promising preliminary calculations have been completed and are on-going to determine how much power can be produced and the optimal band gaps for each system.

The choice of which system to use will depend on how the crops respond to different shading characteristics and how much light deprivation and which portions of the spectrum they can tolerate being deprived of. It will also depend on the relative percentage of diffuse to direct light in a geographic location as the systems in Fig. 1(a)-(b) are better suited for areas with more diffuse light. We explore these options and make some general conclusions on system selection.

References

[1] Solar Projects Sow Tension, Wall Street Journal, Wall Str. J. (2017) A3.

[2] D.J.C. MacKay, Solar energy in the context of energy use, energy transportation and energy storage., Philos. Trans. A. Math. Phys. Eng. Sci. 371 (2013) 20110431. doi:10.1098/rsta.2011.0431.

[3] S. Ong, C. Campbell, P. Denholm, R. Margolis, G. Heath, Land-Use Requirements for Solar Power Plants in the United States - Technical Report NREL/TP-6A20-56290, 2013.

[4] State Profiles and Energy Estimates, (n.d.). http://www.eia.gov/state/.

[5] Major Land Uses, (n.d.). https://www.ers.usda.gov/data-products/major-land-uses/.

[6] T. Ulavi, T. Hebrink, J.H. Davidson, Analysis of a hybrid solar window for building integration, Sol. Energy. 105 (2014) 290–302. doi:10.1016/j.solener.2014.03.006.

[7] E. Gençer, C. Miskin, X. Sun, M.R. Khan, P. Bermel, M.A. Alam, R. Agrawal, Directing solar photons to sustainably meet food, energy, and water needs, Sci. Rep. (2017).

498203 Embodied Phosphorus in Interstate U.S. Food Transfers: Sustainability Implications for Food-Energy-Water Nexus

Abstract:

Modern agriculture relies heavily on nitrogen, phosphorus, and potassium based fertilizers to reduce the yield gap and sustain current crop yields. Recently inorganic phosphorus fertilizers, an essential input to agriculture have garnered attention as they are primarily obtained from non-renewable and geographically concentrated rock phosphate sources. Phosphorus content in the soil is heavily skewed across the world with many regions facing water quality issues due to nutrient overloading in surface waters, while some face nutrient deficiency and soil degradation issues. Food trade is essential for supplying diverse and large quantities of food items to feed the growing population at global, national, and regional scale. Food trade also presents a pathway for transferring large volume of virtual nutrients and associated environmental impacts. The spatial heterogeneity of food production systems can increase or decrease the consumption based environmental impact of food trade in a particular region. At domestic level, vast quantities of food is traded between states resulting in a complex interstate food trade network. Although, estimates of United States (U.S.) centric domestic production and international trade of embodied phosphorus are available, we lack an understanding of interstate flux of embodied phosphorus for the U.S. To avoid unintended consequences and ensure optimal outcomes for a resilient food supply system, a holistic systems approach to studying interconnected resources is required. We model the origin and destination of food and embodied phosphorus flows to quantify trends in regional phosphorus use, identify hotspots, and propose interventions for enhancing the fertilizer use efficiency and overall environmental sustainability of the U.S. food system.

Using publicly available disparate datasets, we develop a food trade and associated virtual phosphorus model for the U.S. To this end, we evaluate over 60 crop items covering commodity groups of cereal grains, and fruits, nuts and vegetables. The interstate food trade network is translated to virtual phosphorus network by using crop yields and phosphate fertilizer application rates from estimates provided by the U.S. Department of Agriculture. Preliminary results indicate that over 655 million tons of food commodities were transferred across the U.S. in 2012 along with 61 million tons of embodied phosphorus fertilizer. Additionally, fruits and vegetables category contributed only 30% by mass for food trade but accounted for half of embodied phosphorus trade compared to cereal grains. We also evaluate virtual phosphorus savings for each individual item to identify phosphorus use efficiency in the US food production system. Additionally, we quantify the embodied energy flows associated with embodied phosphorus trade and discuss the inorganic phosphorus impact on water quality. Furthermore, we apply a network theory framework to understand the structure, robustness, and environmental sustainability of the considered trade networks. From a network perspective, our previous work has found that interstate food trade forms a well-connected, robust network with a majority of states participating in a high volume of trade. We characterize interstate virtual phosphorus network through series of higher order network measures to understand robustness and resilience patterns in virtual phosphorus trade. This will provide a necessary framework to evaluate series of interventions in a highly interconnected food trade system.

495017 Comparing Beccs and DAC for Climate Change Mitigation: The Water-Land-Energy Nexus

Abstract:

The Paris Agreement signaled global commitment to limit average global temperature rise to 2°C and to pursue efforts to achieve 1.5°C. The IPCC AR5 cites negative emission technologies, particularly bio-energy with carbon capture and storage (BECCS), as critical to achieving this[1]. BECCS' ability to deliver negative emissions has been questioned with many uncertainties around land-use change emissions. Furthermore, large-scale biomass use poses several biophysical and socioeconomic risks such as exacerbating scarce water resources, biodiversity loss, and large opportunity costs of land-use for biomass. Capturing carbon dioxide (CO_2) directly from the atmosphere, *direct air capture* (DAC), presents an alternative source of negative emissions. Established capture technologies can be used to extract it from the atmosphere, seemingly without the significant natural resource requirements of BECCS. However, the low concentration of CO_2 in air necessitates a high energy cost of capture.

In this work, a comparison of the strengths and weaknesses of each technology based on key performance metrics (land, water and energy requirements) has been done. A negative emissions target of 12 Gt_{CO2}/y was used; this is the central estimate of what is required from RCP2.6[1] and has been cited as the target beyond which BECCS is no longer economically feasible. A *'farm-to-NET plant'* model of BECCS accounting extensively for the biomass supply chain and power plant specifications (coal co-firing proportions, capture rates) has been built to determine the overall GHG emission balance of the technology. Data from DAC processes proposed in the literature and existing pilot projects have been used to quantify the potential land, water and energy requirements of a large-scale system.

For BECCS, high variability in carbon balance outcomes was observed depending on the feedstock type and source, and assumptions on indirect land-use changes. Depending on decisions taken throughout the supply chain, a BECCS facility could take over 50 years to begin removing CO_2 from the atmosphere, or it could start removing CO_2 from day one[2]. Stark contrasts in resource requirements to meet the mitigation target were found for BECCS and DAC. Depending on the conditions of its deployment, a BECCS plant operating at 100% load factor could need 363-2100 Mha of land and 4-16 Tm^3/yr of water dedicated.

DAC's needs were 3-5 orders of magnitude lower at 0.04-3.3 Mha and 180 km^3/yr[3], thus posing little to these resources. It is important to note that DAC's water requirements are very sensitive to relative humidity and hence location of the plant[3]. Furthermore, the land area associated with the electricity used for DAC process was not accounted for, and could be significant when considering low carbon power, such as solar and wind.

BECCS' status as an energy-positive mitigation option was found to be untrue in some scenarios. Its annual energy output in the case of US-sourced Switchgrass could be as high as 22 EJ/yr, however up to 37 EJ/yr could be required if Brazil-sourced Miscanthus is the choice biomass. DAC energy costs, however, are still significantly higher at 81-274 EJ/y, even when the lower-bound from literature is assumed.

Surplus resource availability, particularly land and water, are highly-dependent on geography. BECCS and DAC have been shown to require contrasting amounts of these. Therefore, the preferred technology must be discussed within the context of a specific location, and in the case of BECCS, biomass source if not locally-grown.

References

1. IPCC. *Climate Change 2014: Mitigation of Climate Change. Working Group III Contribution to the Fifth Assessment Report of the Intergovernmental Panel on Climate Change* (2014). doi:10.1017/CBO9781107415416

2. Fajardy, M. & Mac Dowell, N. Can BECCS deliver sustainable and resource efficient negative emissions? *Energy Environ. Sci.* 0–42 (2017). doi:10.1039/C7EE00465F

3. Socolow, R. *et al.* Direct Air Capture of CO 2 with Chemicals Panel on Public Affairs. *Am. Phys. Soc. - Panel Public Aff.* 100 (2011).

499908 Food Manufacturing: The Shift from Centralised to Distributed Production

Abstract:

Projections for global food demand for a growing world population indicates an increase of 70% by 2050 (United Nations, 2017). The competitiveness among the food producers to satisfy production demand and consumer expectations has led to the search of new alternatives and innovations to make food products more attractive for the consumers, thus increasing company profits (Ingram et al., 2013).

With the growing concern about climate change, the consumer expectations are not limited to the relation quality-price, but also they include a low environmental impact of the product (Kemp et al., 2010).

As in other industries, the food manufacturing has benefited from the economy of scale, where the mass production allows the reduction of the manufacturing cost. However, a centralised production leads to a complex distribution chain, where the distances travelled for the product distribution increase and with them the cost and the CO_2emissions related to transportation.

Another production alternative is distributed manufacturing with shorter distances among the facilities (i.e. farm-factory, factory-distribution centre), thus the transportation cost and hence environmental impact are reduced. Nevertheless, other factors can affect the environmental impact of a production-distribution system, e.g. the carbon footprint of the raw materials.

The aim of this work is to analyse from an economic and environmental perspective several manufacturing-distribution scenarios, and identify the factors with greater influence in the choice between a centralised manufacturing and a more distributed one. For this purpose, a novel framework we call the *honeycomb model* is presented for the techno-economic and sustainability assessment of different manufacturing scenarios. This framework can be applied for the estimation of the optimum configuration that minimises the cost and CO_2emissions of the manufacturing-distribution chain, as well as perform studies of sensitivity of the optimum number of processing plants to certain parameters/characteristics of the system. Here, the tomato paste production is used as case study.

The results indicate that increments in the area of the system analysed and the product demand favour a more distributed manufacturing, while the origin of the raw materials and their carbon footprint could lead to the centralisation of the production.

Sensitivity analysis of the optimum manufacturing configuration provides important information about when a more distributed manufacturing is preferred over a centralised one. This information can be useful for further design of strategies to shift a centralised manufacturing-distribution configuration to a more distributed one when the purpose is to boost local economies.

This project is part of the Centre for Sustainable Energy Use in Food Chains (CSEF) from RCUK.

References

K. Kemp, A. Insch, D.K. Holdsworth, J.G. Knight. 2010. Food miles: Do UK consumers actually care? Food Policy 35:504â€'513.

J.S.I. Ingram, H.L. Wright, L. Foster, T. Aldred, D. Barling, T.G. Benton, et al. 2013.Priority research questions for the UK food system. Food Sec. 5:617â€"636.

United Nations. Available in: http://www.un.org/en/development/desa/news/population/2015-report.html. Date accessed: February 18th, 2017.

494133 Understanding Biomass Value Chains and the Environment-Food-Energy-Water Nexus through Whole-Systems Analysis and Optimisation

Abstract:

Biomass has the potential to provide access to a sustainable supply of resources such as energy, chemicals and materials. Since the production of biomass, i.e. plant matter such as forestry, agricultural crops and residues, requires a large amount of land and water, competes with food production and can affect the ecosystem (e.g. deforestation, leaching of fertilisers and pesticides into water bodies), there is a delicate balance between all of these elements. Together, these are known as the environment-food-energy-water nexus and all of the benefits associated with the use of biomass to meet our requirements for energy and other products must be considered along with the potential impact to the nexus as a whole. Therefore, it is important not only to understand how best to exploit the great potential of biomass but also to understand the detailed interactions among all of the elements in the nexus.

The BEFEW (Biomass and the Energy-Food-Environment-Water Nexus) project funded by the Newton Fund and EPSRC (Grant No. EP/P018165/1) is examining the many different possibilities for producing energy and high-value products from biomass that also maintain the balance in the nexus. Detailed maps of existing land use and soil quality, water availability, quality and tolerance to contamination, and climate projections in terms of temperature and rainfall are used in order to understand the potential for growing biomass. Different processing facilities that convert the biomass to energy or other chemicals are researched and a comprehensive database of such technologies, including a site suitability analysis is developed. Thus, a detailed and rich representation of the influence, on the whole of the nexus, of the types of biomass grown, where they are grown, the types of technologies used (and therefore the products that will be made from biomass) and where they are located can be built. A mixed-integer linear programming (MILP) model is developed in order to determine the combinations of crops grown and technologies used to convert biomass to energy and high-value products (thus forming a biomass value chain) that provide the greatest benefit with the lowest impact on the nexus. Using the model, the efficient and robust value chains that are synergistic with the nexus and have the potential for deployment over a wide range of scenarios with different combinations of key socio-economic and environmental performance indicators and constraints are identified. In this conference, the latest progress in the BEFEW project will be presented, with a focus on the MILP model and case studies.

505039 Innovative Approaches to Achieving FEW Nexus Goals that can Influence Near-Term & Long-Term Decision Making

Abstract:

The world population is estimated to reach to 9.7B by 2050, resulting in increased demand for resources and increased production of waste. As a result, today's waste disposal systems and approach to waste management will not be adequate for future applications. Therefore, a systematic approach to sustainable food, energy and water management is needed to meet increased demand for resources and efficient waste utilization systems. Currently, the food, energy and water (FEW) sectors use technologies developed to address the needs of a given sector (*i.e.* energy, water, food or agriculture), and wastes from each sector are usually managed separately. Production systems underlying these sectors have traditionally treated pollution and waste as externalities that are diluted into the ambient environment, and separate infrastructure systems have consequently been developed with unnecessary inefficiencies that are increasingly problematic. Because neither landfilling nor industrial composting provide ultimate reutilization of organic waste, and neither can achieve both energy and nutrient recovery in one process from organic waste, we need to transition from a "Traditional Linear Waste Disposal" approach to a sustainable "Closed Loop Resource Recovery" approach. In addition, finding efficient and affordable solutions to avoid GHG emission leakages from landfills and composting facilities through the use of Anaerobic Digestion infrastructure will be a key element of a sustainable system for future waste management. Thus, exploring how FEW systems can efficiently be optimized to (1) repurpose or recycle organic waste products, (2) internalize traditional externalities, and (3) integrate wastes with resource inputs across systems, by deploying integrative management strategies is essential. In addition, integrating other relevant disciplines that bring the necessary economic, social and educational perspectives to addressing this problem, will achieve broader impacts across the system.

497155 Optimizing Drinking Water Disinfection: Balancing Corrosion, Byproduct Formation, and Pathogen Removal

Abstract:

Currently, treatment plants most commonly disinfect drinking water with free chlorine to remove pathogens. Free chlorine, however, is prone to forming carcinogenic disinfection byproducts from reactions with organic matter. Ill-considered application of disinfectants and complications from cleanup of their byproducts, recently resulted in catastrophic corrosion and discharge of lead into the water supplies of Washington D.C. and Flint, Michigan. To avoid a recurrence of such mistakes, the strategy for drinking water disinfection in vulnerable, lead-containing supply networks was reconsidered. Specifically, we evaluated the merits and drawbacks of replacing free chlorine with chlorine dioxide in systems with lead infrastructure.

Free chlorine, the most commonly used disinfectant in drinking water treatment, is a powerful oxidant that readily reacts with viral and bacterial pathogens, but also with other water constituents, including dissolved organic matter and metallic infrastructure surfaces. Reactions with organic matter produce hazardous byproducts, while reactions with infrastructure surfaces can increase the concentration of metals, including lead, in drinking water. Some water treatment facilities have substituted free chlorine with chloramines, an alternative disinfectant, to meet disinfection byproduct regulations set by the Environmental Protection Agency. However, chloramines produce their own set of toxic byproducts and have lower disinfection efficiency than free chlorine. Removal of disinfection byproducts and the precursors to said byproducts can exacerbate corrosion in the metal pipes of water distribution systems.

Unlike free chlorine, chlorine dioxide has a significantly lower propensity to form organic byproducts, and it is as potent, and in some cases, more potent than free chlorine for the inactivation of bacteria and viruses. In a properly managed dose, chlorine dioxide can deliver effective disinfection of drinking water[1]. Chlorine dioxide is more expensive to use for drinking water treatment than free chlorine and it can decay in water to form chlorite and chlorate, two potentially hazardous inorganic byproducts. It has been assumed that chlorine dioxide will behave similarly to free chlorine with regards to lead corrosion mechanisms because of their similar oxidation reduction potentials, but there have been few studies to confirm.

Using three different lead minerals, a series of experiments were performed to measure the kinetics of chlorine dioxide decay and byproduct formation, considering chlorine dioxide concentration, pH, presence of relevant anions, and presence of natural organic matter. Bicarbonate/carbonate, sulfate, silicate, bromide and phosphate are common anions in drinking water systems and were all considered in this study for the following reasons. The chloride to sulfate mass ratio has been shown to influence lead leaching[2]. The bromide to bromate pathway is a concern for drinking water standards. Phosphate is commonly added to drinking water to control corrosion[3]. Finally, natural organic matter has also been shown to be an important variable when analyzing corrosion dynamics and chlorine dioxide reactions[4].

Spectrophotometry was used to standardized chlorine dioxide concentrations at 359nm. Ion chromatography, by a Dionex 5000 Reagent-Free ion chromatograph, was used to measure byproducts (chlorite and chlorate) and specified anions. All experiments were batch reactions, conducted in the dark under constant agitation of a stir plate. The influencing factors on the kinetics of chlorine dioxide decay and byproduct formation in the presence of lead minerals will be presented and discussed. Our data will be contrasted with the results of published results on other metals including nickel, copper, and iron.

Aging lead infrastructure will continue to be an issue in the United States. The replacement of distribution systems could take decades and at high environmental, economic, and social costs due to our poor understanding of corrosion and byproduct formation mechanisms. There is a demand for alternative disinfectants that have the capability to protect society from waterborne diseases without polluting our environment with lead and toxic byproducts. This work is intended to provide guidance, on the viability of chlorine dioxide as a disinfectant, to avoid costs and dangers associated with lead corrosion.

References

(1) Gagnon, G. A.; O'Leary, K. C.; Volk, C. J.; Chauret, C.; Stover, L.; Andrews, R. C. *J. Environ. Eng.* **2004**, *130*(11), 1269–1279.

(2) Edwards, M.; Triantafyllidou, S. *J. / Am. Water Work. Assoc.* **2007**, *99*(7).

(3) Dodrill, D.; Edwards, M. *Am. Water Work. Assoc.* **1995**, No. C, 74–85.

(4) Liu, C.; Von Gunten, U.; Croue, J. P. *Environ. Sci. Technol.* **2013**, *47* (15), 8365–8372.

496642 Enhanced Oxidation of Toxic Organic Pollutants By Ferrate(VI) and Its Modifications in Aqueous Solution

Abstract:

Iron in oxidation state of +6, also known as ferrate(VI) (Fe(VI); $Fe^{VI}O_4^{2-}$), is receiving great attention as a multipurpose green chemical with a wide range of applications in different research areas such as organic synthesis, water and wastewater treatment, disinfection of viruses and iron batteries. In recent years, there has been an increasing interest in the application of Fe(VI) in water and wastewater treatment because of its unique oxidation, disinfection and coagulation properties (Fe(VI) is reduced to Fe(III) initiating coagulation). Although a lot of work related to the oxidation of pollutants by Fe(VI) has been reported, limited work related to the activation of Fe(VI) to decrease the required dose of the oxidant has been done so far. Fe(VI) could be activated by acid (e.g. HCl and HNO_3) to significantly enhance the oxidative transformation of caffeine (psychostimulant; CAF), acesulfame K (artificial sweetener; ACE) and atenolol (Î²-blocker; ATL) in aqueous solution, at a relevant to wastewater treatment pH range of 8.0-9.0 and room temperature. Almost complete removal of CAF and ACE by acid-activated Fe(VI) was achieved at Fe(VI) to CAF/ACE molar ratio of 8.0 and HCl to Fe(VI) molar ratio of ~1.5, in seconds to minutes. Interestingly, only ~60% removal of CAF/ACE by Fe(VI) (non-activated Fe(VI)) was achieved in hours. ATL was not removed by Fe(VI) at Fe(VI) to ATL molar ratio of 10.0. Importantly, around 30% ATL removal by acid-activated Fe(VI) was observed at the same Fe(VI) to CAF molar ratio (10.0) and HCl to Fe(VI) molar ratio of ~1.5. The effect of individual anions, cations and natural organic matter usually present in wastewater on the oxidation of CAF by acid-activated Fe(VI) and Fe(VI) (non-activated Fe(VI)) has been also investigated. Results showed that acid-activated Fe(VI) oxidation of CAF is not affected by ions. Interestingly, a negative effect on the removal of CAF by Fe(VI) (non-activated Fe(VI)) in the presence of calcium, magnesium and sulfate ions was observed. The oxidative transformation of CAF by both Fe(VI) and activated Fe(VI) decreased in the presence of natural organic matter. However, activated Fe(VI) could mineralize some of the other organic components present in natural organic matter indicating higher oxidation capacity of Fe(VI) upon activation. All in all, activated Fe(VI) technology is more efficient (lower dose of Fe(VI) required) and faster for treating water than Fe(VI) (non-activated), and it is not affected by common ions present in wastewater. The authors suggest that the findings of this study will be of great interest to the scientific community dealing with oxidation of pollutants in water and will initiate the investigation of the oxidation of a range of compounds by activated Fe(VI).

488490 Cyanide Recovery from Barren Solution Using UV Photodissociation and Gas-Filled Membrane Technology

Abstract:

A process was developed to recover cyanide from barren solution containing simple and metal cyanide complex. The technology includes pretreatment of metals and oxyanions, UV-light-aided photodissociation of metal-cyanide complex, and Gas Filled Membrane (GFM). Using a 200w medium pressure UV light at 254nm wave length, more than 85% cyanide ions (CN^-) were recovered from soluble iron cyanide complex at pH between 10.0 and 10.5 within one hour of reaction. The effluent from the UV reactor was filtered and pH was adjusted between 5.0 and 5.5 in a leak proof container, prior to pumping to the GFM that separated hydrocyanic acid (HCN^0) from water. Liquid solutions were pumped on both sides of the membrane. One side contained hydrocyanic acid solution, and 1.0 N sodium hydroxide solution was recirculated on the other side of the membrane. Within one hour, more than 95% of the cyanide was recovered in the sodium hydroxide solution. The impacts of pH, UV dosage, sequestering agent dosage, and reaction time on the photodissociation efficiency will be presented. Furthermore, the impact of membrane flux rate on cyanide recovery will be discussed.

493939 Removal of Phenol from Wastewater of Petroleum Refinery By Using Advanced Oxidation Process

Abstract:

Advanced oxidation processes (AOPs) are promised techniques for treatment of wastewater which containing refractory contaminants. In the present study an experimental work and kinetic study have been carried out to investigate performance of Lab-scale trickle bed reactor for phenol degradation of wastewater effluent from a petroleum refinery in Baghdad. Several process variables are effecting on the system performance. These selected variables are: initial concentration of phenol (5 to 15 mg/L), temperature (120 to 160 °C), pressure (0.2 to 0.9 Mpa), air flow rate (0.01 to 0.03 L/min), pH of solution (3 to 10), liquid flow rate (0.6 $x10^{-3}$ to $1.66x10^{-3}$ L/min), and flow modes are down-flow and up-flow. The experimental set-up is mainly consists of a stainless steel reactor of dimensions (13 mm ID x 160 mm L), packed with 5 gram activated carbon (0.8 mm particle diameter) as catalyst. The optimum experiments were obtained by using 3-level, 4-factor Box-Behnken design (BBD), with aid of modified response surface methodology (RSM).Results showed that phenol degradation was enhanced by increasing of temperature, increasing of pressure, and increasing of gas flow rate, while initial concentration of phenol and liquid flow rate would give negative images. High removal of phenol at about 97% was obtained at optimum operating condition of; LHSV=7 h^{-1}, temperature=160°C, oxygen partial pressure= 0.9 MPa, and phenol concentration= 5 mg/L. Down-flow mode has been proven more effective than up-flow mode for reactor performance. Kinetics of the reaction behaved as pseudo first order reaction with respect to phenol concentration,and with 0.6 order with respect to oxygen solubility. Activation energy was equal to77.7 kJ/mol, and pre-exponential factor was equal to ($1.826*10^9$ L/kg-cat .h). It was observed that deactivation constant (k_{do}) was equal to ($2.9*10^{11}$ min $^{-1}$),was about (114.43 kJ/mol). When catalyst deactivation had been taken into account, the oxygen order was equal to 1.67 and the frequency factor was about ($4.2*10^9$ L/kgcat.h), also the activation energy was (95.35 kJ/mol). The kinetic parameters (i.e., k and E) were estimated at each operating temperature by using statistical analysis method. Catalyst effectiveness and deactivation rate were also estimated. Reasonable agreements have been found when compared the simulated results with the experimental data.

492840 The Role of Oxygen Vacancies of CeO_2 Supported Catalyst on Catalytic Performance of the CO_2 to CO

Abstract:

The performance of a catalyst depends on many factors. It is usually hard to discuss these effects separately due to their interwoven relationship. Here, we focus on the catalytic activity of nanosized Cu supported on cerium oxide by the space-confined method and mesoporous silica, whereby Cu nanocatalysts are deliberately controlled under the same loading, particle size and coordination number to highlight the interactions caused by different supports. It is interesting to note that the redox temperature of cerium oxide is dramatically decreased from >600°C to <300°C by CuCeOx solid solution so that its conversion is 18% with 100% selectivity approaching the equilibrium conversion 23% of the CO_2 to CO at 1 bar and 300°C the by flow reactor.

Oxygen vacancies generated by the reducible cerium oxide were found to be able to contribute more carbonate with bidentate type absorption. The space confined method was capable of generating high Cu dispersion (~50%) from nano-size CuCeOx solid solution. The Cu maintains the CeO_2 support to keep the oxygen vacancies recycling by hydrogen spill over behavior at 300°C when the reaction of CO_2 to CO oxidized the catalyst. However, similar metal-support was not seen in the mesopeorous silica system.

We have demonstrated that the active sites locate in the proximity between support and reduced nano-sized metal by synchrotron analysis, TEM mapping, in-situ FTIR and catalytic reaction. Furthermore, the work presented illustrates the significant role of oxygen vacancies in the metal-support interaction and how the potential of oxide supported nanocatalysts can be further explored and tuned by for the space confined method and thermal treatment for CO_2 to CO.

Advanced Ni-CeO$_2$ Nanocatalysts for Chemical CO$_2$ Recycling

T. Smith, T. Stroud, E. Le Saché, H. Arellano-Garcia, T.R. Reina[*]

Department of Chemical and Process Engineering, University of Surrey, Guildford, GU2 7XH, UK

[*]Corresponding author: t.ramirezreina@surrey.ac.uk

Introduction

Carbon dioxide is one of the major issues facing the world currently, the adverse effects on the environment that CO$_2$ emissions cause are well documented. Major steps are being taken by industry to mitigate these effects, with a proportion of the CO$_2$ produced being captured each year. This scheme to reduce carbon emissions into the atmosphere is part of a worldwide initiative, whereby the carbon captured can be stored, but there is still a lot of progress to make within this field before this is a truly viable option. There is however, another exciting alternative which is the chemical upgrading of CO$_2$ to obtain fuels and chemicals [1] [2]. This way, the depletion of CO$_2$ emissions is accompanied by an extra motivation in terms of the generation of added-value products.

Among the different alternatives for CO$_2$ utilisation, dry reforming of methane is a promising route leading to the production of syngas (H$_2$ and CO) [3]. This is a highly useful and valuable intermediary, which can be used in processes like the Fischer-Tropsch Synthesis to create long chain hydrocarbons, such as diesel [4].

The issue that arises in the dry reforming process, is the deactivation of the catalysts used in the reaction of the feed chemicals. This deactivation is due to sintering of the active phase and/or carbon deposition [3] [5]. In order to overcome these problems noble metal based catalysts have been employed but their cost make them unsuitable for a real application.

The aim of this work is to develop a highly active dry reforming catalyst based on Ni. Considering the disadvantages of Ni in terms of deactivation, we have developed multicomponent materials based on monometallic and bimetallic Ni/CeO$_2$/Al$_2$O$_3$ catalysts with excellent features for reforming reactions. In this study, all the catalysts contain 10 %wt. Ni and 20 %wt. CeO$_2$. As for the bimetallic formulations, Pt and Sn have been selected as promoters with a nominal loading of 0.3 and 0.8 % wt. respectively.

Experimental

The catalysts were prepared by wet sequential impregnation using alumina (Sasol) and the relevant precursors (Sigma Aldrich) employing a Rotovap, with a session of drying and calcining between each step of the impregnation. The drying was performed at 80°C and was left overnight, the calcination was performed at 700°C for 4 h.

The dry reforming of methane was carried out at atmospheric pressure in a quartz tube reactor. 0.1 g of catalyst was used per reaction supported upon a reactor bed of quartz wool. The catalyst was reduced in situ at 800°C for 1 h, in a 100 mL/min flow containing 20% H$_2$ and 80% N$_2$. The reaction was then conducted at 700°C for over 20 h, in a 100 ml/min flow containing 12.5% CH$_4$, 12.5% CO$_2$ and 75% N$_2$ (WHSV = 60,000 mL g^{-1} h^{-1}). The gas products of the reaction where analysed using an ABB AO2208 gas analyser. The spent catalysts were recovered and characterized.

All the fresh catalyst samples have been characterized fully, by XRD, H$_2$-TPR and N$_2$ adsorption. The spent samples have also been characterized fully, post reaction, by XRD, Raman and SEM.

Results and Discussion

The textural properties of the sample catalysts are shown below in **Table 1**. What can be observed from the data is that the specific surface area as well as the pore volume of the catalyst is reduced by the addition of ceria. However, the addition of Sn and Pt does not affect the textural properties of the catalysts. The decrease in the surface area can be attributed to ceria species covering the surface of Al$_2$O$_3$, this is corroborated by the decrease in pore volume.

Table 1: *Textural properties of the catalyst samples.*

Catalyst	BET (m^2 g^{-1})	Pore Volume (cm^3 g^{-1})
Ni/Al	164	0.42
Ni/CeAl	116	0.28
SnNi/CeAl	120	0.29
PtNi/CeAl	118	0.28

The XRD data shown below in **Figure 1** provides insight into the chemical structure of the prepared catalyst samples. The absence of any crystalline peak attributed to Ni species indicates a good metallic dispersion and small particle sizes. However, the presence of NiAl$_2$O$_4$ cannot be fully discarded since the spinel peaks overlap with those of the bare Al$_2$O$_3$. In addition, the presence of ceria fluorite type structures is evidenced in the XRD patterns. As for the bimetallic samples, the XRD depicted for PtNi/CeAl is comparable to that of the Ni/CeAl base catalyst except for the small PtO peak at $2\Theta = 40°$, while the SnNi/CeAl catalyst shows a very similar XRD pattern to Ni/CeAl.

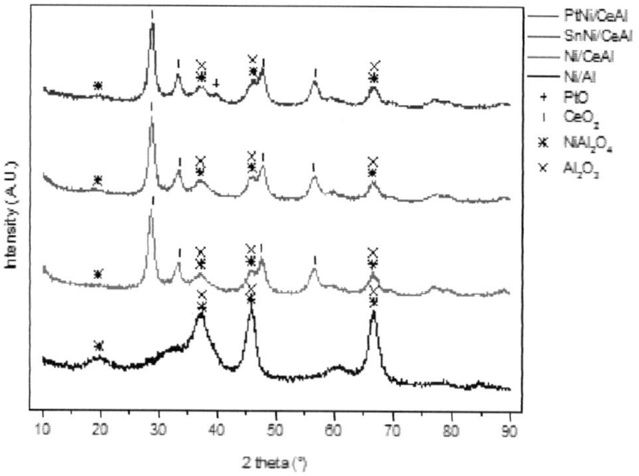

Figure 1: *XRD patterns of the catalysts.*

The redox properties of the catalysts are shown in the H$_2$-TPR profiles displayed in **Figure 2**. All the profiles present, contain a peak at high temperatures (650-850°C), this is due to the bulk reduction of the NiAl$_2$O$_4$ spinels, which due to their structure require a high temperature to be reduced. The later agrees with the XRD data. The three catalysts containing ceria exhibit a shoulder at 850°C as well as much larger H$_2$ consumption peak than the base Ni/Al catalyst. This is attributable to the bulk ceria reduction occurring in the temperature range 700-900°C. Also, a reduction process at 250°C on the PtNi/CeAl profile was identified and attributed to the reduction of PtO.

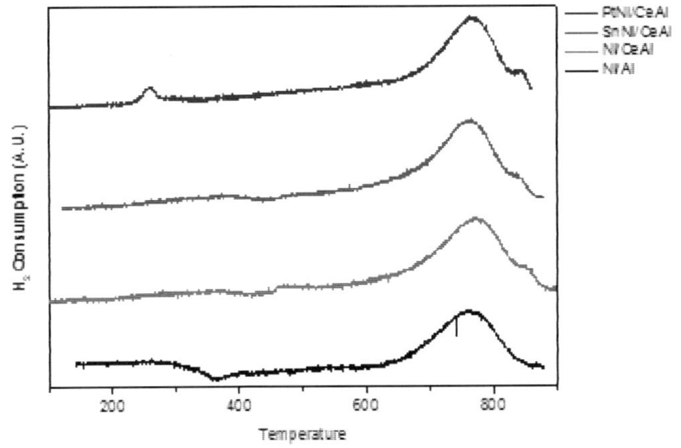

Figure 2: H_2-TPR profiles of the catalysts.

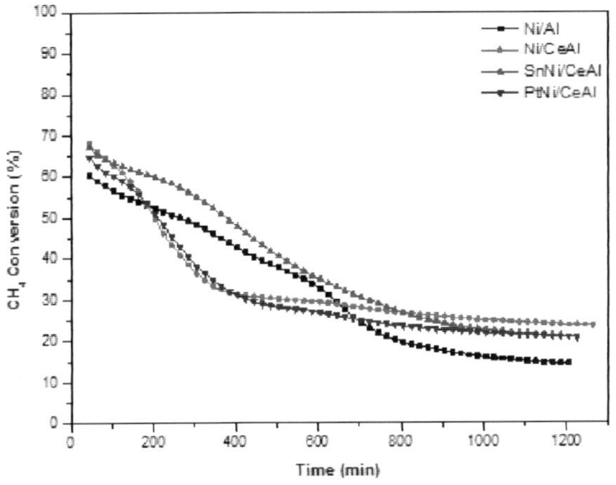

Figure 3: CH_4 conversions of the catalysts.

Figure 3 and **Figure 4** below show the CH_4 and CO_2 conversions for all the catalyst after 20+ h on the reaction stream. It can be seen that the addition of ceria helps to significantly stabilise the catalyst due to its high oxygen storage capacity (OSC), in comparison the Ni/Al conversions are observed to drop especially after 10 hours. This is attributed to the formation of hard carbon on the surface of the catalyst blocking the accessibility of the molecules to the active sites. The effect the promoters have on the conversions differs, the Sn promoted catalyst has higher conversions to begin with before stabilising at similar conversions to the other ceria based catalyst. This has been attributed to the smaller particle size and higher particle dispersion the Sn promotion causes. By contrast the addition of Pt as a promoter hardly affects the catalytic performance of the Ni/Ce/Al sample. The Sn promoted catalyst also had the highest H_2/CO ratio to begin with, of 0.97, before stabilising at a similar value to the other catalysts of around 0.75.

The characterisation performed post reaction on the spent catalysts (not shown for sake of briefness) concur with the conversion hypothesis. Large concentration of hard carbon were deposited on the Ni/Al while smaller concentrations and softer carbon haven been found on the ceria and ceria-tin promoted catalysts.

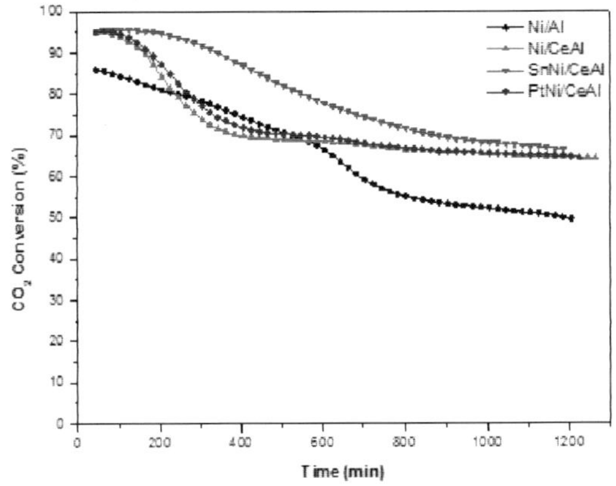

Figure 4: CO_2 conversions of the catalysts.

Conclusions

This work evidences that multicomponent Ni based catalysts can be efficiently utilised for the chemical CO_2 recycling via dry reforming. High levels of CO_2 conversion and good H_2/CO ratios have been achieved in general. The bimetallic Ni-Sn system supported on CeO_2-Al_2O_3 was the most promising material among the studied catalysts. The excellent behaviour of this catalyst is due to its improved redox properties and the Ni-Sn synergy which prevents active phase sintering and carbon poisoning.

Further studies are currently being undertaken by our group to discern the effect of the operating conditions as well as the long-term stability of these advanced materials, in order to explore their potential application in real CO_2 utilisation processes.

References

1. Pires JCM, Martins FG, Alvim-Ferraz MCM, Simoes M. Recent Developements on Carbon Capture and Storage: An Overview. Chemical Engineering Research and Design. 2011; 89(9): p. 1446-1460.
2. Leung DYC, Caramanna G, Maroto-Vler MM. An Overview of Current Status of Carbon Dioxide Capture and Storage Technologies. Renewable and Sustainable Energy Reviews. 2014; 39: p. 426-443.
3. Pakhare D, Spivey J. A Review of Dry (CO2) Reforming of Methane over Noble Metal Catalysts. Chemical Society Reviews. 2014; 43: p. 7813-7837.
4. de Araujo GC, de Lima SM, Assaf JM, Pena MA, Fierro JLG, Rangel MdC. Catalystic Evaluation of Perovskite-Type Oxide LaNi1-xRuxO3 in Methane Dry Reforming. Catalysis Today. 2008; 133: p. 129-135.
5. Charisiou ND, Siakavelas G, Papageridis KN, Baklavaridis A, Tzounis L, Avraam DG, et al. Syngas Production via the Biogas Dry Reforming Reaction Over Nickel Supported on Modified with CeO2 and/or La2O3 Alumina Catalysts. Journal of Natural Gas Science and Engineering. 2016; 31: p. 164-183.

497365 One-Step Synthesis of Transition Metal Doped Titanium Dioxide Catalysts for Oxygen Removal

Abstract:

Carbon dioxide (CO_2) capture and sequestration have been considered as promising techniques to alleviate the emission of greenhouse gas. However, it is difficult to cost effectively deploy them in conventional coal-fired combustion plants due to the diluted CO_2 concentration in the flue gas, requiring additional methodologies for concentrating the stream. An oxy-fuel combustion technique which uses pure oxygen (O_2) to obtain a concentrated CO_2 stream has been developed to solve this problem, and has obtained a sequestration-ready gas stream. However, this gas stream contains O_2 as well as CO_2, which can cause damage to the gas infrastructure through corrosion. To overcome this shortcoming, catalytic O_2 removal system with hydrocarbons has been recently applied. In this study, we develop 1) a one-step synthesis of transition metal doped catalysts and 2) an efficient catalytic O_2 removal system with methane (CH_4). To achieve this, we first synthesize different transition metal (Cu, Fe, Ir, Ni, Mn, Rh, Ru) doped titanium oxide (TiO_2) catalysts in a flame aerosol reactor. Compared to liquid phase synthesis methods such as wet impregnation and sol-gel, flame synthesis is a gas phase technique that can produce homogeneous nanoparticles and can be scaled up for high throughput production. The different catalysts are characterized by BET, X-ray diffraction, scanning electron microscopy and transmission electron microscopy to relate nanoparticle characteristics to performance. Second, we test the synthesized catalysts for O_2 removal with CH_4 to optimize the catalyst. Third, we optimize the synthesis conditions (dopant concentration (0-5 wt.%), fuel flow rate (0.8-1.8 LPM)) to get the highest O_2 removal efficiency.

494289 Carboxylation of Propylene Oxide to Propylene Carbonate

Abstract:

Polycarbonates, a class thermoplastic polymer, have wide ranging applications in automotive & transportation, construction, packaging, consumer goods, optical media, electrical & electronics and medical devices. Conventionally, manufacturing of polycarbonates involves phosgenation of bis-phenol-A, which has drawbacks due to toxic phosgene and corrosive products like HCl in stoichiometric amounts. Alternative non-phosgene route for polycarbonates involves carboxylation of propylene oxide as a key step. This is also an example of CO_2 utilization for value added chemicals. Cyclic carbonates are also widely used as aprotic solvents, anti-foaming agents, antifreeze, plasticizers, and monomer for various commodity polymers. Several studies reported carboxylation of epoxides employing either homogeneous or heterogeneous catalysts. Reported heterogeneous catalysts include metal oxides, zeolites, polymer supported quaternary onium salts, ion exchange resins, and polymer supported ionic liquids. However, the main challenges in this reaction are low catalyst activity, and high CO_2 pressure. In this study, we systematically investigated the influence of solvents on ion exchange resin catalyzed carboxylation of propylene oxide with the aim of enhancing the activity and selectivity. Intrinsic kinetics of carboxylation of propylene oxide has also been studied in a batch slurry reactor to understand the effect of reaction conditions and develop a suitable rate equation.

The experiments for solvent effects and kinetic studies were carried out in a stirred pressure reactor with 100 cm^3 capacity with provisions for control of agitation speed, temperature and sampling of liquids. In these experiments, the CO_2 pressure in the reactor was kept constant by continuous supply through a CO_2 reservoir using a constant pressure regulator such that the temporal reaction progress was followed from the pressure decrease in the reservoir. At the end of each experiment, liquid products were analyzed for PO and PC to assess the material balance. The effect of solvents and the order of activity is: Methanol > Ethanol > Isopropyl alcohol > Dimethyl formamide > N-methyl-2-pyrrolidone > Propylene carbonate > Dimethyl carbonate and it was found that using methanol as a solvent rate of carboxylation were enhanced by a factor of 3. The polar protic solvents (such as linear alcohols) are strong nucleophiles, strong hydrogen-bond donors and acceptors and interact strongly with the electron-deficient electrophiles, epoxides and CO_2. In contrast, polar aprotic solvents are weak nucleophiles compared to the polar protic solvents resulting in low activity. These conclusions, based on CO_2 solubility measurements in methanol and PC solvents along with kinetic studies of PO carboxylation, will be presented and discussed.

Improving the Performance of Sn Promoted Ni/Al_2O_3 Catalysts for the Dry Reforming of Methane

*T. Stroud, T. Smith, E. Le Saché, H. Arellano-Garcia, T.R. Reina**

Department of Chemical and Process Engineering, Faculty of Engineering and Physical Sciences, University of Surrey, Guildford, UK

Corresponding author: t.ramirezreina@surrey.ac.uk

Introduction

Prevention of CO_2 reaching the upper atmosphere is a topic of great interest in the world today. Co-currently, green energy production and chemical feedstock sources are being highly researched, where combining these issues represents the most efficient solution. Dry reforming of methane (DRM) represents one such solution:

$$CH_4 + CO_2 \rightleftharpoons 2CO + 2H_2$$

$$\Delta H_{298K} = +247 \text{ kJ/mol} \qquad \Delta G_{298K} = +170 \text{ kJ/mol}$$

The products of this reaction (known colloquially as syngas) are a highly useful precursor for a number of chemical processes, such as the production of methanol and synthesis of liquid fuels via Fischer-Tropsch (FTS).

Given the endothermic nature of the reaction, high reaction temperatures are required to achieve favourable conversions of the reactants to syngas. Unfortunately, elevated temperatures also promote the formation of solid carbon via a number of further side reactions [1]. This carbon then deposits on the surface of the catalyst which (alongside metal sintering) causes the catalyst to deactivate.

Hence an effective catalyst must resist sintering and be resilient towards the formation of carbon upon the surface; whilst also being relatively inexpensive, such that it can be realistically employed within the wider industry. Nickel has previously been identified as a viable active metal to use, whilst an alumina (Al_2O_3) support helps disperse the Ni over a wider surface area, improving the activity over that of just Ni alone [1].

However, this still suffers from severe carbon deactivation and hence there remains potential for improvement. The aim of this work is to develop advanced catalysts for chemical CO_2 recycling via DRM using tin and ceria as promoters of Ni/Al_2O_3

Experimental

Catalysts were prepared by the method of sequential impregnation, where a base of 10 wt.% Ni was first prepared by adding the appropriate amount of $Ni(NO_3)\bullet6H_2O$ (Sigma-Aldrich) to an alumina (SASOL) base, using acetone (Sigma-Aldrich) as the solvent. This was then thoroughly mixed for 30 minutes, before evaporating the acetone off with the aid of a vacuum pump. The sample was then further air dried in an oven overnight at 80 °C, after which calcination was performed at 700 °C for 4 hours. The resultant Ni/Al catalyst was employed as a reference material in this paper. Following the same preparation method, 3 promoted catalysts were prepared: Sn 0.02 mol.% Ni/Al, Sn 0.04 mol.% Ni/Al, Sn 0.02 mol.% Ni/Ce-Al. The tin was added using a $SnCl_2$ (Sigma-Aldrich) precursor and the ceria by $Ce(NO_3)_2\bullet H_2O$ (Sigma-Aldrich) aiming for a nominal CeO_2 loading of 20 wt.%.

Experiments were run in a quartz tube reactor at atmospheric pressure, utilising quartz wool as a reactor bed. An electric fired oven was used to ramp up (15 °C/min) and maintain the temperatures used for reduction (800 °C) and reaction (700 °C). Reduction was performed in a total flow of 100 ml/min for 1 hour in an atmosphere consisting 20 % H_2 (BOC) and 80 % N_2 (BOC).

Each reaction was run for over 20 hours, using 0.1 g of catalyst and a total inlet flow of 100 ml/min, corresponding to a weight hourly space velocity (WHSV) of 60, 000 ml/(g-cat h). Of the inlet flow, 12.5 % was CO_2 (BOC), 12.5 % CH_4 (BOC) and the remainder being that of N_2.

All fresh catalysts were characterised by means of XRD, H_2-TPR and N_2 physisorption (BET) analysis. Selected spent catalysts also underwent SEM and Raman.

Results and Discussion

H_2-TPR analysis (not shown here) confirms that a reduction temperature of 800 °C is sufficient to reduce the metal oxides to their elemental forms, whilst the XRD data (**Figure 1**) shows the Ni particles have been well dispersed. The increase of Sn loading as a promoter appears to have little effect on the crystalline structure of the samples, whereas cerium shows typical diffraction peaks of CeO_2 fluorite phases.

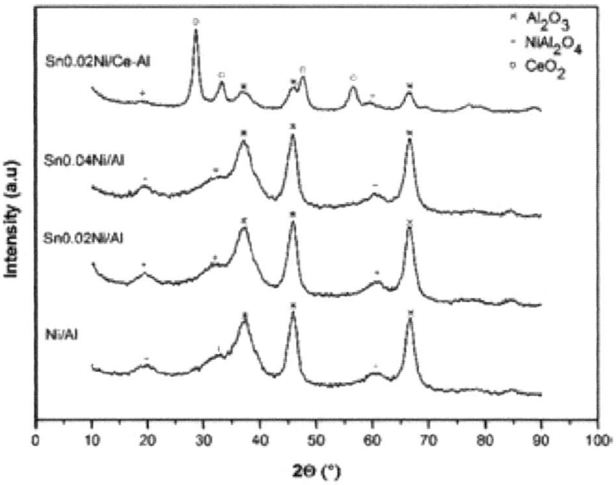

Figure 1 XRD profiles of the sample catalysts

N_2 physisorption results (**Table 1**) show the addition of Sn does not affect the textural properties of the reference Ni/Al material. However CeO_2 reduces the surface area and volume of the pores, which is down to the CeO_2 nanoparticles covering the surface of the primary support (Al_2O_3).

Table 1 Results of N_2 physisorption analysis on fresh catalysts

Sample	S_{Bet} (m^2/g)	V_{Pore} (cm^3/g)	D_{Pore} (nm)
Ni/Al	164	0.42	10.1
Sn0.02 Ni/Al	173	0.42	9.8
Sn0.04 Ni/Al	172	0.42	9.8
Sn0.02 Ni/Ce-Al	120	0.29	9.6

In terms of catalytic performance, both methane and CO_2 conversions (**Figure 2** and **Figure 3** respectively) are improved by the addition of Sn, showing less deactivation with time when compared with the un-promoted Ni/Al catalyst. Initially, both the un-promoted and 0.04 Sn sample appear to perform better than the 0.02 Sn catalyst, but this quickly changes around the 500 minute mark for CH_4 conversion and 450 minute mark for CO_2. Increasing the quantity of Sn from 0.02 mol.% to 0.04 mol.% exhibits a decrease in performance with time. This is theorised to be down to the extra Sn blocking the active sites [2], therefore reducing performance; hence a ratio of 0.02 mol.% of Sn was selected as the optimum to be further improved upon.

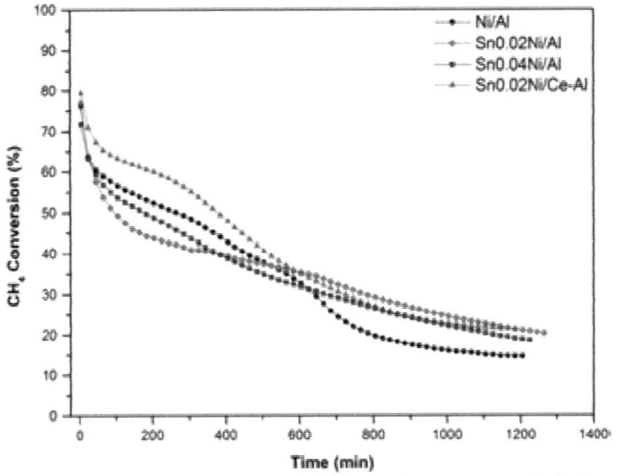

Figure 2 Methane conversions for the catalysts over a period of approximately 20 hours

Ceria is an excellent redox promoter due to its high oxygen storage capacity [3]. Therefore by adding CeO_2, the amount of surface oxygen available increases [1]. This helps with the removal of carbon species from the catalyst surface, increasing the lifespan and improving activity. This has been observed in the case of the multicomponent catalyst (Ni-Sn/Ce-Al), with conversions being greater for longer periods of time in the case of methane. An impressively high peak of 95 % CO_2 conversion can be seen, with a seemingly stable 66 % conversion after approximately 20 hours of on-stream operation. Lower conversions of methane (around 20 % after 20 hours) in comparison to CO_2 are to be expected, due to decomposition of CH_4 being the rate determining step and also the primary depositor of carbon [4].

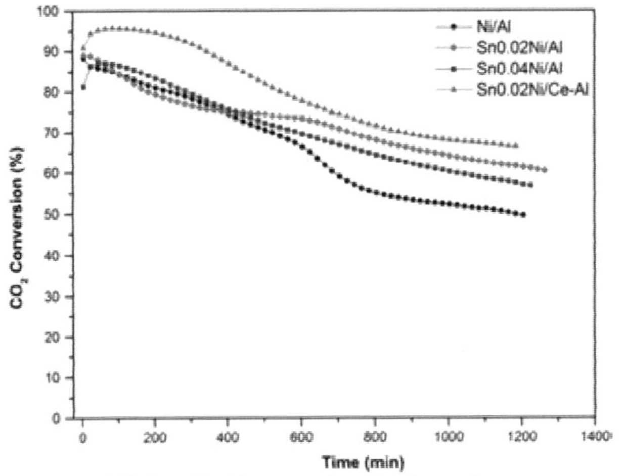

Figure 3 Carbon dioxide conversions for the catalysts over a period of approximately 20 hours

For the H_2/CO ratio (**Figure 4**), the Ce doped sample appears the best candidate early on, peaking at a ratio of 0.97 after *ca.* one hour of operation. However this eventually decreases down to the same level (a ratio of around 0.75) of the un-promoted and 0.04 Sn samples after 20 hours. The Sn0.02 Ni/Al catalyst shows a slightly more favourable ratio of 0.81 after 20 hours, showing the addition of tin is also effective at improving the H_2/CO ratio; with the optimum ratio again being 0.02 mol.%.

With the addition of ceria, a slight drop in performance is observed when compared with the 0.02Sn promoted catalyst. At the end of the experiment, a higher conversion of CO_2 is present for Ce, with a similar methane conversion. This therefore results in a comparatively greater abundance of CO being present in the product stream (as the H_2 is only formed by methane decomposition), giving a lower ratio. The observed trend could be due to the ability of ceria to catalyse the reverse water shift reaction, RWGS ($CO_2 + H_2 \rightleftharpoons H_2O + CO$) which explains greater CO_2 conversion and the smaller H_2/CO ratio for the ceria promoted sample.

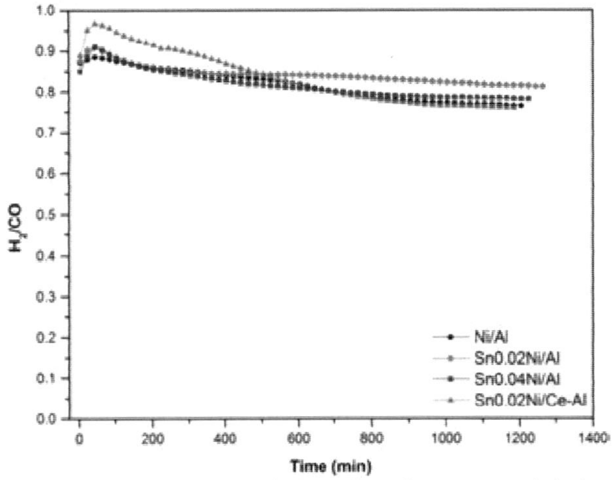

Figure 4 Syngas product ratio for each catalyst over a period of approximately 20 hours

Conclusions

The addition of tin as a promoter to a nickel alumina catalyst shows a marked improvement in conversions of both carbon dioxide and methane to syngas. However, adding too large amounts of tin can have a detrimental effect on catalytic performance by restricting access of reactants to the active nickel sites. It has been found that 0.02 mol.% is an optimum tin loading.

Addition of ceria to the alumina support increases resistance to carbon formation on the surface, exhibiting higher conversions for a longer period of time. In parallel, ceria aids the RWGS reaction boosting the CO_2 conversion. When included together, tin and ceria work synergistically, resulting in a very promising catalyst for dry reforming reactions.

References

1. Charisiou N.D., Baklavaridis A., Papadakis V.G., Goula M.A. *Syn Gas Production via the Biogas Reforming Reaction Over Ni/MgO–Al2O3 and Ni/CaO-Al2O3 Catalysts.* Waste and Biomass Valorization. 2016; 7: p. 725-736.

2. Hou Z., Yokota O., Tanaka T., Yashima T. *Surface Properties of a Coke-Free Sn Doped Nickel Catalyst for the CO2 Reforming of Methane.* Applied Surface Science. 2004; 233: p. 58-68.

3. Reina T.R., Ivanova S., Centeno M.A., Odriolzola J.A. *Boosting the Activity of a Au/CeO2/Al2O3 Catalyst for the WGS Reaction.* Catalysis Today. 2015; 253: p. 149-154.

4. Wei J., Iglesia E. *Isotopic and Kinetic Assessment of the Mechanism of Reactions of CH4 With CO2 or H2O to Form Synthesis Gas and Carbon on Nickel Catalysts.* Journal of Catalysis. 2004; 224(2): p. 370-383.

498418 The Effects of Aging on Steady State Ammonia Oxidation over Pt/Al_2O_3

Abstract:

Three-way catalyst (TWC) light-off (LO) performance is critical in cost-effectively meeting emission standards. Improvements in catalyst LO should include catalyst modifications that minimize inhibition effects of exhaust species such as CO and hydrocarbons (HCs), lower precious group metal (PGM) loading, or use a lower cost PGM such as Pd. Alumina (Al_2O_3) and zirconia (ZrO_2) supports provide high surface area and stability. Furthermore, ceria (CeO_2) provides the key benefit of oxygen storage, compensating for deviations from stoichiometric TWC operating conditions by supplying oxygen for CO and HC oxidation under rich transient conditions. Inclusion of ceria in PGM catalysts enhances CO and HC conversion and minimizes self-inhibition; this improved activity is mainly attributed to ceria containing more efficient oxygen activation sites at the PGM-support interface relative to other support materials such as Al_2O_3 [1-4].

Monolith catalysts having washcoats with 1 wt% Pd, 3 g/in^3 loading, 400 cpsi and formulations Pd/Al_2O_3, Pd/CZO, and $Pd/CZO/Al_2O_3$ (17 wt% CZO, balance Al_2O_3) were synthesized and compared. Catalyst performance experiments were conducted to study the role of Pd and support compositions and loadings on inhibition during oxidation of CO, acetylene (C_2H_2), and propylene (C_3H_6) under near-stoichiometric conditions ($\lambda = 1.01$). Transient and steady-state bench scale reactor studies with simulated exhaust gas mixtures were conducted to extract oxidation kinetics, compare LO behavior, and develop a predictive model for understanding and optimizing catalyst performance.

Catalyst activities were compared via temperature-programmed oxidation experiments. Figure 1 compares CO LO curves using feed concentrations of 0.5% CO and 1% CO. The Pd/CZO catalyst was tested when "fresh" (initial tests after synthesis) and 10+ months later ("deactivated"). Using all three catalysts, LO temperatures increase with feed CO concentration, showing that CO oxidation is self-inhibiting. With increasing CZO content, LO temperatures decrease, demonstrating the promotional effects of using ceria.

Figure 2 shows C_2H_2 LO (feed concentrations of 375 ppm C_2H_2 and 750 ppm C_2H_2) and C_3H_6 LO (feed concentrations of 250 ppm C_3H_6 and 500 ppm C_3H_6) using the deactivated Pd/CZO catalyst. Like CO, C_2H_2 and C_3H_6 are self-inhibiting. For the same concentrations of carbon in the feed mixture (750 ppm C_1 or 1500 ppm C_1), C_2H_2 oxidation occurs at higher temperatures than C_3H_6 oxidation.

Exhaust hydrocarbons propylene and acetylene reveals some interesting differences during co-oxidation with CO. The individual LO features of 1% CO and 750 ppm C_2H_2 in Figures 1 and 2 are compared to their co-oxidation (1% CO + 750 ppm C_2H_2) in Figure 3. CO lights off at higher temperatures in the mixture, i.e. C_2H_2 inhibits CO oxidation. During co-oxidation, C_2H_2 oxidation is inhibited by the presence of CO below

~15% conversion, but is otherwise improved compared to oxidation of C_2H_2 alone. Whereas 1% CO lights off before 750 ppm of C_2H_2 during individual oxidation, the reverse is true during co-oxidation. This is in contrast to earlier findings of CO + C_3H_6 trends in which was found that CO inhibits C_3H_6; i.e. propylene conversion follows CO light off (see below). It is instructive to compare the behavior of the two hydrocarbons at the same C1 concentration. Though the same feed concentration of 1% CO + 1500 ppm C_1 (from HCs) is used, C_3H_6 and C_2H_2 co-oxidation with CO exhibits different trends in terms of LO temperatures and order of species oxidation. As shown in Figure 4, using the deactivated Pd/CZO catalyst, 1% CO lights off before 500 ppm C_3H_6 both during individual oxidation and mixture LO. During co-oxidation, 500 ppm C_3H_6 is 50% converted at ~ 246 °C, compared to ~ 266 °C for 750 ppm C_2H_2.

Differential kinetics studies were conducted to quantify reaction orders and activation energies. The reaction order with respect to CO was found to be approximately −1 using Pd/Al_2O_3 and Pd/CZO/Al_2O_3 and 0 to slightly negative using Pd/CZO, consistent with literature values [5-7]. The activation energies for CO oxidation were found to be 134-213 kJ/mol using Pd/Al_2O_3 and Pd/CZO/Al_2O_3, and 25-70 kJ/mol using Pd/CZO. Using all three catalysts, the reaction order with respect to C_3H_6 was found to be approximately −0.89 to −0.75 and the activation energy in the range of 72-105 kJ/mol, approximately half of that observed for CO oxidation and similar to literature values [5]. Using the Pd/CZO catalyst, the reaction order with respect to C_2H_2 was found to be approximately −0.52 to −0.96 and the activation energy of C_2H_2 oxidation approximately 73 kJ/mol.

Steady-state and transient CO, C_2H_2, and C_3H_6 oxidation experiments using fresh Pd/Al_2O_3, Pd/CZO, and Pd/CZO/Al_2O_3 monolith catalysts were conducted to compare LO behavior and kinetic parameters (i.e. reaction orders and activation energies). CO LO temperatures decreased with increasing proportion of ceria used in the TWC, and C_2H_2 and C_3H_6 LO studies using Pd/CZO demonstrated self- and mutual-inhibition during species oxidation. Kinetics studies demonstrated further the self-inhibiting oxidation behavior.

Ongoing work involves the use of a mechanistic-based kinetic model incorporated into a low-dimensional monolith model [8-9]. The global reactor models are used with CO, C_2H_2, and C_3H_6 oxidation on Pd with steps involving ceria. The 1+1 dimensional model formulation comprises coupled species balances with transverse-average equations having external and internal mass transfer coefficients to verify experimental LO behavior during individual and co-oxidation using catalysts and feed mixtures of different formulations. Finally, experiments involving C_2H_2 oxidation on Pd/Al_2O_3 and Pd/CZO/Al_2O_3 catalysts and other hydrocarbons including toluene are being conducted to investigate inhibition during LO, extract kinetic parameters for oxidation reactions, and provide a basis for validating modeling simulations.

Figure 1: CO conversion (%) versus monolith temperature using Pd/Al_2O_3, $Pd/CZO/Al_2O_3$, and Pd/CZO catalysts for feed concentrations of 0.5% and 1% CO.

Figure 2: C_2H_2, C_3H_6 conversion (%) versus monolith temperature using deactivated Pd/CZO for feed concentrations of 375 ppm C_2H_2, 750 ppm C_2H_2, 250 ppm C_3H_6, and 500 ppm C_3H_6.

Figure 3: CO, C_2H_2 conversion (%) versus monolith temperature using deactivated Pd/CZO for feed concentrations of 750 ppm C_2H_2 and/or 1% CO.

Figure 4: CO, C_3H_6 conversion (%) versus monolith temperature using deactivated Pd/CZO for feed concentrations of 500 ppm C_2H_2 and/or 1% CO.

References

[1] Fernandez-Garcia, M.; Martinez-Arias, A.; Iglesias-Juez, A.; Hungria, A.B.; Anderson, J.A.; Conesa, J.C.; and Soria, J. *App. Cat. B* 2001, **31**, 39-50.

[2] Martinez-Arias, A., Fernandez-Garcia, M.; Iglesias-Juez, A.; Hungria, A.B.; Anderson, J.A.; Conesa, J.C.; and Soria, J.. *App. Cat. B* 2001, **31**, 51-60.

[3] Sharma, S.; Hegde, M.S.; Das, R.N.; and Pandey, M. *App. Cat. A* 2008, **337**, 130-137.

[4] Harmsen, J.M.A.; Hoebink, H.B.J.; and Shouten, J.C. *Chem. Eng. Sci.* 2001, **56**, 2019-2035.

[5] Yao, Y-.F. Y. *J. Catal.*, 1984, **87**, 152-162.

[6] Cant, N.W.; Hicks, P.C.; and Lennon, B.S. *J. Catal.* 1978, **54**, 372-383.

[7] Rainer, D.R.; Koranne, M.; Vesecky, M.; and Goodman, D.W. *J. Phys. Chem. B* 1997, **101**, 10769-10774.

[8] Joshi, S.; Harold, M.P.; Balakotaiah, V. *AIChE J.,* 2009, **55,** 1771-1783.

[9] Raj, R.; Harold, M.P.; Balakotaiah, V. *Chem. Eng. J.*, 2015, **281**, 322-333.

494315 Two-Step Selective Bromination of Thiophenic Compounds for Gasoline Desulfurization Under Mild Conditions

Abstract:

The removal of trace amount of refractory thiophenic compounds from fuels has raised intense concern globally. The development of less energy-intensified desulfurization process is under great demand. In this work, we proposed a new approach of two-step selective bromination of thiophenic compounds for gasoline desulfurization under mild conditions. The Ag modified MoO_3 nanorods (AMO) were synthesized and characterized by XRD, HRTEM and N_2 adsorption, and evaluated in selective bromination of thiophenic compounds. Results showed that (a) thiophenic compounds in model gasoline can be selectively brominated over AMO catalyst with the addition of Br_2, and their bromination production can be removed selectively by followed-up extraction. (b) At the optimized Ag loading of 10%, the apparent rate constants for T and BT reached as high as $95.8h^{-1}$ and $142.7h^{-1}$, respectively under ambient conditions, much higher than those cases in oxidative desulfurization, adsorptive desulfurization, etc. Product analyses showed that the portion of dibromo-thiophenes increased dramatically with increasing loading of Ag, due likely to the formation of $MoAg_2O_4$. (c) The bromination selectivity of thiophenic compounds reached four orders of magnitude than those of aromatic compounds in fuel. Meanwhile, the competition from olefins can be strategically suppressed by HBr pretreatment. (d) With the integration of the 2nd-step of extraction, desulfurization capacity of thiophenes reached as high as 30 mg-S/g-cat, while the integrated system showed excellent bromination desulfurization capacity of thiophene at ~260 mg-S/g-cat under high fuel-to-catalyst mass ratio of 5000. The adsorption of various bromides over $MoAg_2O_4$ was simulated to elucidate the desulfurization mechanism. The fast kinetics, high desulfurization capacity and selectivity of the proposed two-step approach under mild conditions make it promising for gasoline desulfurization.

Keywords: desulfurization, gasoline, selectivity, Ag modified MoO_3

494394 Improvement of Low Temperature Activity for CO Oxidation on Ceria Catalyst By Transition Metal Substitution

Abstract:

Recently, reduction of exhaust gas has been an important issue with increasing the sales of automobiles. CO reduction in various exhaust gases is particularly critical because it damages people and environment. A ceria-based catalyst that can reduce automotive emissions instead of three-way catalysts have been developed. When the temperature is lowered, however, the efficiency is rapidly decreased. In order to improve the activity of ceria catalyst, we synthesized transition metal (TM)-doped ceria. In this study, we examined four transition metal (TM=Cu, Co, Mn, and Fe) doped and their co-doped ceria. TM-doped ceria ($Ce_{0.95}TM_{0.05}O_{2-\text{î}'}$) were synthesized by EDTA-citrate complexing method. By means of this method, we obtained highly pure and homogeneous nano-powder. The nanoparticles were then characterized by X-ray diffraction (XRD), scanning electron microscopy (SEM), and energy dispersive spectroscopy (SEM-EDS). To clarify the positive effects of doping, the surface area was evaluated by Brunauer-Emmett-Teller (BET) measurement. It turns out that the TM-doped ceria powders had higher specific surface area than that of undoped ceria, and correspondingly, the larger specific surface area enhanced the catalytic activity. Our density functional theory results demonstrated that this is attributed to the different interactions between dopant and support. Our results will be helpful for effectively exploring the TM dopant candidates of ceria and determining their optimal concentration for the enhanced catalytic activity.

494685 Evaluating Combined Heat and Power Deionization Systems for Efficient Water Reuse at Thermoelectric Power Plants

Abstract:

Efforts to increase both the United States and global water supply has primarily taken place through seawater and brackish water desalination. According to the International Desalination Association, the United States capacity for desalination increased in 2013 by 50% from 4 million to 6 million m^3/day. Unlike most areas of the world, nearly 77% of all desalination in the United States comes from the treatment of brackish water. Brackish water has lower salt concentrations (~<10 g/l) than seawater (35 g/l) and thus the energy consumed to remove salt ions in theory is much less than seawater desalination. Current technologies utilized for seawater desalination, reverse osmosis (RO) and Membrane distillation (MD), require a significant amount of energy regardless of the type of water (brackish or seawater). Furthermore, these technologies have approached their thermodynamic energy limits (1-3 kWh $m^{\text{â€“3}}$), indicating that further energy savings cannot be achieved. RO and MD are energy intensive technologies because they operated through extracting the main component water (H_2O) from the salt (NaCl). Alternatively, capacitive deionization (CDI), an emerging separation-based technology, operates by adsorbing the minority component (NaCl) from the feeding brackish water. Traditional CDI cells operate in a cyclic process where the capacitive electrodes are brought in contact with a concentrated salt solution (e.g. brackish water) and charged. Once the charged electrodes are fully saturated with salt ions, the electrodes are regenerated through discharging the electrodes in a brine solution. Analogous to a P-V diagram for traditional heat engines, the enclosed area of the charge versus voltage represents the energy that is consumed during this charging-discharging process. Until recently the CDI three step intermittent process has been relatively untouched in terms of literature and system operation. In this study, we explore the effect of temperature on salt removal with CDI. Experimentally, we use traditional electrochemical characterization methods, cyclic voltammetry (CV) and electrochemical impedance spectroscopy (EIS), to evaluate the capacitance and resistance changes with temperature. Here it was found that through increasing the operating temperature from 25°C to 45°C, the specific capacitance of carbon electrodes increased from 34.60±2.15 to 42.56±3.57 F/g, and the cell resistance decreased from 20.20±2.97 to 15.57±2.52 ohm. Additionally, the salt adsorption capacity decreased from 13.82±1.13, to 6.58±1.01 milligram NaCl/mg carbon. This indicates that although the electrochemical performance of CDI can be increased with temperature, the ion storage within the electric double layer is hindered at higher temperatures.

493557 Biochar Amendments for Increased Crop Yields: How Can Biochars Improve Crop Nutrient Availability?

Abstract:

Background:

Maintaining a sustainable food supply for the foreseeable future without excessive environmental degradation will require novel engineering approaches for remediating the human disruption of the nitrogen cycle. Over the past several decades, food production has been able to keep pace with human population growth only thanks to the development of new high-yielding crop varieties grown with the help of fertilizers. Thus, vast amounts of energy are consumed every year to produce the approximately 200 Mt of ammonia needed to make the nitrogen (N) fertilizers used by farmers around the globe.

Unfortunately, more than half of the N fertilizer applied to a field is not available for plant growth due to losses caused by surface runoff, leaching into surface and ground water, or volatilization. Because of these losses, the increased use of nitrogen fertilizers has been linked to a variety of water pollution problems ranging from the expansion of the hypoxic zone in the Gulf of Mexico caused by eutrophication to the contamination of wells and groundwater with nitrate N. Moreover, denitrification reactions convert nitrates to nitrous oxide, a major greenhouse gas and nitrogen oxide, a major atmospheric pollutant and significant contributor to the depletion of atmospheric ozone.

The amendment of soils with biochar has been heralded as a sustainable method for increasing crop yield and preventing pollution problems caused by fertilizer runoff. However, the available experimental results are highly variable [1-3]. While some studies reported significant beneficial effects with up to 60% higher yield after biochar addition, other studies reported that biochar amendments either had no effect or even resulted in up to 30% lower agricultural productivity [3]. One of the most important reasons for this variability is that biochar is not a single entity. Dozens of biomass feedstocks are used to produce biochars in multiple types of reactors under varying temperature and oxygen conditions, leading to thousands of biochars with widely varying properties. To complicate things even more, soil properties, climate conditions (like rainfall and temperature), plant requirements and many other parameters vary from application to application. As a result, such studies and meta-analyses have not provided much insight into the fundamental mechanisms that modulate the environmental performance of biochar amendments.

Our Previous Work on this Problem:

To begin bridging this knowledge gap, we focused our recent research on the mechanisms that control the ability of biochars to improve crop nutrient availability, one of the main mechanisms proposed to explain crop yield increases in biochar-amended soils [2].

As a first step in this direction, we have recently developed and tested a mathematical model that uses first principles to describe nutrient transport in soils amended with biochar [4]. Using a simple Langmuir isotherm to describe nutrient adsorption, our simulation results revealed that nutrient transport in biochar-amended soils is modulated by a complex interplay between the physical mechanisms governing flow in porous media and nutrient transport, and the chemical mechanisms modulating nutrient adsorption of the pore surface of biochar solids. Nutrient transport rates through the topsoil decreased when biochars with the proper adsorption capacity and affinity were mixed with the soil. This was due to the differential dynamics of the adsorption and desorption processes. Nutrient adsorption was significantly faster than desorption, with biochar essentially acting as a slow (or controlled) release medium for the adsorbed nutrient [4].

Current Work and Summary of Results:

The present study extends our previous work with a systematic study of the adsorption and desorption dynamics on a much broader class of biochars. We report the development of a mathematical model that describes the response of biochar particles to a pulse of nutrient that simulates a single fertilizer application. Nutrient adsorbs on the biochar particle when the fertilizer solution flows past the particle and starts desorbing when the fertilizer solution is replaced by pure water. Two nutrient transport mechanisms are considered: (a) diffusion through liquid-filled pores and (b) surface diffusion. We assume that nutrient adsorption follows a Sips isotherm whose parameters can be varied to describe adsorption on both homogeneous and heterogeneous solids. The Freundlich and Langmuir isotherms commonly used to describe adsorption of ammonium or nitrates on biochars are special cases of the Sips isotherm. The highly nonlinear partial differential equations describing this transient adsorption/desorption problem are discretized using high-order finite differences and the resulting system of first order differential equations are integrated with a variable-step, variable-order integrator based on numerical differentiation formulas.

As noted previously, a "good" biochar should be able to quickly adsorb the nutrient when it is present in the liquid flowing around it, but it should desorb it much more slowly in order to act as slow (or controlled) release medium. This is the mechanism through which a "good" biochar will improve crop nutrient availability [2, 4].

Our simulations clearly show that the ability of a biochar to slowly release the adsorbed nutrient depends on a complex interaction of (a) external mass transfer, (b) intraparticle diffusion (both pore and surface diffusion), and (c) adsorption dynamics (affinity constant and adsorption capacity).

The interaction between external mass transfer and intraparticle pore diffusion is a key factor for modulating the release of nutrient that has adsorbed on the biochar. Our system is characterized by low values of the Biot number, the dimensionless number that represents the ratio of external mass transport rate over the intraparticle diffusion rate. This is because of the low interstitial water velocities (10-90 cm/day) in biochar-amended soils. Low Biot numbers are one of the key factors responsible for the long desorption tails that characterize the slow release of nutrients from "good" biochars.

Our simulations reveal that all the processes mentioned above can significantly affect the nutrient release rates. For example, surface diffusion accelerates nutrient desorption from small (2 mm in diameter) and highly porous biochar particles (0.6 total porosity). But, surface diffusion will slow down desorption from larger (4-8 mm in diameter) and less porous particles, even though the latter biochar particles may have exactly the same adsorption isotherm parameters. Adsorption affinity and site heterogeneity can also interact with porosity and particle size to modulate nutrient desorption. Finally, surface diffusion again improves the ability of a biochar to act a slow release medium at high interstitial velocities.

By revealing that the response of biochar particle to a pulse of nutrient is governed by complex interactions among external mass transfer, intraparticle diffusion and adsorption, this study provides new insights into the fundamental mechanisms that influence the environmental performance of biochar amendments. It may also explain why biochars that share some key properties (e.g. carbon content) or produced at the same pyrolysis temperature can potentially have widely different impacts on crop yields, as has been repeatedly reported in the literature [2,3]. Much better characterization of biochar properties (porosity, adsorption isotherm, intraparticle transport mechanism) and more precise control of amendment or soil parameters (like particle size, biochar aging, soil permeability) are necessary before we can fully elucidate the mechanisms of biochar action.

References:

[1] Spokas, K. A.; Cantrell, K. B.; Novak, J. M.; et al. "Biochar: A Synthesis of Its Agronomic Impact beyond Carbon Sequestration." *J. Environmental Quality* **41** (2012) 973-989.

[2] Jeffery, S; F.G.A. Verheijena; M. van der Velde; A.C. Bastos. "A quantitative review of the effects of biochar application to soils on crop productivity using meta-analysis." *Agriculture, Ecosystems and Environment* **144**(2011) 175–187.

[3] Crane-Droesch, A; S. Abiven; S. Jeffery; M.S. Torn. "Heterogeneous global crop yield response to biochar: a meta-regression analysis." *Environmental Research Letters* **8**(2013) 044049.

[4] Sun, S; C. E. Brewer; C. A. Masiello; K. Zygourakis. "Nutrient Transport in Soils Amended with Biochar: A Transient Model with Two Stationary Phases and Intraparticle Diffusion" *Industrial and Engineering Chemistry Research*, **54** (2015) 4123–4135.

501354 Moisture Retention in Emulated Soil Micromodels: Development and Performance of Sustainable Agriculture Biotechnology

Abstract:

Soil bacteria living near plant roots secrete exopolysaccharides (EPS), a polymeric substance which can help modulate soil moisture and promotes plant growth. EPS functions in several different ways: (i) EPS behave as hydrogel, swelling during wet soil conditions and remaining hydrated during dry soil conditions; (ii) EPS promotes aggregation of soil particles and thereby promotes water retention in dry soil conditions through stronger capillary forces; and (iii) EPS alters surface properties of soil, creating more water repellent surfaces to inhibit rapid re-wetting in wet soil conditions. EPS functions in concert with the physical microstructure of real soil; therefore, it is necessary to study EPS performance in physically realistic micro-structures. Here we show pore-scale changes in moisture distribution within emulated soil micromodels that have systematically varied physical microstructures and surface water repellencies, and can be loaded with well-defined pore solutions. Experimental results showed dilute EPS solutions took twice as long to dry compared with pure water in identical micromodels. Microstructure geometry was found to influence drying rate over a range of pore saturation values as well as control the spatial distribution of liquid and gas phases within unsaturated soil micromodels. Deionized water took four times longer to completely dry from micromodels with higher water repellency surfaces compared with micromodels with lower water repellency surfaces. These results illustrate the various mechanisms whereby EPS impacts the spatial distribution and evaporation resistance of soil moisture. We provide a systematic method to evaluate performance of natural EPS solutions in modulating soil water and promoting resiliency of terrestrial ecosystem; we anticipate our approach may speed the development of sustainable agriculture biotechnology to enable food production using less water.

498287 Molecular Modeling and Simulation Studies of the Structural and Energetic Evolution during Dehydration of Food Systems

Abstract:

From a molecular point of view, foodstuffs are mostly comprised of biopolymeric molecules that form highly porous structures in which solvents, solutes, additives, and other natural or artificial ingredients are accommodated. Because of the complexity of the food structures and compositions and because of the sensitivity of the pore structures and properties of food to physical and chemical changes arising externally or internally, foodstuffs are one of the most complex types of soft materials and are inclined to exhibit significant structure-property relationships in the physicochemical, functional, processing, and nutritional aspects of food science and engineering. In this respect, water (moisture) is a major factor in determining the properties of foods in various stages as it is not only a major component of food materials but also acts as space filler for stabilizing food structures, solvent for dissolving additive molecules, medium for heat and mass transfer, and participant in the chemical and/or biological reactions in food systems. .

One of the important ways for preparing and preserving foods is dehydration (water removal), which is a process that requires energy to be supplied to enable water molecules to break loose from their interactions and escape from the porous structures of the foods. The process of food dehydration and the preservation of dehydrated foods can thus be directly related to the interactions of water molecules with food macromolecules, solute ions, and additive molecules, because these interactions dictate the energy requirement for water removal during food dehydration as well as the structures of foods before, during, and after dehydration. Moreover, the magnitudes of these interactions depend on the densities and intermolecular distances of mutually interacting molecules, rendering the water molecules in foods to exist in different energetic states and making single system-wide thermodynamic parameters (e.g, water activity) inadequate in representing the diverse states of water molecules in food porous structures and in elucidating the microscopic and macroscopic structural evolution of food materials caused by dehydration. Capable of tackling these challenges is the molecular dynamics (MD) method which employs molecular-based interaction potentials to construct a model system that provides a satisfactory physicochemical representation of the real system and allows various structural and dynamic properties to be studied at a molecular level. Recently, we have employed MD modeling and simulations to construct an amylose-based food material and study the dehydration process in a continuous dynamic manner by periodically examining the energetics of all water molecules at the water-vapor interface and removing the one with the highest energy from the model food system. It was found that the strong interactions of water molecules with the food polymer stabilize the porous structure of the food material and makes the dehydration energy requirement to be greater than that involved in pure water vaporization. They also cause the water interaction energetics to be highly nonuniform in all three spatial dimensions. As dehydration proceeds, the surface tension of the descending water interface coupled with strong water-polymer interactions exert downward stresses on the food polymer to induce polymer conformational compaction and food structural reduction.

We further added two functional food ingredients to the model food systems in our MD simulation studies. The first one is potassium sorbate, which is a commonly used food preservative against the growth of bacteria, fungi, molds, and yeast in foods. The second one is fructose-1,6-diphosphate (FDP), which is a bioprotective agent and a promptly available high energy substrate and can promote growth and improve health. The stability of the food porous structures was found to be enhanced by the presence of the food additive molecules which function as additional fillers in the food pore space and interact strongly with the polymer chains and water molecules of the food material.Â During dehydration, the descending water level tends to accumulate the additive molecules, in particular sorbate with a nonpolar hydrocarbon moiety, at the vapor-liquid interface, which helps to form dense clusters of food polymer chains at the interface, reduces the level of structural reduction, but increases the energy requirement for water removal.Â The additive molecules in the interior regions of the food structures have strong tendencies to stay close to food polymer and hence exhibit nonuniform density distributions as well. Extensively dehydrated food porous structures were found to have substantially improved stability due to the water molecules remaining in the pore structure that effectively â €œbindâ€ the food polymer and additive molecules together. Which makes it very difficult to recover the original pore structures of the food systems that existed before the commencement of dehydration. The findings of these MD studies suggest that the original pore structures could be reduced less during dehydration and could be recovered more effectively by rehydration, if (i) the food material has a higher polymer density, (ii) solute or additive molecules have elongated shapes to act as cross-linkers, and (iii) the effects of water surface tension can be minimized or eliminated using special dehydration methods such as freeze drying. The MD modeling and simulation approach developed in this work could form a basis for systematic evaluation of food porous structures and their structural and energetic evolution and mechanisms in molecular detail and for providing scientific explanations to the macroscopic trends and empirically observed phenomena during food dehydration.

493109 Metallic Membranes for N_2 Separation and NH_3 Production

Abstract:

The availability of nitrogen, phosphorus, and water are the three main factors that limit our ability to produce enough food to feed the growing population of the planet. The nitrogen cycle is one of the most significant biogeochemical cycles on Earth, as nitrogen is an essential nutrient for all forms of life. Although freely available in the atmosphere as dinitrogen (N_2), access to fixed forms of nitrogen constitutes, in many cases, the most limiting factor for plant growth. The industrial production of ammonia for fertilizers via the current Haber-Bosch process is an energy-intensive process that consumes 1-2% of the worldâ€™s annual energy supply. For these reasons, the need for advanced catalytic methods for the reduction of N_2 to ammonia remains a requirement for sustainability in the food, energy and water systems cycle.

The aim of this work is to explore the potential of metallic membranes for N_2 separation with the final intent to produce NH_3. Based on a preliminary theoretical investigation using density functional theory, the Group V transition metals (e.g., vanadium (V), niobium (Nb) and tantalum (Ta)) show strong affinity toward N_2. Moreover, from solubility and diffusivity values taken from the literature, iron (Fe) is a suitable fit for this application. Therefore, V, Nb, Ta and Fe foils with a thickness of 40 Î¼m have been chosen as membrane materials in this study.

Permeation tests with pure gases (He, N_2, CO_2 and CH_4) have been performed to characterize the membranes in terms of the N_2 permeating flux and ideal selectivity at different temperature and pressure conditions, varying from 400 to 600 Â°C and from 2.0 to 60 bar, respectively. Moreover, scanning electron microscopy (SEM) and an electron microprobe analyzer (EMPA) have been used to investigate the effect of the different operating conditions on the membrane surface. From the experimental results, all of the membranes show infinite selectivity towards N_2 permeation at each operating condition investigated and V showed better performance with respect to Nb, Ta, and Fe.

The V membrane is housed in the membrane reactor where a H_2 stream is used as the sweep gas to promote the NH_3 reaction. Lower pressure than the conventional Haber-Bosch process is used. Specifically, an operating pressure of 60 bar, instead of 200 bar (Haber-Bosch process), and a temperature of 500 °C are used as operating conditions to produce NH_3. The performance of the membrane reactor in terms of NH_3 conversion is, hence, investigated.

485442 Capacitive Deionization of Brackish Water for Irrigation and Energy Storage

Abstract:

Irrigated agriculture water represents the bulk of demand for water in most of the countries suffering from acute water scarcity in the Near East and North Africa, as well as Mexico, Pakistan, South Africa, and large parts of China and India. Greenhouses are emerging solution. Current testing in Mozambique and Sierra Leone has demonstrated significantly reduced water use. Yet even a few gallons per day per greenhouse is challenging because many local water sources are brackish. Capacitive deionization (CDI) is a sustainable, energy efficient, and cost effective technology for removal of low or moderate salinity typical of brackish water as compared to technologies that remove the water instead. Indigenous production of the carbon for the electrodes holds the key to such technology being adopted as a source of irrigation water, simultaneously facilitating the use of such using renewable energy sources. Biochars have recently been explored as electrode materials in CD, and can be easily produced from local plant-based agricultural waste materials.

The goal of this study is thus to identify biomass process conditions to enable the best performance as electrode material in CDI. Studies on CDI so far have been targeted towards achieving the best adsorption capacity, which directs towards the use of highly processed and tailored carbons. This is mostly achieved through chemical activation processes using toxic or corrosive and expensive materials which require deionized water for subsequent washing of the carbon, for achieving better CDI performance. Hence, with the effort directed towards making the process viable for use locally in regions with acute water scarcity, a change in the approach to tailor the electrode material for use in CDI is proposed. This approach is to determine the best possible material utilization by ensuring maximum possible contact between the liquid and the biochar surface (wettability) and by generating the least Ohmic loss thus resulting in highest electrical efficiency.

499583 Gas-Particle Partitioning of Alkyl Nitrates from Anthropogenic Alkanes

Abstract:

Oxidation of volatile organic compounds (VOC) produces secondary organic aerosol (SOA), increasing the overall fine particle mass and aggravating the associated negative health effects. Studies have shown that long-chain anthropogenic VOCs, despite their low ambient abundance relative to biogenic VOCs, can contribute significantly to the overall SOA mass. Typical anthropogenic emission sources, such as oil and gas extraction sites, tend to be environments rich in nitrogen oxides (NO_x) and therefore favorable for the formation of alkyl nitrates (AN), which can partition to the particle phase and therefore contribute to SOA. The gas-particle partitioning of AN is known to aid NO_x cycling and transport, contributing to the regional nature of NO_x pollution. The gas-particle partitioning behavior of SOA derived from anthropogenic VOCs remains poorly characterized.

Here we report results from environmental chamber experiments conducted to investigate SOA formation from select linear, branched, and cyclic C_8, C_{10}, and C_{12} anthropogenic alkanes under high NO_x conditions. SOA formation was significantly higher for cyclic alkanes, consistent with previous findings. Aerosol volatility was determined using a thermodenuder and an aerosol evaporation kinetics model. The partitioning behavior of ANs was determined based on particle-phase measurements and total AN concentration estimated using an updated SAPRC box model with carbon-bond-based reaction mechanisms. The molecular composition of gas- and particle-phase products was determined using a Filter Inlet for Gases and AEROsols (FIGAERO) coupled to a High-Resolution Time-of-Flight Chemical Ionization Mass Spectrometer, where evidence of oligomer formation was observed. Results from this study can help improve air quality models with respect to SOA loading, NO_x cycling, and NO_x transport.

492530 Why Would Cloud Condensation Nucleus Activity Linearly Change with O/C? Assessing the Role of Volatility, Solubility, and Surface Activity

Abstract:

Aging of atmospheric organic aerosol (OA) has been studied in terms of average molecular properties such as elemental ratios (O/C, H/C), volatility, and hygroscopicity. Correlations between O/C and hygroscopicity (or cloud condensation nucleus activity), represented by κ, are a computationally efficient approach to estimate the impact of aerosol aging on cloud formation and climate; however, previously reported correlations between these two variables are largely empirical and vary widely in their slopes and extrapolations to high O/C values. This study proposes a theoretical framework that bridges elemental ratios, volatility, solubility, and κ of OA. The framework estimates inherent κ based on molecular formulas of organics composed of carbon, oxygen, and hydrogen that partition to condensed phase, and then it estimates apparent κ considering solubility distribution parameterized by O/C. This paper applied the new framework to the two-dimensional volatility basis set (2D-VBS) and found that distribution of O/C and molecular size plays a key role in determining apparent κ. Another finding is that for highly soluble organics, κ is dictated by gas-particle partitioning and it is unlikely for κ of organics to go beyond 0.3 in typical ambient organic material loading. Sensitivity analysis showed that presence of surface-active compounds may slightly enhance apparent κ when solutes are not entirely dissolved, but it would not change the major conclusions. Although inherent uncertainties in various parameters prevent exact underpinning of activation mechanisms, this framework provides plausible explanations of why κ would and would not correlate with O/C depending on product distribution in the 2D-VBS space. The framework opens up new opportunities to evaluate two-dimensional representations of organic aerosol aging using κ, a significant advancement from the current empirical linear fits to κ and O/C.

498962 Bimodal Aerosol Size Distributions from Laboratory Atomization/Evaporation

Abstract:

Bi-modal aerosol size distributions have been observed in laboratory studies using atomized droplet generation followed by diffusion drying. While there is evidence of multi-component distributions forming from various component separations within droplets, our research has observed similar effects from single-component aerosols. The bi-modal effect is observed via Scanning Mobility Particle Sizing (SMPS) analysis. This effect is controllable by altering the concentration of the initial atomized solution between 0.1 and 10 grams per liter, by altering the flowrate of the aerosol stream through a series of dryers, and by altering the drying amount and rate via the number of and saturation of water within diffusion dryers. In this research, we have investigated both single- and multi-component solutions using both organic and inorganic constituents dissolved in water. We have done a series of experiments generating aerosols by varying the above factors such that the final aerosol size distribution can be predictably controlled for single- or bi-modality. The resulting aerosols have been collected with both PIXE impactors and Nano-Aerosol Samplers and have been imaged with both Scanning Electron Microscopy (SEM) and Atomic Force Microscopy (AFM) in order to understand their true particle shapes and dimensions. Furthermore, we have used X-ray Diffraction (XRD) to analyze the collected aerosols for crystal polymorphic structures and compared the XRD graphs against known structures to determine the extent to which meta-stable polymorphic forms of organic crystals are occurring. Results of all of these experiments and analysis will be presented and explained. One key conclusion from these studies is that aerosol atomization and evaporation producing particles on the order of 100 nanometers in diameter actually produces crystal structures and can be modified to produce both stable and metastable polymorphs of these crystals.

499859 Oxidized Organic Compounds Formed from Chlorine-Initiated Oxidation of Toluene

Abstract:

Oxidation of small aromatic compounds can affect the formation of ozone and organic aerosol. Most of the recent literature has focused on their reaction with hydroxyl radicals. However, recently, Cl-initiated photochemistry has been shown to significantly affect ozone formation in urban cities. In this work, we quantify the formation of secondary organic aerosol (SOA) from Cl-initiated photo-oxidation of small aromatic compounds including toluene, m-xylene and 1, 2, 4-trimethylbenzene under both low NOx and high NOx conditions. A high resolution time-of-flight chemical ionization mass spectrometer (HR-Tof-CIMS) was used to measure gas-phase products formed during each experiment. Major products observed during the Cl-initiated photo-oxidation of toluene included benzaldehyde, benzyl alcohol, benzyl hydroperoxide, phenol, cresol, epoxide, methylbutanedial, methylhexanedial, benzaldehyde, butanedial and bicyclic peroxy radical (Figure 1). Toluene-Cl photochemistry also produces secondary OH radicals, and typical toluene-OH products were also observed in these experiments.

Toluene-OH oxidation has been studied in the past, and it has been difficult to form highly oxidized SOA in environmental chamber experiments. In the work reported here we achieved SOA oxidation states as high as 0.24 using Cl-initiated photo-oxidation reactions, which occur at a faster timescale than OH-initiated reactions. The SOA oxidation states were inferred from mass spectra measured by an Aerosol Chemical Speciation Monitor (ACSM) and a High Resolution Aerosol Mass Spectrometer (HR-AMS) and were consistently higher than those reported previously. The addition of NO_x resulted in a higher organic aerosol oxidation state. Overall, our results indicate that toluene-Cl reaction under both low and high NOx conditions can yield highly oxidized organic aerosol and that this reaction pathway has the ability to alter SOA formation in the atmosphere.

Figure 1: Panel A shows major gas-phase products from toluene-Cl and toluene-OH pathways as measured by the HR-Tof-CIMS; Panel B shows common gas-phase products from both pathways as measured by the HR-Tof-CIMS; Panel C shows uncorrected organic aerosol formation as observed from the HR-AMS. Panels A, B and C all represent data obtained from low NO_x-toluene-Cl system. The thick black line represents additional injection of chlorine during the experiment.

501209 Hydrolysis of Organic Nitrates Formed from Oxidation of Biogenic Volatile Organic Compounds

Abstract:

Organic nitrates play an important role in the cycling of NO_x and secondary organic aerosol (SOA) formation, yet their formation mechanisms and fates remain highly uncertain. Results from ambient studies suggest that organic nitrates have a relatively short lifetime, though corresponding laboratory data are limited. One further reaction of organic nitrates in the particle phase is hydrolysis in the presence of particle water. In this work, we investigate the formation and hydrolysis of organic nitrates from the photooxidation and nitrate radical oxidation of biogenic volatile organic compounds, a-pinene and b-pinene. Experiments are conducted in the Georgia Environmental Chamber facility (GTEC) under dry and humid conditions in the presence of different types of aerosols seeds. Experiments are also design to probe different peroxy radical pathways (RO_2+HO_2 vs RO_2+NO_3). Speciated gas-phase and particle-phase organic nitrates are continuously monitored by a Filter Inlet for Gases and AEROsols High Resolution Time-of-Flight Chemical Ionization Mass Spectrometer (FIGAERO-HR-ToF-CIMS). Bulk aerosol composition is measured by a High Resolution Time-of-Flight Aerosol Mass Spectrometer (HR-ToF-AMS). A large suite of highly oxygenated gas- and particle-phase organic nitrates are formed rapidly in photooxidation experiments in the presence of NO_x as well as from NO_3 oxidation. We find that the oxidation pathway (photooxidation vs. nitrate radical oxidation) plays a more important role than precursor hydrocarbon (a-pinene vs. b-pinene) in the extent of organic nitrate hydrolysis. The dynamics of the speciated organic nitrates over the course of the experiments will also be discussed. Results from this chamber study can be used to interpret organic nitrate fates under humid ambient conditions with the presence of aerosol water.

493674 Analytical Model for Vapor Condensation on Soot Agglomerates

Abstract:

Soot is a major environmental pollutant which deteriorates air quality and affects human health. Moreover, soot nanoparticles serve as condensation nuclei for atmospheric aerosols, light absorbers and scatterers, impacting the climate [1,2]. Soot nanoparticles are fractal agglomerates of graphitic spheres mixed with organic or inorganic products of combustion. Both the optical properties of soot in the atmosphere and the transport properties of soot in human respiratory tract strongly depend on the morphology of soot agglomerates [3,4]. Therefore, in order to assess the negative impacts of soot on both climate and human health one has to know the microstructure of soot nanoparticles.

After their formation, the soot agglomerates are exposed to other vapors that condense on the agglomerates surface. In some cases vapor condensation does not affect the soot morphology, however, often it induces restructuring of soot agglomerates, so that they collapse into globules [5,6]. While some studies suggest that the tendency to restructure depends on the surface tension of condensing liquid [7], other experiments show that even solid coatings can cause restructuring [8]. A theory of this process is still lacking.

Restructuring of soot agglomerates involves two aspects: mechanics of the agglomerates and kinetics of vapor condensation on agglomerates surface. Here we focus on the latter. When considering vapor condensation on agglomerate of spheres, there are two distinct regimes of condensation: liquid can form a film on a surface of the spheres or it can fill the gaps between the spheres ("capillary condensation"). We hypothesize that two regimes of condensation correspond to the two scenarios in mechanical evolution: when vapor condenses uniformly, it is unlikely to induce significant mechanical stresses and cause restructuring; but when it condenses in the gaps, the capillary forces at the joints of the spheres can collapse of the agglomerate.

The kinetics of vapor condensation on a spherical surface is well-known [9]. Here we describe the condensation of a vapor into a pendular ring in the gap between the adjacent spheres. We found an analytical expression for the amount of condensed liquid as a function of time. This result allowed us to find a non-dimensional parameter which value determines realization of one of the two condensation/restructuring scenarios. Our model not only supports well our experimental observations, but also can be used to rationalize most of the published experimental results from other groups [10-12].

References

1. Bond, T. C., Doherty, S. J., Fahey, D., Forster, P., Berntsen, T., DeAngelo, B. J., Flanner, M. G., Ghan,

assessment. J. *Geophys. Res. Atmos.*, 118(11):5380–5552
2. Shiraiwa, M., Selzle, K., and Poschl, U. (2012). Hazardous components and health effects of atmospheric aerosol particles: reactive oxygen species, soot, polycyclic aromatic compounds and allergenic proteins. Free Radical Res., 46(8):927–939.
3. Xue, H., Khalizov, A. F., Wang, L., Zheng, J., and Zhang, R. (2009). Effects of dicarboxylic acid coating on the optical properties of soot. *Phys. Chem. Chem. Phys.*, 11(36):7869–7875
4. Broday, D. M. and Rosenzweig, R. (2011). Deposition of fractal-like soot aggregates in the human respiratory tract. *J. Aerosol Sci.*, 42(6):372–386.
5. Weingartner, E., Baltensperger, U., and Burtscher, H. (1995). Growth and structural change of combustion aerosols at high relative humidity. Environ. Sci. Technol., 29(12):2982–2986.
6. Mikhailov, E. F., Vlasenko, S. S., Kiselev, A. A., and Ryshkevich, T. I. (1998). Restructuring factors of soot particles. *Ivz. Atmos. Ocean. Phys.*, 34:307–317.
7. Schnitzler, E. G.; Gac, J. M. and Jäger, W. (2017) Coating surface tension dependence of soot aggregate restructuring *J. Aerosol Sci., 106*:43-55
8. Chen, C., Fan, X., Shaltout, T., Qiu, C., Ma, Y., Goldman, A., and Khalizov, A. F. (2016). An unexpected restructuring of combustion soot aggregates by subnanometer coatings of polycyclic aromatic hydrocarbons. *Geophys. Res. Lett.*, 43(20).
9. Seinfeld, J. H. and Pandis, S. N. (2006). *Atmospheric chemistry and physics: From air pollution to climate change*. John Wiley & Sons, 2nd edition.
10. Ma, X.; Zangmeister, C. D.; Gigault, J.; Mulholland, G. W. and Zachariah, M. R., Soot aggregate restructuring during water processing. (2013) *J. Aerosol Sci.* 66:209-219.
11. Lu, Z. F.; Hao, J. M.; Hu, L. H. and Takekawa, H., The compaction of soot particles generated by spark discharge in the propene ozonolysis system. (2008) *J. Aerosol Sci.*, 39 (10):897-903.
12. Miljevic, B.; Surawski, N. C.; Bostrom, T. and Ristovski, Z. D., Restructuring of carbonaceous particles upon exposure to organic and water vapours. (2012) *J. Aerosol Sci.*, 47:48-57.

498959 Implications of High Accuracy Thermochemical Kinetics for H + CH_3 (+M) â‡„ CH_4 (+M) on Combustion Models

Abstract:

CH_3, the simplest alkyl radical, is an extremely stable intermediate that persists in large concentrations even in high temperature chemically reacting systems such as flames. Radical-radical recombination with H-atoms (R1: H + CH_3 â‡„ CH_4) is the dominant process for CH_3 removal in high temperature combustion and flame processes. As such, the net result of this reaction is to terminate reactive H-atoms and CH_3 radicals in flames. Numerous studeies have been conducted to directly measure k_1, however, k_1 has only been measured at low temperatures (< 700 K) while all data at high temperatures (> 1300 K) relevant to the pre-flame regime have been measured in the reverse direction. Hence, reconciling k_1 and k_{-1} requires accurate equilibrium constants spanning an extended range of temperatures. In the present work, we include all relevant anharmonic corrections to calculate an accurate Active Thermochemical Tables (ATcT) partition function for CH_3. The resulting nonrigid-rotator-anharmonic-oscillator partition function is used to determine thermochemical parameters (heat capacity, entropy, and enthalpy increment values) for CH_3 and K_{eq} values for reaction R1 over a wide temperature range (200-6000 K). With the newly evaluated equilibrium constant, literature experiments and theory (based on two-dimensional master equation calculations with a first principles energy and angular momentum transfer kernel) are used to obtain an accurate representation of the kinetics for the title reaction. In particular, the present work provides fits to k_1(T,P) for N_2, Ar, and He as colliders for direct use in combustion modeling. The results of the new rate constant evaluation for R1 has a dramatic impact on combustion modeling. In particular, we assess the impact of incorporating anharmonic thermochemistry for CH_3 and the updated fits for k_1 (and k_{-1}) in current widely used literature models for simulations of CH_4/air and n-heptane/air laminar flame speeds and find that flame speed predictions can be affected by as much as 50%.

This work was supported by the Office of Basic Sciences, Division of Chemical Sciences, Geosciences, and Biosciences, U. S. Department of Energy, under Contract No. DE-AC02-06CH11357.

496829 Estimation of Thermodynamic Properties of Polycyclic Molecules By a Linear Regression Model

Abstract:

Though recent advances in *ab initio* methods have opened the door toward accurate calculations of molecular thermochemistry, large-scale theoretical studies are still often limited, at least initially, to the use of empirical methods to rapidly screen out unimportant species, so only the important species are the subject of CPU-time intensive quantum chemistry calculations. Among the empirical methods developed in the past decades, the Benson group additivity scheme is one of the quickest and most convenient methods to determine thermodynamic properties of molecules without needing the 3D molecular structures. It has achieved great success in accurate prediction of thermochemistry of small molecules and has been widely adopted in modeling software for on the fly prediction of thermodynamic parameters. However, because this additivity scheme simply uses the properties of individual chemical groups independently to calculate the entire property for a molecule, the contribution of the overall molecular structure to the property is usually not taken into account. This problem manifests itself for strained structures and can cause significant errors for polycyclic molecules with fused rings. Therefore, the application of the additivity method is often restricted to simple chemical systems without the presence of polycyclic species.

To address this issue, we have developed an improved model to predict the thermochemistry of polycyclic species.Â A regularized linear model was chosen to resemble the simplicity and interpretability of the additivity method. However, instead of following the definitions of the chemical groups of Bensonâ€™s scheme, we generated a comprehensive list of identifiers containing local (atoms, bonds, and angles) and/or nonlocal (rings) structural information and used these identifiers to compose feature vectors for molecules.Â The identifiers carrying nonessential information were eliminated by L1 regularization during model training so that those remained in the final model represent an optimum set of chemical units for the calculation of the thermodynamic property of interest with the contribution of the cyclic structures. These chemical units are human interpretable and conceptually equivalent to the chemical groups defined in Bensonâ€™s scheme but are selected by the model objectively without human intervention.

For a training set of 25,716 cyclic and polycyclic organic molecules made up of C, H, and O atoms, 408 identifiers were found to be needed for the calculation of formation enthalpies. The transferability of the trained model was validated on an independent test set of 2,858 molecules with a mean absolute error of 1.88 kcal/mol.

490830 Hydrogen Binding Versus Dissociation during Soot Formation: Insights from Reactive Molecular Dynamics

Abstract:

It has been proposed that two reactions are prominent in the earliest stages of soot nanoparticle formation; hydrogen abstraction via collision between a growing PAH (polyaromatic hydrocarbon) cluster and a hydrogen atom (yielding H_2 and a dehydrogenated cluster) and carbon addition via the binding of an alkane/alkene/alkyne chain to the cluster. We examine the first part of this process via reactive molecular dynamics using REAXFF C-H potentials. In simulations, cluster structures determined using the previously developed Atomistic Model for Particle Inception (AMPI) code of Violi and coworkers are equilibrated (at a predefined temperature) and a hydrogen atom with a specific velocity and impact parameter is driven towards the cluster. Depending upon the velocity, impact parameter, and collision orientation, we find that pure hydrogen abstraction is rare and instead at low velocities hydrogen binds to clusters. At increasing velocities, hydrogens rebound from the cluster surface and can drive hydrogen facilitated collision induced dissociation (CID) of the cluster. CID events yield smaller alkane/aklene/alkyne chains, which may serve to bind other clusters. Integration over all impact parameters and velocities for collisions (in accordance with the Maxwell Boltzmann distribution) yields the hydrogen binding rate coefficient and hydrogen induced dissociation rate coefficient for a given cluster structure. We discuss how these rates vary with temperature and cluster size.

499595 Effect of Composition Variation of Syngas Mixtures on NO_x Formation at High Pressure

Abstract:

Increasing interest in syngas combustion in gas turbines has led to highlighting the environmental feasibility, in particular the control of nitrogen oxide (NO_x) emissions. Trace amounts of NO_x species may be present during combustion and lean premixed (LPM) combustion method is one of the effective NO_x control technologies for gas turbines. Despite the proven feasibility of syngas, these fuels face variation in their composition due to different feedstock and production methods. Syngas can contain small amounts of hydrocarbons, such as methane, ethane and ethylene, and the presence of hydrocarbon species in the fuel can affect the NO_x emissions in the post-flame region. A detailed investigation of high-pressure syngas combustion is required to provide both practical information regarding NO_x emissions from gas turbine systems and fundamental understanding of reaction pathways to verify and develop chemical kinetic models.

In this study, a high-pressure burner facility capable of operating at pressures up to 15 bar, was designed and fabricated to simulate syngas combustion in a gas turbine. Syngas consisting mainly of H_2 and CO was blended with components such as nitric oxide (NO), nitrogen dioxide (NO_2) and trace amount of hydrocarbons prior to combustion at high pressure. Hot combustion products were supplied to the high pressure vessel by a flat flame burner operating with lean, premixed syngas-air mixtures with various H_2/CO ratios ranging between 0.25-1.0 and equivalence ratios between 0.5-1.0. Detailed NO_x speciation measurements in the post-flame region were conducted for different fuel compositions at various pressures up to 10 bar using Fourier transform infrared (FTIR) spectrometer. The experimental data generated from this project can be used to validate/develop kinetic models to predict NO_x emissions.

Numerical modelling of industrial burners for reduction of NO_x emissions using flamelet methods in combination with a newly developed postprocessor for fast and accurate emission prediction

DDI Werner Pollhammer, K1-MET GmbH, Linz, Austria
DI Dr. Christoph Spijker, Montanuniversitaet, Leoben, Austria
DI Michael Koller, Ebner Industrieofenbau GmbH, Leonding, Austria
Univ.-Prof. DI Dr. Harald Raupenstrauch, Montanuniversitaet, Leoben, Austria

Abstract

Current trends of research on combustion systems show increased interests on prediction of pollution. To meet the demands of modern combustion systems such as industrial burners detailed studies of their behaviour is needed. One of these pollutants which are commonly of interest is NO_x. The aim of the work is to develop a numerical toolbox that is able to predict NO_x emissions of industrial burners by numerical simulation using Computational Fluid Dynamics (CFD).

A set of numerical equations is solved for flow, temperature, pressure and species. Using CFD it is possible to model a combustion system using different approaches like flamelet methods or detailed chemical reaction systems. To predict the emission of NO_x from a flame a detailed chemical reaction system is necessary such as the GRI 3.0. mechanism [1]. The problem of using this mechanism is that it can be applied only for small geometries to avoid long calculation time or processing costs. Using a detailed chemical reaction system for every species a few chemical reactions have to be solved. Using the GRI 3.0. mechanism 325 reactions for 53 species have to be solved [2]. Also with modern computers it is not affordable to model industrial burners using this approach which would calculate NO_x emissions well.

The second main possibility for simulation of burners is to use flamelet methods [3][4]. Choosing this approach the combustion system is pre-calculated and reaction data will be stored in so called lookup tables. During simulation this reaction data will be accessed by reading them from the lookup tables. This approach allows fast calculation of temperature and flow field of gas burners as well as fast chemical reactions like formation of CO_2, H_2O or CO. This is the standard procedure for simulation of industrial burners for analysis of flame shape and temperature fields. The problem is that the described approach is not applicable for slow chemical reactions such as formation of NO_x. Therefor a detailed chemical reaction system is necessary. The problem is that on the one hand there is the detailed chemical reaction system which would predict NO_x emissions very well but takes too long for calculation. On the other hand there is the flamelet method which is fast but delivers only for flow fields, temperature fields as well as some species fields right results but models the formation of NO_x not very well. The idea now is to combine the benefits of both approaches and to postprocess the results out of the flamelet calculation using a detailed chemical reaction system. This means temperature and flow field are calculated by the fast flamelet calculation. After convergence of this calculation the solution for the flow field and temperature field as well as for some species fields are known. Subsequently for slow chemical reactions such as NO_x production, which cannot be predicted by the flamelet method, the postprocessor calculates the reactions using the already existing temperature and flow field from the flamelet calculation. This combination enables a much faster calculation of NO_x formation in the flame and combines the speed of flamelet method with the accuracy of using detailed chemical reaction systems with reasonable calculation costs [5].

Model description

For a more precise description of radiation effects it was necessary to implement the Discrete Ordinate Model (DOM) into the flamelet solver of OpenFOAM [6]. The standard radiation model is only valid for free stream flames with a constant surrounding wall temperature. The radiation model was used in combination with a Weight Sum of Gray Gases Model (WSGGM) which calculates the optical density of the gas by applying a mixture of temperature dependent absorption coefficients for calculation of the radiation beams [7].

For flamelet calculation a steady laminar flamelet for non-adiabatic combustion was used out of literature [3][4]. By defining a range for enthalpy defects in flamelet generation heat transfer in the system and out of the system can be taken into account. Flamelets were generated for different scalar dissipation rates which takes mixing time scales into account and for different enthalpy defects.

For modelling of turbulent effects the k-ε-model was used with a slight change according to literature data. For a more precise prediction of flame acceleration the constant $C_{2\varepsilon}$ was modified to 1.8. [8]. Out of data from Sandia Flame D boundary conditions for turbulence were generated for turbulent kinetic energy k and dissipation of it named ε [1].

The NO_x postprocessor uses a detailed chemical reaction approach the so called partially stirred reactor model implemented in OpenFOAM for calculation of chemical reaction [9]. It takes chemical reaction kinetics, for instance out of GRI 3.0., as well as turbulent mixing effects into account [2]. The solver uses a conservation equation for each species and chemical reaction is implemented as a source term in every species equation. This means for chemical reaction in every mesh cell the whole reaction mechanism consisting of hundreds of Arrhenius equations has to be solved.

To reduce the consumption of calculation power a special strategy of solving chemistry was implemented. As a matter of fact that the system seeks for a steady state solution that means after a certain time species composition will not chance any more in the cells. While post processing there is no change of temperature and turbulence because these fields are calculated by the previous flamelet calculation and they are constant during post processing. The reaction rates of chemistry then only depend on species concentration. If steady state is nearly reached in a cell and the species concentration varies just a bit there is no big difference between reaction rates of the previous and actual time step. An algorithm which analyses these changes was implemented. Using this algorithm the chemical reaction source terms were tabled from time step to time step. If the change in species is lower than 2 percent of the relative concentration of every species in a cell then the source terms of the previous time step is used. Using this algorithm about 95 percent of chemistry can be accessed by tabling and only 5 percent of the cells have to be calculated. This leads to huge saving of computational power and to a big speed up in simulation.

To avoid unphysical states there is also a check if one of the species concentrations goes under zero which is physically incorrect. If this occurs then chemistry will be calculated in the next time step. To avoid pseudo steady state solutions every 200 time steps the whole domain will be calculated.

The model validation was done by means of a comparison with measurements on the Sandia Flame D [1]. Therefor a two dimensional mesh consisting of 200 000 hexahedra was generated using blockMesh and analysed in terms of mesh independency. The benefit of validation on Sandia Flame D is that temperature, velocity, turbulence and species decomposition is also known at certain points in the flame. This means it is feasible to analyse the whole flame shape and its position dependent characteristics and not only the properties at the outlet. For reliable results it is important to be able to predict the temperature maximum very well. Comparison of species decomposition shows very good agreement in terms of oxidizer consumption and CO_2 formation [1]. Using the newly developed postprocessor a very accurate description of NO_x formation is possible which fits well to literature data [1].

Model validation on Sandia Flame D and results

First the flamelet model was used to describe the temperature, velocity and turbulence field of the flame. The following Figure 1 shows the solution of the Sandia Flame D case and its temperature field.

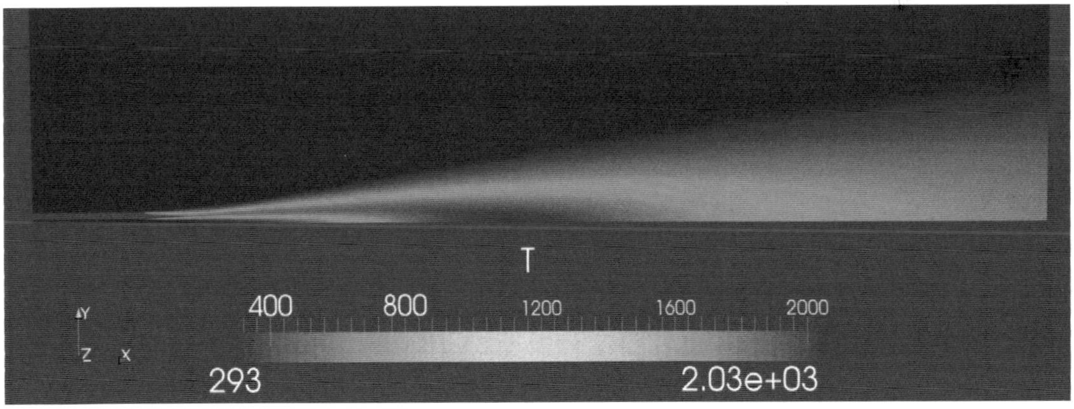

Figure 1: Temperature distribution in Sandia Flame D calculated using flamelet model

This simulation results were then compared to literature data [1] which is displayed in Figure 2. As it can be seen is that the temperature along the rotational axis fits very good to measurements. This is very important for an accurate prediction of nitrogen oxide formation.

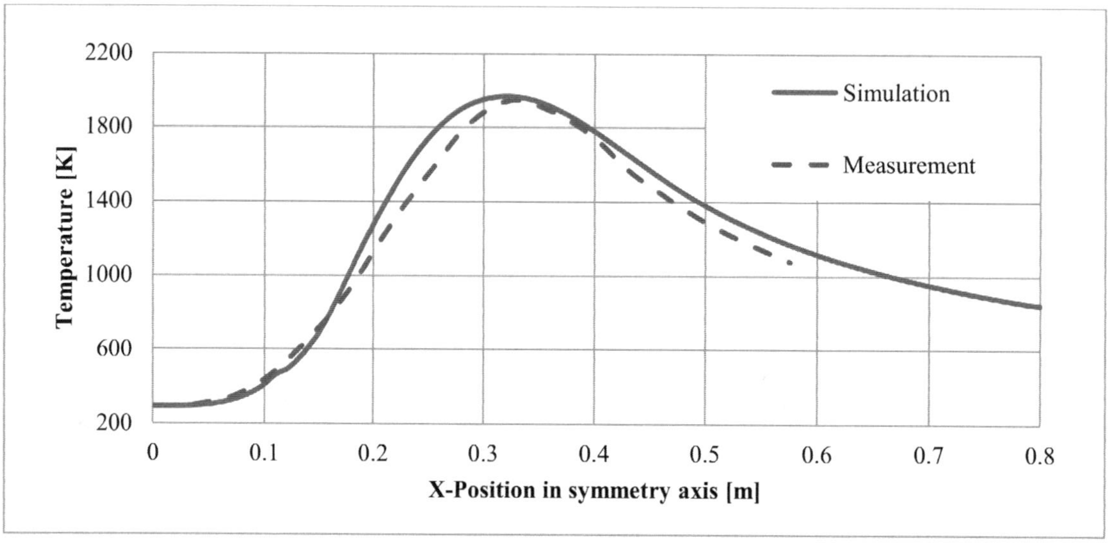

Figure 2: Plot of flame temperature comparing simulation and measurement data from Sandia Flame D [1]

Using this data for temperature, velocity, turbulence and species which can be predicted by flamelet the post processing was performed to calculate species like NO, N_2O, NO_2 which cannot be predicted accurate enough by flamelet. For this species a detailed chemical reaction mechanism was used. In this case the simulations were performed using GRI 3.0. [2]. The following Figure 3 shows the distribution of NO in the Sandia Flame D with three sections through the flame. At these positions a comparison with measurement data was made which is displayed in Figure 4.

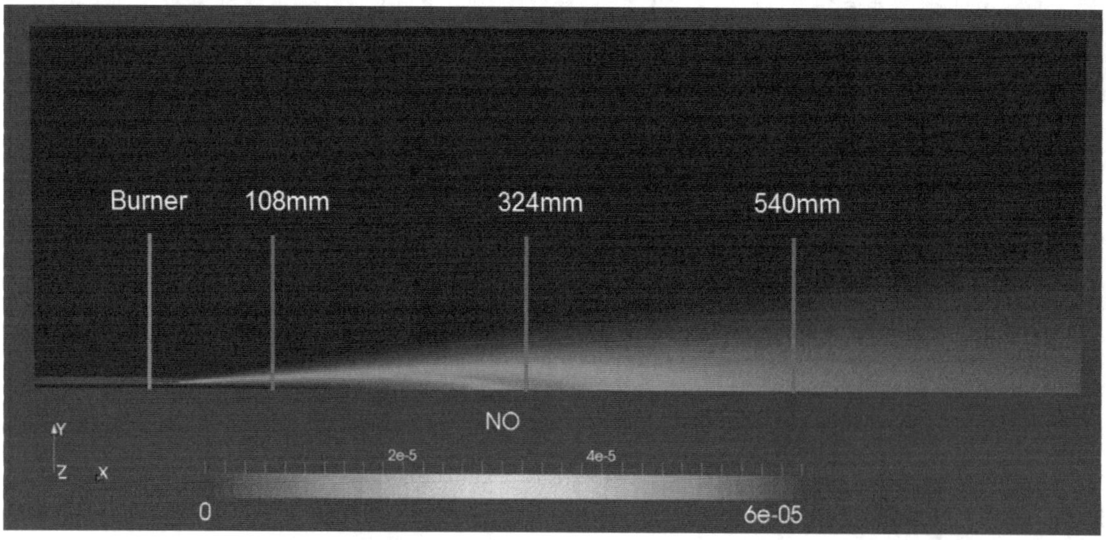

Figure 3: NO field of Sandia Flame D after convergence of postprocessor

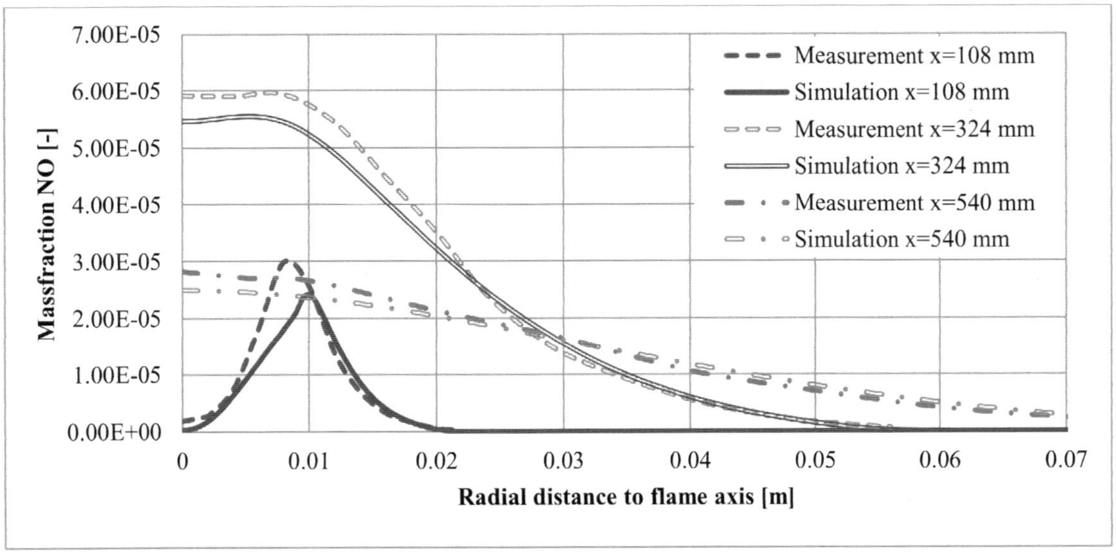

Figure 4: Comparison between simulation and measurement at three cuts through Sandia Flame D as shown in Figure 3 [1]

The result of the first validation is that the generated postprocessor is able to calculate nitrogen oxide formation in Sandia Flame D very well. It predicts slightly lower concentration but the accordance is good in the whole flame. So the developed model can be used for detailed investigations on nitrogen oxide formation in industrial burners.

Model validation on test chamber and results

A second model validation was done by means of an industrial test where a burner was operated in a test chamber. Measurements of wall temperatures by thermocouples and a detailed exhaust gas analysis were performed. Out of the measurements of 20 thermocouples at different positions on the furnace wall a temperature profile was generated to implement it as boundary condition for the flamelet calculation. This lab test was modelled using the developed simulation toolbox. A mesh of the whole test geometry was generated using cfMesh [10]. Out of mass flow and temperature measurements boundary conditions for simulation were calculated. The first validation for the model was a comparison between exhaust gas temperature in the lab test and simulation. These data were then used to start the NO_x post processing. Therefor the developed transient postprocessor was used to calculate the formation of NO_x and various other species which are not predicted well by the flamelet method. At that point it is only necessary to calculate the chemical reaction source terms without solving flow and temperature fields which are now constant in this calculation and already known from the flamelet calculation. This way the simulation is much faster and calculation of NO_x formation now is possible in affordable time using the detailed chemical reaction mechanism with the benefit of a much more accurate description of NO_x formation. For the whole geometry a mesh independency study was performed on the one hand for the flamelet model and on the other hand for the postprocessing. The mesh was generated with a cell size which delivers mesh independent results. For the test chamber with a height of 2.3, a width of 1.7 and a depth 2.5 meters the mesh consists of 5.26 million polyhedral cells.

Figure 5 shows the result of the post processing of the test chamber simulation.

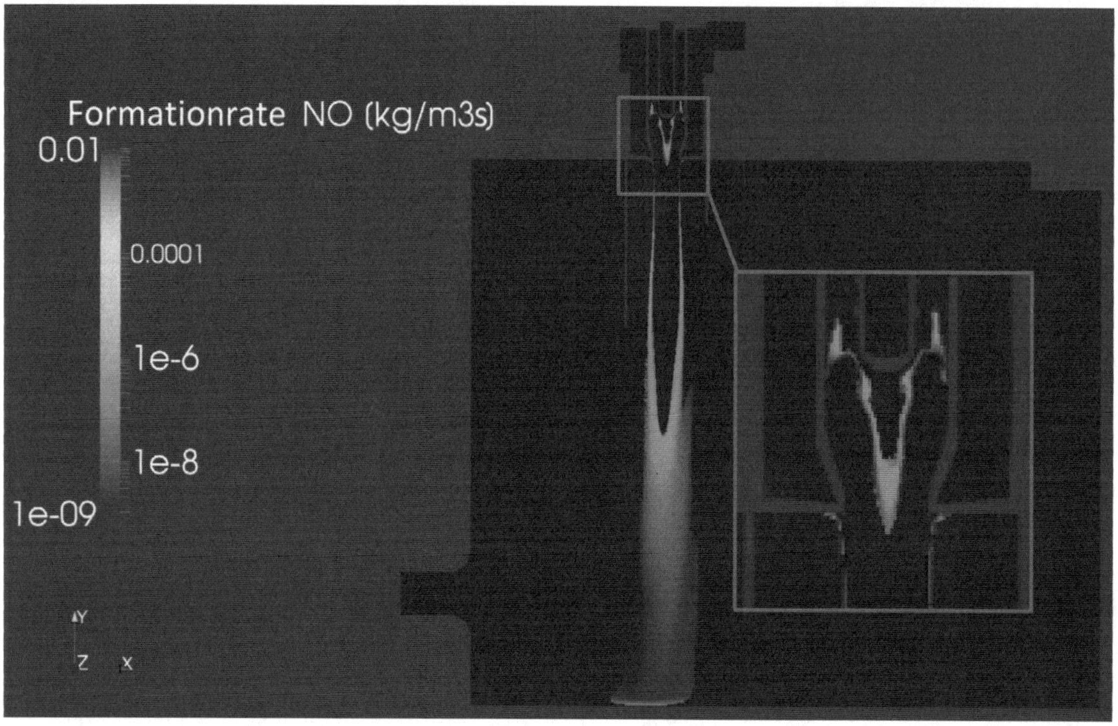

Figure 5: Areas of NO formation in test chamber, calculated using developed postprocessor

This test was performed twice. One test was made with air at ambient temperature and one with preheated air with 199°C.

Table 1: Comparison between simulation results and measurements in test chamber

Formation of NO_x	Simulation [ppm]	Measurement [ppm]
Cold burner air with 28°C	63	65
Preheated burner air with 199°C	78	92

As it can be seen in Table 1 there is also a good accordance for the postprocessor with measurements in the big test chamber. The case with air at 28°C shows the best accordance. The second lab test using preheated air shows the right trend by increasing NO_x formation in terms of the cold air case but shows lower NO_x emissions like the measurement at the test chamber. This effect of lower NO_x concentrations can also be seen in Figure 4 that the model predicts lower nitrogen oxide emissions than the measurements. Important is that the model predicts relative improvements and trends. So now the model was used to perform a geometry study and to test different improvements of the burner geometry by changing the geometry and using the postprocessor for analyzation.

Outlook

Now the model is used to analyse the formation zones of nitrogen oxides in the burner. The next steps are further improvements of the code efficiency and to find more strategies to reduce the calculation time. After some improvement steps of simulation a prototype of the new developed burner will be build a new chamber test will be performed to check these improvements also in a real test.

Acknowledgements

The authors gratefully acknowledge the funding support of K1-MET GmbH, metallurgical competence center. The research program of the competence center K1-MET is supported by COMET (Competence Center for Excellent Technologies), the Austrian program for competence centers. COMET is funded by the Federal Ministry for Transport, Innovation and Technology, the Federal Ministry for Science, Research and Economy, the province of Upper Austria, Tyrol, and Styria and the Styrian Business Promotion Agency.

References

[1] Barlow R, Frank J. *Piloted CH4/Air Flames C, D, E, and F – Release 2.1.* Available at: http://www.sandia.gov/TNF/DataArch/FlameD/SandiaPilotDoc21.pdf; 2007 [accessed 07.04.2017].

[2] Smith GP, Golden DM, Frenklach M, Moriarty N, Eiteneer B, Goldenberg M, Bowman T, Hanson RK, Song S, Gardiner WC, Lissianski, Jr., Vitali V., Qin Z. *GRI Mech 3.0.* Available at: http://www.me.berkeley.edu/gri_mech/; 1999 [accessed 11.04.2017].

[3] Cuoci A, Frassoldati A, Faravelli T, Ranzi E. Formation of soot and nitrogen oxides in unsteady counterflow diffusion flames. *Combustion and Flame* 2009;156:2010–22, doi:10.1016/j.combustflame.2009.06.023.

[4] Cuoci A, Frassoldati A, Faravelli T, Ranzi E. Kinetic Modeling of Soot Formation in Turbulent Nonpremixed Flames. *Environmental Engineering Science* 2008;25:1407–22, doi:10.1089/ees.2007.0193.

[5] Spijker C. *Unsteady Laminar Flamelet Modellierung zur Beschreibung von Mündungsmischbrennern.* Masterthesis, Montanuniversitaet. Leoben; 2010.

[6] Thynell ST. Discrete-ordinates method in radiative heat transfer. *International Journal of Engineering Science* 1998;36:1651–75, doi:10.1016/S0020-7225(98)00052-4.

[7] Barlow RS, Karpetis AN, Frank JH, Chen J-Y. Scalar profiles and NO formation in laminar opposed-flow partially premixed methane/air flames. *Combustion and Flame* 2001;127:2102–18, doi:10.1016/S0010-2180(01)00313-3.

[8] Pfeiler C, Spijker CJ, Raupenstrauch H. CFD als Werkzeug in der Industrieofentechnik. *Berg Huettenmaenn Monatsh* 2011;156:347–52, doi:10.1007/s00501-011-0018-z.

[9] OpenFOAM. *OpenFOAM 2.4.0.* Available at: https://openfoam.org/; 2017.

[10] cfMesh. *Professional Meshing Solutions - cfmesh.com.* Available at: http://cfmesh.com/; 2017 [accessed 11.04.2017].

491775 A CFD Model-Based Optimization of a Low NO_X Burner in Integrated Process System

Abstract:

Although process burners are frequently used in many industries such as refeinring, petrochemical and chemical processes, there is a serious concern about emission control of these equipment. Recently, the deveopment of new numerical approaches such as computational fluid dynamics (CFD) has led to significant improvements in emission control of industrial processes. The proper analysis and testing leads to more efficient combustion, less physical testing and lower cost for full scale equipment burners.

In this study, we used CFD simulations to design, develop and optimize a low NOx burner which produces heat for methane steam reforming. The CFD method helped to design a burner with angled air holes arranged in a swirl pattern. This optimized design increased the mixing near the burner plate that led to lower peak flame temperature and therefore lower NOx emissions. The designed burner was built and experimentally tested and showed a major NOx reduction in steam reforming hydrogen generation process. In addition, reformer performance improved significantly due to the better temperature distribution by the new CFD optimized burner.

496078 Mapping the Stable Region of Operation of Homogenously Charged Compression Ignition (HCCI) Engine for Methane and DMF Fuel Blend

Abstract:

Abstract:

The emission legislations are transmuting to be rigid in accordance to reduce the emissions from the IC engines. Research has been focused to ameliorate both Spark Ignition and Compression Ignition engine in order to make them more efficient. Over the previous decade, major research has been focused on Homogenous Charge Compression Ignition (HCCI) engines. This engine has the potential to consummate the eminent feature of Spark Ignition engine, with no emission of NOx and Soot leading to cleaner combustion, the best of CI engine with higher efficiency and low fuel combustion along with lower carbon dioxide emissions[1].

The focus of this work is to map the stable regions of operation of HCCI engine using Methane and Dimethyl Furan blend as fuel. DMF is comparable to gasoline since it has an energy density of 40% higher than ethanol. Also, Burning DMF requires 33% less air compared to the same amount of gasoline since its stoichiometric air to fuel ratio is 10.72. The Simulations of HCCI engine were carried out in CHEMKIN, where the model used was a closed transient CSTR[2]. The main objective of the study was to depict the stable regions of operation of HCCI engine with and without Exhaust gas recirculation. EGR refers to a part of the exhaust stream recycled back to the engine inlet along with the fuel mixture[3].

Parameters such as EGR Temperature, EGR percentage and equivalence ratio were varied to study the effect on the sustainable emissions at the exhaust for CO, CO_2 and NOx. Detailed Kinetics of DMF obtained from NCBI was used. Initial understanding of the working of HCCI engine was done using Methane as source fuel, keeping EGR temperature and HCCI engine temperatures as a constant at 600K and 447K and the equivalence ratio were varied from 0.5 to 0.7 and the EGR percentage were varied from 0 to 30%. The in-cylinder temperature, pressure and species profile were obtained as a function of Crank Displacement angle (CAD).

The Figure (1) shows the effect of equivalence ratio on the in-cylinder temperature, pressure and species profile for a engine temperature of 447K and EGR temperature of 600K for 30% EGR. Increase in equivalence ratio states the presence of higher amount of fuel where as smaller ratios represent deficit of fuel in the mixture. It could be observed that Peak temperature and pressure increases with increase in equivalence ratio. The mole fraction of CO increase with increase in equivalence ratio, this is because of the lack of oxygen available for complete combustion. In case of Thermal NO production rate, it increases with increase in equivalence ratio due to increase in temperature which leads to increased production of NO at higher temperatures. It was observed from the initial study using Methane as a fuel that, for Lesser exhaust emissions, the fuel must be lean with a higher percentage of EGR.

Fig 1: Variation of In-cylinder temperature, pressure, mole fraction of CO, CO_2 and Thermal NO production rate with Equivalence ratio

The study was further extended to DMF as an additive to Methane and the parameters such as EGR percentage, Equivalence ratio and EGR temperature were varied to find the stable region of operation of HCCI engine for Methane and DMF blend. The effect of Equivalence ratio on the in-cylinder temperature, pressure and species properties for 0% EGR for an engine temperature of 447K and EGR temperature of 600K were studied initially. The figure (2) shows the effect of Equivalence ratio on Peak CO_2 mole fraction in the cylinder for different percentage of Methane and DMF blend. It could be observed that the peak mole fraction of CO_2 increases with increase in Equivalence ratio and also increases with increase in blend percentage. The latter is due to the increase in the carbon availability in the cylinder with increase in fuel.

Fig 2: The Variation of Peak Mole Fraction of CO_2 with percentage of the DMF blend for different Equivalence ratios

As a next step, it is intended to study the effect of EGR percentage and EGR mixing temperature on Methane and DMF blend as shown in the Table 1 and analyse the Peak in-cylinder temperature, pressure and species properties for the different parameters combinations and map the stable region of operation of HCCI engine for Methane and DMF fuel blend.

Table 1: Parameters to be estimated for the Blend

Blend	**Parameters**
Methane & DMF blend	EGR Percentage – 0%, 10%, 20% & 30% EGR temp – 400K, 500K & 600K Equivalence ratio – 0.5, 0.6 & 0.7 Blend Ratio – 0%, 50% & 90%

Keywords:

Exhaust Gas Recirculation, DMF, Crank Rotation angle.

References:

1. Harrison R. M. ed. (1996) Pollution, Causes, Effects and Control. 3rd Ed. Cambridge, The Royal Society of Chemistry.

2. Xu H., Liu, M., Gharahbaghi, S., Richardson, R., Wyszynski, M., Megaritis, T., "Modelling of HCCI Engines: Comparison of Single-zone, Multi-zone and Test Data", SAE 2005-01-2123

3. Heywood JB, Internal Combustion Engine Fundamentals, Mc Graw Hill; 1988

498234 Low-Temperature Catalytic Gasification of Particulate Waste Materials

Abstract:

Americans generate more than 250 million tons of waste each year. Most of the items discarded go un-recycled, are incinerated, or end up in landfills. With landfill reaching capacity and energetic and environmental impacts of incinerators, waste management and resource utilization have attracted the attention of engineers and scientists worldwide. One such alternative is waste gasification, a process that breaks down carbon-based materials, into synthetic gas.

Common long-chain polymers found in municipal waste, wastewater treatment plants, as well as space exploration waste materials are polyethylene, cellulose, and nylon, among others. The catalytic gasification process is studied in a low-to-mid temperature/high pressure. One of the advantages of this process, typically referred to as Wet-Thermal Catalytic Oxidation (WTCO), is the low temperature and energy footprint requirements as compared to more classical routes such as incineration. Proof-of-concept experiments are reported. These results demonstrate the potential of this technology as a sustainable approach for waste management.

This project is focused on the catalytic gasification of cellulose as it has proven to present different challenges. Indeed, unlike that for low melting temperature substrates, cellulose remains in solid phase. This creates a heterogeneous, a particulate or slurry system, reacting environment where interaction must be promoted between catalyst, substrate, and oxygen. This is not an easy feat, and it requires first a characterization of the fluid dynamics of the particulate system. The fluid dynamics of the WTCO reacting environment in a high pressure stirred reactor presents a formidable problem to assess for design and optimization purposes. A multiphase flow exhibiting stratification of particulates in a reacting environment presents several technical challenges. In this paper, we highlight how experiments and computational fluid dynamics (CFD) can be used as a powerful tool for characterization and process optimization.

An imperative stage in ensuring a successful scale-up and design for this project required characterization of the mixing patterns, particle dynamics, and transport phenomena associated with the degradation (gasification) of cellulosic substrates. Several diagnostic challenges presented by the multiphase flow of the gasification reaction were analyzed. In this paper, the modelling of the gasification of cellulose particles in a pressure stirred reactor using a finite element Multiphysics simulation environment (COMSOL) is used for CFD characterization.

A summary of complementary studies on the kinetics of the Sabatier reaction along with models for the water-gas shift reactions are used to describe catalytic kinetic in the gas phase. A phenomenological model integrating the CFD characterization with transport phenomena, thermodynamics, and gas-phase kinetics is demonstrated. The results clearly demonstrate the potential of catalytic gasification as a sustainable waste management alternative.

499736 Experimental Analysis of Catalytic Gasification of Waste Polymers

Abstract:

(1) Title:

Experimental Analysis of Catalytic Gasification of Waste Polymers

(2) [Undergraduate] Student Researchers:

- Samuel O. Sanya
- UC Obiako
- Aliandra Barbutti
- Stephen Reeves
- Eric M. Lange

(3) Faculty Mentor:

- Dr. Jorge Gatica

(4) Abstract

Catalytic gasification, or "gasification", is one effective method which can be applied to promote low-temperature conversion of solid waste to energy. This research focuses on advancing the knowledge of a catalytic gasification process as a potential in-situ resource utilization and waste management alternative. This research has significance in a variety of engineering applications, but it is of particular relevance towards reducing landfill waste or as an in-situ resource generation system for space exploration beyond LEO.

This process evolves through a reaction mechanism consisting of two liquid-phase oxidation reactions of long-chain polymers, complemented by two gas-phase reactions. Using a laboratory grade reactor, the gasification of two model substrates, polyethylene and cellulose, was examined in the presence of ruthenium and platinum catalysts.

Gaseous products consisting of hydrogen, methane, carbon monoxide, carbon dioxide, were quantified by Gas Chromatography (GC). Calibration of the GC using chromatographic standards enabled the formulation of an equation to calculate the composition of the gaseous products for different reaction times. Fundamental reactor design equations along with stoichiometry calculations were then used for determinations of percent polymer gasified as well as selectivity of the gasification process.

Additional characterization of the solid and liquid residuals was performed using Differential Scanning Calorimetry (DSC), Thermo-Gravimetric Analysis (TGA), and Scanning Electron Microscopy (SEM). Quantification of the DSC spectra was used to correlate the thermal characterization of the residues with the unconverted (or non-gasified) substrate after the reaction was quenched. The SEM provided information on the microstructure of the residues, their atomic composition, and preliminary assessment of the possibility of catalyst recovery.

Preliminary results suggest that catalytic gasification provides potential control for the liquid-phase oxidations. Under appropriate conditions, oxidation was found to follow the incomplete oxidation pathway which resulted in significant yields of hydrogen and methane, which are potentially high value products as they can be used for energy recovery.

Comparisons between the gasification rates for the two substrates are drawn. Similarly, comparative studies of different catalysts are also presented. The results are used in conjunction with a kinetic model for the gas-phase to formulate a methodological model for the gasification of solid waste. The results will be presented along with a preliminary extension for a continuous gasification process.

487237 Intensification of Enzymatic Hydrolysis of Cellulose to Glucose By High Frequency Ultrasound: Optimization of Process Parameters and Effects on Enzyme Stability

Abstract:

The conversion of lignocellulosic biomass for biofuels and biorefinery applications is limited due to the cost of pretreatment to separate or access the biomass's three main components, cellulose, hemicelluloses, and lignin. Ultrasound and cavitation-assisted technologies are green techniques that have the potential to enhance the separation and hydrolysis of lignocellulosic materials for use in biofuel production and biorefineries, through physical and/or chemical mechanisms. We have studied the enzymatic hydrolysis of cellulose using a novel multifrequency ultrasonic reactor, and optimized glucose yield at 50°C and 5.2 pH. The best cellulose hydrolysis was observed at 611 kHz and 104 W. The effects of ultrasound and enzyme on the surface and structure of the cellulose were studied using BET, SEM, X-Ray Diffraction (XRD) and Scanning Helium Ion Microscopy (SHIM). The effect of ultrasound on the enzymes (cellulase + cellobiase) was also preliminarily observed from High Pressure Liquid Chromatography (HPLC) and FTIR analyses. It appeared that ultrasound did not affect the specific activity of the complex structure of the enzyme macromolecules at the treatment conditions reported in this study. Up to 70 % glucose yield was obtained with prolonged time (40 – 50 h). A synopsis of our current research projects and some results of work on cavitation-enhanced intensification and optimization of the conversion of cellulose to glucose will be discussed. An overview of studies of ultrasound effects on enzyme stability and future research work will also be presented.

REFERENCES:

Adewuyi, Y.G. Intensification of Enzymatic Hydrolysis of Cellulose Using High-Frequency Ultrasound. In: Intensification of Bio-Based Processes, A. Gorak and A. Stankiewicz (editors), Royal Society of Chemistry (RSC), **2017**. In press.

Adewuyi, Y.G.; Deshmane, V.G. Intensification of Enzymatic Hydrolysis of Cellulose Using High Frequency Ultrasound. An Investigation of the Effects of Process Parameters on Glucose Yield. Energy & Fuels. **2015**, 20, 4998 – 5006.

Deshmane, V.G.; **Adewuyi, Y.G**. Mesoporous Nanocrystalline Sulfated Zirconia Synthesis and Its Application for FFA Esterification in Oils. Appl. Catal. A: General. **2013**, 462-463, 196-206.

Deshmane, V.G.; **Adewuyi, Y.G**. Synthesis and Kinetics of Biodiesel Formation via Calcium Methoxide Base-Catalyzed Transesterification Reaction in the Absence and Presence of Ultrasound. Fuel, **2013**, 107, 474-482.

Mahamuni, N.N.; **Adewuyi, Y.G**. Application of Taguchi Method to Investigate the Effects of Process Parameters on the Transesterfication of Soybean Oil Using High Frequency Ultrasound. Energy & Fuels, **2010**, 24, 2120-2126.

Mahamuni, N.N.; **Adewuyi, Y.G.** Optimization of the Synthesis of Biodiesel via Ultrasound-Enhanced Base-Catalyzed Transesterification of Soybean Oil Using a Multifrequency Ultrasonic Reactor. Energy & Fuels. **2009**, 23, 2757-2766.

502968 Energy Performance of an Integrated Biomass Gasification and SOFC Combined Cycle Power Plants with CO_2 Capture

Abstract:

Torrefied biomass is a potential alternative fuel to replace raw biomass or coal in power plants. This study develops a hybrid power plant by integrating torrefied biomass co-gasification (TBCG) with solid oxide fuel cell (SOFC) systems. The calcium looping (CaL) CO_2 capture process is also implemented into the hybrid power system. Two configurations for the proposed hybrid power plant are designed and compared each other, namely, pre-SOFC and post-SOFC. Thermodynamic analysis is adopted to examine the performance of hybrid power generation system using Aspen Plus. Three different scenarios of the energy efficiency indexes are taken into account to evaluate the performance of hybrid power plant. The simulation results show that the maximum net system efficiency of pre-SOFC and post-SOFC are 40.73 and 36.56%, respectively. In order to improve the maximum net system efficiency of hybrid power generation system, the heat recovery approach is performed. Through the heat integration design, the maximum net system efficiency of pre-SOFC and post-SOFC are 45.72 % and 44.63 %, respectively. Overall, pre-SOFC is superior to post-SOFC in terms of energy efficiency. However, post-SOFC can pursue the hybrid power plant with zero CO_2 emissions.

499868 Pyrolysis of Food Waste: Thermal Deconstruction Rates in a Regular Calcination Oven Vs. an Induction Heating Reactor

Abstract:

The wastes generated by food courts consists mainly of organic residues, paper, cardboard, plastic, wood, glass bottles and aluminum cans which are usually send to landfill. This type of food court waste (FCW) is generated in large quantities and it could be used as alternative primary energy source due to its large organic content. The purpose of this research was to obtain thermal degradation profiles through a thermochemical conversion process â€“ pyrolysis â €“ of FCW in two reactor configurations. The composition of FCW was determined to be 25.11% paper, 14.62% plastic, 20.39% cardboard, 33.40 organic residue, 6.17% glass, 0.02% metal, and 0.28% wood. The pyrolysis processes were performed in two pyrolysis reactors, one heated conventionally and one heated via induction heating in the temperature range of 300-800Â°C, in an inert environment on 30 g sample size. It was determined that the process temperature has a significant effect on the thermal degradation profiles, with the higher temperatures having significantly faster degradation rates in both conventional and induction heating reactor. As the temperature increased, the amount of char residue decreased, and its stabilization occurred in much shorter times. Degradation rates were used to determine the activation energies and pre-exponential molecular constants using Arrhenius-type relationships for both types of reactors.

499573 Hydrothermal Carbonization of Wastes for Simultaneous Nutrient Recovery and Energy Capture

Abstract:

Hydrothermal carbonization (HTC) is a process that converts wet biomass into carbon-rich, energy dense solid fuel, called hydrochar, and an aqueous product. The process is done in hot, compressed water, typically in a temperature range of 180 °C - 260 °C. The thermodynamic properties of water change greatly in the subcritical region, and as a result, subcritical water behaves as a non-polar solvent with high ionic activity. When subjected to HTC, biomass undergoes rapid hydrolysis and produces solid hydrochar, as well as an aqueous product containing organic acids, sugars, and sugar derivatives [1].

Industrialized societies produce large waste streams with high moisture content, such as wastewater solids, animal manure, food processing wastes, and some industrial wastes. HTC can readily produce a carbon-rich, hydrophobic solid hydrochar from these waste streams. Most nitrogen compounds (derived from protein) are found in the aqueous stream, while much of the phosphorous is tightly bound in the hydrochar matrix [2]. Ideally, the nutrients can be captured in the aqueous stream, while the carbon (energy) is retained in the hydrochar. Recently, it has been shown that the presence of Ca, Mg, and Fe, as well as pH and reaction temperature, all affect the extraction of P and N from swine manures [3].

In this study, we evaluate the effects of reaction temperature, reaction time, pH, and especially addition of specific metal oxides on HTC of cow manure. We will show that the energy can be most effectively concentrated in the hydrochar at high reaction temperature with short reaction time. Under most reaction conditions, the nitrogen is readily partitioned to the aqueous phase. However, phosphorous capture requires longer residence times and addition of MgO, CaO, or Fe_2O_3.

References

[1] J. G. Lynam, M. T. Reza, W. Yan, V. R. Vásquez, and C. J. Coronella, "Hydrothermal carbonization of various lignocellulosic biomass," *Biomass Convers. Biorefinery*, vol. 5, no. 2, pp. 173–181, 2015.

[2] M. T. Reza, A. Freitas, X. Yang, S. Hiibel, H. Lin, and C. J. Coronella, "Hydrothermal carbonization (HTC) of cow manure: Carbon and nitrogen distributions in HTC products," *Environ. Prog. Sustain. Energy*, vol. 35, no. 4, pp. 1002–1011, Jul. 2016.

[3] U. Ekpo, A. B. Ross, M. A. Camargo-Valero, and L. A. Fletcher, "Influence of pH on hydrothermal treatment of swine manure: Impact on extraction of nitrogen and phosphorus in process water," *Bioresour. Technol.*, vol. 214, pp. 637–644, Aug. 2016.

496722 Using Fly Ash As pH Adjustment for Efficient Phosphorus Immobilization and Reutilization from Swine Manure in Hydrothermal Treatment

Abstract:

Concentrated animal feeding operations (CAFOs) are an important part of the U.S. economy, providing the country with a substantial low-cost source of consumable meat and dairy products such as beef, pork, poultry, milk, and eggs. However, CAFOs are becoming a greater public/government concern due to the amount of animal manure (waste) generated by such facilities, resulting in both air and water pollution. The United States Department of Agriculture (USDA) estimates that 335 million tons of dry manure waste is annually generated by CAFOs. This waste stream contains significant quantities of nutrients, accounting for nitrogen and potassium totals of approximately 6.17 and 1.85 million metric tons, respectively.

Application of manure to pasture and crop land is the most prevalent CAFO waste management practice. This practice can be beneficial as it improves soil quality by replenishing important nutrients for plant health. However, the amount of applied manure along with nutrient mobility leads to runoff into local watersheds, causing a host of local/regional public health and economic issues. Nutrient runoff (especially P) leads to growth of harmful algal blooms, which has led to the loss of water supply to U.S. cities (due to elevated microcystin levels), loss of aquatic life, and over $100 million in economic damage since 2008.

Due to its distributed nature, nutrient runoff is one of the most prevalent and challenging environmental issues in the U.S. To address this issue, Ohio University (OHIO) is developing a supercritical water (SCW)-based modular process to be operated at CAFO sites in order to recover the nutrients as a solid product. This nutrient-rich solid product may be used as a fertilizer supplement for field application, with significantly reduced potential for runoff into local watersheds. In addition, the process produces a methane-rich gas product, which may be utilized at the CAFO site or upgraded into pipeline-quality natural gas. SCW provides an advantageous media for recovery of nutrients from animal manure; its lower fluid density and hydrogen bonding strength promotes precipitation of solid P components, and provides a gasification medium for undigested carbonaceous species.

In this research, basic environment was evaluated at various temperatures to determine whether it has impacts on nutrient immobilization and phosphorus mobility of biochars. For a more controlled basic environment, 0.1 M NaOH was used to replace DI water; and for a more economical approach, fly ash was used for adjusting pH. After hydrothermal treatment, nutrient elements were found to be immobilized into the solid phase more efficiently with addition of basic materials, especially with addition of fly ash. In addition to being more efficient in immobilizing nutrient elements, fly ash also altered crystal phases formed at different temperature treatments which thus results in different profile of phosphorus mobility of the biochars. Results to be presented in this presentation include experimental results such as nutrient immobilization as a function of process conditions, post-trial analyses of biochar products (including XRD and SEM/EDS), and biochar phosphorous mobility.

494564 Recovery of High-Value Chemicals from Organic Waste: Economic Potential and Logistical Issues

Abstract:

Hexanoic (C6) and caprylic Acid (C8) are chemicals of high value due to their versatility, high demand, and relatively high energy density [1]. Studies have shown that such chemicals can be recovered from different types of organic waste streams such as biomass waste, food waste, and municipal solid waste; but such studies have not evaluated economic effectiveness [1-4]. In this work, we present supply chain models to study the economic potential and logistical issues associated to the recovery of C6 and C8 from different kinds of organic waste. We conduct a case study using real data from the State of Wisconsin.

In our supply chain framework, we consider four different sources of organic waste (urban centers, wastewater treatment plants, large dairy farms, and landfills), three different types of waste products (cow manure, sludge residues from wastewater treatment, and food waste) and five different types of waste processing technologies. We assume that 2-4% of the chemical oxygen demand (COD) in the waste streams can be converted as C6 or C8 [5,6]. The constraints in the model involve product balances and conversion constraints, capacity constraints, and logical constraints associated to technology sizing and placement. The objective function is to maximize total profit which includes capital and transportation costs and revenue from the recovered product. We conduct the following studies: a) trade-off analysis of profit against total amount of treated waste in the state, b) effect of uncertainty on yield factors and on prices of waste, C6, and C8, c) effect of waste mixing, and d) comparison of profitability from C6/C8 against the popular biogas (C4) recovery alternative.

Preliminary results indicate that, under current market conditions, it is profitable to recover C6 and C8 from organic waste and that much higher mobility can be achieved compared to C4. The results can be refined by conducting experimental studies that resolve uncertainties in key yield factors and technology costs.

References:

[1] Angenent L T, Richter H, Buckel W, et al. Chain elongation with reactor microbiomes: open-culture biotechnology to produce biochemicals. Environmental science & technology, 2016, 50(6): 2796-2810.

[2] Grootscholten T I M, Strik D, Steinbusch K J J, et al. Two-stage medium chain fatty acid (MCFA) production from municipal solid waste and ethanol. Applied energy, 2014, 116: 223-229.

[3] Kucek L A, Xu J, Nguyen M, et al. Waste conversion into n-caprylate and n-caproate: resource recovery from wine lees using anaerobic reactor microbiomes and in-line extraction. Frontiers in Microbiology, 2016, 7.

[4] Grootscholten T I M, dal Borgo F K, Hamelers H V M, et al. Promoting chain elongation in mixed culture acidification reactors by addition of ethanol[J]. Biomass and Bioenergy, 2013, 48: 10-16.

[5] Neves L, Oliveira R, Alves M M. Co-digestion of cow manure, food waste and intermittent input of fat[J]. Bioresource Technology, 2009, 100(6): 1957-1962.

[6] Ge S, Usack J G, Spirito C M, et al. Long-term n-caproic acid production from yeast-fermentation beer in an anaerobic bioreactor with continuous product extraction[J]. Environmental science & technology, 2015, 49(13): 8012-8021.

498471 Optimization of Biorefinery Production Chains and Decision-Making through Sustainability Evaluation: A Biojet Fuel Case Study

Abstract:

The aviation industry is a large contributor to the environmental pollution. For example, during the course of a roundtrip flight from Paris to New York, 2.2 Tons of CO2 are released per passenger into the atmosphere; a number that is roughly equivalent to sixty five 100km trips in a medium sized car (2). Clearly, there is a need for solutions that help mitigate these emissions, and advocacy groups such as the Air Transport Action Group (ATAG) push for actions towards more sustainable operations. According to the International Air Transport Association (IATA) sustainable aviation fuels derived from biomass (i.e. biojet fuels) "can reduce the overall carbon footprint by around 80% over their full lifecycle" (3), and actively supports their development.

Previous contributions have addressed the production of biojet fuel; as examples Crawford et al. (4) and Klein-Marcuschamer et al. (5) have analyzed the economic viability of biojet fuel production under different geographical and technological contexts, while Klein-Marcuschamer et al. (5) have studied the air emissions for its production. Going a step forward, Agusdinata et al. (6) have recognized that multiple actors/ stakeholders (e.g. national/international policymakers, biorefineries, farmers) are involved in the development of biojet-fuels, each of which has differents objectives to consider in their decision-making.

In this work, we combine mathematical programming tools (optimization) and the GREENSCOPE approach developed by the U.S. Environmental Protection Agency (7) to assess the sustainability of production of biojet fuels from the perspective of the different actors that participate in the overall production chain. The GREENSCOPE approach allows for the quantification of process sustainability in four main areas: Efficiency, Energy, Economics, and Environment; in here, each E is characterized by a series of metrics that indicates the performance of a particular process. Optimization is applied to find the processing pathways that provide the best performance for each E, and for clusters of the indicators that represent the preferences of the different stakeholders. Multi-objective optimization is then used to find the Pareto frontier that describes the trade-offs among their optimal solutions.

As a case study, we consider the production of biojet fuels from three different organic material sources: eucalyptus, pine, and macauba, through three main technological pathways: fast pyrolysis, hydrothermal liquefaction, and gasification followed by Fischer-Tropsch synthesis. In all cases, the benefits of adding a follow-up cracking step of long-chain (hydrocarbon) residues, is also considered.

The views expressed in this presentation are those of the authors and do not necessarily represent the views or policies of the U.S. Environmental Protection Agency.

References

1. Brasseur, G.P., M. Gupta, B.E. Anderson, S. Balasubramanian, S. Barrett, D. Duda, G. Fleming, P.M. Forster, J. Fuglestvedt, A. Gettelman, R.N. Halthore, S. Jacob, M.Z. Jacobson, A. Khodayari, K. Liou, M.T. Lund, R.C. Miake-Lye, P. Minnis, S. Olsen, J.E. Penner, R. Prinn, U. Schumann, H.B. Selkirk, A. Sokolov, N. Unger, P. Wolfe, H. Wong, D.W. Wuebbles, B. Yi, P. Yang, and C. Zhou "Impact of Aviation on Climate: FAA's Aviation Climate Change Research Initiative (ACCRI) Phase II.", Bull. Amer. Meteor. Soc., 2016, 97, 561–583.

2. Simulations done on the basis of the tool available at https://co2.myclimate.org/. Accessed on April 16, 2017.

3. IATA website: http://www.iata.org/whatwedo/environment/Pages/sustainable-alternative-jet-fuels.aspx. Accessed on April 16, 2017.

4. Crawford, J. T., Shan, C. W., Budsberg, E., Morgan, H., Bura, R. and Gustafson, R., "Hydrocarbon bio-jet fuel from bioconversion of poplar biomass: techno-economic assessment", Biotech. for Biofuels, 2016, 9 (1), 141.

5. Klein-Marcuschamer, D., Turner, C., Allen, M., Gray, P., Dietzgen, R. G., Gresshoff, P. M., Hankamer, B., Heimann, K., Scott, P. T., Stephens, E., Speight, R. and Nielsen, L. K., "Technoeconomic analysis of renewable aviation fuel from microalgae, Pongamia pinnata, and sugarcane". Biofuels, Bioprod. Bioref., 2013, 7: 416–428.

6. Agusdinata, D. B., Zhao, F. and DeLaurentis, D. A., "Sustainability of Biojet Fuels: A Multiactor Life Cycle Assessment Approach", IEEE Potentials, 2012, 31(1), 27-33.

7. Ruiz-Mercado, G. J., Gonzalez, M. A. and Smith, R. L., "Sustainability Indicators for Chemical Processes: III. Biodiesel Case Study", Ind. & Eng. Chem. Res., 2013, 52, (5), 6747-6760.

502474 Eco-LCA Methodolgy for Sustainability Assessment of Bio-Wwts Focused on Energy Recovery

Abstract:

Introduction. The energy potential of agro-industrial effluents receiving biological treatment is often not much exploited, particularly in emergent regions because this alternative energy is forced to compete disadvantageously with conventional energy sources from an economic point of view. Consequently, smart strategies are required to prove the environmental/economic potential of biological wastewater treatment systems (Bio-WWTs) depolluting agro-industrial effluents committed with improved energy recovery. In this context, the following contribution introduces a methodological framework for refined sustainability assessment of Bio-WWTs treating agro-industrial effluents. It combines life cycle assessment (LCA) and ecological economics methods, specifically the emergy accounting approach. Results of a first proof of concept of this methodology are presented here using a Bio-WWTs of the poultry sector and a Bio-WWTs treating beer production effluents as case studies.

Methods. The methodology is developed as follows:

(i) Obtaining of Net Environmental Benefits Indicators-(NEB) by means LCA. The analyses includes avoided or shifted conventional energy in the borders of Bio-WWTs (Godin et al. 2012.) by the recovery of biogas from the Bio-WWTs operation. For each case study the NEB indicators were calculated for two life cycle scenarios considering different energy recovery routes in order to transform these NEBs in Environmental Influence Indicators (IAM). Application of the analytical identification method enables obtaining a reduced set ofIAM (Grunwald-Rosch, 2011).

(ii) Identification of potential co-benefits or "add-values" related with changes in the ecosystem services offer, which are promoted by differences between life cycle scenarios. These "add-values" are specifically appraised by emergy methods (Zhang Y, 2010). Then a set of normalized add-value indicators (NAV) are proposed to assess the "eco-economic" dimension of sustainbility.

Results. A set of Sustainable-Development Indicators (SDIs) has been obtained and defined as NAV/IAM ratios, similarly to ecoefficiency indicators. These SDIs are related to: CO_2-capture capacity, resources utilization, water pollutants and airborne emissions dilution. An overall sustainability indicator NAV_T/IAM_P has been also tailored from the contribution of single SDIs. Charts as presented in figure 1 were obtained and used to identify sustainability conditions.

Discussion and conclusions.A methodology unique for sustainability assessment of Bio-WWTs focused on energy recovery has been developed. This is based on integration of LCA/emergy methods (Eco-LCA) and allows to build representative SDIs for Bio-WWTs. These SDIs are dimensionless quantities because emergy (NAV) and environmental (IAM) terms are normalised values ranging from -1 to 1. Positive values of SDIs indicate "additionality", potential "add-value" or a co-benefit obtained, which is promoted by reducing a direct environmental impact (positive IAM). A strong non-additional condition as far as an individual SDI results from both emergy and environmental terms minus numbers. A weak non-additional condition is produced when the emergy term is negative but the respective environmental is not. Results show that sustainability requires synergy among different SDIs and is not only dependent of climate change control/mitigation, which is the more generalized and accepted sustainbility criteria for wastewater treatment systems. Bio-WWTs focused on energy recovery increase resources utilization, this is an "emergy cost" which is "not additional" but it is required to promote additionality through other SDIs. The study shows that sustainability of Bio-WWTs with higher energy availability or production (beer case) is more dependent on the contribution to climate change control than Bio-WWTs with lower energy potential (poultry case). The baseline choice and the application of the NEB approach are key aspects to prove additionality of Bio-WWTs by means the outlined methodology.

References.

1. Godin, D., Bouchard, C., Vanrolleghem, P.A. 2012. Net environmental benefit: introducing a new LCA approach on wastewater treatment systems. *Water Science and Technology*, **65**(9), 1624-1631.
2. Grunwald, A., Rosch, C. 2011. Sustainability assessment of energy technologies: towards an integrative framework. *Energy, Sustainability and Society*, **1**(1), 3.
3. Meneses-Jacome, A., Osorio-Molina, A., Parra-Saldivar, R., Gallego-Suarez, D., Velasquez-Arredondo, H.I., Ruiz-Colorado, A.A. 2015. LCA applied to elucidate opportunities for biogas from wastewaters in Colombia. *Water Science and Technology*, **71**(2), 211-219.
4. Wang, X., Liu, J., Ren, N.-Q., Yu, H.-Q., Lee, D.-J., Guo, X. 2012. Assessment of Multiple Sustainability Demands for Wastewater Treatment Alternatives: A Refined Evaluation Scheme and Case Study. *Environmental Science & Technology*, **46**(10), 5542-5549.
5. Zhang, Y., Singh, S., Bakshi, B.R. 2010. Accounting for Ecosystem Services in Life Cycle Assessment, Part I: A Critical Review. *Environmental Science & Technology*, **44**(7), 2232-2242.

501481 Sustainability Assessment Water Resources Use Mining in Colombia, By the Emergy Analysis

Abstract:

Due of its great climate diversity, Colombia has an abundance of natural resources and possesses a wide range of mining products, including coal, gold, platinum, nickel, emerald, limestone, among others that are extracted at a smaller scale. This extraction is carried out in two ways: a formal large-scale extraction and other smaller-scale, traditional and artisanal; the latter lacks adequate and defined technology, usually in some cases it is considered an illegal activity, making it unsafe, unprofitable, uncompetitive and environmentally unsustainable.

In the last decade, there has been a significant increase in both the volume and value of production of primary goods, particularly in the mining and hydrocarbons sector, which has positively and significantly affected the countryâ€™s gross domestic product (GDP) (FEDESARROLLO, 2012), currently being one of the most dynamic of the Colombian economy. The GDP from this sector has grown from 5.2 to 10.3 trillion pesos between 2000 and 2011 (Colombian Ombudsman, 2010), although in the last two years (2013 - 2014) this activity has gone into recess.

Thus, according to the large-scale mining sector (LSM), in 2011 mining represented 24.2% of exports; 2.4% of GDP; 20% of the total foreign direct investment; 650 billion pesos in construction of infrastructure; 2.6 trillion pesos in purchases from domestic suppliers, 65 billion pesos of investment in social responsibility and 178 billion pesos in environmental responsibility (LSM, 2012). This constitutes evidence for the mining sector in Colombia to be considered one of the main economic engines, since it does not only contribute with employment creation (there were 836000 in 2011, between direct and indirect jobs), but also with investments in infrastructure, public utilities, and social and environmental management (LSM, 2012). In royalties, the mining sector contribution to the country is 16.5%; 13% of it comes from coal companies (1.2 trillion pesos a year).

This constitutes evidence for the mining sector in Colombia to be considered one of the main economic engines, since it does not only contribute with employment creation (there were 836000 in 2011, between direct and indirect jobs), but also with investments in infrastructure, public utilities, and social and environmental management (LSM, 2012).

However, it is impossible to avoid discrepancies and political, economic, environmental and social polemics around this productive activity in the country, since mishandled mining practices lead to: social impacts, since many of these unformalized activities allow child labor, make use of intensive techniques in unskilled labor, generating poor quality jobs with low levels of industrial, social, and health security for miners, it is usually developed in remote and deprived areas; furthermore, it generates conflicts between large mining industries due to concession contracts awarding. Regarding political and economic impacts, a minimum working capital and scarce financial resources for investment are evidenced, and therefore an improper handling of inputs and royalties from this activity is presented. In a greater relevance, there are the irreversible environmental impacts such as destruction of ecosystems, damage to protected areas, air pollution, and ***pollution and impact on the availability of water resources***. Globally, water demand is driven primarily by the population and economic growth, and it is anticipated that in the coming decades, it will increase considerably, while supplies will remain almost constant or decreasing by factors such as climate change, overpumping and pollution of aquifers (Drelich, 2012). It is not unknown that the primary input in gold extraction is water; the water that is used in this sector comes from a variety of sources, and sources and quality are the main areas of controversy and debate in this productive sector. ***The main problem of the mining industry is to generate confidence in the development of a responsible, sustainable and transparent strategy in the use of water*** (Drelich, 2012).

Undoubtedly, the flow of resources from this activity is increasing and projections show that the growing trend will continue, since the international scene shows an important dynamism of the mining market. This situation could be exploited by Colombia, given its geographical location and its geological potential (FEDESARROLLO, 2012); it is estimated that US$250 billion will be invested in mining projects in Latin America between now and 2020. Thus, the mining sector gains importance among public policies, not only due to the high revenues generated by its operation, but also because of the social and environmental impact of the sector activity. Hence the importance of seeking for solutions, methodologies and sustainable alternatives to assimilate the challenge of turning the mineral wealth in an opportunity for development, and thus respond to future generations about how resources from non-renewable assets were invested without affecting renewable resources.

This highlights the fact that, despite the importance that the mining sector has taken, the debate about the true benefits and costs, or the environmental, economic, energy and social problems of the sector's activity has intensified. It is misleading to decide whether mining offers a significant potential or a threat to society without a methodological tool that allows predicting how sustainable is this productive activity, which represents a profitable business for many, which provides social and economic welfare (ultimate goal of the implementation of development projects of this magnitude), and for other sectors of the society symbolizes the opposite, considering as innefective and undervalued all corrective and preventive measures that have been implemented by the government agencies or those regulatory entities that have some kind of control in the field of sustainable management of the processes of extraction and use of non-renewable resources, including products that are generated in such activity, in response to the environmental awareness that has been growing in recent years.

Based on the concept of sustainability as "the compatibility between energy, economics (maximum performance) and environmental aspects" (Redclif, 1987), all development projects especially those that threaten the environmental integrity such as the exploitation of natural resources and mining processes, should be focused on being an economic use alternative that provides an energy yield with acceptable environmental burden. That is, this highly demanding production activity in the country (including byproducts generated in it) should be given a particular management by means of methodological tools to harmonize the economic with the energy and environmental aspects; since these three factors can not be assessed independently, but they must be analyzed together and in an interrelated way, in order to give management on a sustainable basis, which avoids the discrepancies noted above. In addition to these tools, it is possible to identify where the weaknesses lie, which apparently make this activity a threat to some sectors of the society.

It is then necessary to establish, among the many valuation methodologies available, those that allow a sustainable assessment of the current management and mining process in Colombia in both artisanal and large-scale mining; likewise identify the gaps in it, whose purpose is to make mining a lever for development not only with good labor, industrial and environmental safety practices, but also going further in its commitment to the country and the city that hosts it. That is, for the management of mining resources, a valuation method should be used thereof, which is bearable and balanced from the economic, energy and environmental levels, as in the current regulations governing this sector a large imbalance is evident between these aspects. That is to say, in many situations mining is presented with a great economic benefit, but it entails irreversible environmental damage or major social problems that trigger collective chaos; or on the contrary, many mining licenses are denied impeding economic growth because of an environmental impact that may be acceptable, discrepancies caused by non-uniformity of a regulatory framework for the mining sector and the lack of methodological tools to assess the sustainability of different productive sectors of society, which and/or balance economic, energy and environmental aspects.

Globally, the possibility of quantifying the sustainability of different production sectors (agriculture, energy, legislative, sanitation) has been subject of study using the method of emergy which allow not only to quantify parameters, but also to analyze the implications and interrelationships of economic and environmental aspects of these long-term processes.

Thus, this research aims to develop a methodological tool; emergy analysis to assess and compare the sustainability of the current small and large scale mining systems in Colombia, including byproducts generated in this process, emphasizing on the **use of water resources**, gaps in these systems which may not make them sustainable, and a uniform methodology that can be used as a road map to regulate that productive sector in a sustainable way, because without doubt, regulation constitutes the source of many of the limitations of the mining sector

494533 Contribution of Fugitive Particulate Matter to Airborne Pollution in Arid Areas: A Case Study for an Urbanized Middle Eastern City

Abstract:

Fugitive particulate matter (fPM) is a substantial source of airborne pollution in dry arid lands such as the Middle East Area (MEA). However, accurate determination of fPM emissions has been an ongoing challenge for the air quality research community because of the induced health effects and the large uncertainty on their determination. This work examines fPM emissions from loose Calcisols (i.e. soils with substantial accumulation of secondary carbonates) and evaluates the accuracy of the existing emission factors to apply for dry and arid conditions. Using own field measurements for airborne particulate matter composition and concentrations, new emission factors were developed for fPM emissions from anthropogenic activities such as construction and traffic-induced. A high resolution emission inventory was assembled and the contribution of fPM to the local air quality, environmental health, and solar energy production was assessed.

499587 Study on Regional Air Quality Impact from a Chemical Plant Emergency Shutdown

Abstract:

Ground-level ozone is the pervasive air pollutants, which is formed by the reaction between VOC and NOx. Chemical Plant emergency shutdown generates huge amounts of VOC and NOx emissions in a short-time period, which might cause significant impacts to local air quality. Unfortunately, such adverse impacts have not been studied quantitatively. Olefin plant is the backbone of the entire petrochemical industry; however, it is also the largest emission producer of the industry. The cracked gas compressor (CGC) is the most critical unit in an olefin production facility. Once the CGC is tripped, the entire chemical plant will fall into the emergency shutdown procedure. Streams from the cracking furnace have to be sent to flare systems for combustion to secure the safety of chemical plant, which will generate intensive and tremendous flare emissions in a very short time (e.g., 0.5 hr in this study). These flare emissions consisted of VOC and NOx may potentially aggravate the regional ozone pollution and seriously damage the human health. Thus, it is important to know the ozone impact caused by tremendous flare emissions from the CGC emergency shutdown of a typical chemical plant.

In this paper, a systematic methodology has been proposed to study the air quality impact from the CGC emergency shutdown of an olefin plant. Through case studies, it demonstrates that emergency shutdown may cause the ozone hourly increment ranging from 4.7 to 61.6 ppb; meanwhile, it differs due to the different of emergency shutdown starting time and the DRE (Destruction and Removal Efficiency) values. This study will be helpful to understand the ozone impact from CGC emergency shutdown of the olefin plant for decision makers on air quality controls in the future.

499860 Source Apportionment and Composition of Fine Particulate Matter in Delhi, India

Abstract:

Chronic exposure to fine particulate matter (PM) leads to premature death and increases the risk of cardiopulmonary disease. Industrialization and urbanization in developing countries like India and China present unique air quality challenges, where dense populations are routinely exposed to high levels of atmospheric pollutants. Pollutant composition and source profiles remain poorly characterized and lack temporal resolution in such settings. A better understanding is required to devise effective control and remediation policies to combat air pollution challenges in megacities such as Delhi, India.

Ambient measurements of submicron aerosol (PM_1, d_p < 1 μm) were carried out at The Indian Institute of Technology Delhi during the winter of 2017. Bulk chemical composition, number, size distribution, black carbon content, and concentration of PM_1 were measured by a suite of instruments, including an Aerosol Chemical Speciation Monitor (ACSM). Results show high PM concentrations dominated by organic species, exceeding 50 μg/m^3 on average during the campaign. Speciated diurnal profiles highlight a unique morning (6:00–9:00) particulate chloride peak, as shown in the figure below. A literature survey was conducted to identify potential local and regional chloride sources. Complimentary laboratory studies were performed to improve chloride quantification. Source apportionment was carried out via Positive Matrix Factorization of ACSM mass spectra. Results indicate significant contribution from hydrocarbon-like organic aerosol (HOA) sources. This study provides time-resolved information on particulate matter concentrations, sources, and characteristics in Delhi, India.

Figure 1: Preliminary diurnal PM_1 profile in Delhi, India from February 2017, as measured by ACSM. Unity collection efficiency was used.

497891 An Improved Hybrid Modeling Framework for Estimation of Human Exposure to Near Roadway Air Pollution

Abstract:

The rapid growth of the worldâ€™s motor-vehicle fleet due to population growth and economic improvement causes a significant negative impact on public health. The currently available air quality modeling approaches can compute the source specific pollutant fate on either a regional or a local scale but still lack effective ways to estimate the combined regional and local source contributions to exposure. Temporal variabilities in human activities and differences in pollutant dispersion pattern in stable and unstable atmospheric conditions greatly influence the exposure. Estimating air pollution exposure from local sources such as motor vehicles while considering all the variables impacting the dispersion make the process computationally intensive.

In this study, we employ a hybrid modeling framework combining a 3-D Eulerian chemical transport model and a pseudo steady state dispersion model to provide improved estimation of near road pollutant concentrations. We use the Comprehensive Air Quality Model with Extensions (CAMx), one of the two major air quality models used by Environment Protection Agency (EPA). For local dispersion modeling, we use R-LINE, a line source dispersion model for near surface releases. We employ this modeling framework and estimate human exposure for a wide variety of primary and secondary pollutants. Exposure to roadway emissions mostly depends on population density and temporal activities. In our study, we quantify exposure considering census tract population density and temporal and spatial variability in concentrations. Our approach using a dispersion model is unique as it uses the mass fraction of the total dispersed pollutant at different receptor points and hence is not dependent on either roadway emissions data or extensive model runs. This approach helps overcome the limitation associated with computational burden of regional models.

489593 Air Filtration of Ultrafine Particles Using PVDF-Peg Hollow Fibers

Abstract:

The particulate matter (PM) air pollution has become a global issue in the last few decades. Various techniques, such as electron spun air filters have been developed to remove PM from the air. However, limited studies focused on filtration of small particles (< 0.3 mm) which are more toxic to human. Compared to conventional fibers filter media, hollow fibers feature of self-supported properties with tunable pore sizes, making hollow fibers a good candidate for filtration processes. Until now, limited studies have worked on particle-laden gas filtration using hollow fibers. This study reveals as the first attempt to apply hollow fibers for air filtration of ultrafine particles. The asymmetric polyvinylidene fluoride - polyethylene glycol (PVDF-PEG) hollow fibers with high gas permeance have been developed. In order to increase the gas permeance of hollow fibers for air filtration, different molecular weight of PEGs was added into spinning dopes. The addition of PEGs facilitate the formation of loosely connected cross-section and porous outer skin, thus enhances the gas permeance for air filtration. Under the inside-out testing mode, all PVDF-PEG hollow fibers display excellent filtration efficiency of 99.999% against polydispersed NaCl particles with a geometric mean size of ~30 nm (Figure 1). This study may provide useful insights for developing hollow fibers for air filtration with the optimal operation conditions.

Particle size distributions arising from vaporized components of coal combustion fly ash – a comparison of theory and experiment

Huimin Liu
North China Electric Power University, Baoding, China
Yueming Wang and **Jost O. L. Wendt**
University of Utah, Salt Lake City, USA

Paper 333f
Presented at 2017 AIChE Annual Meeting
Minneapolis, MN
October 29 – November 3, 2017.

The problem

- Prediction of ash particle size distributions arising from coal combustion
- Focus on ash constituents that undergo:
  - Vaporization
  - Nucleation
  - Coagulation
  - Growth
- Restrict modeling to prediction of temporal evolution of particle size distribution due to coagulation only
- Compare to measurements from a laboratory combustor that simulates practical conditions

THE UNIVERSITY OF UTAH

2

Ash partitioning during coal combustion

Sarofim and co-workers at MIT

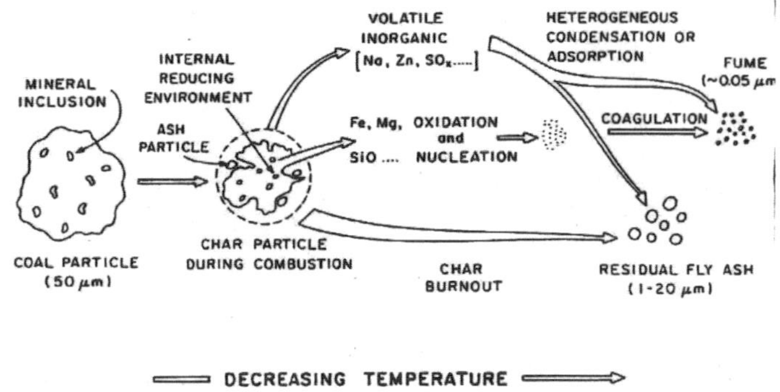

Submicron particles with diameter less than 0.6μm are formed by vaporization of inorganic (metal) constituents, followed by nucleation and coagulation.

3

Microscale diffusion phenomena during coal particle combustion

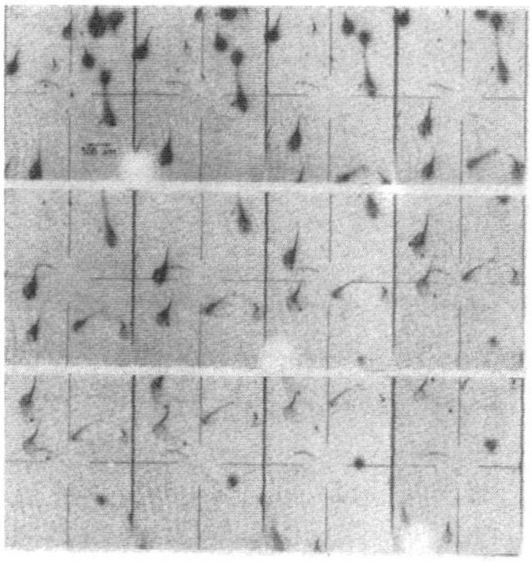

HIGH-SPEED CINEPHOTOGRAPHS OF SINGLE PARTICLE COAL COMBUSTION IN SANDIA REACTOR

Individual coal particles exhibit local microscale diffusion effects surrounding each particle during combustion. Photograph by Pohl et al, Sandia Laboratories

Question:

1. Does nucleation of vaporized ash constituents and subsequent coagulation occur within the diffusion layers shown here or after mixing with the total flue gas?
2. Can we use coagulation theory and experimental data to address this?

4

Coagulation theory

- Solution of General Dynamic Equation (GDE) using a sectional model.
- Coagulation only

$$R_{Coagulation} = \frac{dn_k}{dt} = \frac{1}{2}\sum_{i+j=k} N_{ij} - \sum_{i=1}^{\infty} N_{ik} = \frac{1}{2}\sum_{i+j=k} K_{ij}(v_{pi}, v_{pj})n_i n_j - n_k \sum_{i=1}^{\infty} K_{ik}(v_{pi}, v_{pk})n_i$$

- Multicomponent aerosol simulation code, MAEROS2 (Fred Gelbard, Sandia Laboratories, 1992, Gelbard and Seinfeld (1979))
  - Coagulation kernels from Fuchs, 1964
  - Conserves mass
  - Methods validated and verified by Gelbard et al (1998) for marine boundary layer
  - In this study, use experimental data to calculate vaporized inorganic metals mass.
  - Assume all vaporized ash metals placed in size "bin" 2-10 nm and then allowed to coagulate until the psd essentially ceased to change.

THE UNIVERSITY OF UTAH

5

EXPERIMENT: 100kW (max thermal) Oxy-Fuel Combustor (OFC)

1. Temperatures, particle and species concentrations similar to practical units.
2. Aerosol dynamics (coagulation, condensation etc) similar to those in practical units.
3. Simulates practical units but allows lower cost and better control.
4. Sufficiently well defined to allow systematic experimentation, elucidation of mechanisms, and modeling of experimental data.

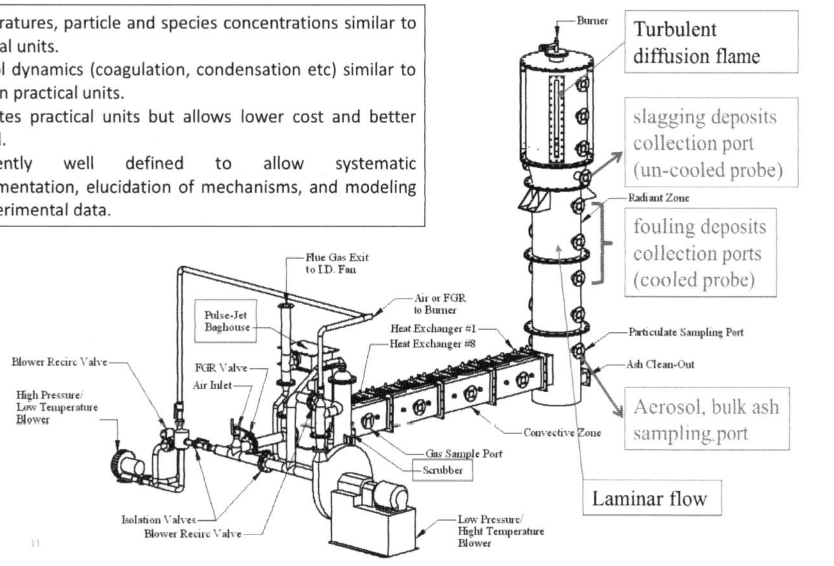

THE UNIVERSITY OF UTAH

11

6

Ash aerosol sampling systems

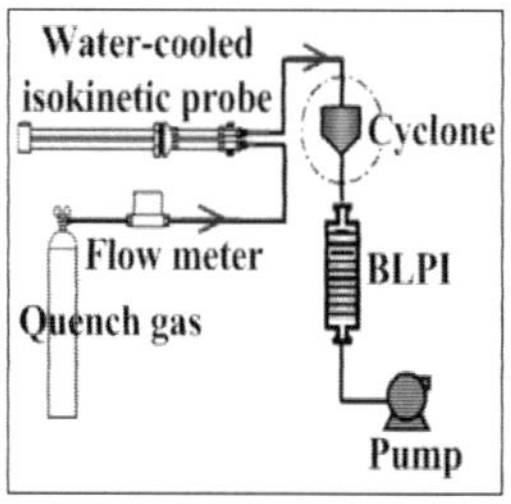

+

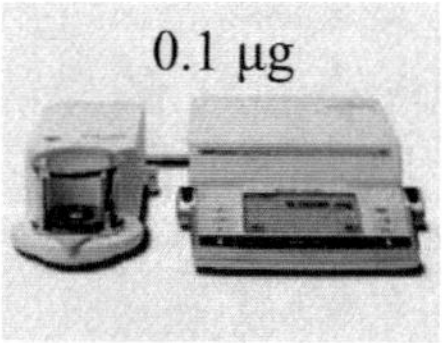

1) Cellulose acetate membrane substrates for chemistry
2) Aluminum foil substrates for PSD

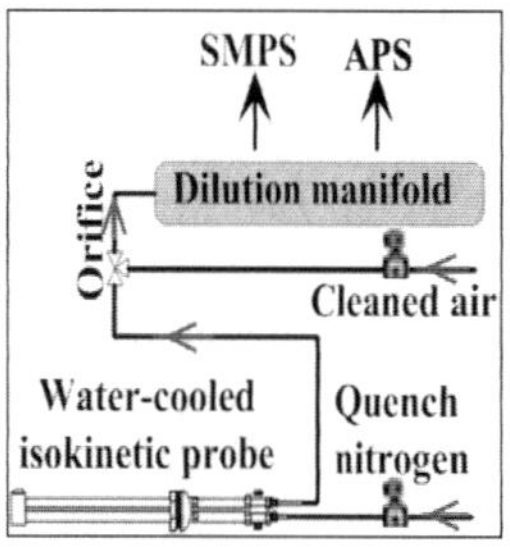

BLPI: 0.0324 - 15.7 μm
SMPS: 0.0143 - 0.6732 μm
APS: 0.532 - 20 μm

THE UNIVERSITY OF UTAH

7

PSD measurements SMPS/APS: 5 Runs Two coals, air and oxy-combustion conditions

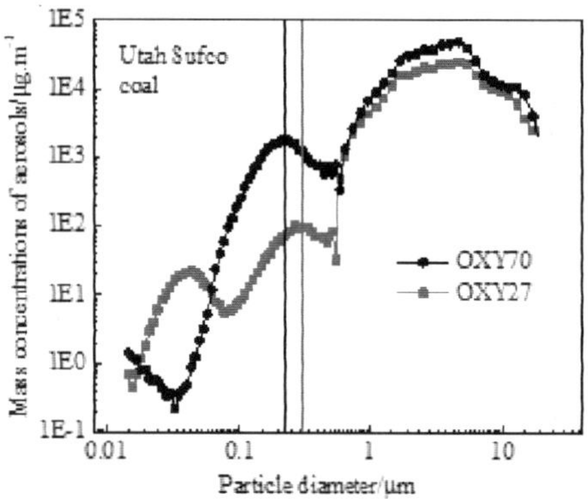

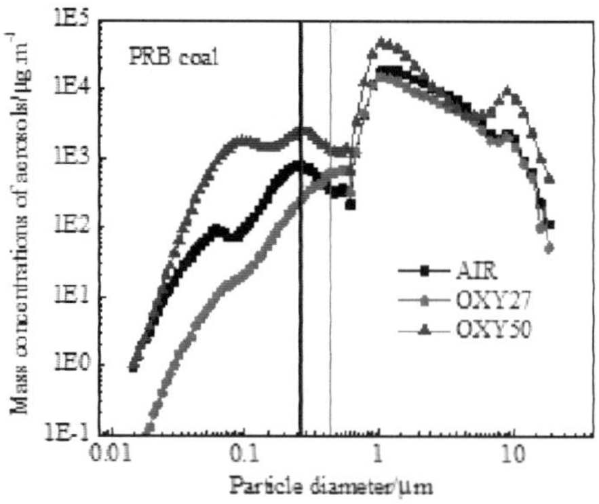

THE UNIVERSITY OF UTAH

Purpose is to predict the size location of the accumulation modes designated by vertical lines shown. Location of the nuclei peak at d_p <0.1 is not discussed here.

8

Models tested

1. MODEL A
   - Assumes coagulation occurs only in the mixed flue gas. State 2 followed by cooling to State 3 and then dilution at sample probe at State 4.
2. MODEL B
   - Assumes coagulation occurs primarily within a space between R and 2R where R is the coal or coal char particle radius. This occurs over a time interval of 0.05s (State 1), then dilution with flue gas occurs (State 2), cooling to State 3, and sample dilution to State 4
3. MODEL B_{mod}
   - Model B with diffusion layer decreased to ~0.2R (instead of R)

THE UNIVERSITY OF UTAH

9

Experiment and model description

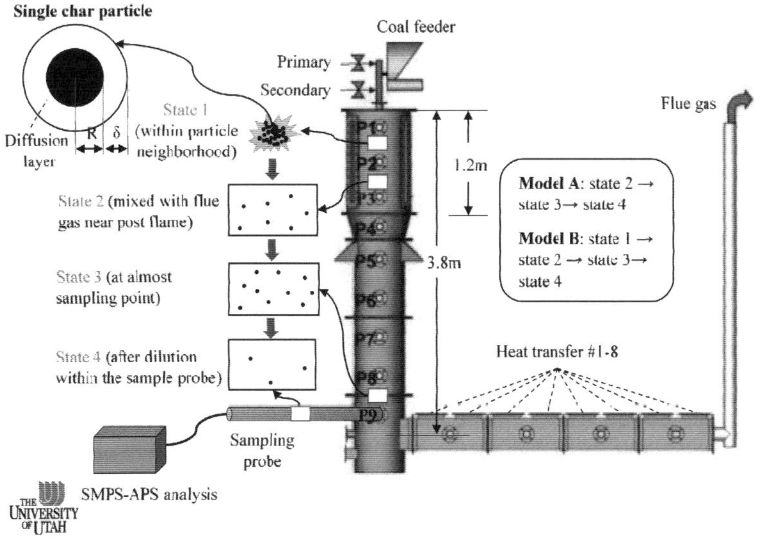

THE UNIVERSITY OF UTAH

Particle size distribution
- SMPS/APS
- BLPI

Particle modes:
- Vaporization(<0.6)
- Fragmentation(>0.6)

Vaporization:
- Nucleation(<0.1)
- Accumulation(0.2-0.5)

Coagulation theory
- GDE
- MAEROS

Assumptions
- Diffusion layer surrounding each char particle
- Thickness δ

RESULTS

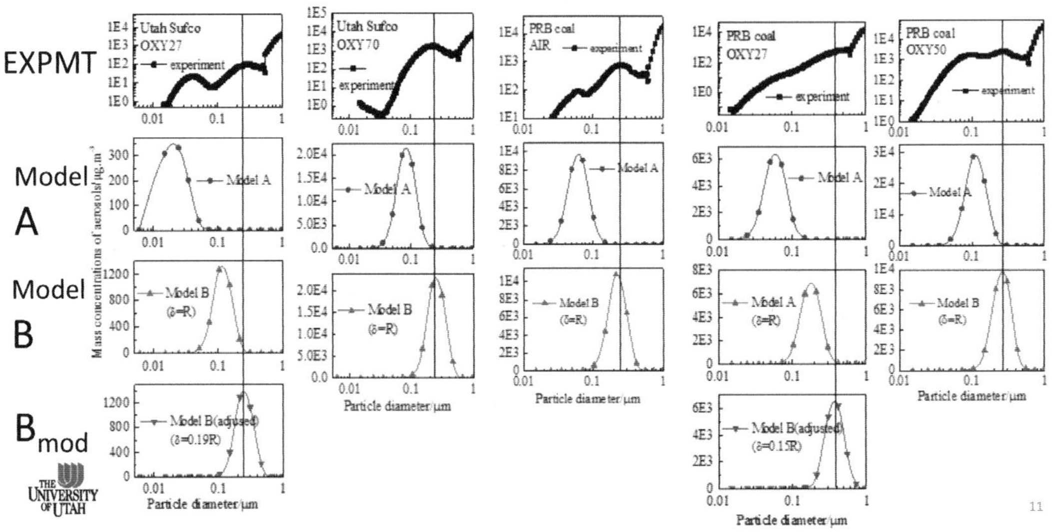

11

CONCLUSIONS

1. Predictions of the vaporized ash particle size distribution MUST account for the microscale diffusion phenomena near individual coal particles. (MODEL B).
2. Coagulation occurs within 0.05s in a volume within 1 radius of the coal particle surface.
3. Suggests that particle size distribution determined primarily by elements , such as Si and Ca, that vaporize in the reduced form within the char particle and nucleate, and coagulate very close to the particle surface.
4. One can use coagulation theory to gain insight into mechanistic details of ash partitioning.

THE UNIVERSITY OF UTAH

12

Acknowledgements

US National Science Foundation Award 1603249 for supporting this research.

China Scholarship Council (CSC) for stipend support for HL

Fred Gelbard, Ph.D., Sandia Laboratories for providing MEAROS2(1992) code.

THE UNIVERSITY OF UTAH

13

500440 Large Eddy Simulation of Soot Formation in Oxy-Coal Combustion

Abstract:

Oxy-Coal combustion is a promising clean coal technology. Typical combustion burns coal in air, resulting in a product mixture of water, nitrogen, and carbon dioxide. Separating carbon dioxide from nitrogen is costly. In oxy-fired combustion, nitrogen is first separated from air and a mixture of coal, oxygen, and (optionally) recirculated flue gases is burned. The resultant products are water and carbon dioxide, which are easily separated for subsequent sequestration. Oxy-fired combustion often occurs at high temperatures and under flow and thermal conditions that differ somewhat from combustion in air. Radiative heat transfer is a key process in such combustion as it is a primary source of heat transfer in steam generation applications. Radiation is present from gaseous species, ash, coal char, and soot, with soot being a significant contributor. Unfortunately, soot formation is a complex process. While much research has been done on soot formation in general, relatively little research has been published on soot formation in coal combustion. In oxy-fire systems with high water and carbon dioxide concentrations, and at high temperature, soot consumption via gasification may be significant. (Normally so consumption is dominated by oxidation reactions.)

We present a series of large eddy simulations in an oxy-coal combustor in which sensitivity to the soot formation mechanism is presented. Differences is flow, temperature, and radiative heat transfer are shown. We compare results using a previously published coal soot model, as well as extensions that include optimized gasification and oxidation rates. An advanced physics-based soot formation from coal tar mechanism is presented.

Ash aerosol and ash deposit formation during high temperature oxy-combustion of petroleum coke

Yueming Wang [a], Xiaolong Li [a], Jost Wendt [a]

[a] Department of Chemical Engineering, University of Utah, Utah, USA

Background

What is petroleum coke

- Peroleum coke (petcoke) is a byproduct of petroleum refining process

Advantages

- Petcoke is available in large amounts with low price
- Petcoke has low ash content and high heating value

Challenges

- The volatile content is very low in petcoke making it hard to be burned efficiently
- The sulfur content is very high and can cause corrosion and pollution

2

Research Objectives

- Assess the combustion efficiency of petroleum coke under pulverized coal furnace
- Uncover the effect of flame temperature on combustion performance of petroleum coke
- Study the formation mechanisms of ash aerosol and ash deposition (fouling)
- Develop the relationship between ash deposition rates and concentration of the ash aerosol

THE UNIVERSITY OF UTAH

3

Oxy-Fuel Combustor (OFC)

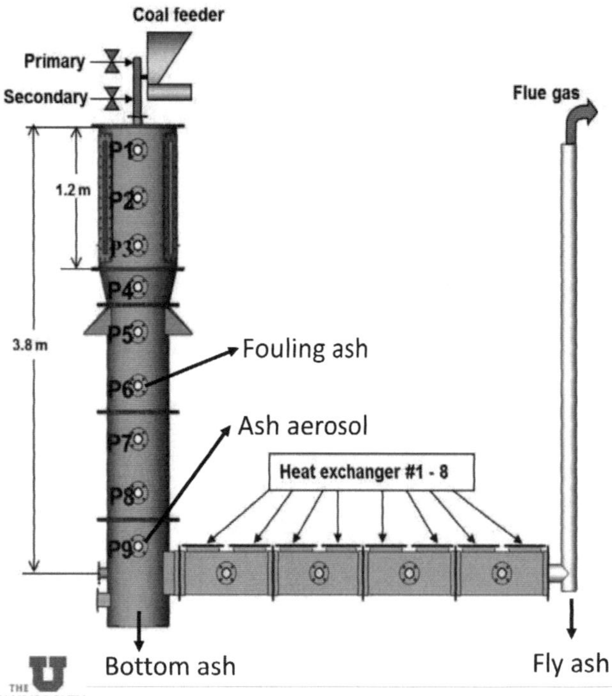

- 100 kW pilot scale
- Self-sustained down-fired
- Systematically controlled
- Temperature profile and particle concentration comparable with full scale units

THE UNIVERSITY OF UTAH

4

Fouling ash sampling system

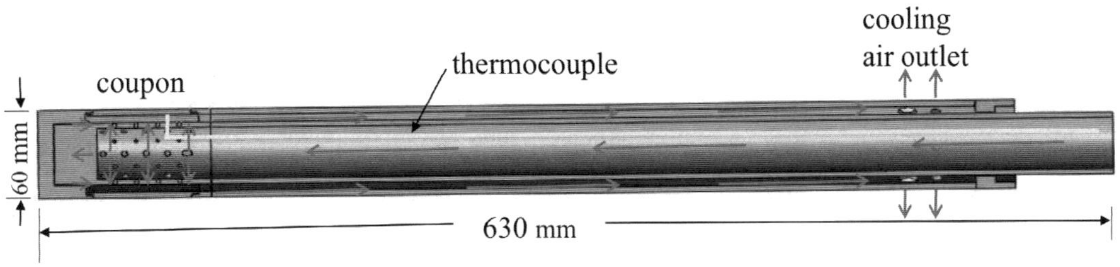

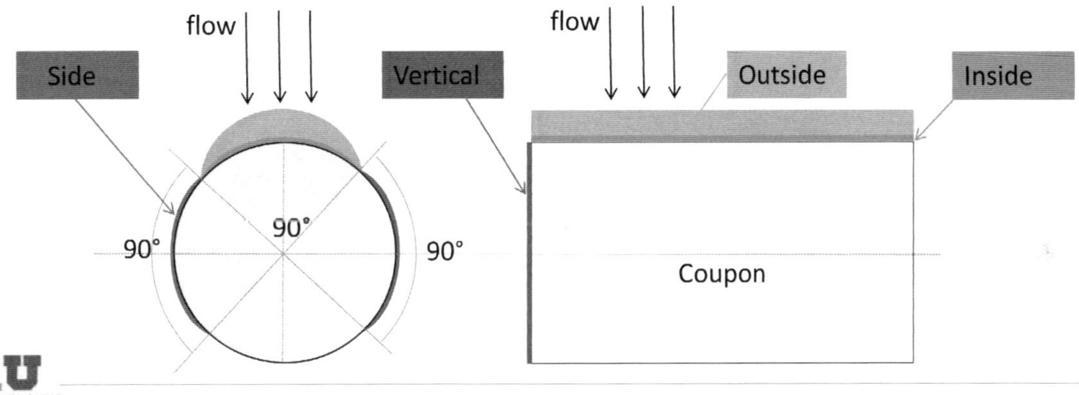

THE UNIVERSITY OF UTAH

5

Ash aerosol sampling system

BLPI:
0.0324 - 15.7 µm

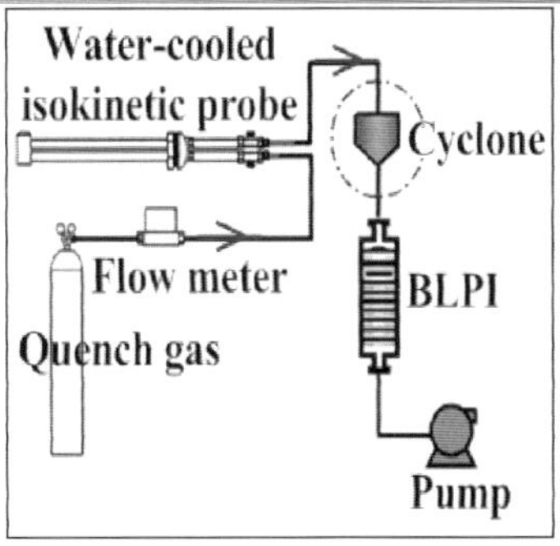

\+

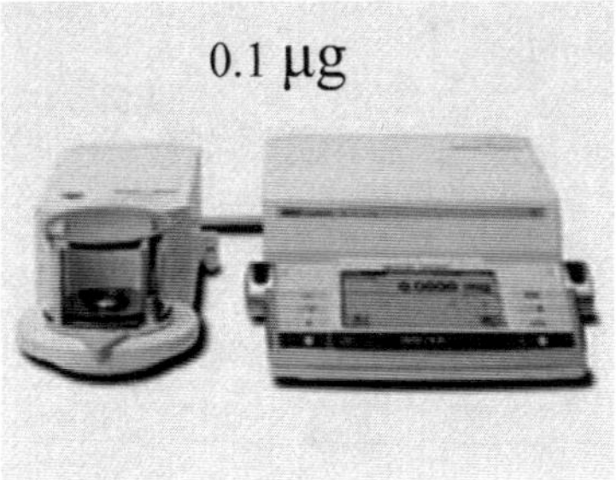

SMPS:
0.0143 0.6732 um
APS:
0.532 - 20 um

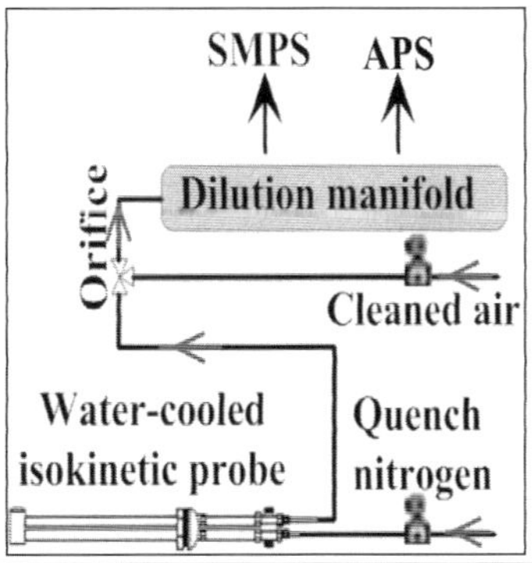

THE UNIVERSITY OF UTAH

6

Combustion parameters

Composition Analysis

ASH (%)	C (%)	H (%)	N (%)	S (%)	O (diff) (%)	H_2O (%)	Volatile (%)	Fixed Carbon (%)	HHV (kJ/kg)
2.99	82.51	6.02	1.71	5.65	0.49	0.57	10.18	86.26	34447.6

Al_2O_3 (%)	CaO (%)	Fe_2O_3 (%)	MgO (%)	K_2O (%)	SiO_2 (%)	Na_2O (%)	SO_3 (%)	TiO_2 (%)	NiO (%)	V_2O_5 (%)
19.4	4.22	7.02	0.66	1.17	46.7	0.72	3.77	0.63	1.26	8.24

Combustion conditions

Tests conditions	OXY70	Air
Feeding rate of petroleum coke (kg/h)	1.36	1.36
Feeding rate of natural gas (kg/h)	1.36	1.36
Excess O_2 (vol%_dry)	3	3
Adiabatic flame temperature (K)	3095	2312
Flow rate of flue gas (m^3/h_standard condition)	11.85	37.73

THE UNIVERSITY OF UTAH

Temperature profile of flue gas

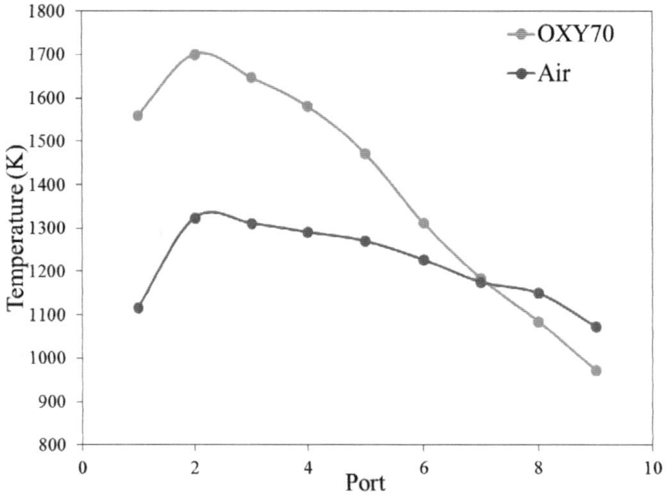

- Peak temperature of OXY70 (1701 K) is much higher than Air (1324 K)
- Peak temperatures for both cases are much lower than adiabatic flame temperatures.

THE UNIVERSITY OF UTAH

Aerosol particle size distribution (PSD)

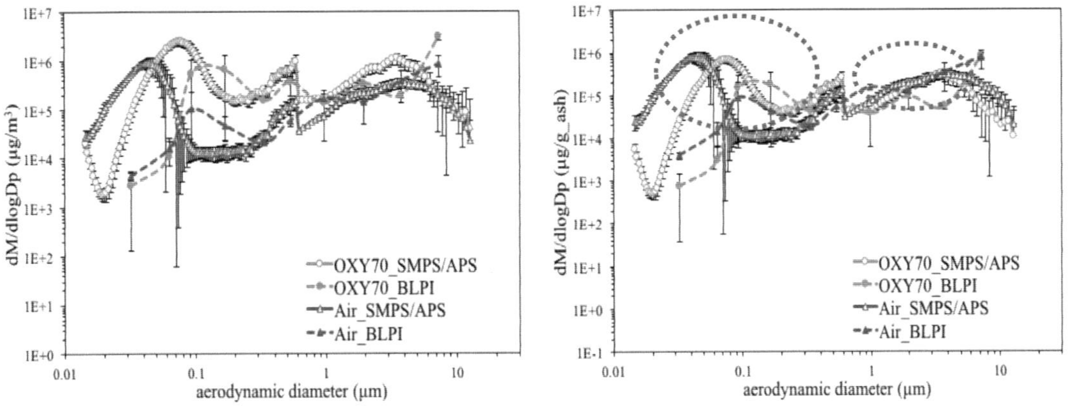

- BLPI data match with SMPS/APS data
- High temperature oxy-firing generates more submicron particles (99.1 mg/g_ash) than air-firing (72.4 mg/g_ash), but supermicron particles are not affected by temperature

Aerosol size segregated composition

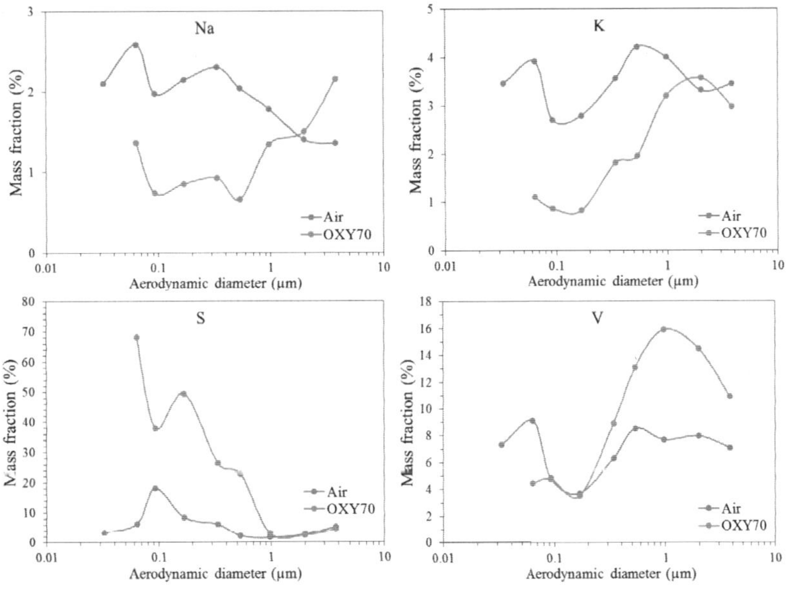

- More alkali metals are scavenged from submicron particles by alumina-silica under high temperature condition

Composition of fouling ash

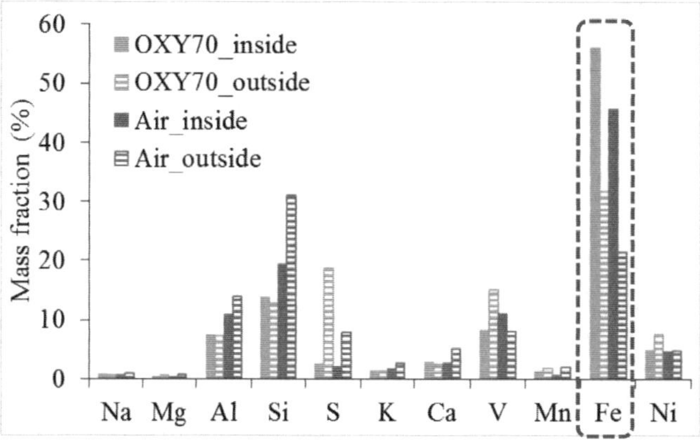

Al_2O_3 (%)	CaO (%)	Fe_2O_3 (%)	MgO (%)	K_2O (%)	SiO_2 (%)	Na_2O (%)	SO_3 (%)	TiO_2 (%)	NiO (%)	V_2O_5 (%)
19.4	4.22	7.02	0.66	1.17	46.7	0.72	3.77	0.63	1.26	8.24

- Composition of deposits are dominated by Fe

THE UNIVERSITY OF UTAH

SEM and EDS analysis for deposits

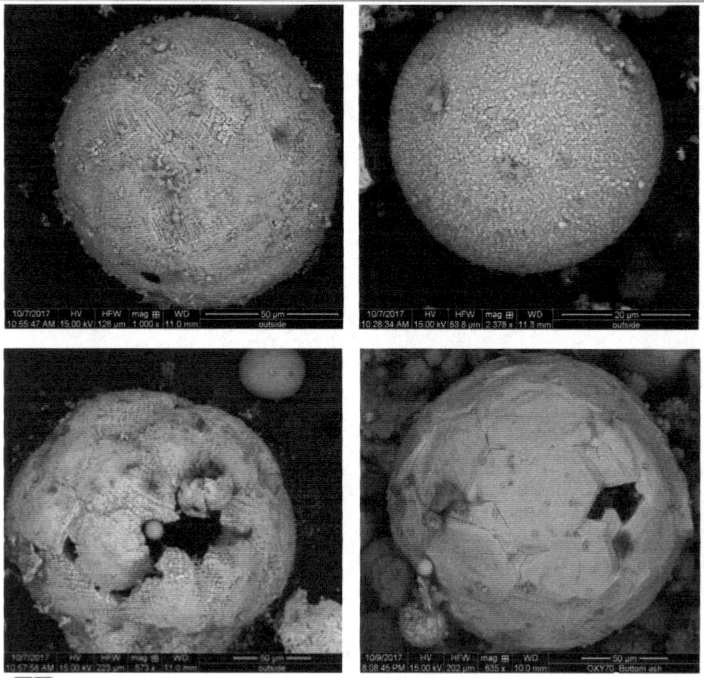

Na	0
Mg	0.17
Al	1.63
Si	2.30
S	1.05
K	0.33
Ca	0.82
V	1.33
Mn	0.61
Fe	**90.01**
Ni	1.75

THE UNIVERSITY OF UTAH

Composition of fouling ash

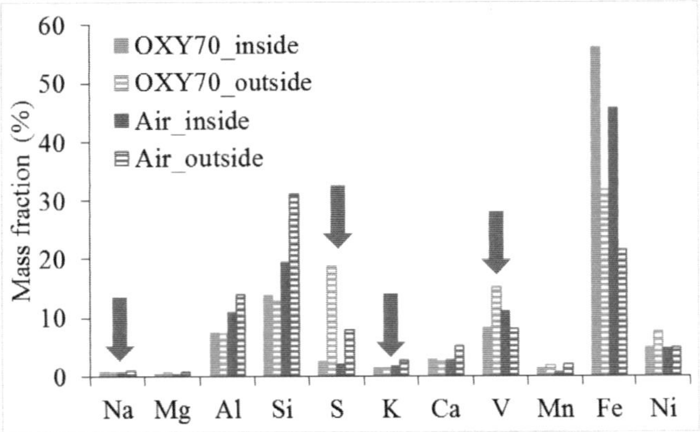

- Composition of deposits are dominated by Fe
- OXY70 deposits have more S and V, but less Na and K than Air, which are consistent with aerosol composition analysis

Sulfur and carbon analysis

Condition	OXY70		Air	
Content	S (%)	C (%)	S (%)	C (%)
outside deposit	3.21	79.0	1.13	28.0
fly ash	3.43	11.9	1.55	13.3
bottom ash	3.12	68.8	1.89	36.8

- Air firing seems surprisingly more efficient to burn petcoke than OXY70
- Sulfur content in OXY70 is higher than Air case

Correlation of deposition rates with PM1

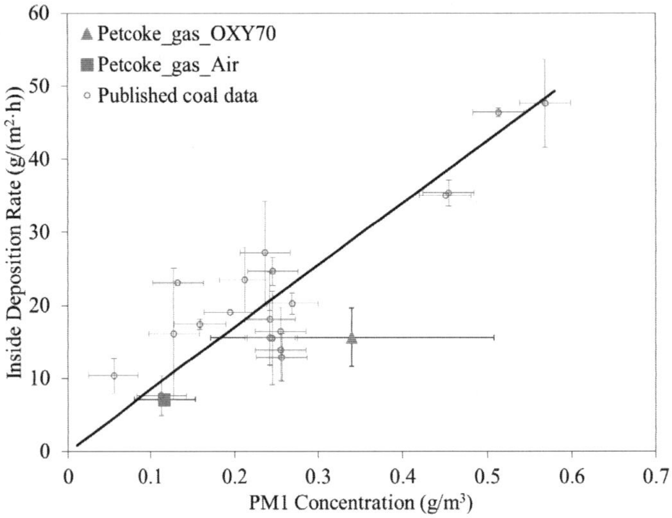

Published data from "Zhan, Z.; Wendt, J. O. L. Energy & Fuels **2017**, 31 (3), 2198–2202."

- Inside deposition rates of petcoke show positive relationship with concentration of PM1, which is consistent with previous findings in coal.

THE UNIVERSITY OF UTAH

15

Conclusions

- More alkali metals are scavenged from ultrafine fly ash by alumina-silica under high temperature condition
- The composition of fouling ash can be predicted by ash aerosol
- Fe is enriched in fouling ash, especially for high temperature condition
- Mixing between oxidant gas and petcoke is more important for carbon burnout than flame temperature
- Inside deposition rates can be predicted by the concentration of PM1

THE UNIVERSITY OF UTAH

16

Acknowledgements

- National Science Foundation under Award 1603249
- Praxiar for contributing the O_2 and CO_2 supply
- Brittany Walker, Dave Wagner, Paulo Perez and Brian van Devener

THE UNIVERSITY OF UTAH

17

Thank you!

THE UNIVERSITY OF UTAH

18

499074 Mercury Oxidation over Cu-SSZ-13 Catalysts under Flue Gas Conditions

Abstract:

Increasing concerns over mercury (Hg) emissions from coal-fired power plants have led the creation of new emission standards such as the recent Mercury and Air Toxic Standards, which requires a 90% reduction of Hg emissions. These new standards have led to the implementation of various novel technologies to control Hg emissions. Selective catalytic reduction (SCR) catalysts have shown promise as bifunctional catalysts for limiting both nitrogen oxide (NO_x) and Hg emissions through simultaneous NO_x reduction and Hg oxidation. The catalytic oxidation of elemental Hg, which is difficult to capture without dedicated Hg capture systems, to Hg^{+2} via HCl enables the capture of Hg^{2+} in the existing wet flue gas desulfurization unit as it is water soluble. Previous studies have investigated the possibility of utilizing current commercial SCR catalysts to fill this role, however these catalysts often have unsatisfactory performance when it comes to reducing Hg pollution to the levels required by current emission standards. To develop the next generation of SCR catalysts, a copper (Cu) exchanged small pore zeolite (Cu-SSZ-13) was investigated for its possible application in coal-fired power plants as a new bifunctional SCR and Hg oxidation catalyst. Cu-SSZ-13 has shown promise as a SCR catalyst for diesel exhaust, but has not been studied in coal-fired power plants.

In this study, the activity of Cu-SSZ-13 was investigated for its ability to oxidize Hg under a variety of flue gas conditions. The Cu-SSZ-13 catalyst was compared to SSZ-13 to understand the effect of Cu in the Hg oxidation mechanism. The Hg oxidation performances of both zeolites were compared to that of a commercial SCR catalyst (V-Ti-W), which served as a benchmark. All catalysts required hydrogen chloride (HCl) to exhibit any Hg oxidation with the zeolites outperforming the commercial catalyst. However, the activity of all catalysts significantly decreased in the presence of either sulfur oxides (SO_x) or SCR reactants (nitrogen oxide (NO)+ammonia (NH_3)) in the flue gas, with SO_x having a greater effect on Cu-SSZ-13 and SCR reactants having a greater one on the commercial catalyst. Additionally, it was determined that Cu was not the active site for Hg oxidation but was necessary for the SCR reaction. In fact, in the presence of SO_2, Cu actually contributed to the deactivation of the catalyst for Hg oxidation, likely due to the formation of sulfates as confirmed by XPS analysis.

502807 Mediated Oxycombustion with Integrated Uncoupled Oxygen Supply Â (MOBIUS)

Abstract:

Combustion systems that circumvent or reduce the energy penalty for carbon dioxide separation are being vigorously investigated. The approach we have pioneered aims at synergistically combining the basic objectives of three such technologies, namely oxy-combustion, chemical looping and liquid anode DCFC, and transforming them into a single lean, thoroughly optimized single unit process - Mediated Oxycombustion with Integrated Uncoupled Oxygen Supply (MOBIUS) - through innovations that overcome the bottlenecks of each of these technologies such as a) controlled combustion in a molten oxide to overcome the issues related to high temperatures (oxycombustion); b) obviating the need of solids transport between reactors (CLC); and c) avoiding operational issues such as electrical resistance, ash separation, impurity tolerance etc (DCFC). The system (simplistic concept in Figure 1) synergistically combines the effective aspects of the above mentioned technologies along with innovations to overcome the hurdles faced by these technologies. In MOBIUS, the solid fuel is combusted in the absence of any nitrogen as in oxy-combustion. However, the carbon is not combusted directly by oxygen but by a metal oxide as in a CLC. Unlike conventional CLC, the MeO_x and Me in the proposed Chemical Looping with Oxygen Uncoupling (CLOU) systems are in the liquid phase to improve the combustion rates and the mobility of the metal and the oxide. The metal is re-oxidized at the walls of the combustor with oxygen that is uncoupled from the reactor wall material. The individual subsystem studies will be presented and the integration of the various subsystems into a single reactor system will be discussed. Results from the subsystem will include the combustion rates, metal oxidation rates, oxygen uptake, transport and release rates under various conditions and composition of the membrane (during the membrane evaluation). The long terms operation of the integrated system will be demonstrated.

497932 Mobility Size and Effective Density of Carbonaceous Aerosols

Abstract:

Mobility Size and Effective Density of Carbonaceous Aerosols

Georgios A. Kelesidis, Eirini Goudeli, Sotiris E. Pratsinis

Particle Technology Laboratory, Institute of Process Engineering,

Department of Mechanical & Process Engineering, ETH Zurich, Switzerland

The mobility size estimation of evolving fractal-like soot aggregates is not trivial, as it depends on their structure as well as the number and size of their constituent primary particles. A scaling law describing the mobility size of agglomerates consisting of monodisperse primary particles in point contact (Sorensen, 2011) overestimates the mobility diameter of mature soot aggregates from diffusion flames for a given number of primary particles (Rissler et al., 2013). Rissler et al. (2013) attributed this difference to the primary particle polydispersity and chemical bonding induced by surface growth (aggregation), neglected in the above scaling law (Sorensen, 2011). Typically population balance models for soot formation estimate the mobility diameter assuming it is equal to the gyration diameter (Yapp et al., 2015).

Here, the Discrete Element Model of nascent soot agglomeration and surface growth by acetylene pyrolysis (Kelesidis et al., 2017) is extended to investigate the dynamics of mature soot formation at longer residence times and different flow regimes than nascent soot in the absence of soot oxidation. The evolution of soot mean gyration, mobility and primary particle diameters as function of the mean number of primary particles per aggregate are elucidated. The common assumption that the mobility is equal to the gyration diameter leads to 40 % underestimation of the nascent soot mobility size and 30 % overestimation of the mature soot aggregate mobility diameter.

Relationships for the soot mobility size and effective density are derived by following the evolution from nascent to mature soot morphology and compared to the correlation of Sorensen (2011) and mass-mobility measurements in quenched diffusion flames (Rissler et al., 2013; Yon et al., 2015). The scaling law for agglomeration in the absence of surface growth (Sorensen, 2011) overestimates the measured mature soot aggregate mobility size by 37 % and underestimates its effective density by 44 %. Accounting, however, for the primary particle polydispersity and chemical bonding or aggregation induced by surface growth reduces the deviation from the mass-mobility measurements down to 7 and 11 % for the mobility size and the effective density, respectively.

References:

Kelesidis, G.A., Goudeli, E., Pratsinis, S.E. (2017) *Proc. Combust. Inst.* **36**, 29-50.

Rissler, J., Messing, M.E., Malik, A.I., Nilsson, P.T., Nordin, E.Z., Bohgard, M., Sanati, M., Pagels, J.H. (2013) *Aerosol Sci. Technol.* **47**, 792-805.

Sorensen, C.M. (2011) *Aerosol Sci. Technol.* **45**, 765-779.

Yapp, E.K.Y., Chen, D., Akroyd, J., Mosbach, S., Kraft, M., Camacho, J., Wang, H. (2015) *Combust. Flame* **162**, 2569-2581.

Yon, J., Bescond, A., Ouf, F.X. (2015) *J. Aerosol Sci.* **87** (2015) 28-37.

490537 Adsorption of Tetracycline Antibiotics in Wastewater Onto Biochar-Based Adsorbents

Abstract:

Tetracycline (TC) is one of the most widely used antibiotics for both human and animal diseases. TC has been detected at the level of ng L^{-1} to mg L^{-1} in domestic wastewater, but 10-100 mg L^{-1} in dairy manure, hospital, and pharmaceutical manufacturing wastewater. Like other antibiotic compounds, release of TC to aquatic ecosystem can significantly cause acute and chronic toxicity and propagation of antibiotic resistance in microbial community. Current treatment methods such as biodegradation, advanced oxidation and membrane filtration have not shown effective elimination of TC in wastewater and water due to low concentration and biorecalcitrant nature of TC.

Compared with other methods, adsorption is a simple and cost effective process without generating any toxic byproducts. Recently biochar is generated as a byproduct from gasification of solid wastes such as sludges, biosolids, agricultural residues and animal manure. Owing to beneficial physical and chemical properties of biochar, it is considered as an emerging adsorbent for eliminating various contaminants.

In this study the dairy manure-derived biochar (BC) was modified by the wet impregnation technique to immobilize MgO to the biochar for enhancing TC adsorption capacities and reducing the inherent functional groups interfering interaction between TC and the biochar. The physical and chemical properties (i.e., surface area, porosity, surface chemistry) of the BC and MgO-BC were characterized.

The adsorption efficiencies of TC were investigated using both the BC and the MgO-BC at the selected conditions. The results showed that the MgO-BC showed higher adsorption capacity of TC (70-95 %) than that that by the BC (59-71%) at the selected conditions. The higher adsorption using the MgO-BC was thought to be due to possible π–π electron donor–acceptor interactions and hydrogen bonding between the MgO-BC and TC. Rather the BC would rely on mainly hydrophobic interaction. The adsorption isotherm study suggested Freundlich isotherm to be a representative model for adsorption of TC onto the MgO-BC. It would indicate monolayer adsorption of TC on the MgO-BC. The adsorption capacities of TC onto the MgO-BC were found to be 2 – 10 mg TC per g biochar from the adsorption isotherm. The adsorption capacities were higher than those by the biochar made from agricultural residues and other surface modification techniques. The adsorption kinetic analysis revealed that the pseudo-second order kinetic model fitted to the experimental results (R^2>0.98) better than the pseudo-first order kinetic model. It also indicated that the adsorption of TC on the MgO-BC would include several mechanisms including diffusion, adsorption, and desorption which are also often found in other carbon-based adsorbents such as activated carbon, graphene and carbon nanotube. Further works will include the detailed mechanisms for adsorption and desorption, the column tests for practical application and the mathematical modeling for engineering design.

500121 Reductive Degradation of Co-Contaminant Medium of TCE and Cr(VI) Using Atomized Iron Powder

Abstract:

Elemental Iron in the form of atomized iron powder (Zero Valent Iron, ZVI or Fe(0)) was used to individually and simultaneously degrade a co-contaminant medium of hexavalent chromium (Cr(VI)) and trichloroethylene (TCE). Under appropriate conditions of initial concentrations, the overall degradation reaction was modeled as a combination of two pseudo-first order reactions. The reaction kinetics, degradation efficiency and reductant demand of both Cr(VI) and TCE were determined. Over a 2-hour period, degradation of TCE and Cr(VI) yielded 85% and 25% reduction, respectively with excess amounts of Fe(0). Over a 5-hour period, TCE percent remained at 85% whilst Cr(VI) degradation increased rapidly to almost 100%. This observation is attributed to: (1) increased polarization and destabilization of the double bond in the TCE leading to cleavage of the Chloride and solubility equilibria; (2) the formation of a passive layer of chromic-ferric mixed oxides and oxyhydroxides on the surface of iron particles, which impeded degradation but dissolved due to increasing acidity to expose fresh reactive sites. The observed pseudo-first order degradation rate constants for individual and combined degradations of the co-contaminant were determined.

504261 Tritium Distribution and Cycling on Savannah River Site

Abstract:

There are no standard doses for organically-bound tritium (OBT) therefore, this research aims to develop and enhance procedures for evaluating OBT levels for various plants and animal receptors. Samples were gathered and historical data were used for this investigation. Assays were performed using the freeze drying process and reverse transpiration respiration methods were used for assessing OBT concentrations. Thus, the research findings provide a basis for establishing OBT safe levels.

In recent years, there has been increased interest in the radiological hazards associated with ^{3}H that has accompanied plans for expansion of nuclear power generation and public concerns associated with non-routine ^{3}H releases at nuclear power facilities. For instance, non-routine spills of ^{3}H that exceeded current the U.S. drinking water standard (DWS) have occurred at 46 of the 65 locations where commercial power plants have been operating.

Hence, the dose associated with ^{3}H, a low energy beta emitter (0.0186 MeV), arises almost entirely from the ingestion pathway. Factors such as biological half-life and isotope discrimination can have a significant impact on ^{3}H dose calculations in risk assessment scenarios. Therefore, research in the last decade has focused on improving our understanding of ^{3}H assimilation in organic tissues, and the development of standardized procedures to measure the level of ^{3}H associated with biological tissues, known as organically bound tritium (OBT).

Moreover, tritium (^{3}H) is second only to radioactive carbon (^{14}C) in terms of the collective radioactivity dose released from nuclear facilities. Tritium accounts for greater than 65% of the atmospheric and greater than 99% of the stream water environmental releases of radioactivity from the Savannah River Site (SRS). For example, multiple ^{3}H sources exist on the SRS, including tritium-containing groundwater plumes that outcrop at various locations along the Fourmile Branch stream, a tributary to the Savannah River. Additionally, the SRS released 24,300 Ci of ^{3}H to the atmosphere in 2014, most of which (>95%) were stack releases of water vapor from the tritium processing facility. Annual release levels from the processing facility depend on several factors, including annual production schedules, with an average annual release level of 31,460 Ci for the last ten years.

Consequently, the SRS provides an excellent setting for evaluating low-level tritium cycling dynamics in the environment, including the transformation of tritiated hydrogen (HT) and tritiated water (HTO) to OBT because there are multiple sites with documented histories of ^{3}H exposure that provide excellent test cases. Thus, the current study will utilize SRS resources to develop and refine monitoring protocols for evaluating organically-bound tritium (OBT) levels in various plant and animal receptor species.

493684 Preparation of Geopolymeric Adsorbent Derived from LD Slag for Removal of Zinc (II) from Waste Water

Abstract:

In this work, LD slag Geopolymer (LDSGP), a porous aluminosillicate geopolymeric adsorbent has been successfully prepared from steel plant LD slag for efficient removal of Zinc ions (Zn^{2+}), one of the most toxic heavy metals present in waste water. LD slag, a high calcium oxide rich solid waste, is generated in huge quantity while producing high quality steel from pig iron using Linz Donawitz (LD) converters. Presently, disposal of LD slag has become a major environmental concern for the industries. So, geopolymerization of LD slag is attempted to reuse it as a low cost adsorbent in waste water treatment by removing Zn^{2+}, thus presenting a solution for both treating a problematic wastewater and turning industrial waste into a valuable material. The geopolymerisation process is a series of chemical reactions that include alkali activation of raw LDS, polycondensation of aluminosillicate and structural re-organisation. LD slag geopolymer (LDSGP) was synthesized in a plastic mould where the colloid mixture of raw LD slag and alkaline activator (sodium hydroxide and sodium silicate (1:1 w/w)) were cured for 3 days at room temperature. The LDSGP was characterized by using XRF technique, BET surface area, SEM & TEM images, PXRD patterns, FTIR spectra and TGA. The BET surface area of LDSGP (30.84 m^2/g) is improved considerably compared to raw LDS (4.85 m^2/g). The PXRD pattern of LDSGP confirms the formation of tetrahedral geopolymer matrix consisting Ca_2SiO_4 and Ca_3SiO_5. The SEM and TEM images of the LDSGP particles reveal the combination of micropetal and cauliflower like structures on the surface. The equilibrium data from batch experiments for the adsorption study of Zn^{2+} ions at 298 K was successfully fitted with the Langmuir isotherm. The maximum adsorption uptake of Zn^{2+}, onto LDSGP, was found to be 84.125 mg/g at 298 K. The adsorption rate data was better correlated to pseudo-second-order model indicating the process to be dominated by chemisorption.

Synthesis of "Sea Urchin"-like Carbon Nanotubes/Porous Carbon Superstructures Derived from Waste Biomass for Treatment of Various Contaminants

Yunjin Yao, Chao Lian, Guodong Wu, Yi Hu, Fengyu Wei
School of Chemistry and Chemical Engineering, Hefei University of Technology, Hefei, China

Abstract

Novel "sea urchin"-like Ni nanoparticles embedded in N-doped carbon nanotubes (CNTs) supported on porous carbon (Ni@N-C) 3D materials derived from waste biomass were prepared via pyrolysis and employed as an environmentally friendly, easy available and cost-effective catalyst for removal of toxic pollutants. The characterizations indicated that Ni^0 catalyzed the growth of intertwined CNTs on carbon layers, affording abundant porous materials and larger specific surface area. With the synergistic effect of embedded Ni^0 nanoparticles, nitrogen doping, hierarchical micro-mesopores, and interconnected CNTs, Ni@N-C displayed a superior catalytic capability for the oxidation of organic pollutants using peroxymonosulfate as an oxidant, and catalytic reduction of toxic Cr^{VI} to nontoxic Cr^{III} by formic acid as a reducing agent. It was found that the pyrolysis temperature affected the composition, morphologies, and catalytic properties of Ni@N-C. Inactive oxidized N species have transformed to the highly active graphitic N, pyridinic-N, and Ni-O-N clusters, thereby improving the catalytic activity. Moreover, Ni@N-C maintained good physicochemical structure and stable activity even after several cycles of reactions. The simple synthetic strategies, 3D structure, and remarkable performance of Ni@N-C composites make them serve as alternative environmentally friendly catalysts for removal of pollutants.

Introduction

Toxic pollutants like reactive dyes and hexavalent chromium (Cr^{VI}) presented in water streams have attracted significant research attention because of their toxicity, slow self-degradation in nature, and huge volume of production from various industries.[1, 2] Unfortunately, conventional technologies such as filtration, liquid extraction, biological treatment, chemical oxidation, and membrane process have some disadvantages including insufficient capacities, slow reaction rates, and low selectivity.[1, 3] Thus, it has become a challenge in the development of effective materials and/or methods with an excellent performance to capture and remove these toxic species from waters.[4]

Recently, much research has focused on carbon-based catalysis, an effective and ideal technique, to eliminate the aqueous contaminants. Carbon materials including metal carbides (such as Ni_xC, Fe_xC, Co_xC *etc.*) and heteroatom (such as nitrogen, phosphorus, sulfur, and boron, *etc.*) -doped porous carbons are ubiquitous in environment-related applications because of their superior properties and wide availability.[5, 6] Although metal carbides are abundant and nonprecious, it is highly challenging to solve the problems of sintering or leaching of metals, which causes irreversible deactivation of the catalysts.[7] Metal nanoparticles (NPs) embedded in carbon shells are highly promising and active materials with high stability, because the carbon shells protect the metal NPs. In our previous studies, Fe embedded in N-doped carbon nanotubes (CNTs) has been demonstrated as the heterogeneous catalyst to generate radicals from peroxymonosulfate (PMS, HSO_5^-) for the oxidative degradation of organic wastes.[8, 9] Bhowmik et al.[10] reported Ni NPs stabilized by reduced graphene oxide as heterogeneous catalysts to reduce toxic Cr^{VI} compounds to nontoxic Cr^{III} by formic acid as a reducing agent. It was supposed to possess the features of fast mass transport channels and abundant active sites.[11] Unfortunately, most reported materials have a low activity or stability. Hence, there is still a need to explore rationally designed architectures as the heterogeneous catalysts in water treatment.

Three-dimensional (3D) carbon-based structures have recently attracted enormous interest, due to their unique properties including high porosity, electrical conductivity, and specific surface area, which are beneficial to efficient diffusion/mass transfer of liquid/solid phase.[12, 13] They are also more easily separated because of their larger size and weaker Brownian motion.[14] Numerous techniques have been developed to prepare porous 3D carbon materials, including templating, combustion, cathodic corrosion, and aerogel formation.[15] For example, Wan et al.[16] synthesized superparamagnetic γ-Fe_2O_3 NPs encapsulated in 3D cellulose aerogels for Cr^{VI} removal. However, the harmful and expensive precursors (graphene or CNTs) or complex equipment involved in these syntheses have impeded their commercialization. The development of more efficient and stable catalysts using renewable precursors in a cost-effective manner is highly desirable for a successful large-scale application.

Nowadays, there is a trend to produce carbon-based materials derived from biomass, as they are very cheap, extensively available, accessible, and recyclable, etc. Many efforts have been made to increase the catalytic performance of biomass-derived carbons. Transition metals (Fe, Co, Ni, etc.) coordinated with nitrogen doping in CNTs from biomass have been extensively studied because of their low cost, long durability and superior catalytic activities. For example, N-doped graphene encapsulating Fe_3C was synthesized by Shi et al.[17] via calcination of biomass/melamine/$FeCl_3$ in N_2 atmosphere. Intrinsic properties and unique structure, such as high chemical stability, surface area, and electrical conductivity, make them extremely attractive as catalytic materials. The incorporation of embedded metal NPs and nitrogen doping in CNTs is an interesting approach to synergistically improve active sites.[15] Thus, developing Ni embedded N-doped CNTs with an excellent activity from waste biomass has motivated our curiosity. In addition, recycling of waste biomass provides extra values for contaminant treatment.

Herein, we demonstrated a scalable strategy for the controlled synthesis of “sea urchin”-like nickel embedded in N-doped CNTs-supported on porous carbon (Ni@N-C) as a 3D catalyst through pyrolysis treatment of a waste carbon precursor, dicyandiamide and $NiCl_2 \cdot 6H_2O$. The catalytic performances of Ni@N-C composites were evaluated by the oxidation of Orange II using PMS as an oxidant, and the catalytic reduction of Cr^{VI} to Cr^{III} by formic acid as the reducing agent. As expected, Ni@N-C exhibited catalytic activity and long-term stability under acidic conditions. The prominent catalytic performance and low production cost render Ni@N-C as promising candidates in green catalysis.

Materials and Methods

Synthesis of Ni@N-C catalysts

A biomass (stillage residue, a waste from the liquor industry) was obtained from Anhui Yingjia Gongjiu Corp, China. Before use, the biomass was washed, dried, crushed, and sieved to 200-300 mesh. The biomass powder was activated with a KOH activation process as reported in our previous study.[18] Typically, the raw biomass was mixed with KOH (biomass/KOH mass ratio as 1/1) and then annealed in high purity N_2 atmosphere at 750 °C for 2 h. The resulted samples were washed with HCl solution and distilled water until neutral filtrate, and then dried at 70 °C for 24 h. The obtained porous activated carbon was denoted as AC.

The process used to prepare Ni@N-C catalysts is shown in Figure 1. Typically, 0.2 g AC and 0.0476 mol dicyandiamide were then immersed into 200 mL aqueous solution of $NiCl_2 \cdot 6H_2O$ (5.0 mM) with stirring for 12 h. Then, the solvent solution was slowly evaporated. Subsequently, the obtained powders were annealed in N_2 atmosphere at 500 °C for 2 h to obtain reduced Ni catalysts and then further annealed at the setting temperature (700, 800, and 900 °C) for another 2 h to obtain Ni@N-C composites. The powder collected was filtered, washed, and dried to get the catalyst materials. The obtained products were denoted as Ni@N-C-T, where T represents the pyrolysis temperature (T = 700, 800, or 900 °C).

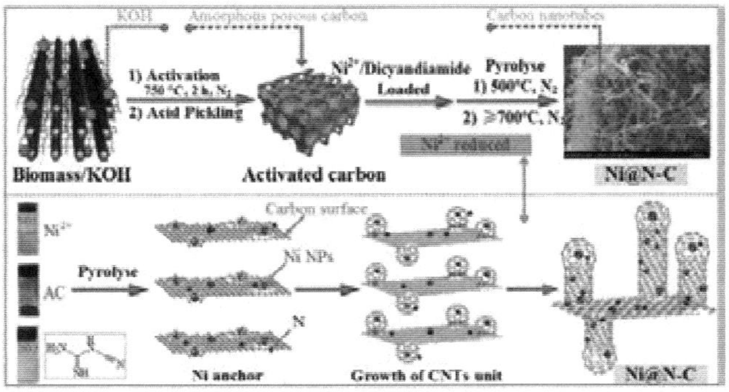

Figure 1. Schematic Illustration for the Synthesis of Ni@N-C Catalysts.

Characterizations

The phase structures of the Ni@N-C catalysts were analyzed by an X'Pert PRO Multi-purpose XRD System using Cu Kα radiation. Raman spectra were recorded at ambient temperature on a confocal Raman microscope (Horiba Jobin Yvon LabRAM Aramis) with 532 nm wavelength laser. Fourier transform infrared (FT-IR) spectra were obtained on a Perkin-Elmer Spectrum 100 spectrometer in the range of 400-4000 cm^{-1}, and X-ray photoelectron spectroscopic (XPS) measurements were performed using an ESCALAB250 instrument. The morphology and elemental composition of the Ni@N-C catalysts were obtained by field emission scanning electron microscopy (FESEM; Hitachi SU8020), field-emission transmission electron microscopy (FETEM, Jeol JEM 2100F), and high resolution TEM (HRTEM) with elemental mapping analysis and energy-dispersive X-ray spectroscopy (EDS). The Brunauer-Emmett-Teller (BET) surface area and pore structure of samples were obtained from N_2 adsorption/desorption isotherms at -196 °C using a Builder 4200 instrument (Tristar II 3020M, Micromeritics Instrument Corp., USA).

Catalytic Evaluation

The catalytic activity for organic degradation was tested in a 250 mL beaker. The catalysts (10 mg) were dispersed into organic dye solutions (200 mL) with the desired initial concentrations. The well-mixed solution was stirred in dark for 120 min to attain adsorption equilibrium, followed by addition of the oxidant (PMS, 0.05-0.25 g/L) for reaction. Later, 2.0 mL of solution samples were taken out at given time intervals, separated with a magnet for analysis. The catalyst stability was checked by recovering the solid through filtration, washing, drying at 80 °C, and retested in the following cycle. In some experiments, the scavenger chemicals were also used specifically to verify the radical species.

The organic concentrations were obtained by a UV-vis spectrophotometer. Chemical oxygen demand (COD) was measured with an analyzer (DR1010, Hach Co., Loveland, CO). The metal ion leaching was carried out using atomic absorption spectroscopy (AAS, AA800, Perkin-Elmer). The reactive radicals were trapped with 5,5-dimethyl-1-pyrroline N-oxide (DMPO, >99 wt %) and measured using electron spin resonance (ESR, JES-FA200, Japan). Samples were collected from the Orange II degradation system in the presence of PMS and Ni@N-C catalyst within different contact time.

For the catalytic reduction of Cr^{VI}, potassium dichromate ($K_2Cr_2O_7$) was the source of Cr^{VI} and HCOOH was the electron donor under an acidic solution. The batch experiments were carried out in a stirred batch-fed beaker at room temperature. Typically, a 200 mL aqueous reaction mixture was prepared by mixing $K_2Cr_2O_7$ and deionized water to make a solution at a desired concentration. Subsequently, the Cr^{VI} reduction was initiated by the sequential addition of Ni@N-C catalyst and HCOOH. Then, a certain volume of the solution was sampled at given time intervals, and then filtered (0.25 μm, PTFE) to remove the catalyst solids. The Cr^{VI} content in the reaction solution was monitored by measuring UV-Vis absorbance of $K_2Cr_2O_7$ at 350 nm. After each experiment, the Ni@N-C composite was magnetically captured on the bottom of flasks, washed with water, and dried before the next catalytic cycle.

Results and Discussion

Catalyst Characterizations

XRD spectra of Ni@N-C composites (Figure 2a) show the diffraction peaks at $2\theta = 26.5°$, attributing to the (002) reflection of graphitic carbon. In addition, the intensity of this reflection increased with pyrolysis temperature, which may be due to the fact that Ni@N-C loses some functional groups at a high temperature.[17] In addition to the characteristic peaks from carbon, the diffraction peaks at around 44.2, 51.7 and 76.2° can be indexed to the (111), (200) and (220) crystalline planes of Ni^0 (PDF no. 65-0380), respectively, suggesting the successful reduction of nickel ions to a metallic state during the pyrolysis process.[19, 20] The Raman spectra contain characteristic wide D and G bands at approximate 1343 and 1577 cm^{-1}, respectively (Figure 2b). The intensity ratio of the D and G bands (I_D/I_G) is used to estimate the degree of graphitization. The ratio values of Ni@N-C-700, Ni@N-C-800, and Ni@N-C-900 are 1.09, 1.05, and 0.93, respectively. This reveals that the degree of crystallinity of Ni@N-Cs increases with increasing temperature.[21]

FT-IR spectroscopy (Figure 2c) indicated the broad peaks centered at 2924 and 3440 cm^{-1}, which were attributed to -CH and -OH stretching vibrations, respectively, while the peaks at 1108 and 1139 cm^{-1} were attributed to C-OH and C-N stretching vibrations, respectively.[22, 23] The peak locating at 1636 cm^{-1} is attributed to N-H vibration coupled with C-N stretching mode.[24] Furthermore, the peak at 636 cm^{-1} typically fits in Ni-O bond vibration. The N_2 adsorption/desorption isotherms are shown in Figure 2d. The shape of the physisorption isotherm is in accordance with type IV hysteresis loop in a P/P_0 range of 0.45−0.98, the typical characteristics of mesoporous structure according to the IUPAC classification.[25]

FESEM images of Ni@N-C composites are presented in Figure 3a-c. It is seen that a large number of intertwined CNTs anchored on the rough surface of irregular shaped carbon support. The proportion and diameter of CNTs increased dramatically with increasing calcination temperature, due to the decomposition of dicyandiamide. Therefore, it is reasonable to speculate that high pyrolysis temperature is more favorable for accelerating the growth of CNTs. FETEM images (Figure 3d) of Ni@N-C-700 confirmed that plenty of CNTs as bamboo-like structures with the length of 100−300 nm and diameter of around 10−15 nm are grown on the outboard of carbon. More importantly, the HRTEM image (Figure 3e) confirms that Ni NPs with the diameter between 8−26 nm are embedded in the carbon layers, which can be ascribed to the catalytic effect of the Ni nano-catalyst during the carbonization process.[19] Metallic NPs are crystalline, and the lattice spacings of ~2.05 and 1.77 Å correspond to those of (1 1 1) and (2 0 0) planes of Ni^0, respectively. The graphitic layers of CNTs grown on the surface of Ni@N-C-700 are not parallel, indicating the existence of abundant defect sites.[32] Taken together, these results indicate the formation of Ni@N-C 3D structures with Ni, N codoped CNTs grown on AC sheets.

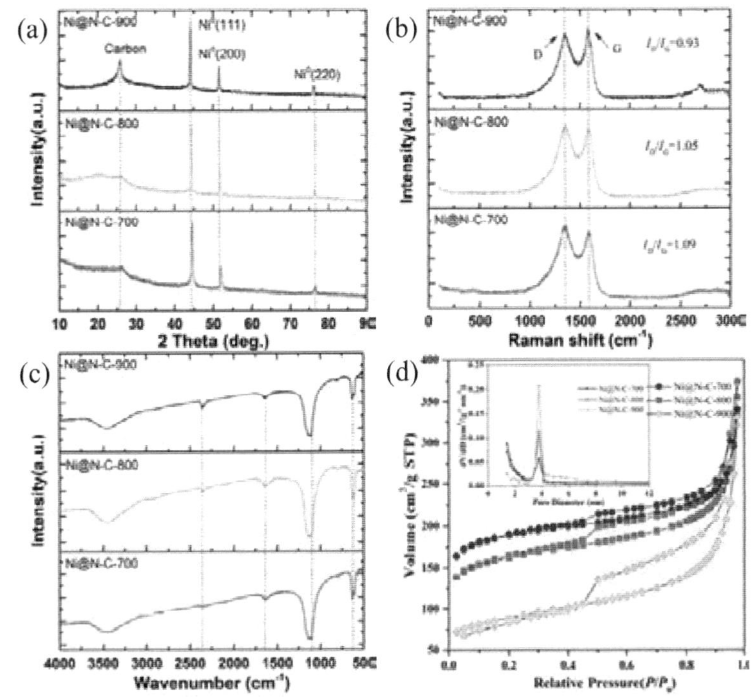

Figure 2. (a) XRD patterns, (b) Raman and (c) FT-IR spectra, (d) Nitrogen sorption isotherms of different Ni@N-C composites with different pyrolysis temperature. Inset: Pore size distribution.

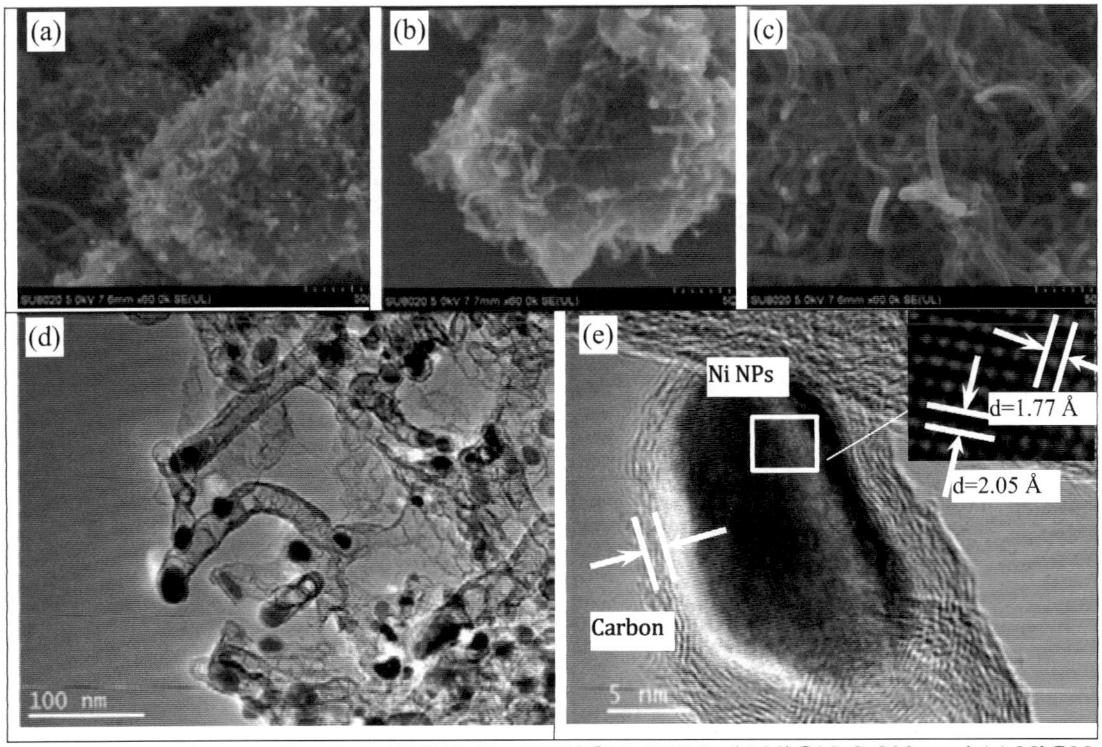

Figure 3. FESEM image of various Ni@N-Cs: (a) Ni@N-C-700, (b) Ni@N-C-800, and (c) Ni@N-C-900. (d) FETEM image, and (e) HRTEM image of Ni@N-C-700.

Catalytic Activity for organics removal

The catalytic performance of Ni@N-C was investigated in catalytic activation of PMS for removal of Orange II. For comparison, CNTs, graphene oxide, Ni^{2+}, Ni^{0} and NiO were also tested. Figure 4a shows that almost no Orange II was removed by PMS alone, meaning PMS itself could hardly oxidize Orange II.[33] As for Ni@N-C performance without the presence of PMS, less than 40% Orange II was removed attributing to adsorption of Orange II onto Ni@N-C. Once the PMS was added, a sharp drop of Orange II concentration occurred over all the Ni@N-C catalysts, suggesting the PMS was activated. Ni@N-C-700 performed the best activity with almost complete removal (~98.2%) of Orange II within 90 min, followed by Ni@N-C-800 and Ni@N-C-900, which removed 96.9% and 68.5% Orange II within 120 min, respectively. These observations suggest that Ni@N-C prepared at lower annealing temperature exhibited greater Orange II removal.[34]The stability of Ni@N-C catalyst was examined by collecting and reusing in five cycles. As shown in Figure 4d, after five cycles, 64.3% of its initial activity was still remained. The concentrations of Ni ions in the reaction medium after the five-recycle usages were quite low by an AAS method. Therefore, it was speculated that the Ni NPs were well protected by the carbon cage. With the increase of the residence time, the adsorption of Orange II on the catalyst surface could decrease accessibility of the active sites and passivate the Ni@N-C surface, resulting in activity deterioration.

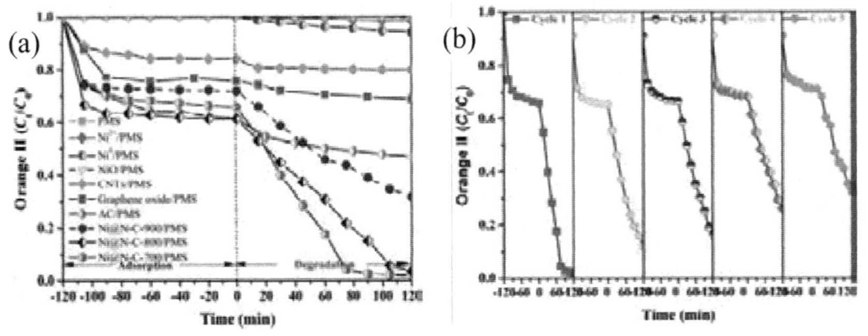

Figure 4. (a) Catalytic performance of various catalysts mediated by PMS. (b) Repeated oxidation of Orange II by the Ni@N-C-700/PMS system. Reaction conditions: [dye] = 20 mg/L, [catalyst] = 50 mg/L, [PMS] =0.10 g/L, T =25 °C.

To verify the existence of radicals for reactions, ESR measurements were performed using DMPO as a trapping agent. As shown in Figure 5a, the quartet lines with peak strength of 1:2:2:1 and hyperfine coupling constant of α_N =1.49 mT and α_H = 1.49 mT could be detected, which accounted for the typical signals of DMPO-$HO^{\bullet}$ adduct.[35] Moreover, the intensity decreased with increasing reaction time. The hyperfine splitting (a_N=0.74 mT and a_H=0.39 mT) was ascribed to 5,5-dimethyl-1-pyrrolidone-2-oxyl or 5,5-dimethylpyrrolidone-2-(oxy)-(1) (DMPO-X) produced via the oxidation of DMPO.[38] A sextet signal with hyperfine coupling parameters (a_H = 2.28 mT, a_N = 1.58 mT) was assigned to the DMPO adducts of carbon-based radicals ($R^{\bullet}$), which might be responsible for the organic decomposition.[39, 40] It is also noted that the signals of DMPO-SO_4 and DMPO-$\bullet O_2^-$ have not been found. This can be ascribed to the weak intensity and superior high adsorption capability of Ni@N-C, which can adsorb not only Orange II but also the radicals.[18]

Thus, from the above results, such a good activity of the prepared Ni@N-C can be attributed to the following factors. Firstly, large amounts of CNTs grown on the C surface with a unique 3D structure not only provide highly active sites but also facilitate electron transport along the basal surfaces of the carbon layer. The anchored configuration of Ni@N-C changes the local work function of the AC sheets, making the surface of the carbon layer more active. Secondly, the N-doped carbon layers protect Ni^0 NPs from aggregation and leaching, making them of high stability and activity in the acid condition. This close interconnection between Ni and carbon species hampers the aggregation of Ni NPs and provides a less path suitable for fast electron transfer. Thirdly, the dicyandiamide and $NiCl_2$ doped into the AC precursor indeed enhance the versatile nitrogen of 3D structures. Pyridinic-N, graphitic N, Ni–N clusters, and N dispersed on the Ni@N-C may be responsible for the higher catalytic activity. Furthermore, the high content of micropores and mesopores from biomass–derived porous carbon precursor provides more active sites. Consequently, with the synergistic effect of embedded Ni^0 NPs, nitrogen doping, hierarchical micro-mesopores, and interconnected CNTs, Ni@N-C exhibits superior heterogeneous catalytic performance.

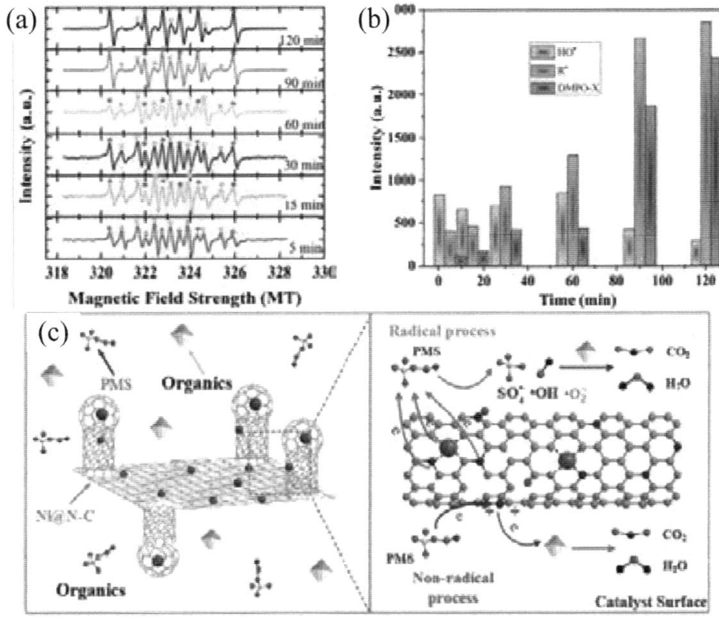

Figure 5. (a) ESR spectra of PMS activation with DMPO as the trapping agent at different reaction time. (b) Comparison of ESR signal intensity.∇ represents •OH adduct and • represents carbon-centered adduct. ◊ represents DMPO-X. (c) Reaction mechanism for organics removal by the Ni@N-C/PMS system. Reaction conditions: [Orange II] = 20 mg/L, [catalyst] = 50 mg/L, T =25 °C, [PMS] =0.10 g/L.

Catalytic Activity for Cr^{VI} Reduction

The catalytic performance of Ni@N-C composites for the transformation of toxic Cr^{VI} to Cr^{III} has been investigated. It was observed that Cr^{VI} reduction did not occur in the presence of only formic acid without a catalyst, which ruled out the possibility of the Cr^{VI} reduction by formic acid solely. When Ni@N-C-800 catalyst was introduced, the absorption peak of Cr^{VI} at 350 nm gradually decreases and finally disappears within 160 min, accompanied with a change in the color from yellow to colorless, indicating the successful reduction of Cr^{VI} to Cr^{III} (Figure 6a). The Cr^{III} formation is verified by adding excess sodium hydroxide to the solution, resulting in a color change from colorless to green due to the formation of hexahydroxochromate (III).[10, 41] After Cr^{VI} reduction reaction, typical Cr XPS peaks were clearly found on the spent Ni@N-C-800 catalyst (Figure 6b). The high-resolution XPS spectra of the Cr 2p region show that Cr $2p_{1/2}$ and Cr $2p_{3/2}$ peaks are located at ~587.2 and ~577.4 eV, respectively. Each of them could be fitted with two strong peaks, the ones at 588.1 and 578.5 eV were assigned to Cr^{VI}, while those at 586.9 eV and 577.2 eV to Cr^{III}, suggesting that Cr^{VI} and Cr^{III} coexisted on the carbon surface. The percentages of Cr^{VI} and Cr^{III} to total Cr species were calculated at 41.2% and 58.8%, respectively.

We further compared the Cr^{VI} reduction performance of different materials, Ni^0, CNTs, graphene oxide, AC, and Ni@N-C at the different annealing temperatures (Figure 6c). The Cr^{VI} reduction rates were found to be in an order of CNTs< graphene oxide< Ni^0< AC< Ni@N-C-700< Ni@N-C-900< Ni@N-C-800. The results clearly show that Ni@N-C-800 is the most active catalyst under our experimental conditions. The change of the annealing temperature remarkably influenced the activity, attributing to the different material structures (such as metal content, nitrogen content, nitrogen type, and CNT structure). To confirm the role of HCOOH, the reduction of Cr^{VI} was also studied using formaldehyde, acetic acid, H_2SO_4, citric acid and HCl as reducing agents(Figure 6d). The Cr^{VI} reduction rates for different reducing agents follow an order of formaldehyde<acetic acid<H_2SO_4<HCl<citric acid<HCOOH, which means that HCOOH can be a potential reducing agent for Cr^{VI} reduction.

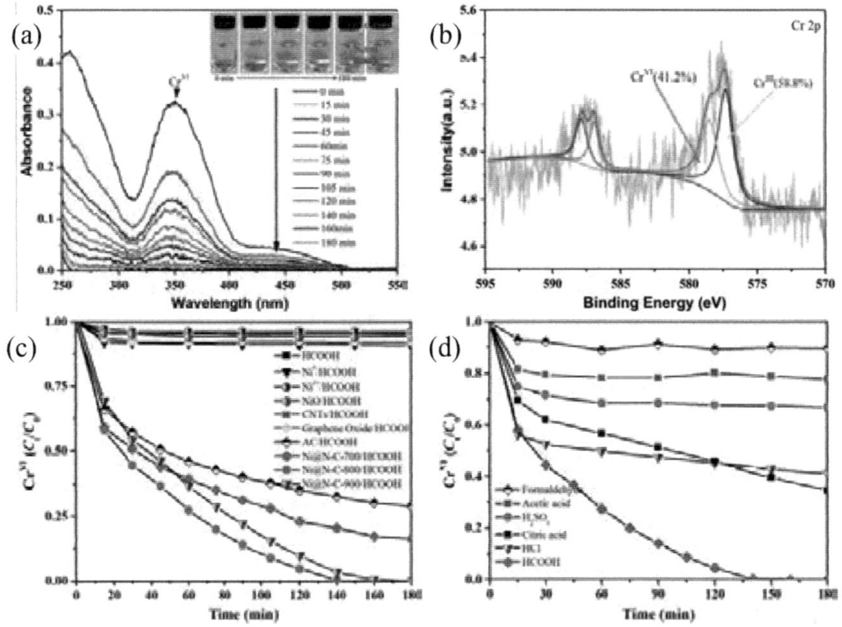

Figure 6. (a) Typical time-dependent UV−vis absorption spectra of Cr^{VI} solution in Ni@N-C-800/HCOOH system. (b) Cr 2p XPS spectra of spent Ni@N-C-800 with the adsorbed Cr^{VI} species. Comparison of the catalytic performance in the presence of (c) different catalysts or (d) different reducing agents. Reaction conditions: [Cr^{VI}] = 10 mg/L, [catalyst] = 50 mg/L, T =25 °C, [HCOOH] =0.117 M.

To check the process viability of Ni@N-C/HCOOH catalytic system, reduction of Cr^{VI} was repeated up to four cycles (Figure 7a). The removal efficiency of Cr^{VI} gradually decreases with increasing cycles because the surface of Ni@N-C is preadsorbed with in situ generated Cr^{III}. These results once again prove that Ni@N-C /HCOOH system is a practical method for Cr^{VI} removal.

Based on the above analyses and discussion, a plausible mechanism for this redox reaction was proposed (Figure 7b). In the solution, potassium dichromate exists mostly in the anionic form as $HCrO_4^-$, $Cr_2O_7^{2-}$, and CrO_4^{2-} ions. Initially, both formic acid and Cr^{VI} ions were firstly enriched on the surface of Ni@N-C by simple electrostatic attraction, where formic acid underwent a dehydrogenation decomposition pathway to produce CO_2 and hydrogen ($H^{\bullet}$).[41, 42] Ni^0 has a strong enrichment capacity for hydrogen molecules, which can enhance the formation of atomic hydrogen.[43] In the presence of electron donating acids, the transformation of Cr^{VI} to less toxic Cr^{III} by the free hydrogen atom occurred. It is noteworthy that Ni@N-C with an excellent electron transfer property could rapidly transfer the electron to reduce Cr^{VI}, which may be responsible for an improved catalytic activity. For Ni^0 NPs, there was no electron-transfer mediator to promote the electron transfer. It might explain pure Ni^0 has a lower reduction level compared to that of Ni@N-C. Hence, Ni@N-C exhibits superior electrochemical activities toward Cr^{VI} reduction due to the abundant active sites and effective electron transfer.

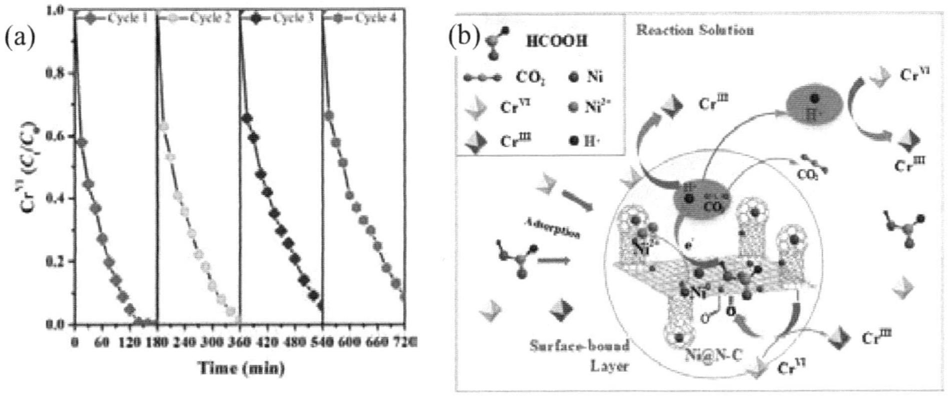

Figure 7. (a) Repeated runs of Cr^{VI} reduction in Ni@N-C-800/HCOOH system. (b) Cr^{VI} reduction mechanism by Ni@N-C/HCOOH system. Reaction conditions: $[Cr^{VI}] = 10$ mg/L, [catalyst] = 50 mg/L, T =25 °C, [HCOOH] =0.117 M.

Conclusion

In this study, we demonstrated a facile, economic, and environmentally friendly method to fabricate novel "sea urchin"-like Ni@N-C 3D composites via an efficient thermal condensation of low-cost biomass, dicyandiamide and $NiCl_2 \cdot 6H_2O$. The physicochemical characterizations including SEM, XRD, and TEM verified that large amounts of CNTs encapsulating Ni NPs had grown on AC sheets, revealing the effectiveness of the synthesis method. Additionally, the as-grown Ni@N-C exhibited excellent catalytic activity and stability for the degradation of organic dyes, and the reduction of aqueous toxic Cr^{VI} to nontoxic Cr^{III}, owing to the rationally designed microporous structure and more active catalytic sites. Detailed results indicated that the pyrolysis temperature affected the composition, morphologies, and catalytic properties of Ni@N-C composites. Furthermore, benefiting from the 3D porous structure, the low cost of the precursors, and green methodology, the obtained highly active Ni@N-C catalysts offer very attractive prospects and could be extended toward practical applications.

ACKNOWLEDGEMENTS

The financial supports by the Anhui Provincial Natural Science Foundation (NO. 1708085MB41), the China Postdoctoral Science Foundation (NO. 2015M570547, 2016T90585), and National Natural Science Foundation of China (Grant 51372062) are acknowledged. A partial support from the Australian Research Council for DP 150103026 is also acknowledged.

The authors declare no competing financial interest.

Reference

1. J. Su, Y. Zhang, S. Xu, S. Wang, H. Ding, S. Pan, G. Wang, G. Li, H. Zhao (2014), "Highly Efficient and Recyclable Triple-shelled Ag@Fe_3O_4@SiO_2@TiO_2 Photocatalysts for Degradation of Organic Pollutants and Reduction of Hexavalent Chromium Ions", *Nanoscale,* February, pp. 5181-5192.
2. Y. Choi, M.S. Koo, A.D. Bokare, D.-h. Kim, D.W. Bahnemann, W. Choi (2017), "Sequential Process Combination of Photocatalytic Oxidation and Dark Reduction for the Removal of Organic Pollutants and Cr(VI) using Ag/TiO_2", *Environ. Sci. Technol.*, March, pp. 3973-3981.
3. D. Xia, Y. Li, G. Huang, R. Yin, T. An, G. Li, H. Zhao, A. Lu, P.K. Wong (2017), "Activation of Persulfates by Natural Magnetic Pyrrhotite for Water Disinfection: Efficiency, mechanisms, and stability", *Water Res.*, January, pp. 236-247.
4. A.V. Desai, B. Manna, A. Karmakar, A. Sahu, S.K. Ghosh (2016), "A Water-Stable Cationic Metal–Organic Framework as a Dual Adsorbent of Oxoanion Pollutants", *Angew. Chem. Int. Ed.*, June, pp. 7811-7815.
5. J. Wang, Z. Wei, Y. Gong, S. Wang, D. Su, C. Han, H. Li, Y. Wang (2015), "Ni-promoted Synthesis of Graphitic Carbon Nanotubes from in situ Produced Graphitic Carbon for Dehydrogenation of Ethylbenzene", *Chem. Commun.*, June, pp. 12859-12862.
6. N. Karousis, I. Suarez-Martinez, C.P. Ewels, N. Tagmatarchis (2016), "Structure, Properties, Functionalization, and Applications of Carbon Nanohorns", *Chem. Rev.*, April, pp. 4850-4883.
7. Z. Wei, J. Wang, S. Mao, D. Su, H. Jin, Y. Wang, F. Xu, H. Li, Y. Wang (2015), "In Situ-Generated Co^0-Co_3O_4/N-Doped Carbon Nanotubes Hybrids as Efficient and Chemoselective Catalysts for Hydrogenation of Nitroarenes", *ACS Catalysis*, July, pp. 4783-4789.
8. Y. Yao, H. Chen, J. Qin, G. Wu, C. Lian, J. Zhang, S. Wang (2016), "Iron Encapsulated in Boron and Nitrogen Codoped Carbon Nanotubes as Synergistic Catalysts for Fenton-like Reaction", *Water Res.*, May, pp. 281-291.
9. Y. Yao, H. Chen, C. Lian, F. Wei, D. Zhang, G. Wu, B. Chen, S. Wang (2016), "Fe, Co, Ni Nanocrystals Encapsulated in Nitrogen-doped Carbonnanotubes as Fenton-like Catalysts for Organic Pollutant removal", *J. Hazard. Mater.*, March, pp. 129-139.
10. K. Bhowmik, A. Mukherjee, M.K. Mishra, G. De (2014), "Stable Ni Nanoparticle–Reduced Graphene Oxide Composites for the Reduction of Highly Toxic Aqueous Cr(VI) at Room Temperature", *Langmuir*, March, pp. 3209-3216.
11. M. Li, Y. Xiong, X. Liu, C. Han, Y. Zhang, X. Bo, L. Guo (2015), "Iron and Nitrogen Co-doped Carbon Nanotubes@hollow Carbon Fibers Derived from Plant Biomass as Efficient Catalysts for Oxygen Reduction Reaction", *J. Mater. Chem. A*, March, pp. 9658-9667.
12. X.-L. Wu, T. Wen, H.-L. Guo, S. Yang, X. Wang, A.-W. Xu (2013), "Biomass-Derived Sponge-like Carbonaceous Hydrogels and Aerogels for Supercapacitors", *ACS Nano*, April, pp. 3589-3597.
13. Z.-Y. Wu, C. Li, H.-W. Liang, J.-F. Chen, S.-H. Yu (2013), "Ultralight, flexible, and fire-resistant Carbon Nanofiber Aerogels from Bacterial Cellulose", *Angew. Chem. Int. Ed.*, February, pp. 2925-2929.
14. H.U. Lee, S.C. Lee, Y.-C. Lee, S. Vrtnik, C. Kim, S. Lee, Y.B. Lee, B. Nam, J.W. Lee, S.Y. Park, S.M. Lee, J. Lee (2013), "Sea-urchin-like Iron Oxide Nanostructures for Water Treatment", *J. Hazard. Mater.*, August, pp. 130-136.
15. M.A. Khalily, H. Eren, S. Akbayrak, H.H. Susapto, N. Biyikli, S. Özkar, M.O. Guler (2016), "Facile Synthesis of Three-Dimensional Pt-TiO_2 Nano-networks: A Highly Active Catalyst for the Hydrolytic Dehydrogenation of Ammonia–Borane", *Angew. Chem. Int. Ed.*, September, pp. 12257-12261.

16. C. Wan, J. Li (2015), "Facile Synthesis of Well-Dispersed Superparamagnetic γ-Fe_2O_3 Nanoparticles Encapsulated in Three-Dimensional Architectures of Cellulose Aerogels and Their Applications for Cr(VI) Removal from Contaminated Water", *ACS Sustainable Chem. Eng.*, July, pp. 2142-2152.
17. J. Shi, Y. Wang, W. Du, Z. Hou (2016), "Synthesis of Graphene Encapsulated Fe_3C in Carbon Nanotubes from Biomass and Its Catalysis Application", *Carbon*, December, pp. 330-337.
18. Y. Yao, J. Zhang, G. Wu, S. Wang, Y. Hu, C. Su, T. Xu (2017), "Iron Encapsulated in 3D N-doped Carbon Nanotube/porous Carbon Hybrid from Waste Biomass for Enhanced Oxidative Activity", *Environ. Sci. Pollut. Res.*, January, pp. 7679-7692.
19. Y. Zhao, J. Zhang, K. Li, Z. Ao, C. Wang, H. Liu, K. Sun, G. Wang (2016), "Electrospun cobalt embedded porous nitrogen doped carbon nanofibers as an efficient catalyst for water splitting" *J. Mater. Chem. A*, July, pp. 12818-12824.
20. H. Song, H. Cui, C. Wang (2014), "Abnormal Cyclibility in Ni@Graphene Core–Shell and Yolk–Shell Nanostructures for Lithium Ion Battery Anodes", *ACS Appl. Mater. Interfaces*, July, pp. 13765-13769.
21. M. Ma, S. You, W. Wang, G. Liu, D. Qi, X. Chen, J. Qu, N. Ren (2016), "Biomass-Derived Porous Fe_3C/Tungsten Carbide/Graphitic Carbon Nanocomposite for Efficient Electrocatalysis of Oxygen Reduction", *ACS Appl. Mater. Interfaces*, November, pp. 32307-32316.
22. F. Luo, Z. Chen, M. Megharaj, R. Naidu (2016), "Simultaneous Removal of Trichloroethylene and Hexavalent Chromium by Green Synthesized Agarose-Fe Nanoparticles Hydrogel", *Chem. Eng. J.*, March, pp. 290-297.
23. C. Wang, G. Shao, Z. Ma, S. Liu, W. Song, J. Song (2014), "Constructing Fe_3O_4 @N-rich Carbon Core-Shell Microspheres as Anode for Lithium Ion Batteries with Enhanced Electrochemical Performance", *Electrochim. Acta*, June, pp. 679-688.
24. S. Li, L. Tang, G. Zeng, J. Wang, Y. Deng, J. Wang, Z. Xie, Y. Zhou (2016), "Catalytic Reduction of Hexavalent Chromium by A Novel Nitrogen-functionalized Magnetic Ordered Mesoporous Carbon Doped with Pd Nanoparticles", *Environ. Sci. Pollut. Res.*, August, pp. 22027-22036.
25. Y. Huang, Y. Liang, Y. Rao, D. Zhu, J.-j. Cao, Z. Shen, W. Ho, S.C. Lee (2017), "Environment-Friendly Carbon Quantum Dots/$ZnFe_2O_4$ Photocatalysts: Characterization, Biocompatibility and Mechanisms for NO Removal", *Environ. Sci. Technol.*, February, pp. 2924-2933.
26. F. Lai, Y.-E. Miao, Y. Huang, Y. Zhang, T. Liu (2016), "Nitrogen-Doped Carbon Nanofiber/Molybdenum Disulfide Nanocomposites Derived from Bacterial Cellulose for High-Efficiency Electrocatalytic Hydrogen Evolution Reaction", *ACS Appl. Mater. Interfaces*, August, pp. 3558-3566.
27. W. Tian, H. Zhang, X. Duan, H. Sun, M.O. Tade, H.M. Ang, S. Wang (2016), "Nitrogen and Sulfur Co-Doped Hierarchically Porous Carbon for Adsorptive and Oxidative Removal of Pharmaceutical Contaminants", *ACS Appl. Mater. Interfaces*, March, pp. 7184-7193.
28. J. Yin, Q. Fan, Y. Li, F. Cheng, P. Zhou, P. Xi, S. Sun (2016), "Ni–C–N Nanosheets as Catalyst for Hydrogen Evolution Reaction", *J. Am. Chem. Soc.*, October, pp. 14546-14549.
29. J.S. Cho, S.Y. Lee, H.S. Ju, Y.C. Kang (2015), "Synthesis of NiO Nanofibers Composed of Hollow Nanospheres with Controlled Sizes by the Nanoscale Kirkendall Diffusion Process and Their Electrochemical Properties", *ACS Appl. Mater. Interfaces*, November, pp. 25641-25647.
30. C. Zhou, J.A. Szpunar, X. Cui (2016), "Synthesis of Ni/graphene Nanocomposite for Hydrogen Storage", *ACS Appl. Mater. Interfaces*, June, pp. 15232-15241.
31. P.K. Boruah, P. Borthakur, G. Darabdhara, C.K. Kamaja, I. Karbhal, M.V. Shelke, P. Phukan, D. Saikia, M.R. Das (2016), "Sunlight Assisted Degradation of Dye Molecules and Reduction of

Toxic Cr(VI) in Aqueous Medium Using Magnetically Recoverable Fe_3O_4/reduced Graphene Oxide Nanocomposite", *RSC Adv.*, January, pp. 11049-11063.

32. T. Chen, B. Cheng, G. Zhu, R. Chen, Y. Hu, L. Ma, H. Lv, Y. Wang, J. Liang, Z. Tie, Z. Jin, J. Liu (2017), "Highly Efficient Retention of Polysulfides in "Sea Urchin"-Like Carbon Nanotube/Nanopolyhedra Superstructures as Cathode Material for Ultralong-Life Lithium−Sulfur Batteries", *Nano Lett.*, December, pp. 437-444.
33. X. Li, Z. Ao, J. Liu, H. Sun, A.I. Rykov, J. Wang (2016), "Topotactic Transformation of Metal-Organic Frameworks to Graphene-Encapsulated Transition-Metal Nitrides as Efficient Fenton-like Catalysts", *ACS Nano*, December, pp. 11532-11540.
34. H. Lee, H.-i. Kim, S. Weon, W. Choi, Y.S. Hwang, J. Seo, C. Lee, J.-H. Kim (2016), "Activation of Persulfates by Graphitized Nanodiamonds for Removal of Organic Compounds", *Environ. Sci. Technol.*, September, pp. 10134-10142.
35. G. Liu, S. You, Y. Tan, N. Ren (2017), "In Situ Photochemical Activation of Sulfate for Enhanced Degradation of Organic Pollutants in Water", *Environ. Sci. Technol.*, January, pp. 2339-2346.
36. X. Cheng, H. Guo, Y. Zhang, X. Wu, Y. Liu (2017), "Non-photochemical Production of Singlet Oxygen via Activation of Persulfate by Carbon Nanotubes", *Water Res.*, February, pp. 80-88.
37. C. Wang, J. Kang, P. Liang, H. Zhang, H. Sun, M.O. Tade, S. Wang (2017), "Ferric carbide nanocrystals encapsulated in nitrogen-doped carbon nanotubes as an outstanding environmental catalyst", *Environ. Sci.: Nano*, pp. 170-179.
38. H. Lee, H.-J. Lee, J. Jeong, J. Lee, N.-B. Park, C. Lee (2015), "Activation of Persulfates by Carbon Nanotubes: Oxidation of Organic Compounds by Nonradical Mechanism",*Chem. Eng. J.*, April, pp. 28-33.
39. B. Jiang, X. Wang, P. Hu, M. Wu, J. Zheng, W. Wu (2016), "Dual Enhancement-inhibition Roles of Polycarboxylates in Cr(VI) Reduction and Organic Pollutant Oxidation in Electrical Plasma system", *Chemosphere*, October, pp. 1611-1617.
40. F. Gong, L. Luo, Y. Yao, D. Dai, W. Lu, W (2016), "Drastic Rate Acceleration Driven by Synergistic Effects: Key Role of Persistent Free Radicals Coupled with Ascorbic Acid in Decomposition of Organic Contaminants by Ferric Citrate", *Chen, Chem. Eng. J.*, November, pp. 440-447.
41. M. Liang, R. Su, W. Qi, Y. Zhang, R. Huang, Y. Yu, L. Wang, Z. He (2014), "Nanocomposite Catalyst with Palladium Nanoparticles Encapsulated in a Polymeric Acid: A Model for Tandem Environmental Catalysis", *Ind. Eng. Chem. Res.*, February, pp. 13635-13643.
42. Y. Guo, D. Wang, X. Liu, X. Wang, W. Liu, W. Qin (2014), "Synthesis and Characterization of the Nickel@carbon Dots Hybrid Material and Its Application in the Reduction of Cr(VI)", September, *New J. Chem.*, pp. 5861-5867.
43. X. Zhou, G. Jing, B. Lv, Z. Zhou, R. Zhu (2016), "Highly Efficient Removal of Chromium(VI) by Fe/Ni Bimetallic Nanoparticles in an Ultrasound-assisted system", *Chemosphere*, July, pp. 332-341.

500591 Technical Economic Analysis of an Intensified Integrated Gasification Combined Cycle Plant Design Featuring Membrane and Adsorptive Reactors

Abstract:

Currently, coal-based power plants are the largest source of electricity both globally and in the United States, and despite there being an abundant supply of low-priced natural gas, the Energy Information Administration (EIA) projects that by 2040, coal-based power plants will still provide 254.1 GWe of the generation capacity of the United States, and will account for approximately 36% of the total electricity generation capacity globally1,2. For this reason, coal-based power plants have begun considering the adoption of carbon capture technology in order to reduce the quantity of CO2 emissions, since each year in the United States 6,870 million metric tons of greenhouse gases (GHG) are emitted into the atmosphere, and CO2 represents over 80% of these GHG emissions3.

Because of the environmental concerns these emissions evoke, several carbon capture technologies have been proposed and integrated into these coal power plants with the goal to reduce the CO2 emissions by at least 90%. The main process utilized for carbon separation and capture employs absorption with amine-type solvents, whereby CO2 is separated through the use of a Dual-Stage Selexol unit4.

In this work, the design of an Integrated Gasification Combined Cycle (IGCC) plant featuring membrane and adsorptive reactors is first developed and then analyzed from an economic viewpoint. The design features membrane and adsorptive reactors that have over 90% carbon capture capability without the use of a Dual-Stage Selexol unit. The resultant intensified design is able to produce power more efficiently than traditional IGCC plants with Carbon-Capture Storage (CCS) technology. The UNISIM software (Honeywell trademark) is used to create an intensified flowsheet for the proposed power production plant design, and heat integration studies are presented. A technical economic analysis (TEA) is carried out on the proposed flowsheet to determine economic feasibility of the novel design. The economics of the intensified designs will be compared to a traditional IGCC plant with CCS to identify the extent of process intensification over the baseline design.

1. Annual Energy Outlook 2013, Department of Energy, U.S. Energy Information Administration, 2013.
2. Annual Energy Outlook 2014, Department of Energy, U.S. Energy Information Administration, 2013.
3. Inventory of U.S. greenhouse gas emissions and sinks: 1990-2014. Washington, DC: U.S. Environmental Protection Agency, n.d. Print.
4. Department of Energy/National Energy Technology Laboratory. Cost and Performance Baseline for Fossil Energy Plants Volume 1b: Bituminous Coal (IGCC) to Electricity Revision 2b â€“ Year Dollar Update. Document Number: DOE/NETL- 2015/1727.

501648 An Electrochemical Method to Remove Aqueous Sulfide from Swine Manure

Abstract:

Sulfide is synthesized by sulfur reducing bacteria during the anaerobic degradation of swine manure. The year-long pit storage of swine manure allows the accumulation of sulfide, which eventually results in a high risk of hydrogen sulfide (H_2S) releasing to the air during the manure pumping. H_2S is an extremely hazardous and toxic gas; and a number of casualties have been reported due to H_2S poisoning in the past several years. In order to minimize the sulfide level in swine manure and create a safe working environment, an effective way is presented in this study. By applying a certain voltage onto low carbon steel electrodes partially immersed in the swine manure storage, a 90% or above sulfide removal can be achieved after 1 to 2 weeks treatment. In our experiment, the optimal voltage was investigated to be 0.7 V, which allowed the majority removal of sulfide, and meanwhile a guarantee of less electrode material consumption. The effects of potential parameters, including manure solid content, electrode surface area, distance between electrodes, treatment temperature, biogas composition in the reactor headspace, and sulfide recovery after shutting down the power, were investigated. This electrochemical treatment has shown a significant aqueous sulfide removal from swine manure. Optimization based on mentioned parameters can further improve the treatment feasibility in reality.

499078 Pilot-Scale Testing of Electrochemical Removal of Hydrogen Sulfide in Deep-Pit Swine Manure Storage

Abstract:

The presence of hydrogen sulfide (H_2S) is a widespread issue in organic waste storage, and this toxic air contaminant causes serious concerns over environment, health and safety. One example is the deep-pit manure storage, a common practice for swine facilities in upper-Midwestern states. It occasionally experiences human or animal fatal overexposure to manure gas that contains H_2S. In this study, we developed an electrochemical-based oxidation method to remove or greatly reduce sulfide generation in the stored liquid manure. The treatment system needed some equipment installation and retrofit of the current storage pit with little maintenance requirement, low electricity consumption, and low chemicals usage. Lab-scale experiments were conducted in simulated swine manure storage, and systematically analyzed sulfide removal kinetics at varying total solids contents, initial sulfide concentrations, pH, applied voltages, electrode surface areas, electrode distances, and temperatures. An on-farm pilot-scale study was then performed to further confirm the treatment performance in close-to-real environment, and provide necessary data for techno-economic assessment. An electrochemical treatment device was designed and installed in two manure pump-out accesses in a testing barn that held about 200 growing-finishing pigs. Each pit access was divided to two compartments, one for control and the other for treatment installation. Perforated sheets of two electrode materials of low carbon steel (LCS) and stainless steel Type AISI 304 (SS) were installed in accesses and tested in 97 daysâ€™ operation. From the monitored H_2S concentration in pit headspace during the treatment, it can be seen that the LCS treatment decreased a substantial amount of H_2S concentration, achieving removal efficiency of 56%-78% depending on the applied voltage of between 0.6 and 0.7 V. SS treatment removed 13%-38% of H_2S. LCS treatment required a lower voltage application but achieved better removal performance, and is more economically feasible compared to SS installation according to techno-economic assessment. However, electrode material cost accounted for 76%-87% of the net present cost of the treatment assuming 20 yrsâ€™ life-time of the treatment device. Cheaper material, lower surface area, or more effective material will have to be optimized in order to decrease the treatment cost.

495035 Economic and Environmental Impact Analyses of Integrated Dehydrogenation and Hydroformylation in Gas-Expanded Liquid Media

Abstract:

Economic and Environmental Impact Analyses of Integrated Dehydrogenation and Hydroformylation in Gas-Expanded Liquid Media

Dupeng Liu,[1,2] Raghunath V Chaudhari[1,2] and Bala Subramaniam[1,2]

[1]Department of Chemical & Petroleum Engineering, [2]Center for Environmentally Beneficial Catalysis, University of Kansas

Lawrence, KS 66047

Increased shale gas production in the U. S. has also enhanced the availability of C_2-C_5 alkanes (the so-called natural gas liquids) as collateral products, and opportunities for their utilization in novel ways. For example, propylene production capacity via propane dehydrogenation (PDH) is slated to grow rapidly over the next several years.[1] However, for a typical PDH process, the per pass conversion is only about 50%. The separation of propane and propylene mixtures is highly energy and capital intensive. Previously, we had reported the beneficial effects of propane expanded liquids (with propylene/propane mixtures as feed) for propylene hydroformylation with Rhodium-based catalysts. At sufficiently high pressures (>10 bar), the pressurized propane/propylene mixture forms a gas-expanded liquid phase that promotes regio-selective hydroformylation of propylene with homogenous Rh complex. In addition, compared with conventional propylene hydroformylation process to produce butyraldehyde using polymer grade propylene as feedstock, the feed in the propane-expanded liquid (PXL) process is 50% cheaper.[3]

In this talk, economic and environmental analyses of the integrated dehydrogenation and hydroformylation in PXL process are presented and benchmarked against simulated conventional hydroformylation processes for which data were obtained from the Ruhrchemie/Rh™ne-Poulenc and Dow processes.[4-5] Preliminary simulation results indicate that capital investment and production cost associated with the PXL process is lower than the conventional processes (by up to 40% and 25%, respectively) due to the elimination of the C_3 separation section. Life cycle analysis (LCA) was also performed based on plant scale simulation of conventional and PXL processes. These analyses have provided guidance in catalyst design and in choosing materials and operating conditions that favor process economics while lessening environmental footprint.

1. Nawaz Z. Light alkane dehydrogenation to light olefin technologies: a comprehensive review[J]. Reviews in Chemical Engineering, 2015, 31(5): 413-436.

2.Nawaz Z. Dynamic modeling of catofin¨ fixed-bed iso-butane dehydrogenation reactor for operational optimization[J]. International Journal of Chemical Reactor Engineering, 2016, 14(1): 491-515.

3. Liu, D., Chaudhari, R.V., Subramaniam, B. Kinetic Modeling of Propene Hydroformylation in Propane-Expanded Liquid with Rh-Based Complexes, 2016 AIChE annual meeting San Francisco, CA.

4. Wattimena F, Heijmen H J. U.S. Patent 4,521,630[P]. 1985-6-4.

5. Devon T J, Phillips G W, Puckette T A, et al. U.S. Patent 5,332,846[P]. 1994-7-26.

492359 Bio-Energy with Carbon Capture and Storage (BECCS): Are Inefficient Power Plants a Better Option?

Abstract:

Negative emissions technologies (NETs) in general, and bio-energy with carbon capture and storage (BECCS) in particular, have been identified as key technologies in the context of limiting climate change to no more than 1.5 – 2°C (Fuss *et al.*, 2014). By actively removing CO_2 from the atmosphere, NETs provide a service, and thus investors would expect an appropriate risk-adjusted rate of return, varying as a function of the quantity of public money involved. In addition to providing negative emissions, BECCS facilities also generate firm low carbon power. However, as the electricity sector is integrating an increasing share of intermittent renewable energy (iRE), such as wind and solar, the dispatch load factor of such BECCS facilities may be small relative to their capacity. This has the potential to significantly underutilise these assets for their primary purpose of removing CO_2 from the atmosphere. In this study, we present a techno-economic-environmental evaluation of BECCS plants with a range of operating efficiencies, considering their full- and part-load operation relative to a national-scale annual CO_2 removal target, electricity price and negative emissions credit. We find that in all cases, a lower capital cost, lower efficiency BECCS plant is superior to a higher cost, higher efficiency facility from both environmental and economic perspectives. We show that it may be preferable to operate the BECCS facility in base-load fashion, constantly removing CO_2 from the atmosphere and dispatching electricity in a demand-following manner. We show that the use of this "spare capacity" to produce hydrogen for, *e.g.*, injection to a natural gas system for the provision of low carbon heating, can add to the overall environmental and economic benefit of such a system. The sensitivity of the model to biomass carbon footprint is also presented. The only point where our conclusions are altered is where the CO_2 emissions associated with the biomass supply chain are sufficiently large so as to eliminate BECCS CO_2 removal (Mac Dowell & Fajardy, 2017).

References

Fuss, S., Canadell, J. G., Peters, G. P., Tavoni, M., Andrew, R. M., Ciais, P., Jackson, R. B., Jones, C. D., Kraxner, F., Nakicenovic, N., Le Quere, C., Raupach, M. R., Sharifi, A., Smith, P. & Yamagata, Y. (2014). Betting on negative emissions. Nature Climate Change, 4 (10), 850–853. https://doi.org/10.1038/nclimate2392

Mac Dowell, N., & Fajardy, M. (2017). Inefficient power generation as an optimal route to negative emissions via BECCS. Environmental Research Letters, at press: https://doi.org/10.1088/1748-9326/aa67a5

APPLICATION OF SOME PREDICTIVE THERMODYNAMIC MODELS FOR INDUSTRIALLY IMPORTANT SYSTEMS

Yizu Zhu
Mohammad Shafiei

AIChE Annual Meeting, Minneapolis, MN

October 31, 2017

CHEMISTRY THAT MATTERS™

THERMODYNAMICS CRITICAL TO INTEGRATED PROCESS ENGINEERING & ECONOMICS

Physical Properties play very important roles in sustainable process engineering and in improving companies' bottom line.

Critical phase equilibrium data are unavailable for compounds under many circumstances:

- Design new products
- Screen conceptual design processes
- Scaleup and commercialize new technologies
- Trouble-shoot & optimize manufacturing plants
- Support techno-economics

Reliable "Predictive Equation Of State" models can be very helpful in tackling the above challenges.

No. 1

IMPACT OF PHYSICAL PROPERTIES ON SYSTEM ECONOMICS

A Close Boiling System

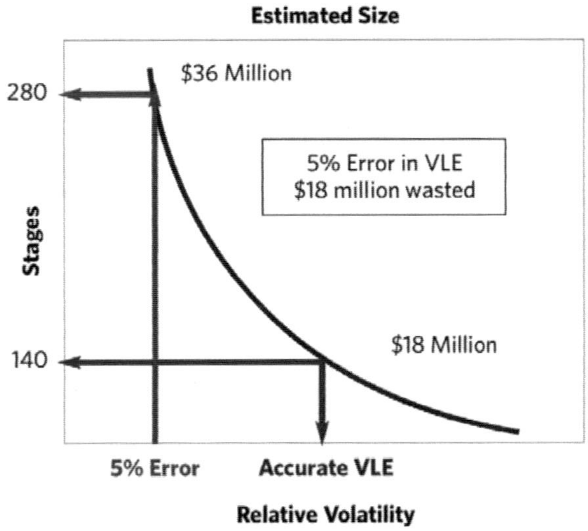

Dhole, et al, Aspen Tech White Paper, 2012

IMPACT OF PHYSICAL PROPERTIES ON SYSTEM ECONOMICS

Ammonia Reaction System using Aspen Plus V8.8

	FEED	PRODUCT		
		"Best available" model with parameters regressed from Data	Using A+ PSRK Prediction	Using Ideal Gas Model
Temperature C	450	450	450	450
Pressure bar	90	150	150	150
Vapor Frac	1	1	1	1
Mole Flow kmol/hr	4	3.272	3.263	3.3
Mass Flow kg/hr	34.06	34.06	34.06	34.06
Volume Flow cum/hr	2.67	1.36	1.36	1.32
Enthalpy Gcal/hr	0.012	0.003	0.003	0.003
Mass Flow kg/hr				
NH3		12.39	12.55	11.92
H2	6.05	3.85	3.82	3.93
N2	28.01	17.82	17.69	18.21
Mass Frac				
NH3		0.364	0.368	0.350
H2	0.178	0.113	0.112	0.115
N2	0.822	0.523	0.519	0.535
Difference%		0.00%	1.28%	-3.82%

The productivity impact is ~17.5$MM/year for a 1,000KTA Ammonia plant (Feb 2017 retail price).

THERMODYNAMIC CHALLENGES AND PREDICTIVE MODELS

- Industry presents challenges for representing
  - Chemically non-ideal systems
  - Involving supercritical components
  - In multiphase regions at extreme temperature and pressures
  - Specialty chemicals and other systems of rare compounds with no VLE-LLE data

- Equation Of State approach is good for high pressure, but until recently, is not-applicable to highly non-ideal chemical systems.

- New Mixing Rules for EOS emerged in the 1990s to represent the interactions in liquid phases of complex mixtures:
  - Predictive SRK (Holderbaum and Gmehling, 1991)
  - Modified Huron Vidal (MHV2) (Dahl, Fredenslund, Rasmussen, 1991)
  - Wong-Sandler (1992)
  - Etc.

No. 4

PENG - ROBINSON EOS

$$P = \frac{RT}{V_m - b} - \frac{a}{V_m(V_m + b) + b(V_m - b)}$$

Mixing Rule: $a = \sum_i \sum_j x_i x_j (a_i a_j)^{0.5}(1 - k_{ij}), \quad b = \sum_i x_i b_i$

$$a_i = \alpha_i 0.45724 \frac{R^2 T_{ci}^2}{P_{ci}}, \quad b_i = 0.07780 \frac{RT_{ci}}{P_{ci}}$$

$$\alpha_i(T) = [1 + m_i(1 - T_{ri}^{0.5})]^2, \quad m_i = 0.37464 + 1.54226\,\omega_i - 0.26992\,\omega_i^2$$

Binary interaction parameter k_{ij} needs to be regressed from phase equilibrium data

No. 5

APPLICATION OF PREDICTIVE EOS MODELS FOR HIGH PRESSURE AND HIGHLY NON-IDEAL SOLUTION SYSTEMS

SRK: $P = RT/(v-b) - a/v(v+b)$

PSRK's Mixing Rule:

$$b = \sum x_i b_i$$

$$a = b[g^E/A_1 + \sum x_i a_i/b_i + (RT/A_1)\sum x_i \ln(b/b_i)]$$

g^E: Excess Gibbs Energy from UNIFAC Group Contribution Activity Coefficient Model

A_1: constant

(Holderbaum and Gmehling, 1991; Dahl and Michelson, 1990)

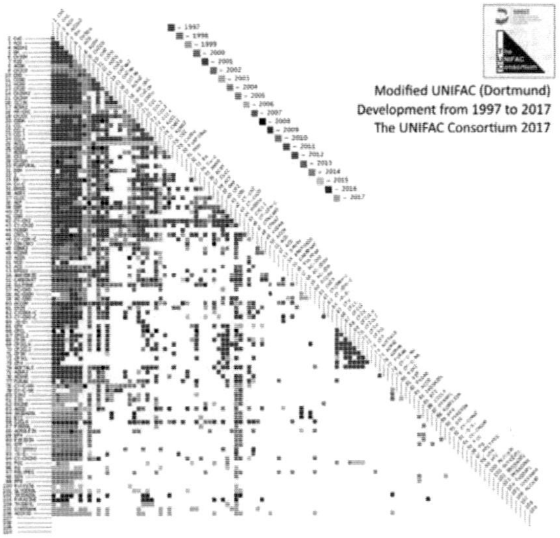

No. 6

KEY FEATURES OF THE NEW MIXING RULES

- A composition-dependent term F(x)
- F(x) is related to Excess Gibbs Energy or Helmholtz Energy
- Allows a single Equation Of State model to represent the phase behavior of:
  - Low Density Fluid (usually by EOS)
  - High Density Fluid (usually by Activity Coefficient Models)
  - Polar Fluid including highly non-ideal LLE systems

No. 7

PREDICTION FOR MEOH-H2S PHASE BEHAVIOR

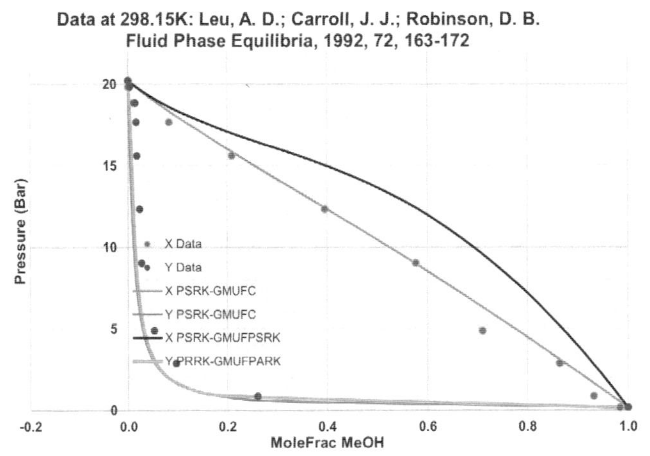

No. 8

METHANE-N-BUTANE P-XY AT HIGH PRESSURE

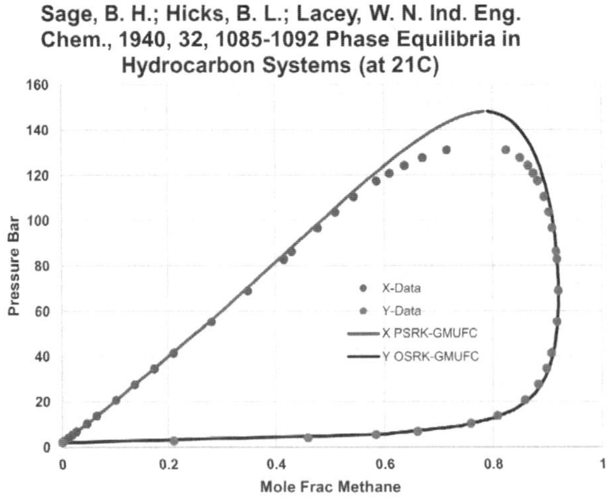

No. 9

EtAc–HEXANE SYSTEM T-XY PHASE DIAGRAM

EtAc-Hexane System at 1 atm
Fernandez, L.; Perez, E.; Ortega, J.; Canosa, J.; Wisniak, J.
J. Chem. Eng. Data, 2010, 55, 5519-5533

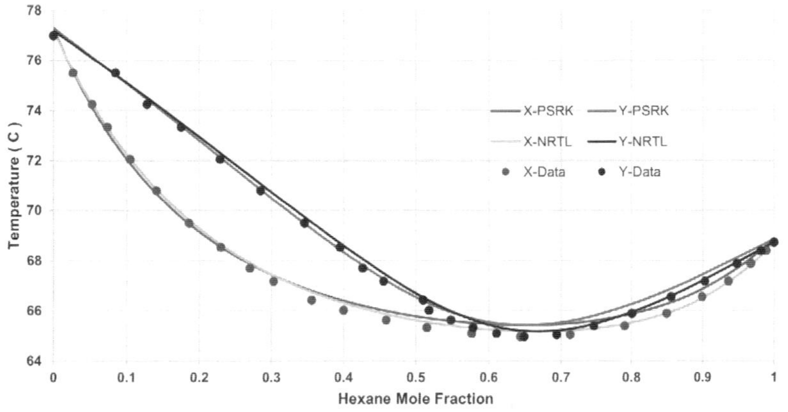

No. 10

PSRK PREDICTION FOR HIGH P, STRONGLY NON-IDEAL SYSTEMS

MEOH-CO2 SYSTEM

P-XY Diagram for Methanol-CO2 System AT 25c Using A+ V8.8)
Ohgaki, K.; Katayama, T. J. Chem. Eng. Data, 1976, 21, 53-5
Isothermal Vapor-Liquid Equilibrium Data for Binary Systems

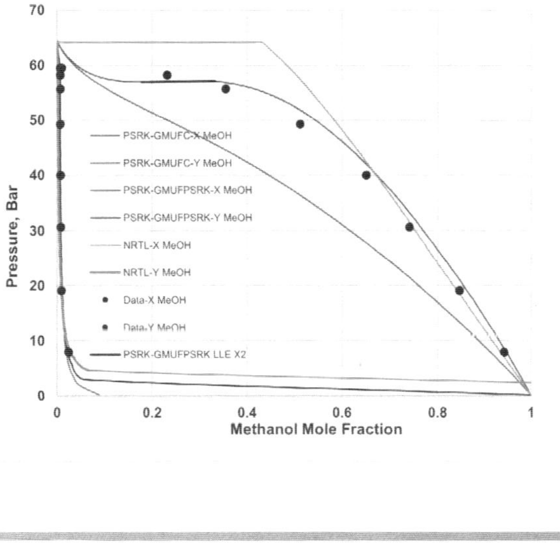

No. 11

CO2-CYCLOHEXANE SYSTEM AT HIGH PRESSURE

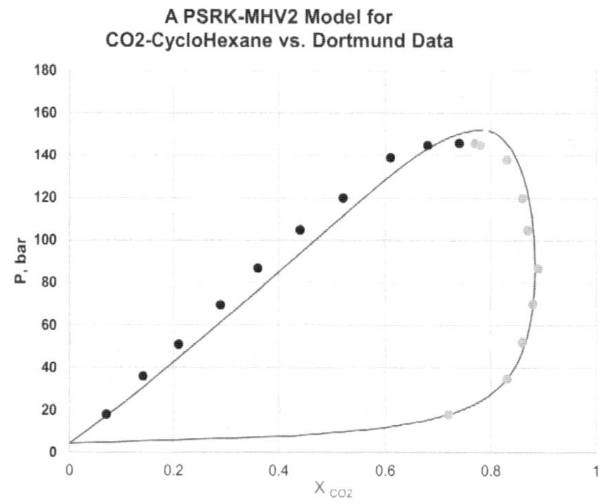

No. 12

NH3-WATER SYSTEM AT TYPICAL PLANT OPERATION CONDITION

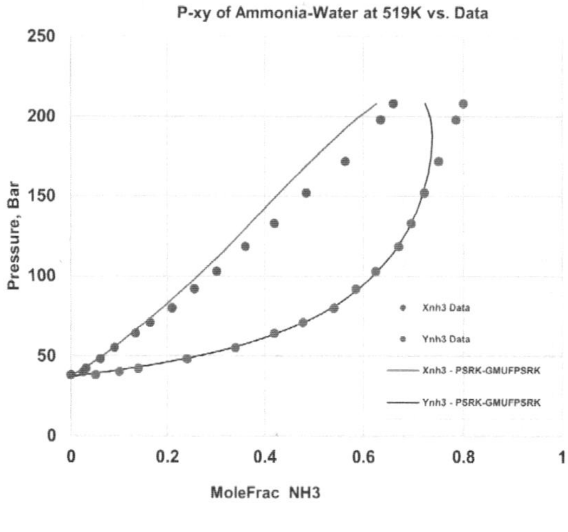

No. 13

N-BUOH – WATER HETEROGENEOUS AZEOTROPE PREDICTION

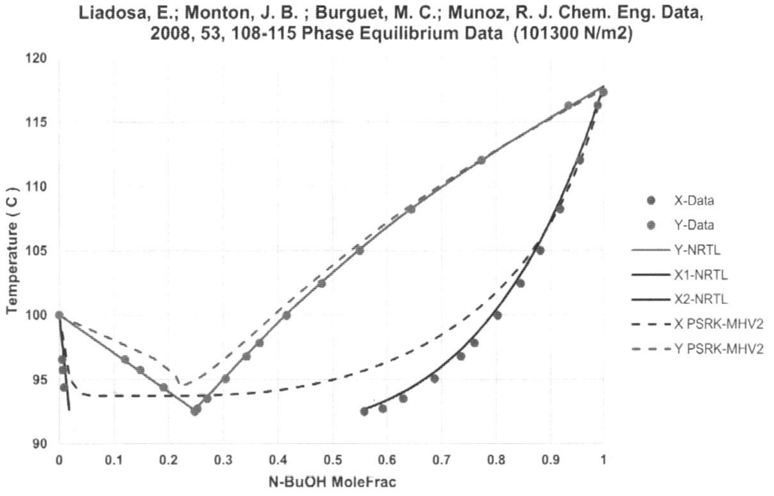

SEPARATION IN AN ETHYLENE PLANT WITH PSRK

	Tower Reflux	Tower Bottom Liquid Rate	Tower Top Temperature
	t/h	t/h	C
Plant Data	12.00	52.30	-98.50
RKS	12.31	51.00	-100.10
PSRK	12.99	52.40	-99.50

TEMPERATURE IMPACT ON HYDROCARBON CONCENTRATION IN RESIN USING POLYSRK FOR PRODUCT QUALITY IMPROVEMENT AND CAPACITY INCREASE

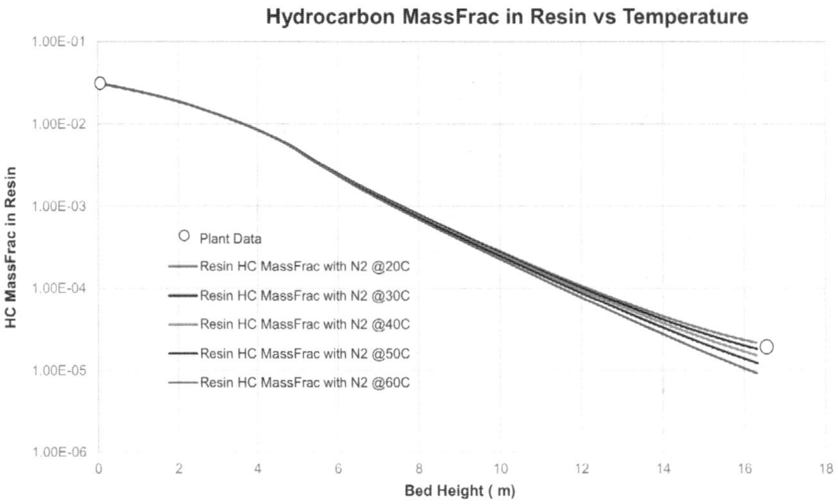

No. 16

CONCLUSIONS

- Physical properties have critical impact to the economics of process design and manufacturing operations.
- Predictive Equation of State models provide a powerful tool in representing highly non-ideal systems under high pressure where data are unavailable and other types of thermodynamic models are unable to handle.
- Predictive EOS models can be used with Excess Gibbs Free Energy of various liquid activity coefficient models available in simulation software such as Aspen Plus where binary interaction parameters are available (such as NRTL, WILSON, UNIQUAC, etc).
- This provides a powerful tool to handle a variety of complex systems where data are sparsely available:
  - Conceptual design of new products and processes
  - R&D
  - Scale-up and commercialization of innovative technologies
  - Manufacturing retrofitting and performance optimization

No. 17

DISCLAIMER

DISCLAIMER: THE MATERIALS, PRODUCTS AND SERVICES OF SAUDI BASIC INDUSTRIES CORPORATION (SABIC) OR ITS SUBSIDIARIES OR AFFILIATES ("SELLER") ARE SOLD SUBJECT TO SELLER'S STANDARD CONDITIONS OF SALE, WHICH ARE AVAILABLE UPON REQUEST. INFORMATION AND RECOMMENDATIONS CONTAINED IN THIS DOCUMENT ARE GIVEN IN GOOD FAITH. HOWEVER, SELLER MAKES NO EXPRESS OR IMPLIED REPRESENTATION, WARRANTY OR GUARANTEE (i) THAT ANY RESULTS DESCRIBED IN THIS DOCUMENT WILL BE OBTAINED UNDER END-USE CONDITIONS, OR (ii) AS TO THE EFFECTIVENESS OR SAFETY OF ANY DESIGN OR APPLICATION INCORPORATING SELLER'S MATERIALS, PRODUCTS, SERVICES OR RECOMMENDATIONS. UNLESS OTHERWISE PROVIDED IN SELLER'S STANDARD CONDITIONS OF SALE, SELLER SHALL NOT BE RESPONSIBLE FOR ANY LOSS RESULTING FROM ANY USE OF ITS MATERIALS, PRODUCTS, SERVICES OR RECOMMENDATIONS DESCRIBED IN THIS DOCUMENT. Each user is responsible for making its own determination as to the suitability of Seller's materials, products, services or recommendations for the user's particular use through appropriate end-use and other testing and analysis. Nothing in any document or oral statement shall be deemed to alter or waive any provision of Seller's Standard Conditions of Sale or this Disclaimer, unless it is specifically agreed to in a writing signed by Seller. Statements by Seller concerning a possible use of any material, product, service or design do not, are not intended to, and should not be construed to grant any license under any patent or other intellectual property right of Seller or as a recommendation for the use of any material, product, service or design in a manner that infringes any patent or other intellectual property right.

No. 18

THANK YOU AND Q&A

No. 19

503648 The Last Quarter Century of Methods and Software for Algorithmic Process Synthesis

Abstract:

Process-Network Synthesis or PNS in short formally defined by Friedler and Fan in 1991 is the constructive procedure of building up complex processes or networks from elementary building blocks in order to achieve overall targets by the help of external and internal resources. Advanced algorithmic methodologies and computational tools developed in the last quarter of century could overcome the combinatorial explosion caused by the large number of alternative combinations of process elements having the potential to achieve the desired overall target facing industrial size problems. Moreover, aiming at finding the global optimum multiple component problems of the process-synthesis traditionally examined sequentially become solvable in integrated way, including reaction path synthesis, separation-network synthesis, heat integration, control design, and scheduling. The presentation aims at reviewing the related methods and software. Latest developments and future challenges are to be highlighted as well.

492362 Near-Term and Sustainable Carbon Dioxide Removal: Is Bio-Energy with Carbon Capture and Storage (BECCS) the Right Answer?

Abstract:

Bioenergy with CO_2 Capture and Storage (BECCS) is commonly regarded as a vital yet controversial negative emissions technology (NET) to meeting our climate goals (Fuss et al., 2014). In this contribution we present a whole-systems analysis of the BECCS value chain associated with the cultivation, harvesting, transport, pre-treatment and conversion in dedicated biomass power stations in conjunction with CCS, of a range of biomass resources â€“ both dedicated energy crops (miscanthus, switchgrass, short rotation coppice willow), and agricultural residues (wheat straw). We explicitly consider the implications of sourcing the biomass from different regions, climates and land types. The water, carbon and energy footprints of each value chain were calculated, and their impact on the overall system water, carbon and power efficiencies were evaluated. An extensive literature review was performed and a statistical analysis of the available data is presented. In order to describe the dynamic greenhouse gas balance of such as system, a yearly accounting of the emissions was performed over the lifetime of a BECCS facility, and the carbon "breakeven time" and lifetime net CO_2 removal from the atmosphere were determined. The effects of direct and indirect land use change were included, and were found to be a key determinant of the viability of a BECCS project. Overall we conclude that, depending on the conditions of its deployment, BECCS could lead to both carbon positive and negative results. The total quantity of CO_2 removed from the atmosphere over the project lifetime and the carbon breakeven time were observed to be highly case specific. This has profound implications for the policy frameworks required to incentivise and regulate the widespread deployment of BECCS technology. The results of a sensitivity analysis on the model combined with the investigation of alternate supply chain scenarios elucidated key levers to improve the sustainability of BECCS: 1) measuring and limiting the impacts of direct and indirect land use change, 2) using carbon neutral power and organic fertilizer, 3) minimising biomass transport, and prioritising sea over road transport, 4) maximising the use of carbon negative fuels, and, 5) exploiting alternative biomass processing options, e.g., natural drying or torrefaction. A key conclusion is that, regardless of the biomass and region studied, the sustainability of BECCS relies heavily on intelligent management of the supply chain (Fajardy & Mac Dowell, 2017).

References

Fuss, S., Canadell, J. G., Peters, G. P., Tavoni, M., Andrew, R. M., Ciais, P., Jackson, R. B., Jones, C. D., Kraxner, F., Nakicenovic, N., Le Quere, C., Raupach, M. R., Sharifi, A., Smith, P. & Yamagata, Y. (2014). Betting on negative emissions. Nature Climate Change, 4 (10), 850â€“853. DOI:Â 10.1038/nclimate2392

Fajardy, M., & Mac Dowell, N. (2017). Can BECCS deliver sustainable and resource efficient negative emissions? Energy and Environmental Science. DOI: 10.1039/C7EE00465F

499278 A Physical Input-Output Model for the Food-Energy-Water (FEW) Nexus in Indiana

Abstract:

One of the major obstacles to understand the challenges of managing the production of food and energy while maintaining water availability and resources is the lack of physical insights into the regional economies. Physical Input-Output Table (PIOT) based models provide insights into physical interaction of various production sectors while taking advantage of structural information about the economy, thus quantifying the interdependency of food and energy production [1]. Furthermore, the environmentally extended form of PIOT allows the studying of environmental impacts from production systems by tracking the emissions to water and air by each production system. While few fully developed physical input output tables have been produced, their application has been demonstrated in measuring land footprints, analyzing the structure of the economy, and testing the resiliency of supply chains, all of which are important in FEW systems [2,3,4]. Hence, we developed an initial PIOT model for agro-based industries in Indiana to analyze the interdependency of food and bioenergy production in the state. In particular, the unidimensional flow of nitrogen was tracked as measure of production from various sectors, since N forms the basis of food production and the optimization of its use is critically important to food systems and ecosystem management. Indiana's agrobased economy is based on corn and soybean production, which together make up over 76.5% of Indiana's agricultural land [5]. Bioenergy is produced as both determining and co-product from the corn processing sectors. To build this model, we generated process models for each production unit and used the results to generate the PIOT as shown in [6]. For agricultural systems, we performed mass balances on production systems. An algorithmic approach was followed to find the statistical sources for information where available, as well as approximation methods from peer reviewed literature. Once the mass balances were completed they served as the basis for process models that fit into the PIOT building approach described in [6]. The structure of PIOT in Indiana is compared with an earlier PIOT developed for Illinois [7].

[1] Hoekstra, R., & van den Bergh, J. C. (2006). Constructing physical input–output tables for environmental modeling and accounting: Framework and illustrations. Ecological Economics, 59(3), 375-393.
[2] Hubacek, K., & Giljum, S. (2003). Applying physical input–output analysis to estimate land appropriation (ecological footprints) of international trade activities. Ecological economics, 44(1), 137-151.
[3] Altimiras-Martin, A. (2014). Analysing the structure of the economy using physical input–output tables. Economic Systems Research, 26(4), 463-485.
[4] Nakatani, J., Tahara, K., Nakajima, K., Daigo, I., Kurishima, H., Kudoh, Y., Matsubae, K., Fukushima, Y., Ihara, T., Kikuchi, Y. and Nishijima, A. (2017). A graph theory-based methodology for vulnerability assessment of supply chains using the life cycle inventory database. Omega.
[5] United States Department of Agriculture National Agricultural Statistical Service. (2017). 2016 State Agriculture Overview: Indiana. Retrieved from https://www.nass.usda.gov/Quick_Stats/Ag_Overview/stateOverview.php?state=INDIANA.
[6] Wachs, L. and Singh, S. (2016, November). Computational Approaches in Systems Modeling for Environmental Impacts of Industries Automating Physical Input-Output Tables (PIOTs) Via Aspen Process Modeling. Poster session presented at the 2016 AIChE Annual Meeting, San Francisco, CA.
[7] Singh S, Compton J, Hawkins, T, Sobota D, Cooter E., "A Physical Input-Output Model for Illinois" In Review

501390 Interventions for Reducing Energy Impacts of Water Embodied in Domestic Food Trade: A Network Perspective

Abstract:

Integrated management of food, energy, and water resources has been advocated to mitigate unintended consequences of managing them in isolation. Food, energy, and water (FEW) systems are highly interdependent, interconnected, and exhibit numerous interactions at distinctive spatial scales. Sustainable management and availability of these resources will require quantitatively mapping their complex interdependencies and understanding their ecology. An integral component of food security and diversity involves trade of food items. Food trade also presents a pathway for transferring embodied resources and emissions associated with food production. For the United States (U.S.), irrigation adds significant value to food and feed production and represents a critical link for studying domestic FEW systems. Specifically, irrigation water withdrawals account for second largest freshwater withdrawals in the U.S. and has substantial associated energy and environmental impacts. We recently developed a network model of interstate food transfers for the U.S. The model was utilized to understand the energetic and greenhouse (GHG) impacts of irrigation embodied in interstate food transfers [1]. This analysis highlighted the high energy and GHG emissions embodied in irrigation associated with domestic food trade. We also proposed relevant network metrics to understand the structure of domestic food trade network and its implications for resilience.

In this work, we propose and evaluate a range of intervention strategies to enhance environmental sustainability and resilience of the domestic food trade network as they relate to irrigation and its embodied energy impacts. We use publicly available datasets to develop weighted and directed networks of interstate food trade and associated environmental impacts. We consider interstate food transfers of over 50 commodities including cereal grains, milled grains, livestock, and meat products. From a network perspective, the weighted food trade network consists of a few key states (Texas, California, Kansas) that control high throughput. We also observe that these influential states depend on water resources facing depletion risks (Colorado river basin, Ogallala aquifer), employ energy and water inefficient irrigation systems, and utilize electricity and diesel-based pumps. We evaluate the impact of technological, behavioral, and policy interventions using our interstate food network model to study their impact on the sustainability and resilience of the domestic FEW nexus. The investigated interventions include changes in irrigation practices, energy mixes, adjustments in diet patterns, and spatial changes in food production patterns. These strategies help in prioritizing and shedding light on interventions that aid in improving sustainability and resilience of the FEW nexus from a systems perspective as opposed to those that offer incremental advantages. The implications of these findings including their potential and limitations for enhancing the sustainability and resilience of the FEW nexus will be described in detail.

Reference Cited:

N. Vora, A. Shah, M.M. Bilec, and V. Khanna. Food-Energy-Water nexus: quantifying embodied energy and GHG emissions from irrigation through virtual water transfers in food trade, *ACS Sustainable Chemistry and Engineering*, 2017, 5(3), 2119-2128.

500119 A General Framework for Process and Utility Networks Synthesis

Abstract:

There are sub-systems in chemical plants that consist of a number of sinks with demand for a certain property and a number of alternative sources that can satisfy these properties to fulfil the sink demands. Examples of such sub-systems are heat exchanger networks [1], fuel gas networks [2], water networks [3], utility networks [4], etc. Properties that are exchanged can be energy in the form of heat as in heat exchanger networks, or in both work and heat form as in the utility networks. Although each subsystem has attracted significant attention, each synthesis problem requires a new superstructure to be postulated whenever a new problem is encountered. Similarly, separation network systems, such as membrane networks [5] and distillation sequences [6], have been a focus of attention, yet a new superstructure representation is needed for each problem. A unified superstructure representation for these different classes of synthesis problems will be highly advantageous. Recently, a systematic framework for process intensification based on building-block superstructure has been proposed based on the idea of dissecting various unit operations into fundamental building blocks [7]. The proposed method can be also used for automatic flowsheet generation.

In this work, we extend the approach to provide a general framework for the synthesis of different networks using a similar representation using building blocks. Here, each block has feed and product streams ensuring interaction with the surroundings and can be viewed as sources and sinks, respectively. Also, streams between blocks are equipped with compressors or expanders depending on the source and sink block pressures. Mass and energy flow within the superstructure are facilitated via block material and energy balances and heating and cooling requirements are decided accordingly. Separation operations, such as gas membrane separations, are represented via two neighboring blocks separated by a semi-restricted boundary. The decisions over the assignment of boundary types between blocks are enabled via use of decision variables. By using all of these features of the building block superstructure, several different process network synthesis problems can be accommodated in a single superstructure with n and m number of blocks in row and column respectively. It should be noted that for different network synthesis problem, the block superstructure size could be reduced to improve the computation performance, i.e, using $2\times m$ superstructure to represent a pooling problem without intermediate pools. The overall process network synthesis problem is formulated as a mixed-integer nonlinear optimization (MINLP) problem where the nonlinear terms are due to splitting, energy balances and/or work calculations and discrete variables are for assigning alternative operations to each block and boundary in the superstructure. In this work, we will demonstrate that the proposed method is capable of unifying the superstructure representations for different class of process network synthesis problems via a range of literature problems.

References:

[1] Yee, T.F., Grossmann, I.E. and Kravanja, Z., 1990. Simultaneous optimization models for heat integrationâ€"I. Area and energy targeting and modeling of multi-stream exchangers.Â *Computers & chemical engineering*,Â *14*(10), pp.1151-1164.

[2] Hasan, M.F., Karimi, I.A. and Avison, C.M., 2011. Preliminary synthesis of fuel gas networks to conserve energy and preserve the environment. Industrial & Engineering Chemistry Research, 50(12), pp.7414-7427.

[3] AhmetoviÄ‡, E. and Grossmann, I.E., 2011. Global superstructure optimization for the design of integrated process water networks. AIChE journal, 57(2), pp.434-457.

[4] Elia, J.A., Baliban, R.C. and Floudas, C.A., 2010. Toward novel hybrid biomass, coal, and natural gas processes for satisfying current transportation fuel demands, 2: Simultaneous heat and power integration. Industrial & Engineering Chemistry Research, 49(16), pp.7371-7388.

[5] Alnouri, S.Y. and Linke, P., 2012. A systematic approach to optimal membrane network synthesis for seawater desalination. *Journal of membrane science*, *417*, pp.96-112.

[6] Giridhar, A. and Agrawal, R., 2010. Synthesis of distillation configurations: I. Characteristics of a good search space. Computers & chemical engineering, 34(1), pp.73-83.

[7] Demirel, S.E., Li, J., Hasan, M.M.F. Systematic Process Intensification using Building Blocks. Computers & Chemical Engineering, http://dx.doi.org/10.1016/j.compchemeng.2017.01.044.

487887 Electrochemical Swing Process for Carbon Capture

Abstract:

Electrochemical Swing Process for Carbon Capture

The capture of carbon dioxide (CO_2) at low concentrations (<1%) is of interest for power generation and other industrial processes, as well as for ventilation of enclosed spaces. We report a solid-state Faradaic electrochemical cell that exploits the reductive addition of CO_2 to quinones for CO_2 capture. The behavior of the cell is modulated by the introduction of CO_2 which changes the reduction of quinone from two one-electron reductions to one two-electron reduction. This results in a change in the potential window of the cell and alters its charge-discharge behavior. An electrochemical cell with a cathode, comprised of a polyanthraquinone-carbon nanotube composite, captures CO_2 upon charging via the carboxylation of reduced quinones, and releases CO_2 upon discharging. The novel cell architecture sandwiches a polyvinylferrocene anode, which serves as both an electron source and sink for the reduction and oxidation of quinones respectively, with two polyanthraquinone cathodes, maximizing the cathode surface area exposed to gas, allowing for ease of stacking of the cells in a device. The bench-scale prototype made demonstrates capture of CO_2 from inlet streams of CO_2 concentrations as low as 0.5% (5000 ppm) at a Faradaic efficiency of >90% and a work of 40-90 kJ per mole of CO_2 captured. This makes it an efficiency improving addition to HVAC systems and cabin environmental control systems aboard aircraft and vehicles.

500966 Opportunities for Industrial CO_2 Capture and Utilization in the US

Abstract:

Carbon dioxide capture from flue gases has largely failed to gain traction due to their prohibitively high cost. One strategy to lower capture costs is to target capture from emissions with higher CO_2 concentration. This presentation focuses specifically on a sub-section of industrial CO_2 emissions, of which there are no carbon-free routes for the formation of products such as cement, glass, iron and other metals. Although these emissions do not represent the majority, unlike the electricity and transportation sectors, they are not replaceable with renewables or bio-energy routes, and in fact, their scale matches well with CO_2 utilization opportunities. A low-cost pathway, including separation, compression and transportation to currently available utilization opportunities has been identified. In particular, our cost model corrects for differences in exhaust composition, flow rate, and geo-specific utilization demand. A regional case-study for the US state of Pennsylvania reveals steel and cement manufacturing as the least cost options. Further, we find that transportation via trucks is generally the low-cost alternative compared to pipeline transport for small volumes on the order of 100 kt CO_2/a. These results are presented in the context of other complexities such as the relative feasibility of extracting CO_2 process emissions from combustion emissions, theoretical maximum commodity outputs by region, and competitive-based commodity trading (regional and international) inspired by localized CCUS efforts.

491206 Improvement of Oxy-Combustion Using Thermodynamic and Exergetic Analysis

Abstract:

Reducing carbon dioxide (CO_2) emissions to the atmosphere is has been the focus of many policy actions directed at mitigating the effects of climate change. Although renewable energies have been widely deployed, there is still a dependence on fossil fuels to satisfy the demands of the grid. Carbon capture and storage (CCS) is generally recognised as providing an essential role to reduce anthropogenic CO_2 emissions from burning fossil fuels for electricity generation. Several technologies have been developed to address this problem, such as oxy-combustion, which also possesses applications in the glass and steel industries. Oxy-combustion is a promising technology that enables the reduction of carbon dioxide (CO_2) emissions from combustion of fossil fuels, such as coal, by more than 90 %. However, the requirement of including an air separation unit (ASU) and a gas processing unit (GPU) introduces an energy penalty to the process, increasing the cost of electricity. This work aims at identifying the potential for improving this technology using a rational approach grounded on the laws of thermodynamics and techno-economic analysis.

Improving the efficiency of oxy-combustion by reducing the parasitic energy consumption of both ASU and GPU is a requirement for this technology to be more competitive in the market. This can be achieved by increasing the separation efficiency of the ASU and GPU. Another way of reducing losses in oxy-combustion is to improve the thermal efficiency of the plant, by increasing the operating conditions to the boiler. However, this would require the use of newer materials, increasing capital expenditure (CAPEX) and make it a riskier project.

Heat integration using low grade heat of compression from the ASU and GPU to pre-heat feedwater is another way of increasing efficiency. This strategy possesses the advantage of using the available resources more efficiently while reducing the number of unit operations, thus reducing CAPEX, and allowing more steam to be used for electricity generation. Acceptance of a technology is highly dependent on the levelised cost of electricity (LCOE) which is influenced by CAPEX and fuel price. This demonstrates that there must be an increase of process efficiency while keeping CAPEX reduced for technological innovation to occur.

An oxy-combustion comprised of a double column cryogenic ASU, an ultra-supercritical boiler with flue gas treatment, and a cryogenic distillation GPU was simulated using Aspen HYSYS. The model developed showed a good agreement with data from IECM program and the values reported by Callide oxyfuel project, with the exception for O_2 and fuel consumption. This difference was found due to the assumption of complete combustion, with no carbon monoxide formation and no heat losses to the environment.

The minimum thermodynamic separation work was calculated for both ASU and GPU and was combined with the theoretical Carnot efficiency to determine the maximum efficiency that an oxy-combustion could achieve. An exergy destruction analysis was performed on each unit operation to quantify the inefficiencies within the process, which would allow for a rational improvement of the process.

In conclusion, the process was improved based on inefficiency minimisation, and then a techno-economic analysis was performed to determine its economic viability. This allowed for an efficiency improvement of 1.4 % (gross) and 3 % (net) as well as a 15 % reduction of CAPEX while decreasing the parasitic power consumption of the ASU from 204 to 197 kWh/t_{O2} and the GPU from 140 to 137 kWh/t_{CO2}. Both reduction of CAPEX and power consumption leads to a decrease of the levelised cost of electricity (LCOE).

The Surface Layer Speciation of CO_2-loaded Aqueous Solutions of Alkanolamines

Naser S. Matin[1], Janice A. Steckel[2], Jesse Thompson[1], Moushumi Sarma[1], and Kunlei Liu[1]

1. Center for Applied Energy Research, University of Kentucky, 2540 Research Park Drive, Lexington, KY 40511, United States
2. National Energy Technology Laboratory, U.S. Department of Energy, Pittsburgh, PA, 15261

UKy-CAER

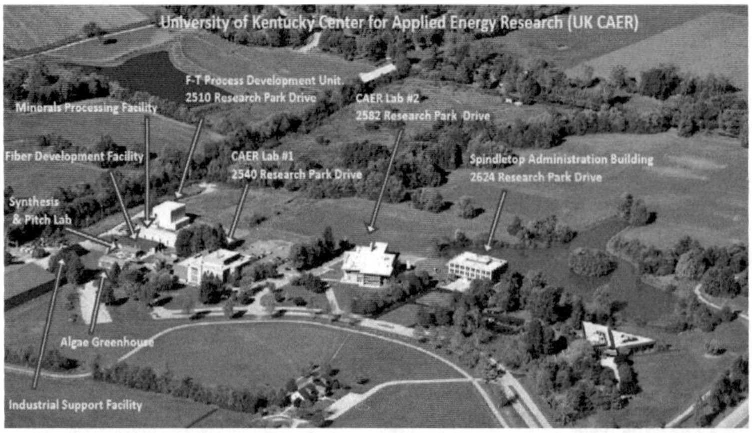

Outline

- ❑ Why Surface Speciation?
- ❑ Surface Behavior of Alkanolamine Solutions
- ❑ Thermodynamic Framework
- ❑ Bulk and Surface Speciation
- ❑ Conclusion

Why Surface Speciation?

Surface layer speciation;

- Has strong impact on the interfacial behavior of a solution.
- Provides deep knowledge and understanding about solutions surface properties.
- Helps in designing proper solvents.
- Is useful in designing proper gas-liquid contactors.
- Provides information about the link between liquid and surface properties.
- Has drawn less attention in amine based CO_2 capture processes.

CO_2 Absorption Rate

Liquid side mass transfer coefficient; $kg' = \{k_2 [Am] D_{CO2}\}^{0.5} / H_{CO2}$

<u>To enhance the CO_2 Absorption rate:</u>

- **The solvent with fast chemical reaction kinetics (k_2)**
  - High absorption kinetics
  - Proper mixed solvents
  - Application of a catalyst
- **Henry's law constant (H_{ij}) decreasing**
  - Different composition of a base solvent
- **Improvement of the CO_2 diffusion in liquid phase**
  - Solution viscosity reduction
  - Physical/mechanical/chemical approaches
- **Amine concentration increasing (optimization)**

The number of amine molecules available at the gas-liquid interface to react with CO_2

- Interfacial/surface properties enhancement

Surface Behavior of Alkanolamine Solutions

Unloaded alkanolamine solutions

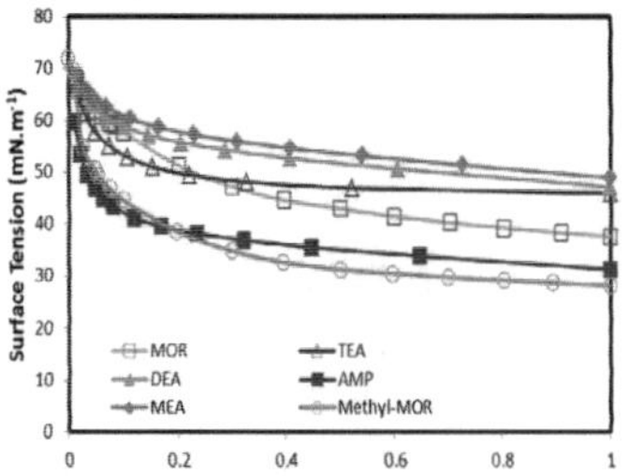

Loaded alkanolamine (MEA 30%)

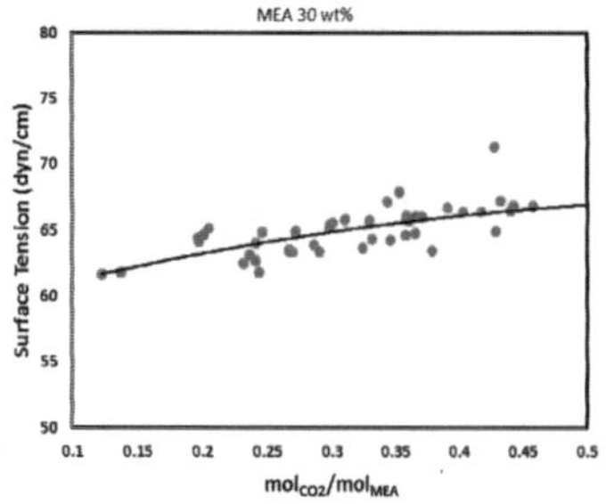

G. Vazquez, et al., J. Chem. Eng. Data 42 (1997) 57.
G. Vazquez, et al., J. Chem. Eng. Data 41 (1996) 806.
P. W. Derks, et al., J. Chem. Eng. Data 50 (2005) 1947.
Y. Maham, et al., J. Chem. Eng. Data 49 (2004) 411.
E. Alvarez, et al., J. Chem. Eng. Data 48 (2003) 32.

CO_2 + **Amine** (aq) →

- HCO_3^- (bicarbonate)
- $CO_3^{=}$ (carbonate)
- $AmCOO^-$ (carbamate)
- AmH^+ (protonated amine)
- Amine (free amine)

How the bulk concentrations of different species can affect the surface properties?

Surface and Bulk Concentrations

m^s : surface concentration

m^b : bulk concentration

Surface concentration (m^s) = *Bulk concentration* (m^b) ?

How can we determine the surface concentration (m^s) ?

Surface Concentration (Speciation) Determination ?

Measurable surface property ⇒ Surface Tension (σ) $\overset{?}{\Rightarrow}$ Surface Concentration

$\sigma = f(I, m_i^s, m_i^b, \Gamma_i)$: solution speciation

Γ_i : Excess of species "*i*" in a unit cross section

I: Ionic strength

m_i^s : concentration of species "*i*" in surface

m_i^b : concentration of species "*i*" in bulk solution

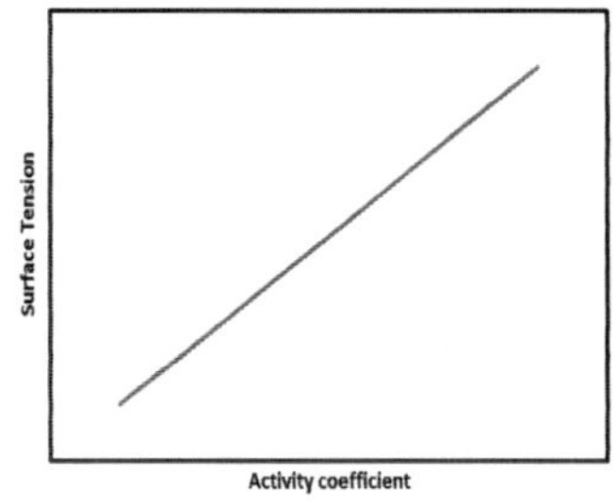

Z. Li, B. C.-Y. Lu, Chemical Engineering Science 56 (2001) 2879.
P. Wang, A. Anderko, and R. D. Young, Ind. Eng. Chem. Res. 2011, 50, 4086.
R. I. Slavchov, J. K. Novev, J. of Colloid and Interface Science 387 (2012) 234.

Theoretical Framework

$$\sigma_{sln} = \sigma_w + f(\text{bulk \& surface compositions}, I, \gamma_i) \quad (1)$$

$$\sigma_{sln} = \sigma_w + \frac{RT}{55.51\, A_w} (\phi^b \sum v_i m_i^b - \phi^s \sum g_i v_i m_i^b) \quad (2)$$

I : ionic strength

γ : activity coefficient

ϕ : osmotic coefficient

σ_w : water surface tension

A_w : partial molar area of water

g : partition coefficient $(g = m^s/m^b)$

Through equation (2), solution surface tension (σ_{sln}) is related to bulk and surface compositions.

Matin, et al., Ind. Eng. Chem. Res. 2017, 56, 5747.
Li, et al., Ind. Eng. Chem. Res. 1999, 38, 1133.

Surface Speciation Estimation

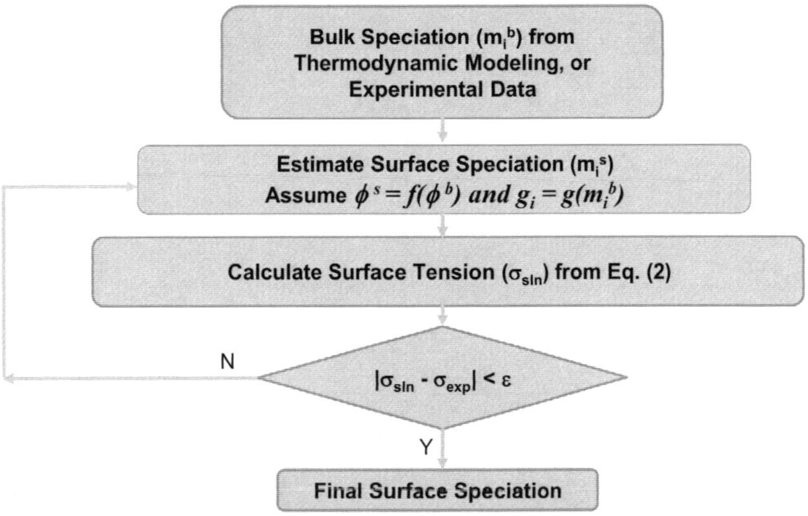

Center for Applied Energy Research AIChE Annual Meeting, Oct. 29 - Nov. 3, 2017, Minneapolis, MN caer.uky.edu 11

Predicted Surface Tension (MEA 30%, 25°C)

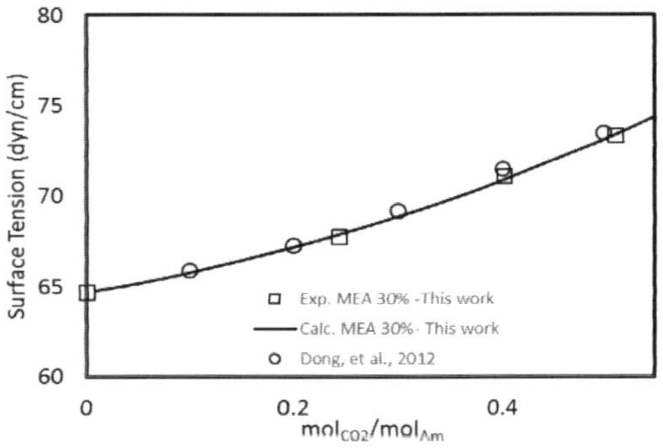

Dong, et al., Science China Chemistry, 55 (2012), 1467– 1473

Center for Applied Energy Research AIChE Annual Meeting, Oct. 29 - Nov. 3, 2017, Minneapolis, MN caer.uky.edu 12

Surface/Bulk Ratio Speciation (MEA 30%, 25°C)

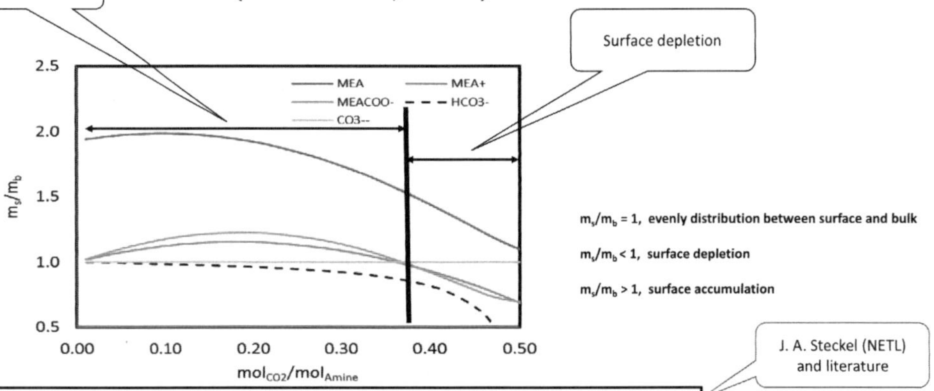

Polarizability: Carbamate (10.2 A^{o3}) > Carbonate (8.1 A^{o3}) > Protonated Amine (5.4 A^{o3}) > Bicarbonate (5.1 A^{o3})
Size: Carbamate > Protonated Amine > carbonate ≈ bicarbonate

Higher size and polarizability ⇒ higher surface tendency

Matin, et al., Ind. Eng. Chem. Res. 2017, 56, 5747.
Krei, et al., "Molecular Thermodynamics of Aqueous Solutions Containing Volatile Weak Electrolytes", Report, Lawrence Berkeley Laboratory University of California, September 1988.
dos Santos, et al., J. Phys. Chem. C 2008, 112, 19431.

Bulk and Surface Speciation (MEA 30%, 25°C)

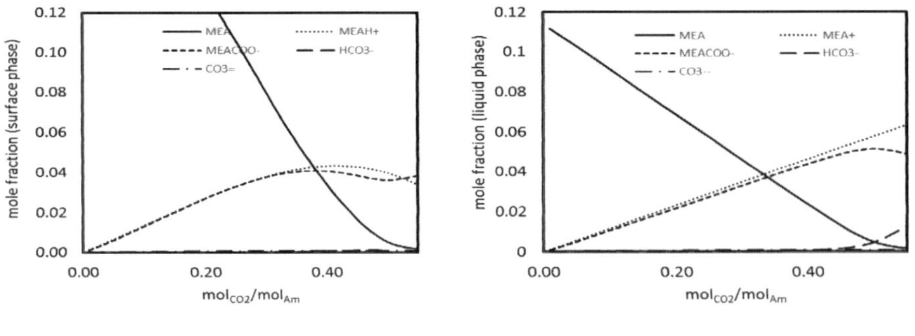

Ionic Strength and Surface Tension

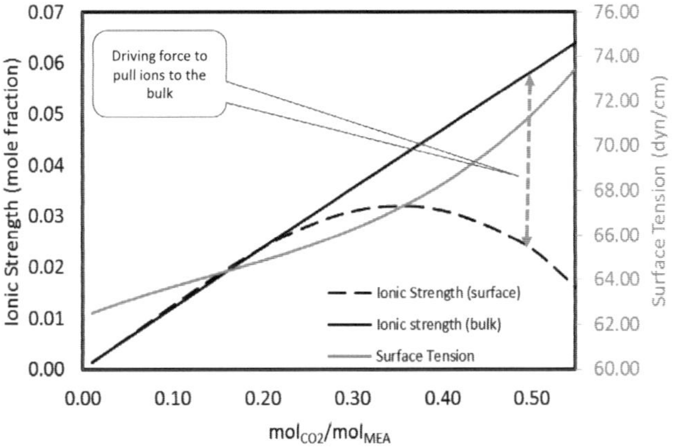

AMP and MEA solutions

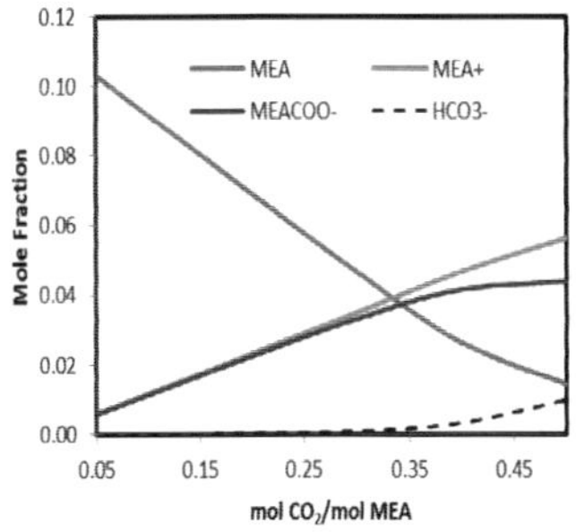

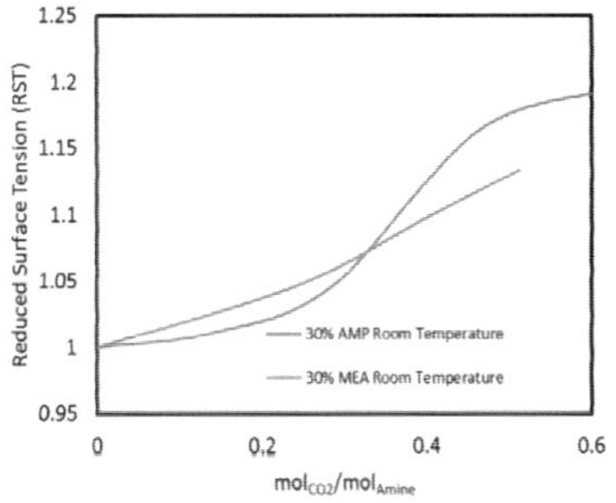

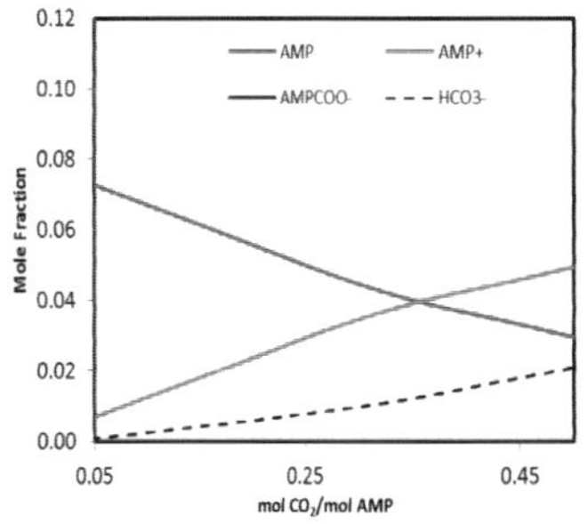

RST = (surface tension at loading α) / (surface tension of unloaded solution)

Solution Behavior Schematic (Theoretical Expectation)

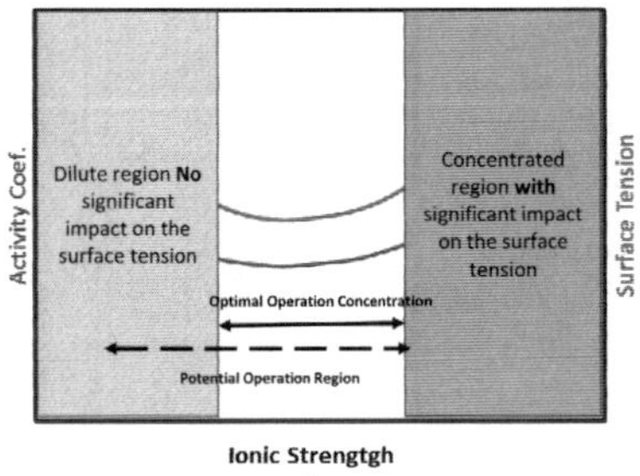

Conclusion

- Thermodynamic platform to estimate the surface phase speciation
- Bicarbonate ion has more impact on the surface tension increasing.
- At high CO_2 loadings all major species show depletion in the surface phase.
- Solvents with more bicarbonate route (hindered, tertiary, carbonated, ..) may need surfactant to decrease their surface tension.
- Higher size and polarizability mean higher surface tendency
- Carbonate polarizability > bicarbonate polarizability $\Rightarrow$ higher surface tendency for carbonate
- Size of MEA carbamate & protonated MEA > size of carbonate & bicarbonate ions $\Rightarrow$ higher surface tendency for carbamate and protonated MEA

Acknowledgement

Carbon Management Research Group Members:

DEDI

DUKE ENERGY

LG&E KU
PPL companies

AEP KENTUCKY POWER
A unit of American Electric Power

EPRI | ELECTRIC POWER RESEARCH INSTITUTE

Thank You

Questions?

Comments?

495771 Post-Combustion CO_2 Capture Using Hexamethylenediamine Activated Aqueous Sodium Glycinate Solvent

Abstract:

One of the most serious problems facing the world today is global warming and consequential climate change. Prevention of the catastrophic global worming requires an effective implementation of CO_2 capture strategy to the major point sources of the CO_2 emission. Among the major point sources, fossil fuel based power plants contributes almost 40% of the anthropogenic discharge. So the main target for implementing the CO_2 emission-reducing strategy is pointed towards this sector. Aqueous amine based absorption-regeneration process for removing CO_2 is a mature technology. Monoethanolamine (MEA), diethhanolamine (DEA) are the conventional solvents for rapid capture of CO_2. However, implementation of the conventional amine solvent for capturing CO_2 from power plant flue gas stream leads to high energy requirement, large solvent circulation rate and high solvent degradation rate, all of which in turn increases the operating and capital cost of carbon capture. These drawbacks are the main motivation for the investigation of the alternative solvents.

CO_2 absorption property of hexamethylenediamine (HMDA) activated aqueous sodium glycinate (SG) solvent is investigated in this work to explore its potential as an efficient alternative solvent for CO_2 capture. The ionic nature of aqueous SG makes it more resistant to oxidative and thermal degradation which is a desired property for the CO_2 capture from the oxygen rich power plant flue gas. It also provides certain other desirable physical properties such as low volatility, higher surface tension, etc. On the other, HMDA, a straight chain diamine (Mondal et al., 2017, 2015) will enhance the solvent characteristics due to its high kinetic and absorption capacity. All these characteristics of the component solvents motivated us to investigate aqueous (HMDA+SG) hybrid solvent.

In this work, the physico-chemical properties (density, viscosity, physical solubility), vapour-liquid equilibrium (VLE) and kinetics of CO_2 absorption in (5 mass% HMDA+ 25 mass% SG+ 70 mass% H_2O) solvent as well as (30 mass% SG+70 mass% H_2O) solvent are measured in the temperature range of 313-333K. These properties are required for the design of absorber-regenerator system for CO_2 capture from power plant flue gas. VLE of CO_2 is measured in a custom made stirred equilibrium cell set up using the method of Park and Sandall (2001). VLE data obtained in this work is presented in terms of CO_2 loading (mole CO_2 absorbed per mole total amine). The experimental VLE data shows a general trend of increasing CO_2 loading with the increase in CO_2 partial pressure. For both the solvent systems, loading capacity decreases with the increase in the temperature which indicate the exothermic nature of the chemical absorption. At 10kPa CO_2partial pressure and 313K temperature, comparison of the hybrid solvent (CO_2 loading = 0.616) with the 30% SG (CO_2 loading = 0.471) shows 20% enhancement of the loading capacity (due to addition of HMDA).

For the kinetic study, a reaction calorimeter (Model: RC1e, Mettler Toledo) set-up is used. Kinetics of CO_2 absorption is explored assuming pseudo-first-order reaction condition. Seconds order rate constant for CO_2 absorption in aqueous 30 mass% SG (592 $m^3mol^{-1}s^{-1}$) is estimated using zwitterion mechanism with fast deprotonation step and compared with the existing literature data (Lee et al., 2007) . For both the solvent systems, overall reaction rate constant is found to increase with the temperature. Overall reaction rate constant for the aqueous 30 mass% SG (2080s^{-1}) and (5mass% HMDA+25mass% SG) (10410 s^{-1}) are compared which shows 5 times enhancement of the absorption rate of the hybrid solvent compared to 30% SG solvent.

So, from the investigation of absorption properties, (HMDA+SG+H_2O) hybrid solvent appears to be a potential solvent with high CO_2 absorption rate and capacity which can be used for the CO_2 capture from the power plant flue gas.

References:

Lee, S., Song, H., Maken, S., Park, J., 2007. Kinetics of CO2 Absorption in Aqueous Sodium Glycinate Solutions. Ind. Eng. Chem. Res. 46, 1578–1583. doi:10.1021/ie061270e

Mondal, B.K., Bandyopadhyay, S.S., Samanta, A.N., 2017. Kinetics of CO2 absorption in aqueous hexamethylenediamine. Int. J. Greenh. Gas Control 56, 116–125. doi:10.1016/j.ijggc.2016.11.023

Mondal, B.K., Bandyopadhyay, S.S., Samanta, A.N., 2015. Vapor–liquid equilibrium measurement and ENRTL modeling of CO2 absorption in aqueous hexamethylenediamine. Fluid Phase Equilib. 402, 102–112. doi:10.1016/j.fluid.2015.05.033

Park, M.K., Sandall, O.C., 2001. Solubility of Carbon Dioxide and Nitrous Oxide in 50 mass % Methyldiethanolamine. J. Chem. Eng. Data 46, 166–168. doi:10.1021/je000190t

499551 Using Intellectual Property to Protect Carbon Capture Innovations

Abstract:

Carbon capture and storage (CCS) technology is increasingly an important tool to meet global emission reduction targets.[1] By the end of 2017, the number of operational large-scale CCS projects is set to rise to over 20, and the CO_2 capture capacity of operational projects will more than double since 2010.[2]

Fueling this growth, companies and universities alike are developing new and more cost-effective CCS technology.[3] Securing strong intellectual property (IP) rights is vital to protecting these innovations. In this session, IP attorneys Charles Collins-Chase and Lauren Dowty will explain what kinds of IP rights are available to protect carbon capture technologies, why those IP rights are valuable in the carbon capture field, how to obtain them, and common pitfalls to avoid. They will also analyze exemplary patents covering CCS technology and discuss case studies concerning CCS patents.

---

1 Global CCS Institute, *The Global Status of CCS 2016*, at 5, 11-12 (2016).

2 *Id.* at 11.

3 *See* Global CCS Institute, *The Global Status of CCS 2014*, at 113-14 (2014) ("Academia is the primary source of publications in general, but corporations are the driving force for patent applications in the area of carbon capture."); *see also ExxonMobil and FuelCell Energy progressing one-of-a-kind carbon capture fuel cell solution*, Energy Factor, https://energyfactor.exxonmobil.com/news/fuel-cell-site/ (Oct. 28, 2016) ("FuelCell Energy and ExxonMobil have announced the site of a pilot plant to test a novel research concept that uses fuel cell technology to capture carbon dioxide. This new technology has the potential to substantially reduce costs and lead to a more economical pathway toward large-scale application globally."); Suzanne McCarron, *Focus on Technology: Finding a way to capture carbon*, Energy Factor (Mar. 7, 2016), https://energyfactor.exxonmobil.com/perspectives/focus-technology-finding-way-capture-carbon/ ("The corporation has a working interest in more than one-third of the world's existing carbon-capture-and-storage capacity. Last year alone, we captured more than 6 million metric tons of CO_2 for storage. . . . In addition to our in-house programs, we have also supported carbon-capture-and-storage research efforts with leading universities such as Georgia Tech, MIT, Stanford, and the University of Texas. That's in addition to technical partnerships we have participated in with the U.S. Department of Energy and the European Commission to evaluate a range of carbon injection and storage technologies."). *But see* Shell, *Shell's Quest Carbon Capture and Storage Project reaches significant milestone*, Global CCS Institute (Oct. 4, 2016) https://www.globalccsinstitute.com/news/institute-updates/shell%E2%80%99s-quest-carbon-capture-and-storage-project-reaches-significant-milestone ("From the outset, any intellectual property or data generated by Quest has been publicly available, in collaboration with the governments of Alberta and Canada to help bring down future costs of CCS and encourage wider use of the technology around the world. This means that others can take the detailed engineering plans, valued at C$100 million, to help build future CCS facilities.")

UNIVERSITY OF BATH

Whole-Systems CO_2 Value Chain Modelling

A closer look at the pathways through syngas

Chris Quarton, c.j.quarton@bath.ac.uk
PhD Student
Department of Chemical Engineering
University of Bath

Supervisor: Dr Sheila Samsatli, s.m.c.samsatli@bath.ac.uk

AIChE Annual Meeting, Minneapolis
31 October 2017

UNIVERSITY OF BATH

A world-class University in a World Heritage City

50 UNIVERSITY OF BATH Established 1966

TOP 5 UK UNIVERSITY The Guardian 2018

TEF Gold

Department of Chemical Engineering

- **Top 3** for course satisfaction *(Guardian University Guide 2018)*
- **Top 5** in the UK for Chemical Engineering overall *(2018 Complete University Guide and Guardian University Guide 2018)*
- **Top 5** for teaching satisfaction *(Guardian University Guide 2018)*
- **89%** of our research is graded as either world-leading or internationally excellent *(Research Excellence Framework 2014)*

UNIVERSITY OF BATH

Outline

1. Context
2. Value Web Model
3. Modelling of UK CO_2 value chains
   - ➢ Model Setup
   - ➢ Results
4. Conclusion

UNIVERSITY OF BATH

Context

- Need to reduce carbon emissions
  - 2050 target of 20% of 1990 levels is enshrined in UK law
- This will require carbon capture
- CCS vs CCU debate is divisive
- Arguments focus around:
  - Scale
  - Sustainability
  - Economics and Efficiency
  - Infrastructure
  - Grid balancing
  - Diversification
- Aim: model both CCS and CCU on same terms to investigate conflicts, synergies, clear winners

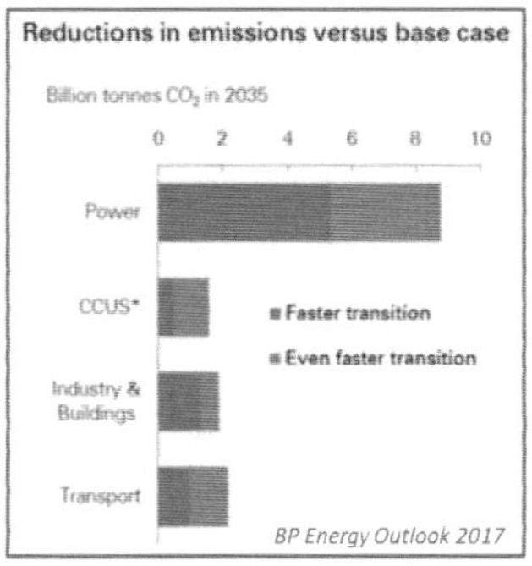

VALUE WEB MODEL

S. Samsatli, N.J. Samsatli (2017). A multi-objective MILP model for the design and operation of future integrated multi-vector energy networks capturing detailed spatio-temporal dependencies, Applied Energy (in press).

Energy systems as value webs

Traditional supply chain model

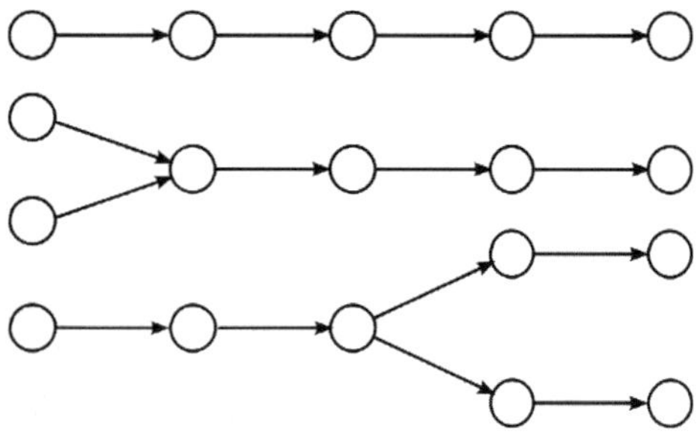

- Linear and unidirectional
- Framework used by many energy system models
- Very limiting for energy systems

Value web model

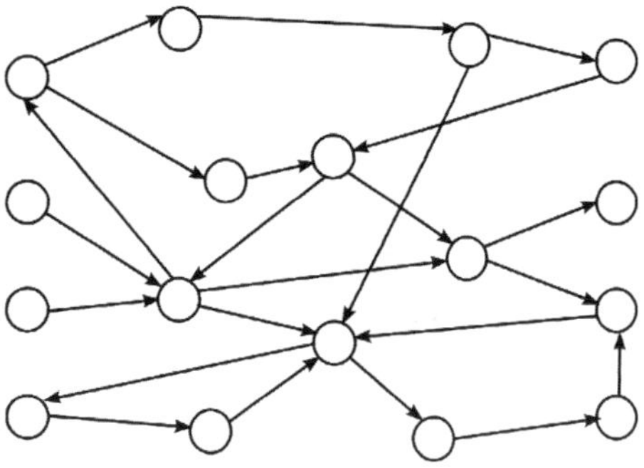

- Ideal framework for energy systems
- Interconvert energy between many different forms
- What is the best form of energy to transport and store?
- Multi-vector energy networks

UNIVERSITY OF BATH

Value Web Model

- Optimal design and operation of systems that utilise primary resources to deliver products and services
- Aids strategic planning and tactical operations (simultaneously)
- Detailed storage and transport representations
- Detailed spatial and temporal representations
- Technology-agnostic
- Data-driven framework (easy to add more networks by simply providing data)
- Can be applied to energy systems, water networks, supply chains for chemicals, pharmaceuticals, food etc., circular economy

UNIVERSITY OF BATH

Problem statement

Given:

- Spatio-temporal demands for resources and energy services
- Spatio-temporal availability of primary energy sources and raw materials
- Characteristics of each technology (e.g. CAPEX, O&M, efficiency, lifetime)

Determine:

- Network design
  - Location, number and capacity of generation/conversion and storage technologies
  - Structure of transport infrastructure network (transmission and distribution)
  - When and where to purchase/install
  - What interactions
- Network operation
  - Which resources to convert, store and transport (how much, where and when)
  - Which technologies to use at different times
  - Transport flows between different regions

Problem statement (cont...)

Subject to:

- Demand satisfaction
- Conservation and other physical laws
- Constraints on resources (e.g. land, water), costs and GHG emissions
- Technological constraints (e.g. tech. availability, build rates)
- Social and political constraints (e.g. siting of specific techs)

Objective:

- Minimise cost
- Minimise environmental impact (e.g. GHG emissions)
- Maximise value
- Any combination of the above

Temporal representation

- Long-term strategic decisions
- Short-term operational issues (intermittency, dynamics of energy storage)

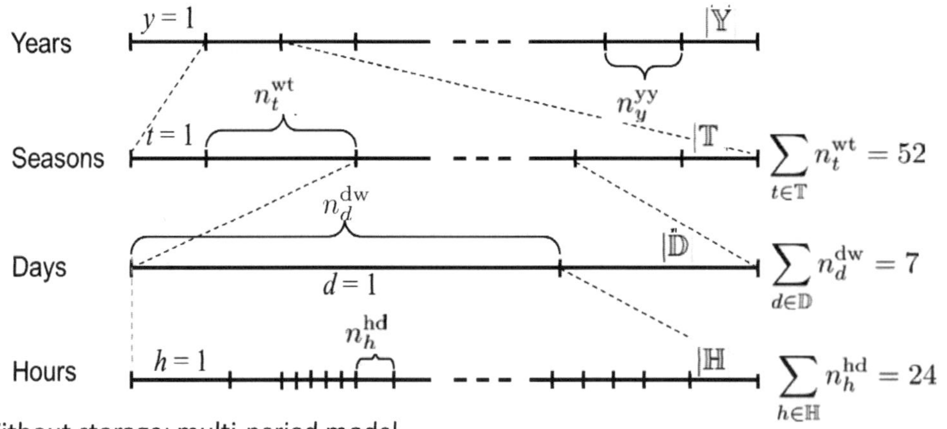

Without storage: multi-period model

With storage: dynamic model; extra variables for initial inventories; extra constraints to link inventories within and between time levels

S. Samsatli, N.J. Samsatli (2015). Computers & Chemical Engineering 80, 155-176, 0098-1354.

Spatial representation

- The region under study divided into a number of zones
- Each zone may:
  - Be of any shape and size
  - Have dynamic demands for various resources
  - Contain some resources that are available in varying quantities
  - Host technologies for conversion and storage of resources
  - Be connected with other cells via transport infrastructures
  - Import or export resources

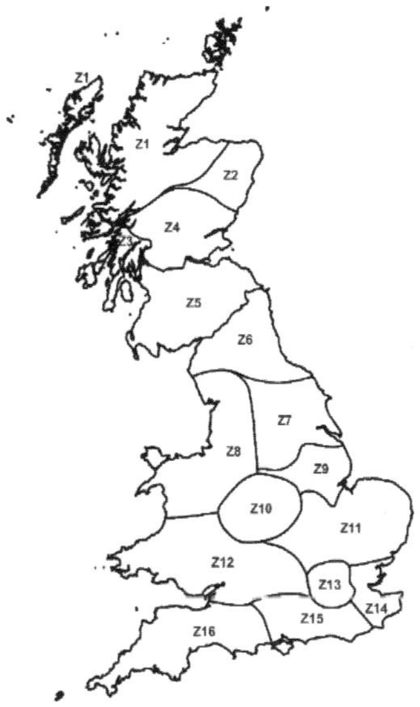

National Grid's SYS study zones

Integrated energy conversion pathways

- Represented using Resource-Technology Networks (RTN)
- Allows representation of "web" pathways
  - convert any resource to any other, including recycles (circular chains)
  - store and transport resources at any stage in the chain
- Consider a wide range of different feedstocks and technologies
  - different operating modes to generate different energy vectors
    - heat, electricity, transport fuel etc.

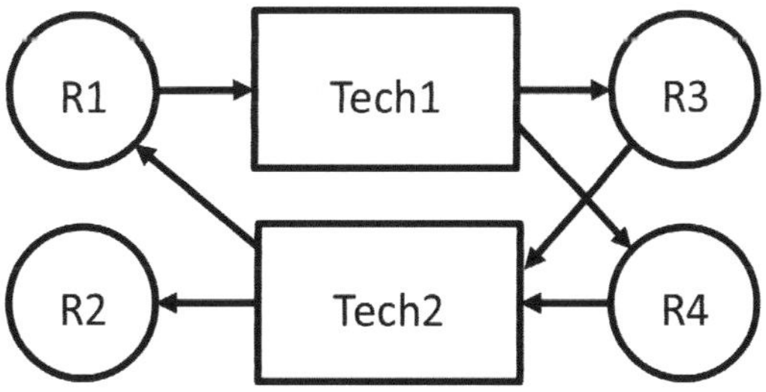

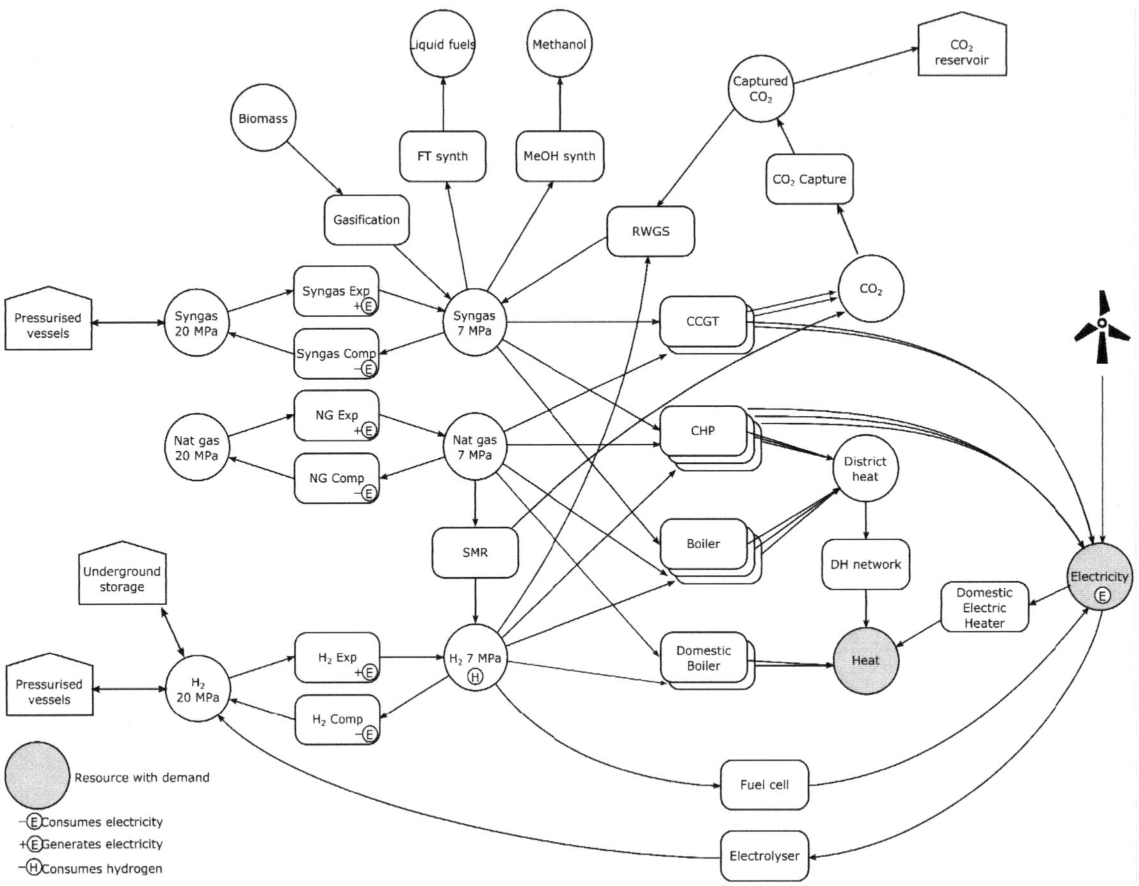

UNIVERSITY OF BATH

Modelling storage

Set of **storage operations** to store resource r_1.

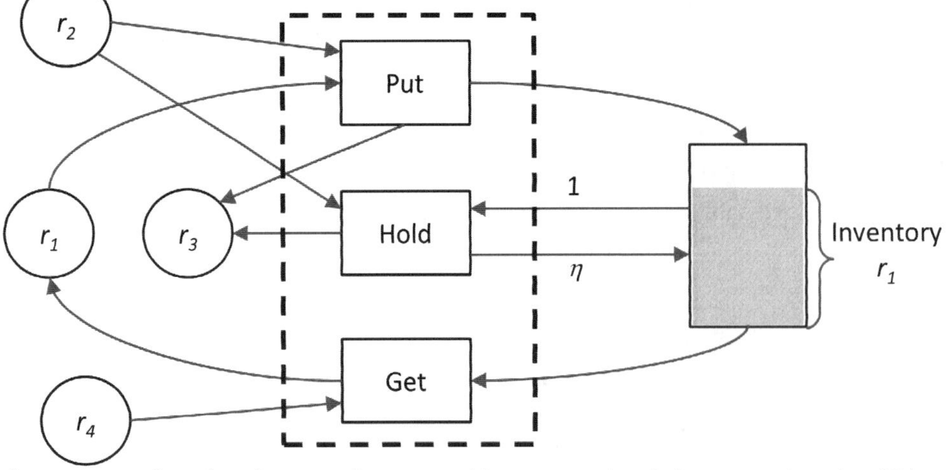

The "put" operation transfers r_1 from the zone to the store, requiring some r_2 and producing some wastes r_3 (e.g. CO_2). The "hold" operation maintains r_1 in storage which could be at less than 100% efficiency, the losses being converted to r_3 ; this operation may also require some r_2. Finally, the "get" operation retrieves r_1 from storage and delivers it to the zone, requiring some r_4.

S. Samsatli, N.J. Samsatli (2015). Computers & Chemical Engineering 80, 155-176, 0098-1354.

Modelling transport

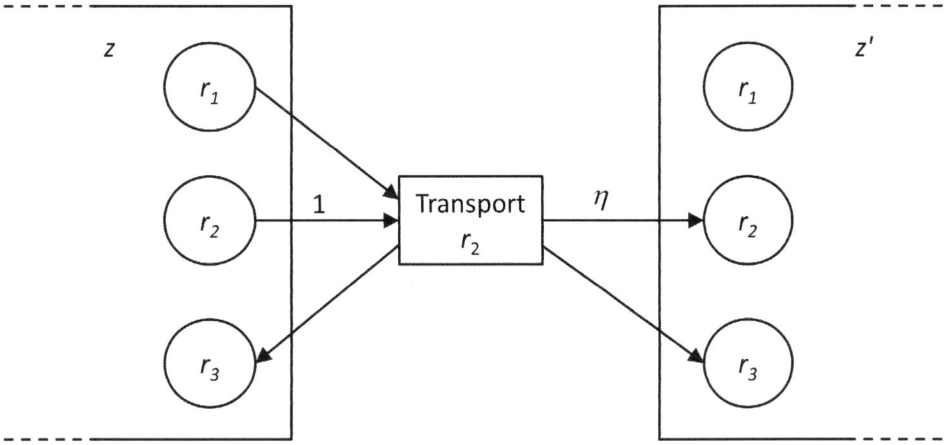

Resource r_2 is transported from zone z to zone z', which requires r_1 from zone z and results in waste r_3 being generated in both zones

S. Samsatli, N.J. Samsatli (2015). Computers & Chemical Engineering 80, 155-176, 0098-1354.

Transmission technologies connect the networks between different zones

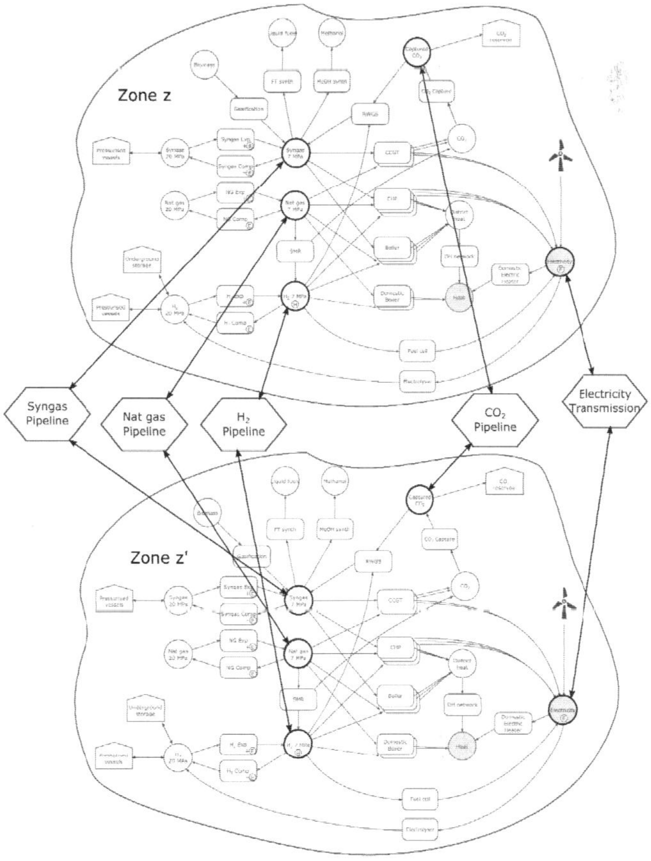

Resource balance and availability

Resource balance:

$$U_{rzhdty} + M_{rzhdty} + P^{\mathrm{h}}_{rzhdty} + P^{\mathrm{d}}_{rzdty} + S_{rzhdty} + Q_{rzhdty} \geq$$
$$D_{rzhdty} + X_{rzhdty} \qquad \forall r \in \mathbb{R},\, z \in \mathbb{Z},\, h \in \mathbb{H},\, d \in \mathbb{D},\, t \in \mathbb{T},\, y \in \mathbb{Y}$$

Resource utilisation: $U_{rzhdty} \leq u^{\max}_{rzhdty} \qquad \forall r \in \mathbb{R},\, z \in \mathbb{Z},\, h \in \mathbb{H},\, d \in \mathbb{D},\, t \in \mathbb{T},\, y \in \mathbb{Y}$

Wind power potential: $u^{\max}_{\mathrm{Elec},zhdt} = 0.5 \times 10^{-6} \eta \rho^{\mathrm{air}} \pi N^{\mathrm{WT}}_{zy} \left(R^{\mathrm{WT}}\right)^2 v^3_{zhdty}$

Biomass potential: $\sum_{hd} U_{\mathrm{Bio},zhdty} n^{\mathrm{hd}}_h n^{\mathrm{dw}}_d n^{\mathrm{wt}}_t \leq A^{\mathrm{Bio}}_{zy} Y^{\mathrm{Bio}}_{zty} \qquad \forall z \in \mathbb{Z},\, t \in \mathbb{T},\, y \in \mathbb{Y}$

Land footprint constraints:

Wind turbines: $\pi \left(5R^{\mathrm{WT}}\right)^2 N^{\mathrm{WT}}_{zy} \leq A^{\mathrm{WT,max}}_{zy} \qquad \forall z \in \mathbb{Z},\, y \in \mathbb{Y}$

Biomass (each zone): $A^{\mathrm{Bio}}_{zy} \leq f_{zy} A^{\mathrm{Bio,max}}_{zy} \qquad \forall z \in \mathbb{Z},\, y \in \mathbb{Y}$

Biomass (GB): $\sum_z A^{\mathrm{Bio}}_{zy} \leq f^{\mathrm{GB}}_y \sum_z A^{\mathrm{Bio,max}}_{zy} \qquad \forall y \in \mathbb{Y}$

Constraints for conversion technologies

Net production of resource r:

$$P^{\mathrm{h}}_{rzhdty} = \sum_{p \in \mathbb{P}^{\mathrm{h}}} \mathscr{P}^{\mathrm{h}}_{pzhdty} \alpha_{rpy}$$
$$P^{\mathrm{d}}_{rzdty} = \sum_{p \in \mathbb{P}^{\mathrm{d}}} \mathscr{P}^{\mathrm{d}}_{pzdty} \alpha_{rpy}$$

Min/max rate of operation:

$$N^{\mathrm{P}}_{pzy} p^{\min}_p \leq \mathscr{P}^{\mathrm{h}}_{pzhdty} \leq N^{\mathrm{P}}_{pzy} p^{\max}_p$$
$$N^{\mathrm{P}}_{pzy} p^{\min}_p \leq \mathscr{P}^{\mathrm{d}}_{pzdty} \leq N^{\mathrm{P}}_{pzy} p^{\max}_p$$

Technology balance: $N^P_{pzy} = N^P_{pz,y-1} + NI^P_{pzy} - NR^P_{pzy}$

Retirements: $NR^{\mathrm{P}}_{pzy} = \sum_{y'} RF^P_{py'y} NI^P_{pzy'}$

Max build rate: $\sum_z NI^{\mathrm{P}}_{pzy} \leq BR_{py}$

- Number of domestic technologies is continuous instead of integers for computational efficiency (i.e. fraction of penetration)

UNIVERSITY OF BATH

Constraints for transport technologies

Net transport of resource r:

$$Q_{rzhdty} = \sum_{z'|\nu_{z'z}=1} \sum_{l \in \mathbb{L}} \left[\left(\bar{\tau}_{lr,\text{dst},y} + \hat{\tau}_{lr,\text{dst},y} d_{z'z} \right) \mathscr{Q}_{lz'zhdty} \right] + \sum_{z'|\nu_{zz'}=1} \sum_{l \in \mathbb{L}} \left[\left(\bar{\tau}_{lr,\text{src},y} + \hat{\tau}_{lr,\text{src},y} d_{zz'} \right) \mathscr{Q}_{lzz'hdty} \right]$$

Max rate of operation: $\mathscr{Q}_{lzz'hdty} \leq \sum_{b \in \mathbb{B}} q_l^{\max} N^{\text{B}}_{bzz'y} |_{LB_{lb}=1 \wedge \nu_{zz'}=1}$

Max infrastructure capacity: $\sum_{l \in \mathbb{L}} \mathscr{Q}_{lzz'hdty} LB_{lb} \leq b_b^{\max} N^{\text{B}}_{bzz'y}$

Infrastructure balance: $N^{\text{B}}_{bzz'y} = N^{B}_{bzz',y-1} + NI^{\text{B}}_{bzz'y}$

Bi-directional links:

$$N^{\text{B}}_{bzz'y} = N^{\text{B}}_{bz'zy} \qquad \forall b \in \mathbb{B} | \beta_b = 1,\ z \neq z' \in \mathbb{Z}$$

UNIVERSITY OF BATH

Constraints for storage technologies

Max charging and discharging rates:

$$\mathscr{S}^{\text{put}}_{szhdty} \leq N^{\text{S}}_{szy} s_s^{\text{put,max}} a_{sz}$$

$$\mathscr{S}^{\text{get}}_{szhdty} \leq N^{\text{S}}_{szy} s_s^{\text{get,max}} a_{sz}$$

Net inflow of resource r due to storage operations:

$$S_{rzhdty} = \sum_{s} \left(\mathscr{S}^{\text{put}}_{szhdty} \sigma^{\text{put}}_{sr,\text{src},y} + \mathscr{S}^{\text{hold}}_{szhdty} \sigma^{\text{hold}}_{sr,\text{dst},y} + \mathscr{S}^{\text{get}}_{szhdty} \sigma^{\text{get}}_{sr,\text{dst},y} \right)$$

Inventory levels:

$$I_{szhdty} = n_h^{\text{hd}} \sum_{r} \left(\mathscr{S}^{\text{put}}_{szhdty} \sigma^{\text{put}}_{sr,\text{dst},y} + \mathscr{S}^{\text{hold}}_{szhdty} \sigma^{\text{hold}}_{sr,\text{src},y} + \mathscr{S}^{\text{get}}_{szhdty} \sigma^{\text{get}}_{sr,\text{src},y} \right)$$

Changes in inventory:

Daily: $\delta^{\text{d}}_{szdty} = I_{sz,|\mathbb{H}|,dty} - I^{0,\text{sim}}_{szdty}$

Weekly: $\delta^{\text{t}}_{szty} = \sum_{d} \delta^{\text{d}}_{szdty} n_d^{\text{dw}}$

Yearly: $\delta^{\text{y}}_{szy} = \sum_{t} \delta^{\text{t}}_{szty} n_t^{\text{wt}}$

Cyclic constraint: $\delta^{\text{y}}_{szy} = 0$

UNIVERSITY OF BATH

Constraints for storage technologies (cont…)

Link inventories between different time levels:

$$I^{0,\mathrm{act}}_{szdty} = I^{0,\mathrm{act}}_{sz,d-1,ty} + n^{\mathrm{dw}}_{d-1}\delta^{\mathrm{d}}_{sz,d-1,ty}$$

$$I^{0,\mathrm{act}}_{sz,1,ty} = I^{0,\mathrm{act}}_{sz,1,t-1,y} + n^{\mathrm{wt}}_{t-1}\delta^{\mathrm{t}}_{sz,t-1,y}$$

$$I^{0,\mathrm{act}}_{sz,1,1,y} = I^{0,\mathrm{act}}_{sz,1,1,y-1} + n^{\mathrm{yy}}_{y-1}\delta^{\mathrm{y}}_{sz,y-1}$$

Shift inventory levels to match average:

$$I^{0,\mathrm{sim}}_{szdty} = I^{0,\mathrm{act}}_{szdty} + \left[\left(n^{\mathrm{dw}}_{d}-1\right)\delta^{\mathrm{d}}_{szdty} + \left(n^{\mathrm{wt}}_{t}-1\right)\delta^{\mathrm{t}}_{szty} + \left(n^{\mathrm{yy}}_{y}-1\right)\delta^{\mathrm{y}}_{szy}\right]/2$$

$$\mathscr{S}^{\mathrm{hold}}_{sz,1,dty} = I^{0,\mathrm{sim}}_{szdty}/n^{\mathrm{hd}}_{1} \qquad \mathscr{S}^{\mathrm{hold}}_{szhdty} = I_{sz,h-1,dty}/n^{\mathrm{hd}}_{h}$$

Storage capacity constraints:

$$s^{\mathrm{hold,min}}_{s} N^{S}_{szy} a_{sz} \le I_{szhdty} \pm \left[\left(n^{\mathrm{dw}}_{d}-1\right)\delta^{\mathrm{d}}_{szdty} \pm \left(n^{\mathrm{wt}}_{t}-1\right)\delta^{\mathrm{t}}_{szty} \pm \left(n^{\mathrm{yy}}_{y}-1\right)\delta^{\mathrm{y}}_{szy}\right]/2 \le s^{\mathrm{hold,max}}_{s} N^{S}_{szy} a_{sz}$$

UNIVERSITY OF BATH

Constraints for CCS and CCU

Rate of utilisation or storage of CO_2:

$$CCUS_{zhdty} = -\sum_{p\in\mathbb{P}^{\mathrm{CCUS}}} \alpha_{p,\mathrm{CO}_2^{\mathrm{cap}}} \mathscr{P}^{\mathrm{d}}_{pzdty}$$

Total impact from carbon credits:

$$\mathscr{I}^{\mathrm{CCUS}}_{iy} = \varsigma \sum_{zhdt} D^{\mathrm{OM}}_{y} V^{\mathrm{CCUS}}_{iy} CCUS_{zhdty} n^{\mathrm{hd}}_{h} n^{\mathrm{dw}}_{d} n^{\mathrm{wt}}_{t}$$

$$V^{\mathrm{CCUS}}_{iy} = \begin{cases} \text{the } CO_2 \text{ price if } i \text{ is cost} \\ 1 \text{ if } i \text{ is } CO_2 \text{ emissions} \end{cases}$$

Objective function

$$\text{Minimise } Z = \sum_{iy} \omega_i \Big(\mathscr{I}_{iy}^{\mathrm{W}} + \mathscr{I}_{iy}^{\mathrm{P}} + \mathscr{I}_{iy}^{\mathrm{Q}} + \mathscr{I}_{iy}^{\mathrm{S}} + \mathscr{I}_{iy}^{\mathrm{w}} + \mathscr{I}_{iy}^{\mathrm{p}}$$
$$+\mathscr{I}_{iy}^{\mathrm{q}} + \mathscr{I}_{iy}^{\mathrm{s}} + \mathscr{I}_{iy}^{\mathrm{r}} + \mathscr{I}_{iy}^{\mathrm{m}} + \mathscr{I}_{iy}^{\mathrm{x}} - \mathscr{I}_{iy}^{\mathrm{Rev}} - \mathscr{I}_{iy}^{\mathrm{CCUS}} \Big)$$

Impacts $\forall i \in \mathbb{I},\ y \in \mathbb{Y}$

	Capital	O&M
Wind turbines	$\mathscr{I}_{iy}^{\mathrm{W}} = \varsigma \sum_{z} D_y^{\mathrm{C}} C_{iy}^{\mathrm{WT}} NI_{zy}^{\mathrm{WT}}$	$\mathscr{I}_{iy}^{\mathrm{w}} = \varsigma \sum_{z} D_y^{\mathrm{OM}} F_{iy}^{\mathrm{WT}} N_{zy}^{\mathrm{WT}}$
Conversion techs	$\mathscr{I}_{iy}^{\mathrm{P}} = \varsigma \sum_{pz} D_{piy}^{\mathrm{C}} C_{piy}^{\mathrm{P}} NI_{pzy}^{P}$	$\mathscr{I}_{iy}^{\mathrm{p}} = \varsigma \sum_{pz} D_y^{\mathrm{OM}} F_{piy}^{\mathrm{P}} N_{pzy}^{P}$
Transport techs	$\mathscr{I}_{iy}^{\mathrm{Q}} = 0.5\varsigma \sum_{bzz'} D_{biy}^{\mathrm{C}} C_{biy}^{\mathrm{B}} NI_{bzz'y}^{B} d_{zz'})$	$\mathscr{I}_{iy}^{\mathrm{q}} = 0.5\varsigma \sum_{bzz'} D_y^{\mathrm{OM}} F_{biy}^{\mathrm{B}} N_{bzz'y}^{B} d_{zz'}$
Storage techs	$\mathscr{I}_{iy}^{\mathrm{S}} = \varsigma \sum_{sz} D_{siy}^{\mathrm{C}} C_{siy}^{\mathrm{S}} NI_{szy}^{S}$	$\mathscr{I}_{iy}^{\mathrm{s}} = \varsigma \sum_{sz} D_y^{\mathrm{OM}} F_{siy}^{\mathrm{S}} N_{szy}^{S}$

Primary resource production $$\mathscr{I}_{iy}^{\mathrm{r}} = \varsigma \sum_{zhdt} D_y^{\mathrm{OM}} \left(RP_{izty}^{\mathrm{Bio}} A_{zy}^{\mathrm{Bio}} Y_{zty}^{\mathrm{Bio}} + \sum_{r \neq \mathrm{Bio}} RP_{rizhdty} U_{rzhdty} n_h^{\mathrm{hd}} n_d^{\mathrm{dw}} n_t^{\mathrm{wt}} n_y^{\mathrm{yy}} \right)$$

Export/import $$\mathscr{I}_{iy}^{\mathrm{x}} = \varsigma \sum_{rzhdt} D_y^{\mathrm{OM}} V_{riy}^{\mathrm{X}} X_{rzhdty} n_h^{\mathrm{hd}} n_d^{\mathrm{dw}} n_t^{\mathrm{wt}} \qquad \mathscr{I}_{iy}^{\mathrm{m}} = \varsigma \sum_{rzhdt} D_y^{\mathrm{OM}} V_{riy}^{\mathrm{M}} M_{rzhdty} n_h^{\mathrm{hd}} n_d^{\mathrm{dw}} n_t^{\mathrm{wt}}$$

Revenue $$\mathscr{I}_{iy}^{\mathrm{Rev}} = \varsigma \sum_{rzhdt} D_y^{\mathrm{OM}} V_{riy}^{\mathrm{Rev}} D_{rzhdty} n_h^{\mathrm{hd}} n_d^{\mathrm{dw}} n_t^{\mathrm{wt}}$$

UNIVERSITY OF BATH

MODELLING OF UK CO_2 VALUE CHAINS

Modelling of UK CO_2 Value Chains

Given:

- Spatio-temporal demands for domestic heat and electricity
- Spatio-temporal supplies of wind, natural gas, and biomass (Miscanthus)
- Carbon capture, storage and utilisation technologies
- What might the optimal 2050 network be?
- 12 timesteps per day, 1 day per season, 4 seasons per year, 1 year type to 2050

Carbon Pathways

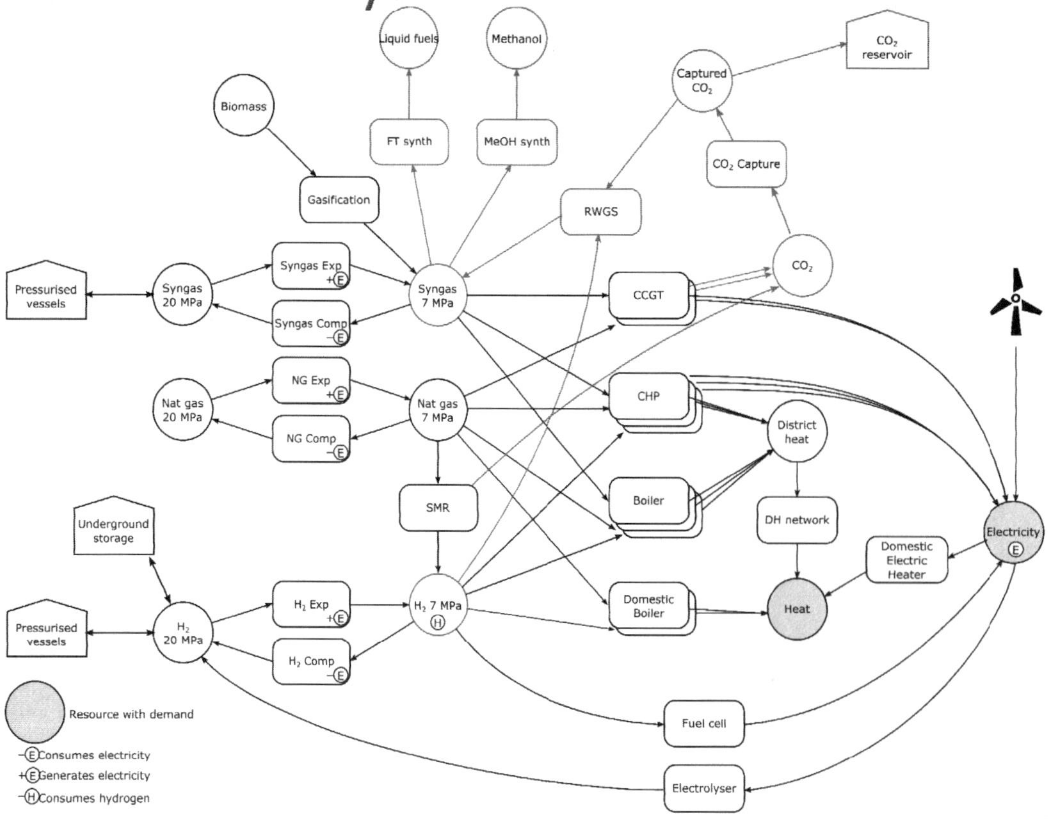

UNIVERSITY OF BATH

Carbon Storage

- 4 potential sequestration sites included
- Sites associated with nearest zone
- Model chooses whether to build CO_2 pipelines between zones

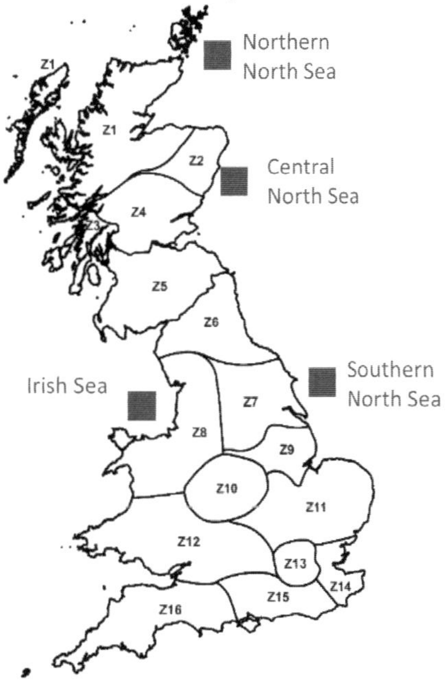

UNIVERSITY OF BATH

Baseline Case: Maximise NPV

- Wind turbines favoured over CCGTs due to long time horizon and production cost of natural gas
- Neither CCU nor CCS is profitable

NPV = £143bn (profit)
CO_2 = 1565 $MtCO_2$

Wind Turbine Installed Capacity (MW @10m/s)

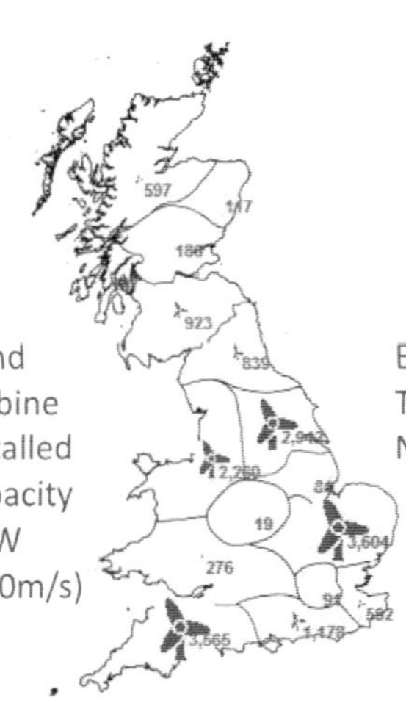

Electricity Transmission Network

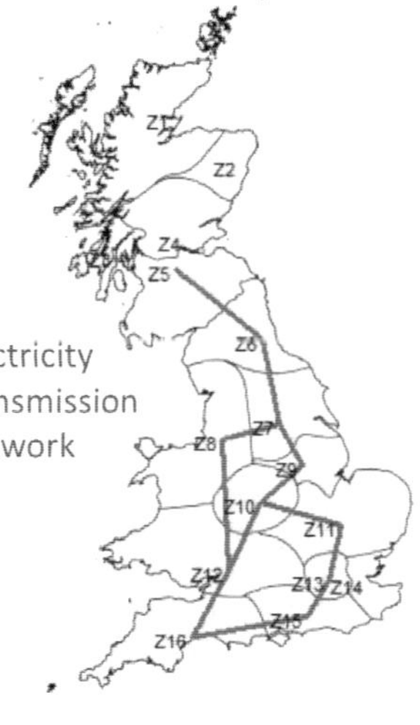

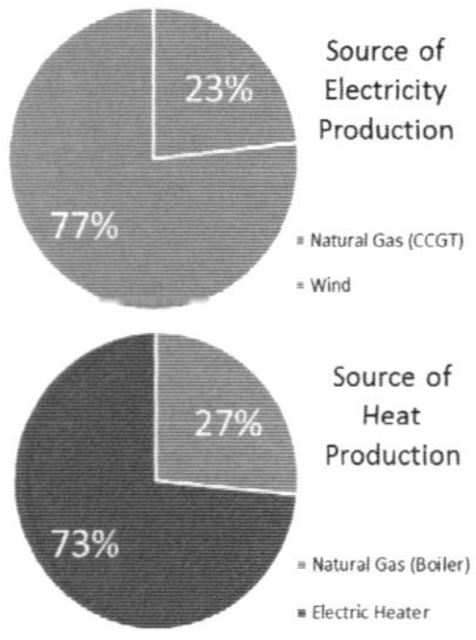

Constraint on Carbon Emissions

- CO_2 constrained to equivalent of 2050 CO_2 target (compared to baseline run)
- Model chooses CCU (MeOH synthesis) – no CCS
- All H_2 produced by electrolysis – some H_2 storage also built
- CO_2 pipeline network built
- CCU focussed in a few zones

NPV = £111bn (profit)
CO_2 = 508 $MtCO_2$

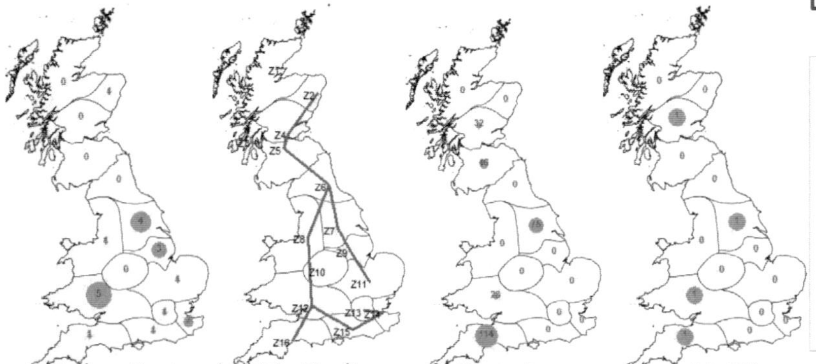

Carbon Capture Facilities | CO_2 Pipeline Network | Electrolysers (70MW) | MeOH Synthesis Plants

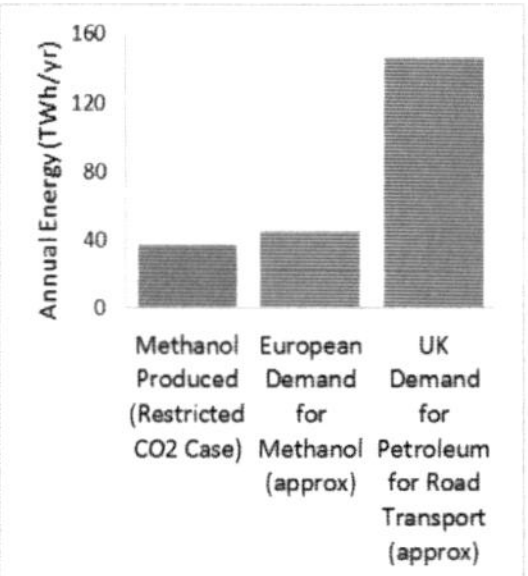

Carbon Credit

- No constraint on emissions
- £100 credit for each ton of CO_2 captured and utilised or stored
- CCS & CCU (MeOH synthesis) both used
- H_2 produced predominantly by SMR - CO_2 and H_2 pipeline networks both built
- Small overall CO_2 benefit given the large credit – due to SMR

NPV = £231bn (profit)
CO_2 = 1291 $MtCO_2$

MeOH plants (green) and Carbon Storage (red)

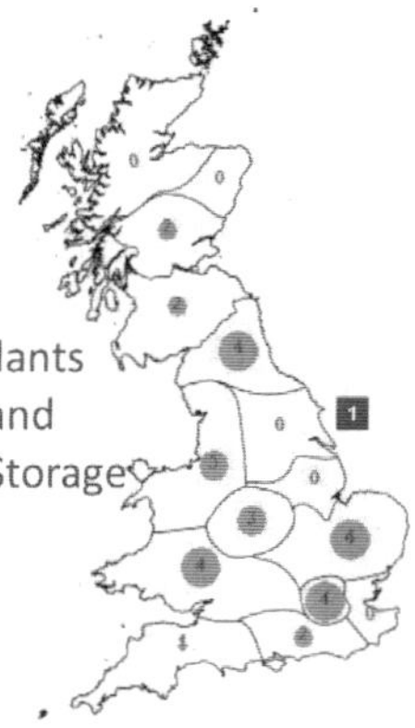

CO_2 Pipeline Network

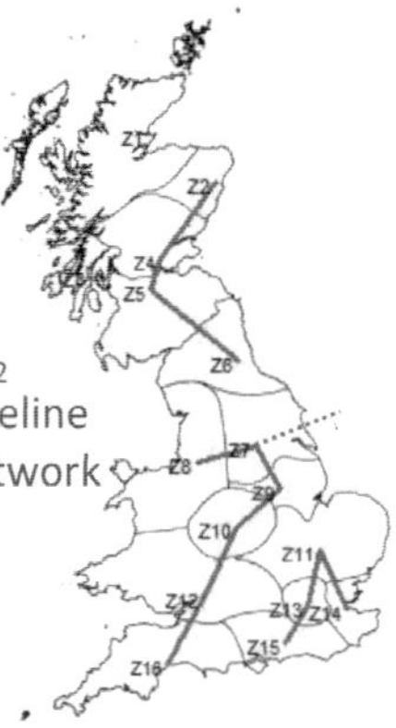

H_2 Pipeline Network

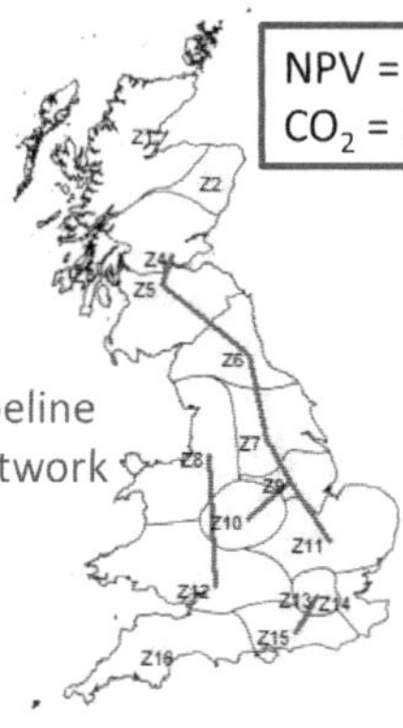

UNIVERSITY OF BATH

Inflated CCU Product Prices

- No carbon credit, no constraint on emissions
- MeOH & Synthetic Hydrocarbons prices doubled (compared to market price)
- CCU (MeOH synthesis) focussed in a few zones – no CCS
- SMR and electrolysis both used for H_2 production
- CO_2 levels are higher than the baseline – due to SMR

NPV = £167bn (profit)
CO_2 = 1617 $MtCO_2$

MeOH plants built

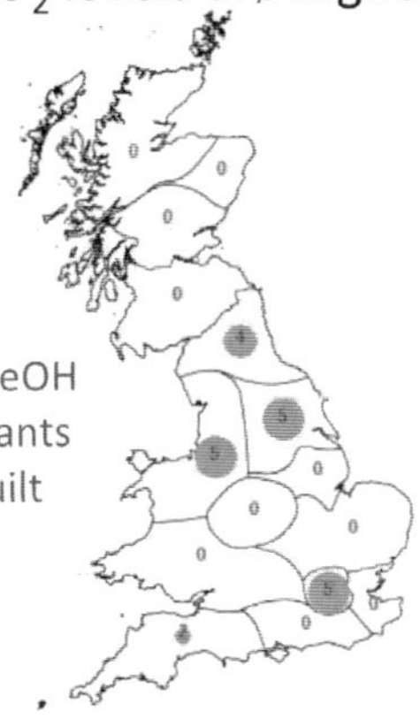

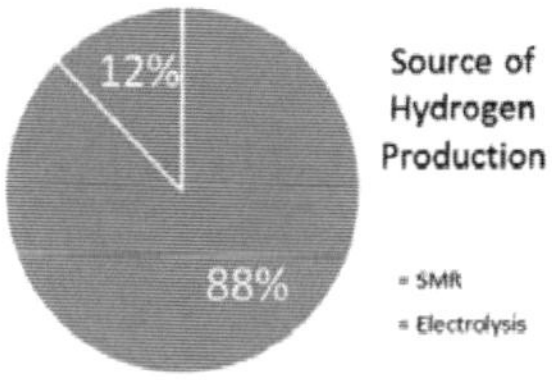

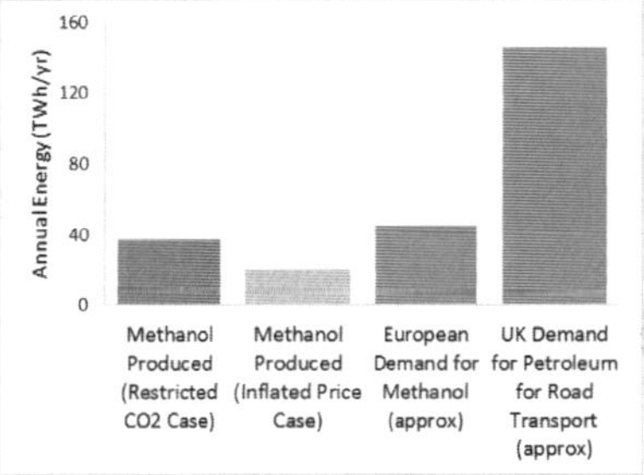

UNIVERSITY OF BATH

Conclusion

- Value Web Model a powerful framework for integrated multi-vector energy networks
  - Spatio-temporal nature matches up primary resource availability and energy service demands which are distributed in space and time
- CCU and CCS both have potential for carbon mitigation
  - CCU has economic advantage over CCS
  - CCU may become economically valid independent of environmental benefits
  - Current CCU products market may be too small but has expansion potential
  - Important to choose policy tools correctly to encourage genuine reductions in total carbon emissions

UNIVERSITY OF BATH

Acknowledgments

We would like to acknowledge the Department for Business, Energy and Industrial Strategy and the Engineering and Physical Sciences Research Council for their support and funding of this work.

Please note that the work presented here represents the views of the authors only and not necessarily those of the sponsors.

EPSRC
Engineering and Physical Sciences Research Council

Department for Business, Energy & Industrial Strategy

500289 Effects of Household Co-Solvents on the Solubility and Oxidation of Trichloroethylene (TCE)

Abstract:

Household co-solvents were chosen to enhance the solubility of trichloroethylene (TCE), a hazardous organic solvent, in water. Two surfactants, Tide and OxiClean, and three alcohols, methanol, ethanol and isopropanol, were selected as co-solvents and analyzed using batch systems. It was determined that isopropanol and OxiClean were the best alcohol and surfactant co-solvents, respectively. 5,300 PPM of TCE dissolved in 500,000 PPM isopropanol/water mixture and 1,600 PPM dissolved in 5,000 PPM OxiClean/water mixture. This is compared to TCE solubility in water at about 1,200 PPM. TCE was totally dissolved by 9,000 PPM of OxiClean. Ternary diagrams also displayed higher miscibility of TCE in water with less required OxiClean concentration, compared to the concentration of other co-solvents tested.

Degradation of TCE by $KMnO_4$ was investigated with and without co-solvents. Experiments were conducted to evaluate simultaneous solubility enhancement and degradation of TCE. While most co-solvents enhanced TCE degradation, OxiClean increased TCE degradation by increasing solubility in the mixture but it also consumed $KMnO_4$. The effects of OxiClean, $KMnO_4$, and TCE on each other were studied. Kinetic relationships between the solvents were determined and an overall rate law was established for the reaction.

These bench scale experiments provide insight into parameters that will effect pump and treat remediation methods for TCE in soil and water contamination sites.

497709 Adsorptive Removal of Phosphorus from Wastewater Using Raw and Engineered Biochars

Abstract:

Releasing high concentration of nutrients (i.e., nitrogen, phosphorus) to environments has caused various environmental problems such as occurrence of algal bloom, eutrophication and lethal level of nitrate in surface and ground water leading to deteriorating water quality and threating human health. Adsorption of nutrients onto various carbon materials among various methods has been actively studied for eliminating nutrients from agricultural and municipal wastewater. Among various adsorbents biochar is a "charcoal-like" carbon material made from gasification or pyrolysis of various solid wastes including biosolids, sludges, agricultural residues and animal manure. Since it has rich nutrients, various functional groups and chemically inert to environments, its application has been extended to soil amendment, water treatment, gas separation, supports for catalysts and fuel sources.

This study focused on recovery of phosphorus from synthetic and dairy wastewater using the dairy manure-derived biochar including raw biochar, CaO-immobilized biochar (CaO-BC) and MgO-immobilized biochar (MgO-BC). The MgO-BC and CaO-BC were prepared by reacting $CaCl_2$ or $MgCl_2$ solution to the raw biochar at ambient temperature under vigorous stirring for 24 h followed by drying at 80 ^oC and calcination at 120 ^oC for 3 h. The adsorption experiments using a synthetic wastewater containing 30 ppm phosphate showed that the CaO-BC removed 89-100% of initial phosphate which was higher than those by the raw biochar and the MgO-BC by a factor of 1.1 to 1.7. When the dairy wastewater (lagoon effluent) was treated by the raw biochar, MgO-BC and Ca-BC, these biochars showed high removal of phosphate (79-90%) and low removal of ammonium.

For better understanding of adsorption characteristics, the effects of pH on adsorption of phosphate were investigated using the CaO-BC and the synthetic wastewater with 30 ppm phosphate. Interestingly the adsorption of phosphate onto CaO-BC was highest at higher pH (7-8) and lower at acidic conditions which was different from others' studies. While the mechanisms associated with adsorption of phosphate will need to be investigated, multiple interactions including ligand complexation, hydrogen bonding and π-π interaction were thought to act on the adsorption of phosphate onto the CaO-BC. The results from the adsorption isotherm experiments supported that Langmuir isotherm was the better-fitted model for adsorption of phosphorus onto the CaO-BC indicating monolayer adsorption of phosphorus onto the biochar. The adsorption capacities of phosphorus onto the CaO-BC were found to be 1 – 6 mg phosphorus per g CaO-BC at the selected conditions. The adsorption capacities were similar and higher than those by the biochars from different feedstock and preparation methods. The adsorption kinetics of phosphorus on the CaO-BC revealed that the pseudo-second order kinetic model was the one to better fit to the experimental results that indicated the chemical interaction-driven adsorption.

Further studies will include the detailed mechanisms of adsorption, thermodynamic analysis associated with adsorption and the column studies for practical application in wastewater treatment.

SIMULATION OF WASTE PLASTIC PYROLYSIS PROCESS USING ASPEN PLUS

Guided by
Dr.Hema Ramachandran
Asst.Professor, TPLC

Jithu Narayanan K
M.Tech T.E.

▶ 1

CONTENTS

▶ 2

INTRODUCTION

- In the present scenario, there is a constant drive for employing profit biased treatment techniques for waste plastics.
- Pyrolysis is such an effective technique done on a beneficiary basis, which involves plastic treatment under anaerobic conditions.
- In case of extensive practical applications, we make use of process simulation for predicting the composition of the yield and optimum process conditions of the pyrolysis reaction.
- Pyrolysis process is simulated and the results of the process simulation are obtained.
- For the simulation purpose Aspen Plus(Version 8.4) software is used.

▶ 3

OBJECTIVE

- To find the optimum process parameters and the optimal yield of the pyrolysis.
- To perform the process with kinetic & equilibrium reactor models.
- To analyze the composition of gas formed by the simulation of thermo-chemical process.
- To compare the output formed by simulation with experimental results.

▶ 4

RESEARCH METHODOLOGY

1. Study on experimental setup
   - Draw the process flow diagram
2. Survey on Literatures
   - Integrate the process parameters and conditions
3. Study on Process simulation
   - To select the simulation package
4. Flowsheet creation in Aspen plus
   a. Selection of blocks
   b. Identification of reactions
   c. Run the simulation
   d. Analysis of output stream

▶ 5

EXPERIMENTAL SETUP

- Consists mainly the reactor
- After the reaction the vapours are passing through a cooling coil, a liquid collecting tank is placed by the end of the coil to collect the liquid pyro-oil
- The non-condensate gases are drawn by a compressor. Compressor outlet is linked with a gas storage facility to accumulate the pyro-gas
- Temperature is a major process parameter deals with the pyrolysis process

Unit	Temperature	Pressure
Reactor	300-500°C	1 atm

Input	Quantity
Poly Propylene	500 gm

▶ 6

Process Flow Diagram

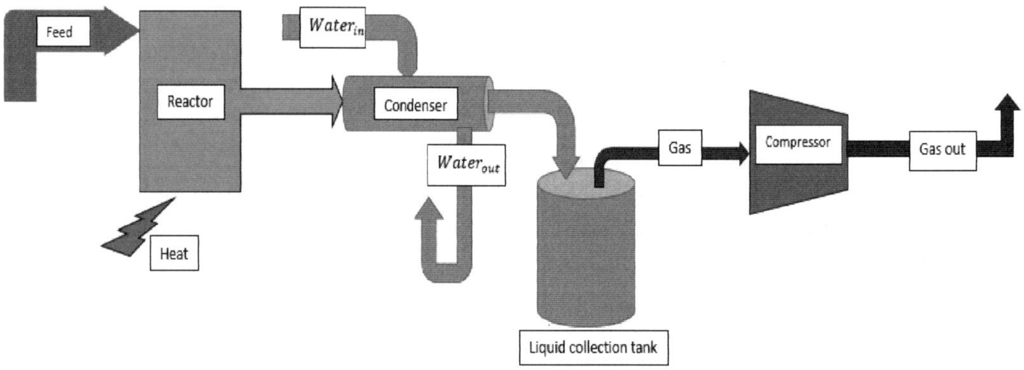

Literature review

Sl. No.	Journal	Author	Remarks
1.	A review on pyrolysis of plastic wastes (Energy Conservation and Management), 2016	*S.D. Anuar Sharuddin et al.*	• Proximate analysis • Oil products • Process parameters & reactor details
2.	Kinetic study of municipal solid waste, (Int.Journal of Hydrogen Energy), 2016	*L.S. Diaz silvarrey et al.*	• Ultimate analysis • Kinetic parameters
3.	Effect of plastic waste types on pyrolysis liquid oil, (Int. Journal of Biodeterioration & Biodegradation), 2016	*R. Miandad et al.*	• Experimental setup • Pyrolysis oil composition
4.	Replacing convention fuels in USA and UK with plastic pyrolysis gases(Energy Conservation and Management), 2016	*S. Honus et al.*	• Weight composition of pyrolysis gases from plastic

Sl. No	Journal	Author	Remarks
5.	Thermal cracking of Virgin and waste plastics of PP, LDPE in a semi batch reactor under atmospheric pressure (Energy & Fuels), 2015	*Guoxun Yan et al.*	• Kinetics • Gas products • Liquid products
6.	Pyrolysis of waste tires: A modeling and parameter estimation study using Aspen Plus (Waste Management), 2016	*H.Y. Ismail et al.*	• Flowsheet model • Aspen plus procedure
7.	Modelling of a Low Temperature pyrolysis process using Aspen Plus (Dev.Chem.Eng.Mineral Process), 1999	*H.M Yan and D.K. Zhang*	• Assumption • Aspen simulation flowsheet
8.	Review of kinetic and equilibrium concepts for biomass modeling by using Aspen Plus (Renewable and Sustainable Energy Reviews)	*A.M.A Ahmed et al.*	• Aspen plus characteristics
9.	Feasibility study for catalytic cracking of waste plastic to produce fuel oil with reference to Malaysia and simulation using Aspen Plus (Environmental progress & Sustainable Energy); 2013	*J.N. Sahu et al.*	• Process flow diagram

▶ 9

Process Simulation

- For better reliability of the system the operating variables have to be optimized and controlled with considerable accuracy.
- The cheapest and most effective technique to qualitatively evaluate the effect of each operating variable and to identify the optimal conditions is through process simulation.
- Conceptual process designs can be tested before they are implemented in full scale.
- Process simulation allows us to predict the behavior of a process by using basic engineering relationships, such as
  - Mass and energy balances
  - Phase and chemical equilibrium relationships
  - Reaction kinetics
- **Required Information**
  - Thermodynamic data
  - Equipment models
  - Operating conditions
- **Output from simulation**
  - Composition analysis
  - Sensitivity studies
  - Optimization runs

▶ 10

Introduction : Aspen plus

- Aspen Plus
  - Aspen Plus is a component of the Process modeling module of Aspen Tech
  - Process modeling tool for steady-state simulation, design, performance monitoring, optimization and business planning for chemicals, specialty chemicals, petrochemicals and metallurgy industries.

- Why Aspen plus?
  - Aspen Plus contains data, physical properties, unit operation models,
  - Built-in defaults, reports and other features and capabilities developed for specific industrial applications

Steps behind process modeling

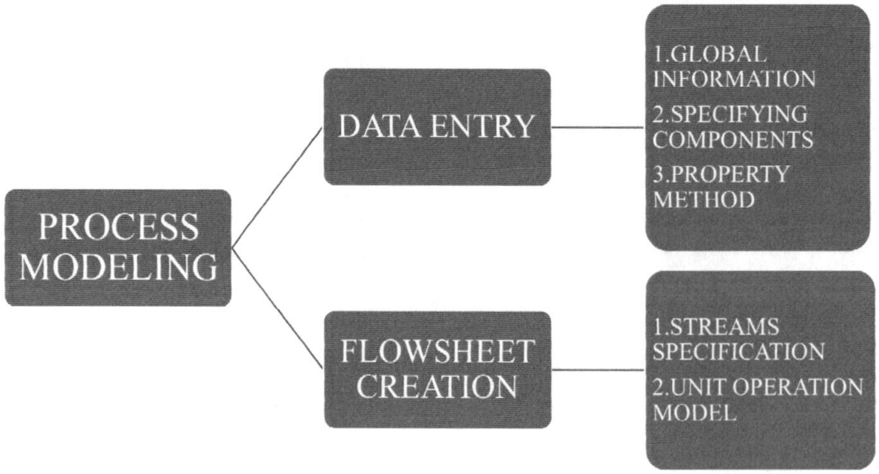

Aspen Plus data entry

- Define the process flow sheet that is being modeled, and to state the purpose of the model.
- This is can be done by **TITLE** and **DESCRIPTION** statements

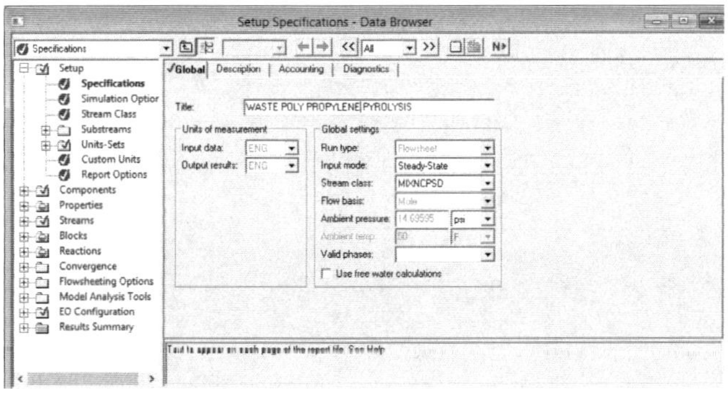

Component specification

- Initially mention the input feed and the possible outcomes of the process in the option of component specification.

Gas	Liquid	
Hydrogen	Styrene	3- methyl cyclo pentane
Methane	Toluene	1-hexene
Ethane	Ethylbenzene	Cyclohexene
Ethane	Benzene	1-octene
Pentene	Xylene	1-nonene
Pentane	Cumene	1-decene
Cyclohexane	Naphthalene	Biphenyl
Cyclopentene	Anthracene	1-heptene
	Di-phenylbenzene	2-methyl 1 1 pentene
	Tri- phenylbenzene	Indene
	1-methyl cyclopentane	

 Ref: Effect of plastic waste types on pyrolysis liquid oil, R. Miandad et al.

Property statement

- Physical properties are satisfied using properties statement
- Polymer type is the process type and the option set PENG-ROB was used in the property method

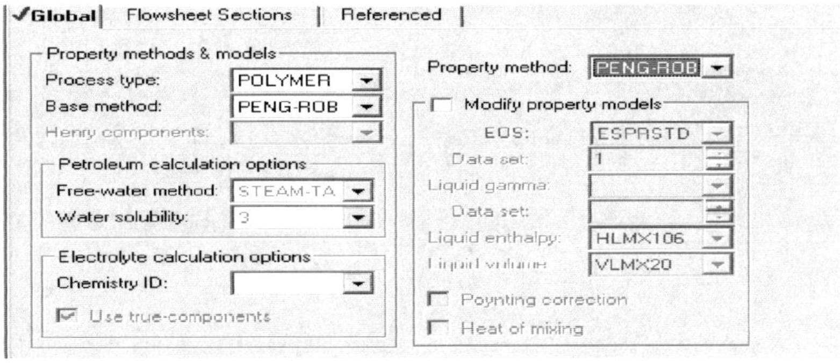

Physical properties set

- For non conventional input quantity the physical property parameters mentioned as,

Property	Model	Attribute requirements
Enthalpy	HCOALGEN	ULTANAL,PROXANAL
Density	DCOALIGT	ULTANAL

Type of plastic	Moisture (wt%)	Fixed carbon (wt%)	Volatile (wt%)	Ash (wt%)
Poly Propylene	0.15	1.22	95.08	3.55

Proximate Analysis (PROXANAL)

Elements	C	H	N	O
Weight (%)	83.25	16.75	0	0

Ultimate Analysis (ULTANAL)

Flowsheet creation

- Initially mentioned the input feed to the pyrolysis setup

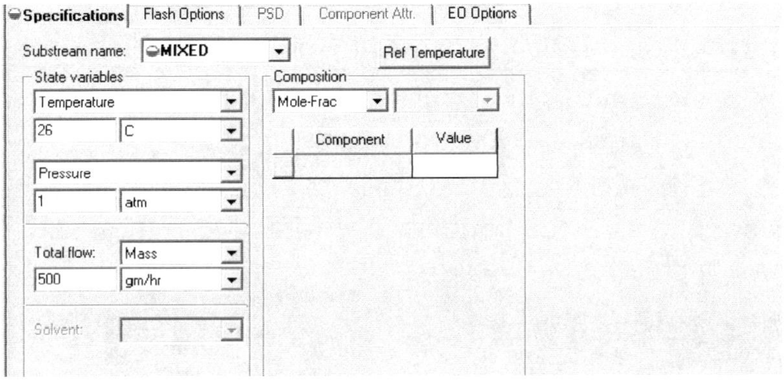

▶ 17

Decompose block

- The Aspen plus yield reactor, RYIELD, is used to simulate the decomposition of feed.
- It is predefined in Aspen Plus library converts the non-conventional solid feed particles into its conventional elemental components.
- Operating range of 300-600°C under 1atm.
- waste poly propylene was decomposed into volatiles and char.

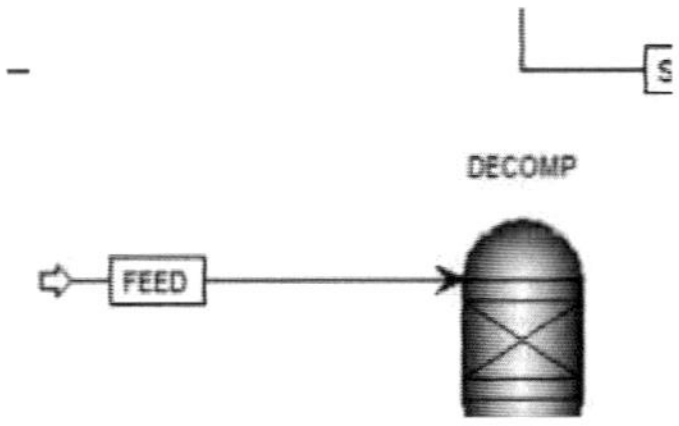

▶ 18

Reactor block

- The block RGIBBS is used to represent the combustion section of the pyrolysis reactor in the flow sheet.
- It provides reaction calculations without providing detailed configuration.
- The block can take one or more input and one or more output streams, and an optional heat input or output streams.
- The block can be used to calculate phase and chemical equilibrium, and allows constraining the equilibrium value with specific heat duty and temperature approach in the Setup.

Temperature	Pressure
300-600°C	1atm

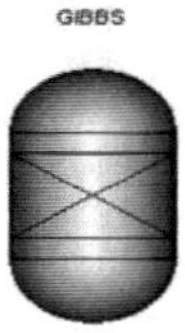

▶ 19

Model	Block	Purpose
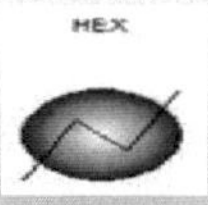	Heat Exchanger	• To cool down the temperature of the pyro-product • Pressure-1atm, temperature - 26°C
	Cyclone	• Used for the separation of solids from the reactor outlet • Split fraction is used for the separation of solids
	Condenser	• Separates the non condensable gas and liquid products • Normal parameters are used
	Compressor	• Compress the outlet gas • Type-Isentropic

▶ 20

Aspen Plus simulation flowsheet for the waste plastic pyrolysis process

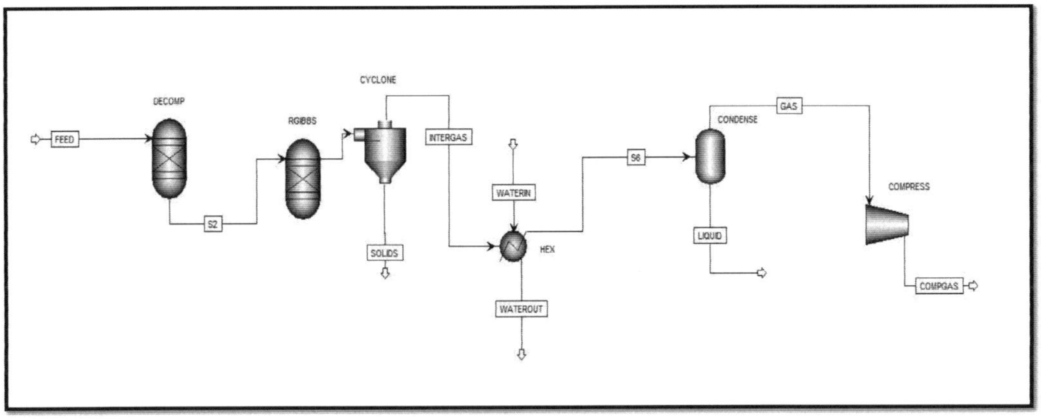

▶ 21

Reaction chemistry

Kinetic

- Kinetic parameter should be known
- Chemical reaction takes place under specified process parameters

Equilibrium

- Only input and output quantities are needed
- Inbuilt operation
- Elemental composition of feed is needed

▶ 22

Kinetic

Sl No.	Product	A (s^{-1})	E (KJ/mol)
1.	2-Methyl-1-Pentene	0.009	1.59
2.	1-Heptene	0.015	1.59
3.	1-Octene	0.07	1.59
4.	Benzene	1.654	33.89
5.	Toluene	7.305	33.89
6.	Xylene	4.476	33.89
7.	Indene	1.278	33.89
8.	Biphenyl	14.201	6.3

Liquid

Sl No.	Product	A (s^{-1})	E (KJ/mol)
1.	Methane	4.877	23.01
2.	Ethene	2.386	23.01
3.	Ethane	0.520	23.01
4.	Propene	0.446	23.01
5.	Propane	0.227	23.01
6.	Butene	0.144	23.01
7.	Pentalyne	0.066	23.01
8.	Methyl Cyclo Pentene	0.009	1.59

Gaseous

E- Activation Energy
A- Pre exponential term(Constant)

Due to certain technical hindrances the model failed to produce output

▶ 23

Equilibrium Model

Assumptions

- All the chemical reactions were assumed to have reached equilibrium within the reactor.
- Only methane, hydrogen, carbon monoxide, carbon dioxide were considered to be present in the product stream.
- The solid products consists only carbon and ash.
- Equilibrium condition facilitates a comprehensive optimization study based on temperature of the pyrolysis process, thereby neglecting the complexities of the pyrolysis hydrodynamics and reaction kinetics.

▶ 24

Output stream

Heat and Material Balance Table												
Stream ID		COMPGAS	FEED	GAS	INTERGAS	LIQUID	S2	S3	S6	SOLIDS	WATERIN	WATEROUT
From		COMPRESS		CONDENSE	CYCLONE	CONDENSE	DECOMP	RGIBBS	HEX	CYCLONE		HEX
To			DECOMP	COMPRESS	HEX		RGIBBS	CYCLONE	CONDENSE		HEX	
Phase		VAPOR	MISSING	VAPOR	VAPOR	MISSING	MIXED	VAPOR	VAPOR	MISSING	LIQUID	LIQUID
Substream: MIXED												
Mole Flow	kmol/hr											
METHANE		1.102863	0.0	1.102863	1.102863	0.0	0.0	1.102863	1.102863	0.0	0.0	0.0
ETHANE		6.89159E-6	0.0	6.89159E-6	6.89159E-6	0.0	0.0	6.89159E-6	6.89159E-6	0.0	0.0	0.0
ETHENE		6.07553E-7	0.0	6.07553E-7	6.07553E-7	0.0	0.0	6.07553E-7	6.07553E-7	0.0	0.0	0.0
PROPANE		2.02568E-10	0.0	2.02568E-10	2.02568E-10	0.0	0.0	2.02568E-10	2.02568E-10	0.0	0.0	0.0
PROPENE		1.73399E-10	0.0	1.73399E-10	1.73399E-10	0.0	0.0	1.73399E-10	1.73399E-10	0.0	0.0	0.0
BUTENE		0.0	0.0	0.0	4.83028E-15	0.0	0.0	4.83028E-15	4.83028E-15	0.0	0.0	0.0
PENTALY		0.0	0.0	0.0	1.57848E-18	0.0	0.0	1.57848E-18	1.57848E-18	0.0	0.0	0.0
MCP		0.0	0.0	0.0	3.86008E-23	0.0	0.0	3.86008E-23	3.86008E-23	0.0	0.0	0.0
MP		0.0	0.0	0.0	8.75248E-24	0.0	0.0	8.75248E-24	8.75248E-24	0.0	0.0	0.0
HEPTENE		0.0	0.0	0.0	6.83368E-29	0.0	0.0	6.83368E-29	6.83368E-29	0.0	0.0	0.0
OCTENE		0.0	0.0	0.0	8.65318E-34	0.0	0.0	8.65318E-34	8.65318E-34	0.0	0.0	0.0
BENZENE		0.0	0.0	0.0	2.62678E-15	0.0	0.0	2.62678E-15	2.62678E-15	0.0	0.0	0.0
TOLUENE		0.0	0.0	0.0	1.28058E-18	0.0	0.0	1.28058E-18	1.28058E-18	0.0	0.0	0.0
XYLENE		0.0	0.0	0.0	6.37658E-23	0.0	0.0	6.37658E-23	6.37658E-23	0.0	0.0	0.0
INDENE		0.0	0.0	0.0	2.63868E-25	0.0	0.0	2.63868E-25	2.63868E-25	0.0	0.0	0.0
BIPHENYL		0.0	0.0	0.0	7.24638E-32	0.0	0.0	7.24638E-32	7.24638E-32	0.0	0.0	0.0
HYDROGEN		8.930226	0.0	8.930226	8.930226	0.0	8.058238	8.930226	8.930226	0.0	0.0	0.0

Stream results of the flowsheet simulation; major outputs are taken for further analysis

▶ 25

Results

Effect of temperature on product yields

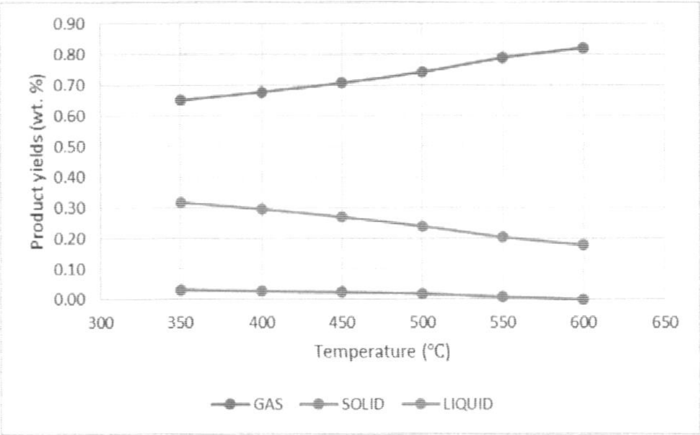

Temperature °C	Gas gm/hr	Solid gm/hr	Liquid gm/hr
350	325.598	15.85714	158.5449
400	339.2943	13.4662	147.2395
450	353.8763	11.73291	134.3908
500	371.4893	9.251487	119.2592
550	394.6382	3.8698	101.492
600	411.0718	0	88.92817

Optimum yield is at temperature 600°C

Sensitivity analysis is a tool for determining how a process reacts to varying key operating and design variables.

▶ 26

Effect of temperature on gas composition

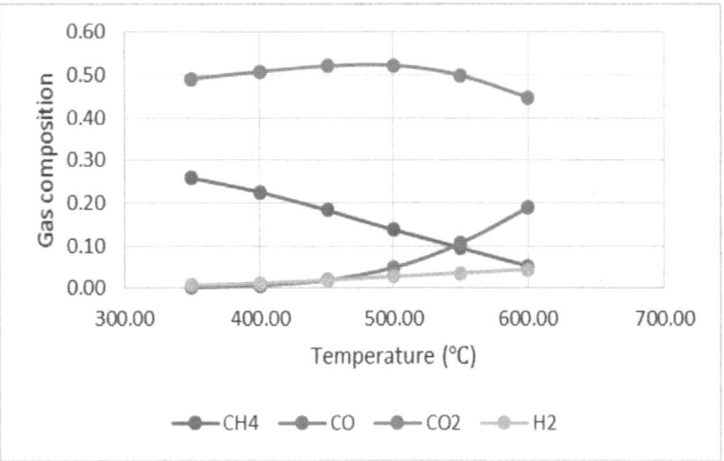

TEMP °C	CH_4 (Wt. %)	CO (Wt. %)	CO_2 (Wt. %)	H_2 (Wt. %)
350.00	0.26	0.00	0.49	0.01
400.00	0.23	0.01	0.51	0.01
450.00	0.18	0.02	0.52	0.02
500.00	0.14	0.05	0.52	0.03
550.00	0.09	0.11	0.50	0.04
600.00	0.05	0.19	0.45	0.04

▶ 27

Effect of pressure on Gas composition

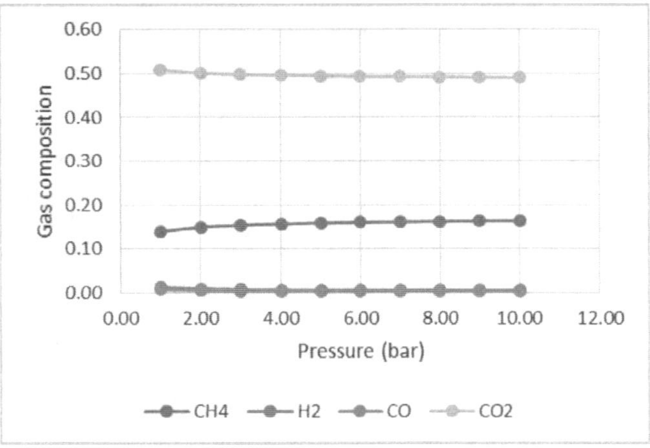

PRESSURE bar	CH_4 (Wt.%)	H_2 (Wt.%)	CO (Wt.%)	CO_2 (Wt.%)
1.00	0.14	0.01	0.01	0.51
2.00	0.15	0.01	0.00	0.50
3.00	0.15	0.01	0.00	0.50
4.00	0.16	0.01	0.00	0.50
5.00	0.16	0.01	0.00	0.49
6.00	0.16	0.01	0.00	0.49
7.00	0.16	0.00	0.00	0.49
8.00	0.16	0.00	0.00	0.49
9.00	0.16	0.00	0.00	0.49
10.00	0.16	0.00	0.00	0.49

▶ 28

Comparison with experimental results

Temperature°C	Component (weight %)							
	Simulation				Experimental			
	CH_4	CO	CO_2	H_2	CH_4	CO	CO_2	H_2
350	0.26	0.00	0.49	0.01	0.40	0.00	0.00	0.15
400	0.23	0.01	0.51	0.01	0.98	0.00	0.00	0.38
450	0.18	0.02	0.52	0.02	1.2	0.00	0.00	0.45
500	0.14	0.05	0.52	0.03	1.6	0.00	0.00	0.70
550	0.09	0.11	0.50	0.04	4.8	0.00	0.00	0.90
600	0.05	0.19	0.45	0.04	5.5	0.00	0.00	1.00

29

Conclusion

- The operating parameters such as temperature and pressure have substantial impact on properties of waste plastic pyrolysis.
- With rise in temperature there is a proportional increase in the gaseous product, which consists predominantly of Hydrogen and Methane.
- The liquid yield shows a decrease in trend with rise in temperature and the solid yield tends to zero.
- This model gives a good initial point for the practical runs to determine the exact operating conditions to manufacture a particular pyrolysis setup.
- It is possible to develop a perfect model only with the knowledge of chemical reaction kinetics of the appropriate process.
- Few variables also affect the yield and properties of the product such as,
  - Reaction chemistry
  - Kinetic parameters

30

Future work

- Inspite of the recent advancements in pyrolysis technology, several issues still remain and there is scope for further process optimization.
- The quality of the work can be further improved by taking into consideration of factors like,
  - Reaction time
  - Equipment specification
  - Presence of catalyst
- This model can be further extended to develop a CNG like model for compressing the filtered synthesized gas from pyrolysis.

31

Reference

- Narendar R. Mitta et. al Modelling and Simulation of a Tyre Gasification Plant for Synthesis Gas Production; 2006.
- Dezhen Chen, Lijie Yin , Huan Wang, Pinjing He, Pyrolysis technologies for municipal solid waste: A review; 2014.
- S.D. Anuar Sharuddin et al. A review on pyrolysis of plastic wastes, *Energy Conservation and Management*; 2016.
- L.S. Diaz silvarrey et al. Kinetic study of municipal solid waste, *Int. Journal of Hydrogen Energy*; 2016
- R. Miandad et al. Effect of plastic waste types on pyrolysis liquid oil, *Int. Journal of Biodeterioration & Biodegradation*; 2016
- S. Honus et al. Replacing convention fuels in USA and UK with plastic pyrolysis gases, *Energy Conservation and Management*;2016

32

- Guoxun Yan et al. Thermal cracking of Virgin and waste plastics of PP, LDPE in a semi batch reactor under atmospheric pressure, *Energy & Fuels*; 2015
- H.Y. Ismail et al. Pyrolysis of waste tires: A modeling and parameter estimation study using Aspen Plus, *Waste Management*; 2016
- H.M Yan and D.K. Zhang Modelling of a Low Temperature pyrolysis process using Aspen Plus, *Dev.Chem.Eng.Mineral Process*; 1999

THANK YOU

INCREASES IN DEMAND FOR FIXED NITROGEN FROM ALTERNATIVE ENERGY AND CARBON CAPTURE SCHEMES

Dr. Luis F. Razon, De La Salle University, Manila, Philippines

Introduction

The world food supply has depended on the supply of synthetic fixed or reactive nitrogen for more than 100 years. The overuse of nitrogen fertilizer has led to numerous negative environmental impacts[1]. In addition, the treatment of nitrogen in sewage leads to increased sewage treatment costs. The manufacture of fixed nitrogen also consumes large amounts of fossil fuels and intrinsically releases large amounts of greenhouse gases. While the link between the food supply, synthetic reactive nitrogen environmental degradation is well known, the scale of synthetic fixed nitrogen requirements for alternative energy and carbon dioxide removal is not often appreciated. The present and possible future uses for reactive nitrogen are diagrammed in Figure 1.

This paper takes projected demands for the year 2050 for alternative energy and carbon dioxide removal schemes from the literature and translates these into projections for fixed nitrogen. A comparison is made with the projected demand for food agriculture. The order-of-magnitude estimates that result from these projections are important when making judgement decisions on investments in new facilities and new technologies. The following sections are arranged in the order of likelihood that the demand for fixed nitrogen for these applications will actually transpire.

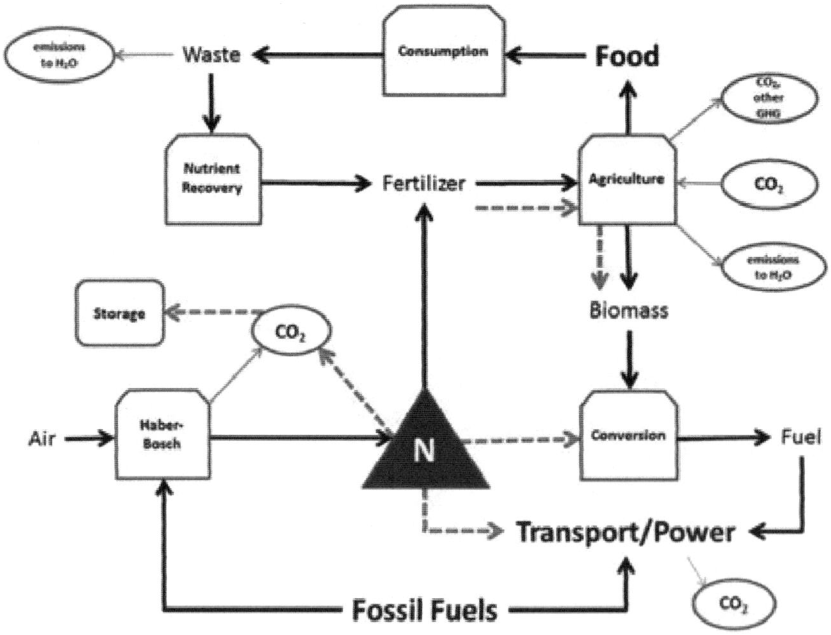

Figure 1. Uses and sources of fixed nitrogen. Solid lines are present uses. Dashed lines are possible future uses.

Demand from Food Agriculture

Fertilizer for food agriculture accounts for about 77% of the global demand for fixed nitrogen[2]. The other 23% is used for explosives, plastics and others. The consumption of N nutrient in 2014 was 109×10^9 kg N. There are many extant projections for food demand available in the literature. The projections are based on population projections and changes in total food intake and food choices based on global and regional scenarios such as those in the "Shared Socioeconomic Pathways" (SSP). These projections predict much higher demands for fixed nitrogen than would be obtained from a simple linear regression. For example, a linear regression of the reactive nitrogen demand from the FAO[3] shows that consumption would be about 187.1×10^9 kg of nutrient N by 2050. On the other hand, Bodirsky et al.[4] analyzed the Shared Socioeconomic Pathway "middle of the road" scenario (SSP2) and arrived at 232×10^9 kg nitrogen (N). Depending on the scenario the estimates of Bodirsky et al. ranged from 188×10^9 kg N under the SSP1 ("sustainability") scenario to the 290×10^9 kg N under the SSP5 ("regional rivalry") scenario. Other estimates, such as those by Galloway et al.[5], are consistent with these estimates. For the rest of the paper, I will use the figure from the SSP2 scenario, 232×10^9 kg N, as a reference figure for comparison.

Demand from Biomass Agriculture and Pretreatment

Concerns about the eventual depletion of fossil fuels and the environmental impact of their use has driven the development and introduction of biofuels such as bioethanol and biodiesel. Reactive nitrogen demand come not only from the fertilizer needed to grow the biomass feedstock but also for pretreatment of lignocellulosic feedstock.

There are not many available estimates of the reactive nitrogen needs for biofuels because of the difficulty in anticipating adoption of biofuels and the feedstocks that might be used[6]. A projection was made by Winiwarter et al.[7] using a representative concentration pathway scenario with modest population growth (RCP 2.6). Using this scenario, Winiwarter et al.[7] projected that the demand for biofuels in 2050 will be about 118 EJ. To supply the biomass for this amount of energy, they estimate that about 41.4×10^9 kg N would be needed for fertilizer. It should be pointed out that the RCP 2.6 scenario is a more optimistic scenario than the SSP2 scenario with smaller population growth because it assumes that measures will be taken such that the radiative forcing values in the year 2100 relative to pre-industrial values is only 2.6 W/m^2. For example, the food fertilizer estimate under this scenario is only 94.4×10^9 kg N as compared to the 232×10^9 kg N estimate for the SSP2 scenario. While this indicates that fundamental differences between the RCP 2.6 and SSP2 scenarios, this estimate is still useful as an order-of-magnitude estimate.

What is not commonly realized is that the reactive nitrogen needs for biofuels are not limited to fertilizer. The present "first-generation" feedstocks for biofuels like corn, sugarcane and palm oil are the same plants used for food. To sustain the expansion of biofuels, it has been proposed that the second generation of feedstocks consist of non-food plants, specifically lignocellulosic feedstocks like wood and switchgrass for making bioethanol. Lignocellulosic feedstocks require pretreatment prior to fermentation to bioethanol. There are several pretreatment methods under development. Ammonia is needed for three of these: ammonia fiber explosion (AFEX), ammonia recycled percolation (ARP) and soaking in aqueous ammonia (SAA)[8]. AFEX appears to be the closest to widespread adoption[8]. AFEX requires about 20g NH_3 per kg of lignocellulosic biomass[9]. In a scenario predicting wide adoption of biofuels to Winiwarter et al.[7] estimate that about 118 EJ of biofuels will be produced and 98% of the bioenergy in 2050 will be from lignocellulosic feedstock. On the other hand, Herve et al.[10] used the IEA Blue Map Scenario to estimate that 90% of biomass feedstock will be lignocellulosic. What is less certain is the means by which the feedstock will be pretreated in 2050. If it is assumed that in 2050, 118 EJ of energy will come from biofuels and that 50-90% of the biofuels are lignocellulosic feedstock pretreated using AFEX, which uses 20g of NH_3 per kg of biomass, then the range for N demand would range from 154×10^9 kg N to 278×10^9 kg N. Therefore, the pretreatment N requirements would be in the same order of magnitude for that of food.

Demand from Atmospheric Carbon Dioxide Removal by Amine Absorption

The rapid accumulation of CO_2 in the atmosphere has led to proposals to actively withdraw CO_2 from the atmosphere for burial in the underground chambers or in the deep ocean rather than to simply rely on limiting carbon emissions[11]. Among these proposals is CO_2 removal by chemical absorption. In this scheme, CO_2 is absorbed by a solvent which is then heated such that the CO_2 is released for further processing and ultimate disposal. The process is already used for CO_2 recovery from fermentation processes for CO_2 reuse in other products like soft drinks. The most commonly used solvent is monoethanol amine (MEA)[12]. Because it is already a mature technology, carbon dioxide removal with monoethanol amine is projected to be a significant contributor to CCS operations. In 2050, the IEA Technology Roadmap for 2013 estimates that 7×10^9t of CO_2 per year will be absorbed by solvent absorption and that "most" of it will be by amine-based absorption[13]. Ideally, all of the amine is regenerated and reused but some is inevitably lost because of evaporation and degradative reactions. About 1.5 kg MEA/t of CO_2 is required as make-up[14]. If 75% of the projected 7000 Mt is absorbed by amine absorption then about 7.9×10^9 kg MEA is required. About 2.2×10^9 kg NH_3 or 1.8 $\times10^9$ kg N would be needed to manufacture this amount of MEA. This is about only a small fraction of the requirement for food, so this application for ammonia will not have a significant impact on reactive nitrogen demand.

Demand from Using Ammonia as a Diesel and Gasoline Substitute

The complete combustion of ammonia releases heat and results in a net increase in the number of moles in the mixture, so it can, in theory, be used as a fuel. The resulting emissions would be carbon-free. The idea is attractive because ammonia has superior properties when compared to hydrogen like higher energy density and lower storage pressure[15]. The US Army tested the concept as early as the 1930's[16]. There is an active NH_3 Fuel Association[17]. Uses of ammonia as a fuel range from internal combustion engines[18] to fuel cells[19] to mixed coal-ammonia power generation plants[20].

Combustion of ammonia should ideally result in N_2 and H_2O. However, incomplete combustion of results in emissions of nitrogen oxides. In addition, ammonia has a low flame speed and therefore ammonia has to be mixed with a secondary fuel if it is to be used in an internal combustion engine[21]. Therefore, only partial replacement of diesel and gasoline would be possible. Nonetheless, it can be seen in the calculations below that the projected demand from enormous.

The data of Kong et al.[18,22] indicate that, to ensure proper performance, 1 kg of NH_3 must be accompanied by 0.31 kg of diesel. On the other hand, in a gasoline engine, 1 kg of NH_3 must be accompanied by 2.5 kg of gasoline. Also, 5.64 kg NH_3 is energetically equivalent to 1 kg of diesel whereas 13.5 kg NH_3 is energetically equivalent to 1 kg of gasoline. Therefore, only 5.42% of the diesel can be replaced with NH_3 while only about 15.6% of the gasoline can be replaced with NH_3.

In 2040, the World Oil Outlook 2016[23] projects that the demand for diesel and gasoline and diesel will be 33.2 mb/d and 28.0 mb/d respectively. A simple cubic regression of the data in the World Oil Outlook 2016, provides a projection for 2050 of 34.2 mb/d and 30.0 mb/d for diesel and gasoline, respectively. This translates to 1.663×10^{12} kg/year for diesel and 1.288×10^{12} kg/year for gasoline.

The requirement for a secondary fuel puts an upper limit to the amount that can be displaced by ammonia. This works out to be a demand of about 0.51×10^{12} kg NH_3/year for the maximum diesel displacement and about 2.72×10^{12} kg NH_3/year for the maximum gasoline displacement. These amounts translate to 0.42×10^{12} kg N/year and 2.24×10^{12} kg N/year for diesel and gasoline, respectively. It is easily seen that these amounts dwarf the amounts necessary for food and the other applications.

Conclusions

The best estimate for each selection is shown in Table 1. The sources of demand for reactive nitrogen are listed in the order of likelihood that the demand may actually come true. Much depends on government policies and development of new technologies. If biofuel subsidies are abandoned, then there would be far less demand for biofuels. If climate change agreements are abrogated or are insufficiently followed, then it is likely that there will not be too much impetus for atmospheric carbon dioxide removal. Finally, the widespread use of ammonia as a transport fuel is purely speculative at the moment. There are many competing technologies including biofuels and electric vehicles. Government policies may very well drive the energy source one way or the other.

Table 1. Summary of possible reactive nitrogen demand in 2050

Source for Demand	**Estimated demand in 2050 (kg N)**
Food Agriculture	232×10^9
Biomass for Bioenergy Agriculture	41.4×10^9
Lignocellulosic Biomass Pretreatment	154×10^9 - 278×10^9
Carbon Dioxide Removal using Amine Absorption	1.8×10^9
Diesel Substitute	0.42×10^{12}
Gasoline Substitute	2.24×10^{12}

The scale of demand for the biomass agriculture and pretreatment are on the same order of magnitude as that for food agriculture. On the other hand, the scale of demand for the substitution for fuels is several hundred times or even a thousand times larger. This is even considering that this is only a partial substitution. Expansion of production and distribution facilities will have to be carefully planned if there is to be widespread adoption of ammonia fuel.

What is clear is that there will be a large and increasing demand for anthopogenic or synthetic nitrogen in the future, even if the demand may be only from agriculture or lignocellulosic biomass pretreatment. The manufacture of such large quantities of reactive nitrogen in the form of ammonia from the Haber-Bosch process will be accompanied by consumption of large amounts of fossil fuels and the release of large amounts of greenhouse gases. Alternatives to the Haber-Bosch process are in early stages of development[24]. More attention needs to be paid to the rapid development of these alternatives.

Acknowledgement

Dr Wilfried Winiwarter is thanked for providing more detailed information from their paper[7].

References

1. Erisman JW, Galloway JN, Seitzinger S, et al. Consequences of human modification of the global nitrogen cycle. *Philos Trans R Soc Lond B Biol Sci*. 2013. doi:10.1098/rstb.2013.0116.
2. FAO. *World Fertilizer Trends and Outlook to 2020*. Rome; 2017. https://www.google.com.ph/url?sa=t&rct=j&q=&esrc=s&source=web&cd=2&ved=0ahUKEwifpcnM88bWAhWLT7wKHaQ0AcMQFggpMAE&url=http%3A%2F%2Fwww.fao.org%2F3%2Fa-i6895e.pdf&usg=AFQjCNGYIMe1yKpnmLGmRqOcner1Es20pA.
3. FAOSTAT. http://www.fao.org/faostat/en/#data/RF. Accessed September 28, 2017.
4. Bodirsky BL, Popp A, Lotze-Campen H, et al. Reactive nitrogen requirements to feed the world in 2050 and potential to mitigate nitrogen pollution. *Nat Commun*. 2014;5(May):3858. doi:10.1038/ncomms4858.
5. Galloway JN, Leach AM, Erisman JW, Bleeker A. Nitrogen: the historical progression from ignorance to knowledge, with a view to future solutions. *Soil Res*. 2017;55(6):417-424. doi:10.1071/SR16334.
6. Searchinger TD, Beringer T, Strong A. Does the world have low-carbon bioenergy potential from the dedicated use of land? *Energy Policy*. 2017;110(December 2016):434-446. doi:10.1016/j.enpol.2017.08.016.
7. Winiwarter W, Erisman JW, Galloway JN, Klimont Z, Sutton MA. Estimating environmentally relevant fixed nitrogen demand in the 21st century. *Clim Change*. 2013;(2005):889-901. doi:10.1007/s10584-013-0834-0.
8. Tao L, Aden A, Elander RT, et al. Process and technoeconomic analysis of leading pretreatment technologies for lignocellulosic ethanol production using switchgrass. *Bioresour Technol*. 2011;102(24):11105-11114. doi:10.1016/j.biortech.2011.07.051.
9. Bals B, Wedding C, Balan V, Sendich E, Dale B. Evaluating the impact of ammonia fiber expansion (AFEX) pretreatment conditions on the cost of ethanol production. *Bioresour Technol*. 2011;102(2):1277-1283. doi:10.1016/j.biortech.2010.08.058.
10. Hervé G, Agneta F, Yves D. Biofuels and World Agricultural Markets: Outlook for 2020 and 2050. In: *Economic Effects of Biofuel Production*. Vol ; 2011:129-162. doi:10.5772/20581.
11. Meadowcroft J. Exploring negative territory Carbon dioxide removal and climate policy initiatives. 2013:137-149. doi:10.1007/s10584-012-0684-1.

12. Shakerian F, Kim KH, Szulejko JE, Park JW. A comparative review between amines and ammonia as sorptive media for post-combustion CO_2 capture. *Appl Energy*. 2015;148:10-22. doi:10.1016/j.apenergy.2015.03.026.
13. IEA. *Technology Roadmap - Carbon Capture and Storage*. Paris; 2013. doi:10.1007/SpringerReference_7300.
14. Rao AB, Rubin ES, Berkenpas MB. *An Integrated Modeling Framework for Carbon Management Technologies. Vol. 1 - Technical Documentation: Amine-Based CO_2 Capture and Storage Systems for Fossil Fuel Power Plants*. Pittsburgh, Pennsylvania; 2004. https://www.cmu.edu/epp/iecm/documentation/tech_04.pdf.
15. Zamfirescu C, Dincer I. Using ammonia as a sustainable fuel. *J Power Sources*. 2008;185(1):459-465. doi:10.1016/j.jpowsour.2008.02.097.
16. Garabedian CG., Johnson JH. *The Theory of Operation of an Ammonia Burning Internal Combustion Engine*. Warren, Michigan; 1966. http://www.dtic.mil/dtic/tr/fulltext/u2/634681.pdf.
17. NH3 Fuel Association | Practical, carbon-free fuel. https://nh3fuelassociation.org/. Accessed September 28, 2017.
18. Gross CW, Kong SC. Performance characteristics of a compression-ignition engine using direct-injection ammonia-DME mixtures. *Fuel*. 2013;103:1069-1079. doi:10.1016/j.fuel.2012.08.026.
19. Afif A, Radenahmad N, Cheok Q, Shams S, Kim JH, Azad AK. Ammonia-fed fuel cells: A comprehensive review. *Renew Sustain Energy Rev*. 2016;60:822-835. doi:10.1016/j.rser.2016.01.120.
20. Japanese utilities team on CO2-reducing tech for coal plants- Nikkei Asian Review. https://asia.nikkei.com/Business/Deals/Japanese-utilities-team-on-CO2-reducing-tech-for-coal-plants. Accessed September 28, 2017.
21. Reiter AJ, Kong S. Combustion and emissions characteristics of compression-ignition engine using dual ammonia-diesel fuel. *Fuel*. 2011;90(1):87-97. doi:10.1016/j.fuel.2010.07.055.
22. Ryu K, Zacharakis-jutz GE, Kong S. Effects of gaseous ammonia direct injection on performance characteristics of a spark-ignition engine. *Appl Energy*. 2014;116:206-215. doi:10.1016/j.apenergy.2013.11.067.
23. OPEC. *World Oil Outlook*.; 2016. doi:10.1190/1.1439163.
24. Razon LF. Is nitrogen fixation (once again) "vital to the progress of civilized humanity"? *Clean Technol Environ Policy*. 2014;17(2):301-307. doi:10.1007/s10098-014-0835-3.

FARADAY TECHNOLOGY, INC.

Pulse/Pulse-Reverse Electrodeposition of Copper Electrocatalysts for CO_2 Reduction to Ethylene

2017 AIChE Annual Meeting
Minneapolis, MN
31 Oct 2017

Brian Skinn[a], Sujat Sen[b], Rajeswaran Radhakrishnan[a], McLain Leonard[b], Stephen Snyder[a], Fikile R. Brushett[b]

[a] Faraday Technology, Inc., Englewood, OH 45315
[b] Dept. of Chemical Engineering, Massachusetts Institute of Technology, Cambridge, MA 02139

BrianSkinn@FaradayTechnology.com

17AIChE
Annual Meeting, Minneapolis, MN

Outline

- Pulse/Pulse-Reverse Electrochemical Processing
- FARADAYIC® ElectroDeposition – Prior Work
- FARADAYIC® Copper Film CO_2 Electrocatalysts
  - Rationale and Electroreactor Concept
  - FARADAYIC® ElectroDeposition & Catalyst Activation
  - Electrocatalyst Performance

FARADAY
TECHNOLOGY, INC.

Electrochemical Processing

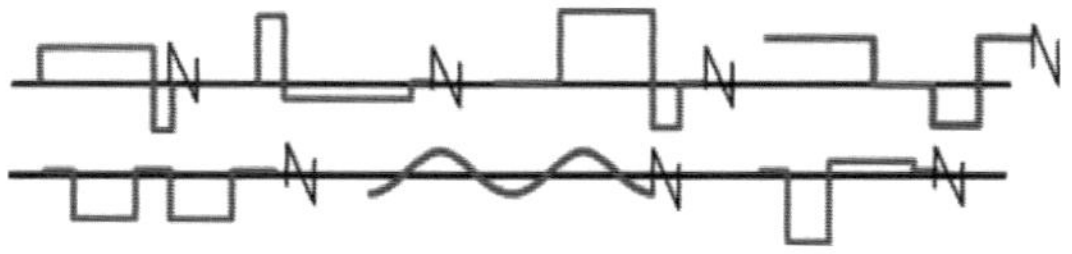

Conventional DC vs Pulse/Pulse-Reverse

Pulsed Waveforms vs DC

Know: *Linearized Concentration Profile*

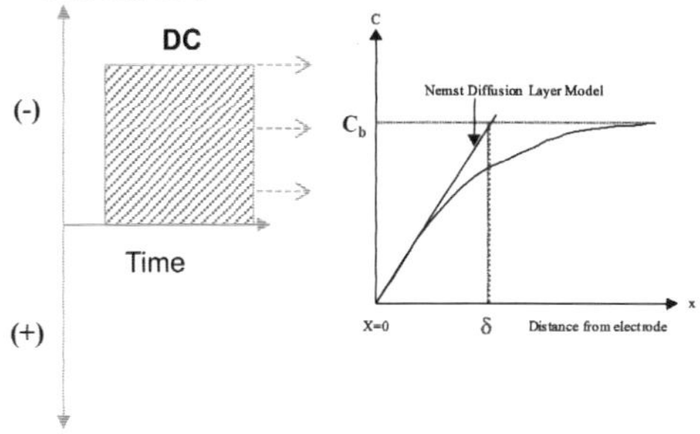

Pulsed Waveforms vs DC

Know: *Linearized Concentration Profile*

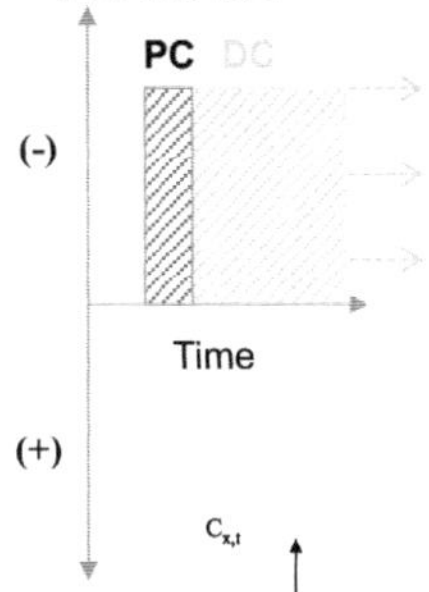

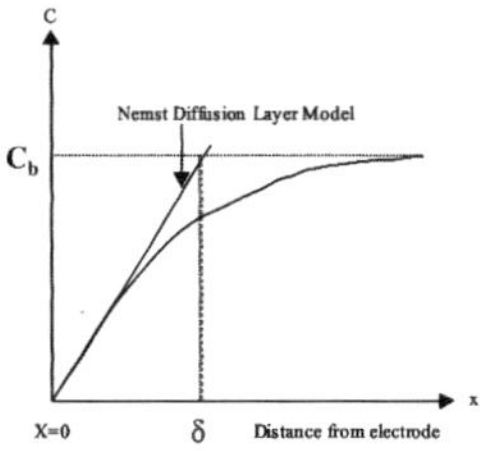

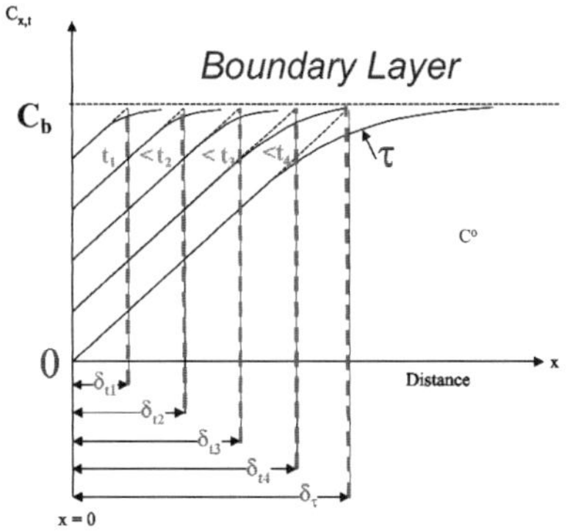

FARADAY TECHNOLOGY, INC.

Pulsed Waveforms vs DC

Know: *Linearized Concentration Profile*

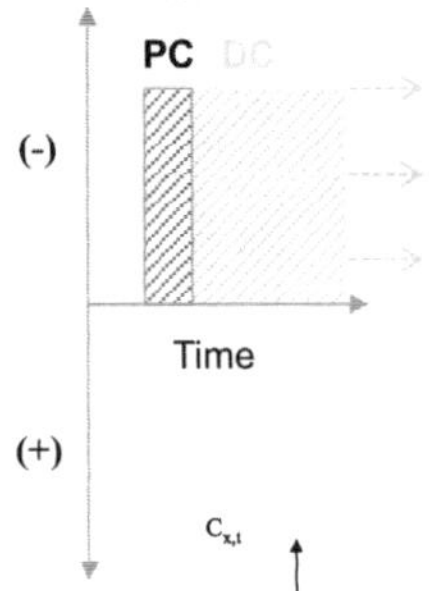

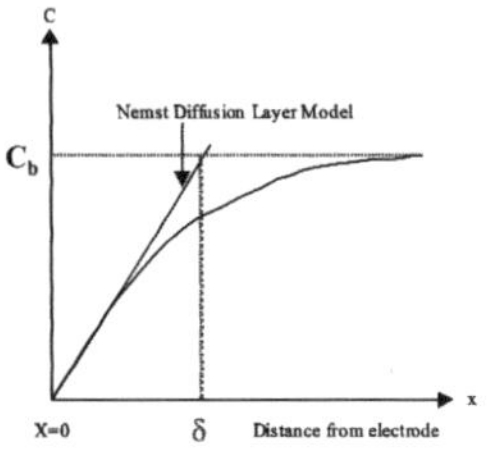

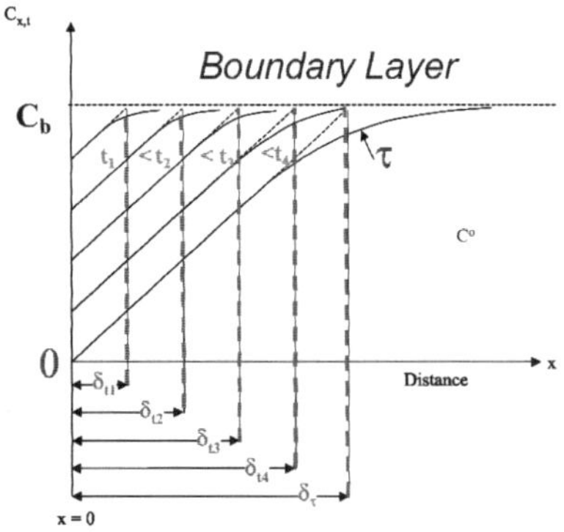

Assume:

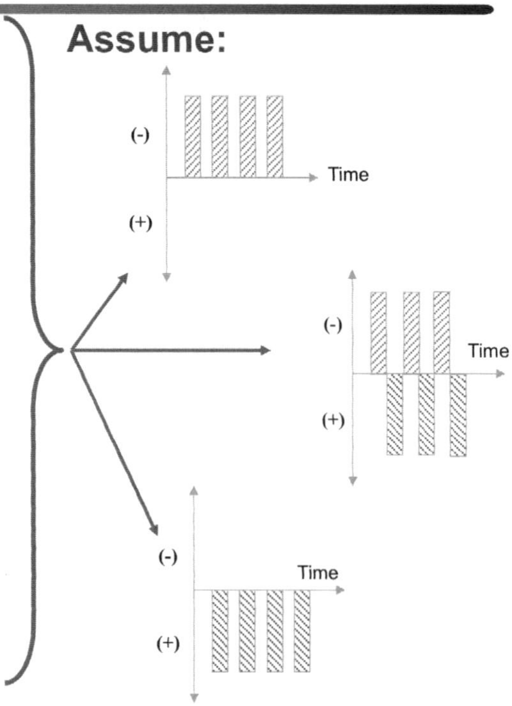

FARADAY TECHNOLOGY, INC.

FARADAY TECHNOLOGY, INC.

FARADAYIC® ElectroDeposition

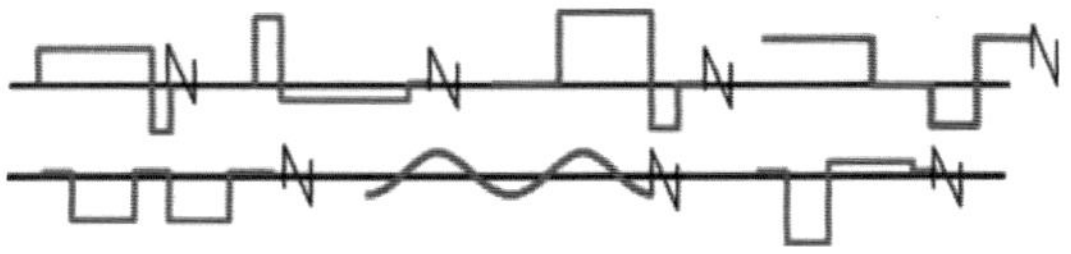

Prior Work

Prior Work – Copper Electrodeposition for PCBs

Standard DC Recipe

$CuSO_4$
H_2SO_4
"Accelerator"
"Suppressor"
"Leveler"
Other Additives

Pulsed Waveform Recipe

$CuSO_4$
H_2SO_4
Trace Cl^-
Trace PEG

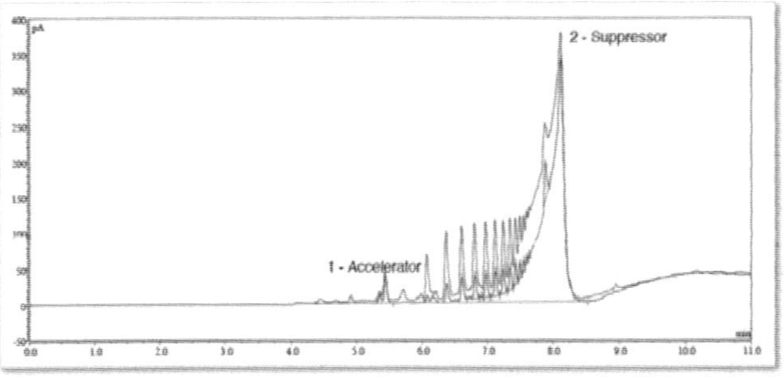

Plante, et al. "Quantization and Characterization of Copper Plating Bath Additives by Liquid Chromatography with Charged Aerosol Detection." Thermo Scientific. http://www.dionex.com/en-us/webdocs/113469-Pittcon12_1820-6_MPlante_PN70008_PlatingBaths.pdf

Z-Interconnects Cu-Plated by Pulse-Reverse FARADAYIC® ElectroDeposition

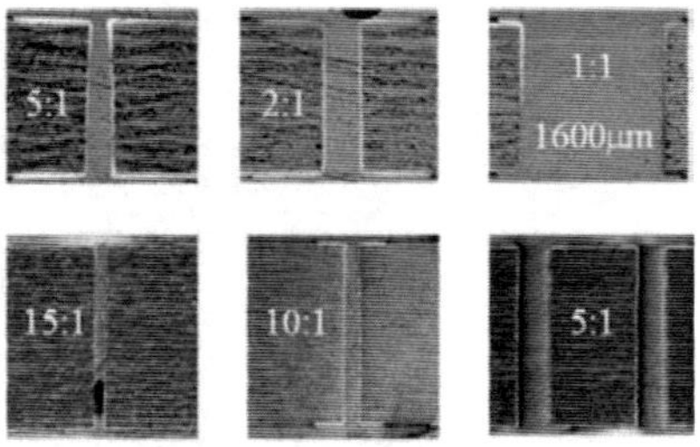

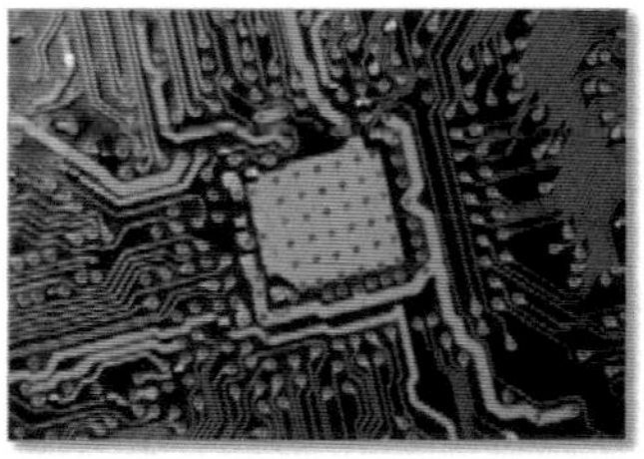

Prior Work – FARADAYIC® Lead-Free Tin

Rotating Disk Plating Tool

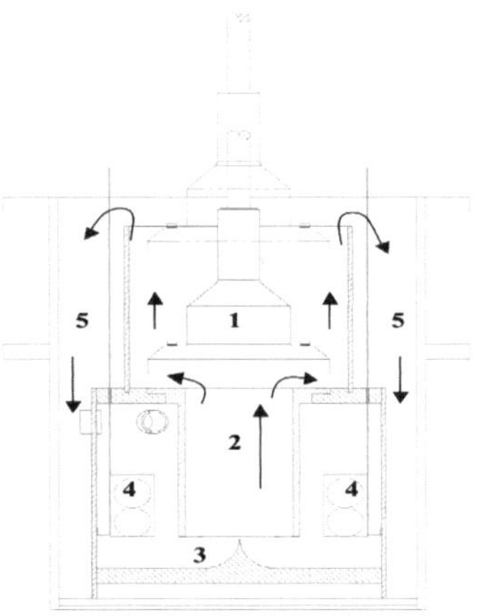

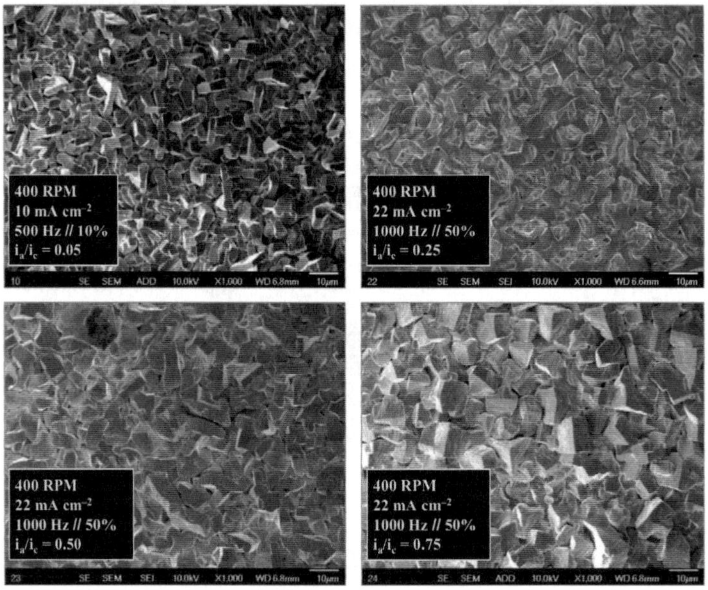

Sn Deposits on Cu Substrates — 1000× Magnification

U.S. Patent 8,603,315 Granted

Prior Work – Sn GDE CO_2 Electrocatalysts

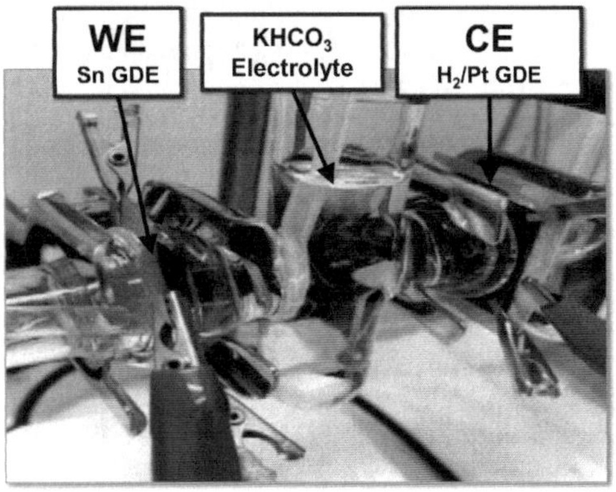

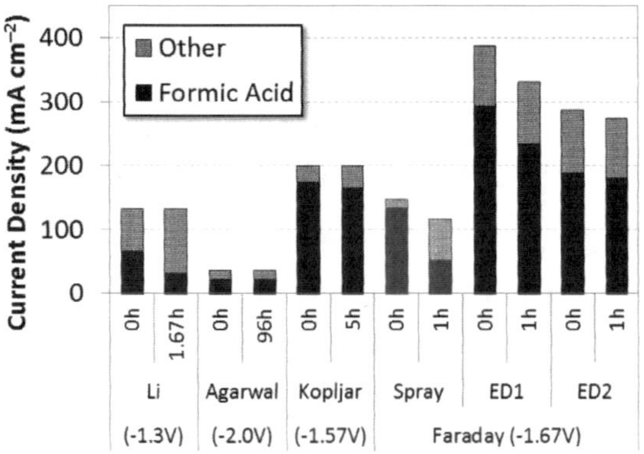

(Half-cell potentials vs SHE)

WE: $CO_2 + 2H^+ + 2e^- \rightarrow HCOOH$

CE: $H_2 \rightarrow 2H^+ + 2e^-$

Cell: $CO_2 + H_2 \rightarrow HCOOH$

Li and Oloman. *J Appl Electrochem* 35: 955, 2005.

Agarwal et al. *ChemSusChem* 4: 1301, 2011.

Kopljgar et al. *J Appl Electrochem* 44: 1107, 2014.

Prior Work – FARADAYIC® ElectroCatalyzation

- Low-loading deposition of Pt fuel cell catalyst
- Beta-scale batch system (NSF)
- Pilot-scale reel-to-reel system (OH)

Batch System **Continuous Reel-to-Reel**

E.J. Taylor *et al., J. Electrochem. Soc.*, 1992, 139 (5), L45; N.R.K. Vilambi *et al.*, U.S. Patent No. 5,084,144, issued 28 Jan. 1992; E.J. Taylor et al., U.S. Patent No. 6,080,504, issued 27 Jun. 2000.

TEM of Pt on Carbon

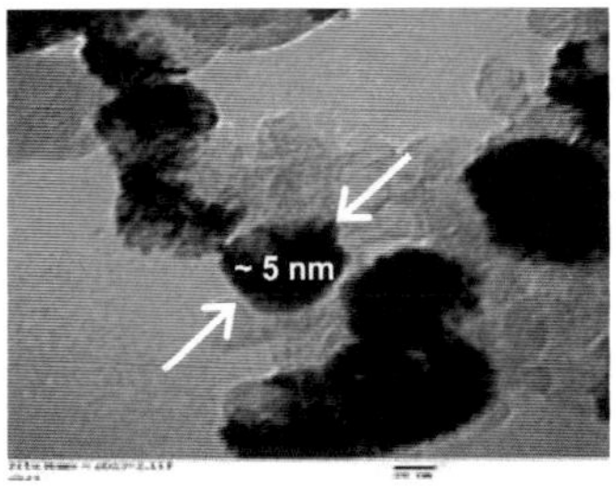

Performance Comparison

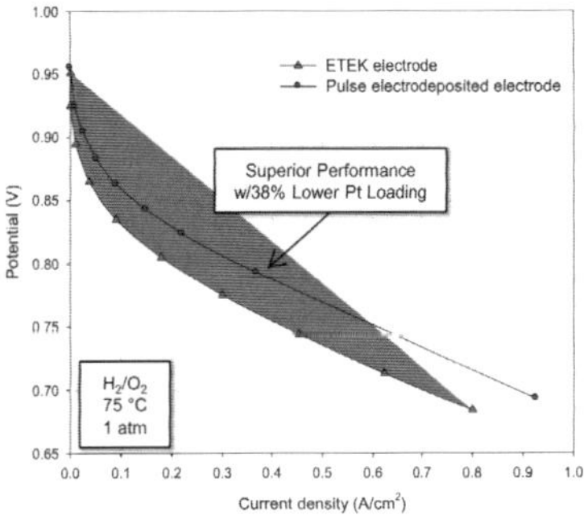

FARADAY TECHNOLOGY, INC.

FARADAY TECHNOLOGY, INC.

FARADAYIC® Copper Film CO_2 Electrocatalysts

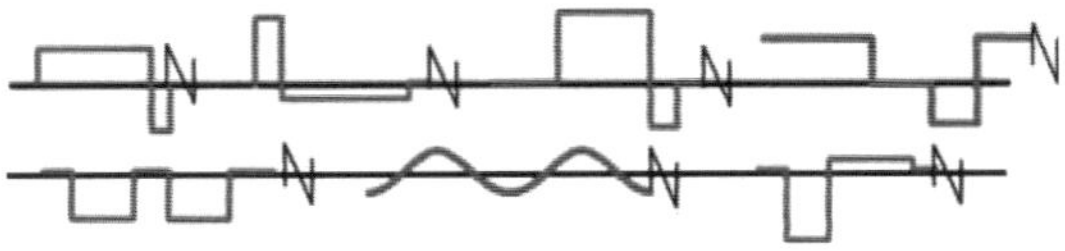

Rationale and Electroreactor Concept

CO_2 Electrocatalysis – Catalyst Selection

Product distribution over single-metal electrocatalysts as determined by Hori et al.
Electrochimica Acta 39(11-12): 1833 (1994).

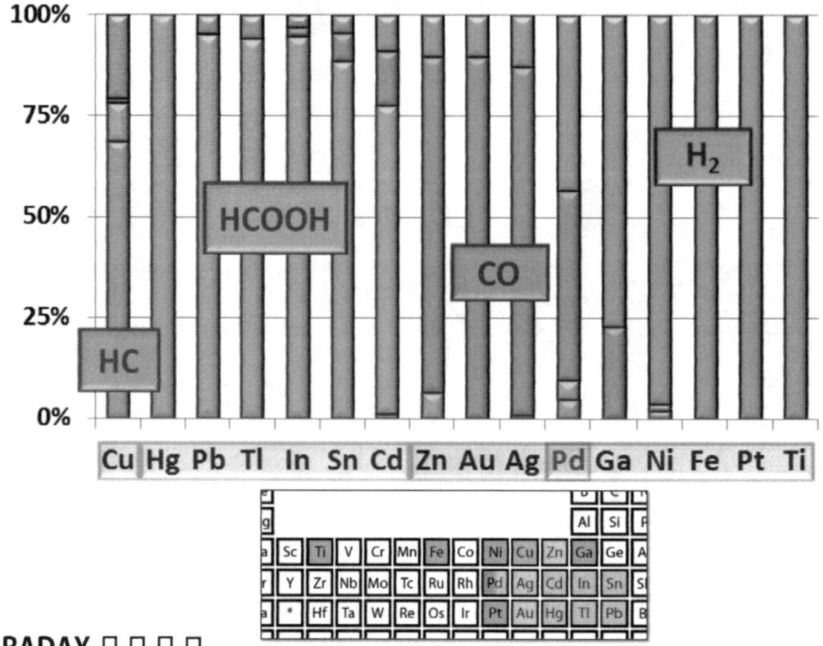

CO_2 Electrocatalysis – Form Factor Selection

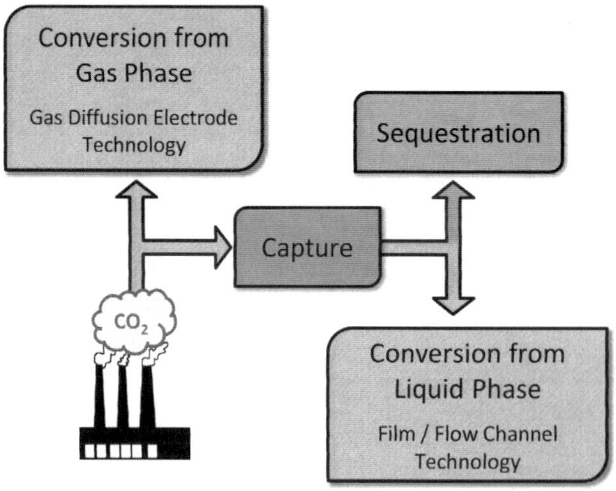

CO_2 Electrocatalysis – Form Factor Selection

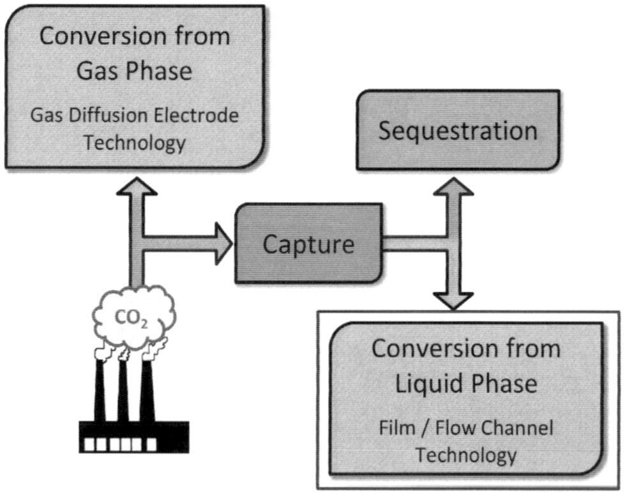

CO_2 Electrocatalysis – Flow Electroreactor Concept

- Flow channels for electrolytes
- CO_2 reduction on Cu catalyst
- O_2 evolution on commercial mixed oxide anode
- Anticipate direct use of CO_2-laden capture media
  - Aqueous amines, etc.

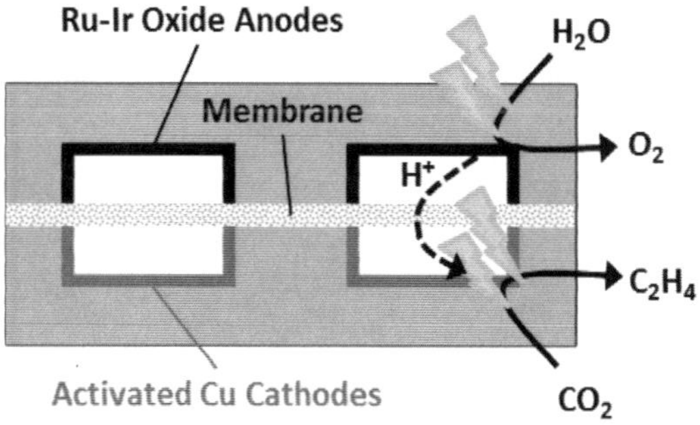

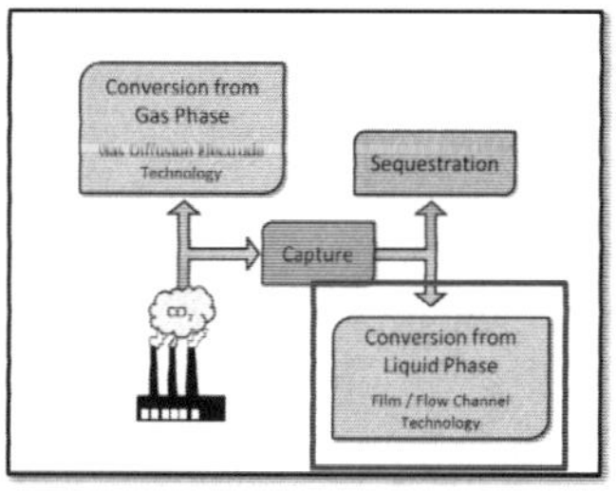

FARADAYIC® Copper Film CO_2 Electrocatalysts

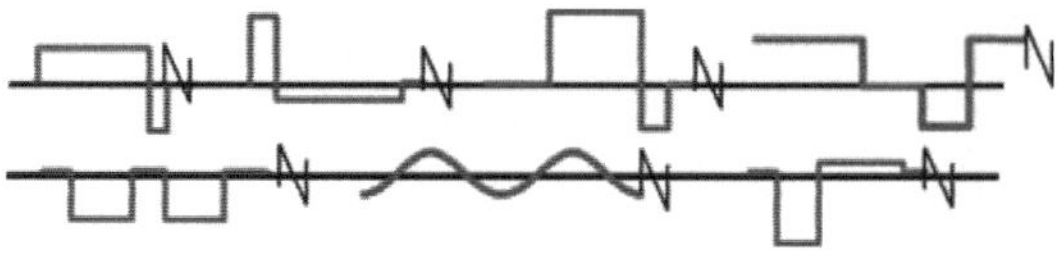

FARADAYIC® ElectroDeposition and Catalyst Activation

FARADAYIC® Copper ElectroDeposition

FARADAYIC® Copper ElectroDeposition

Catalyst Activation

Deposit Oxidize Reduce

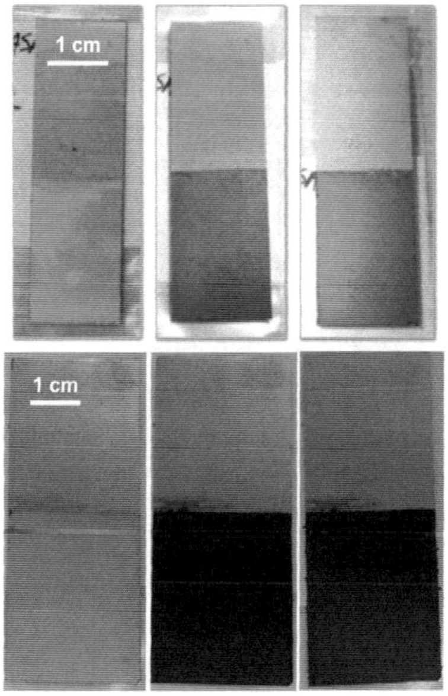

Modified from Li & Kanan, J Am Chem Soc 134: 7231 (2012)

Oxidize

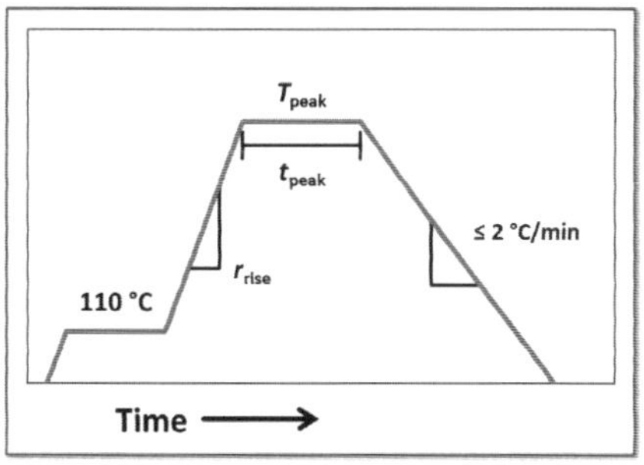

Reduce

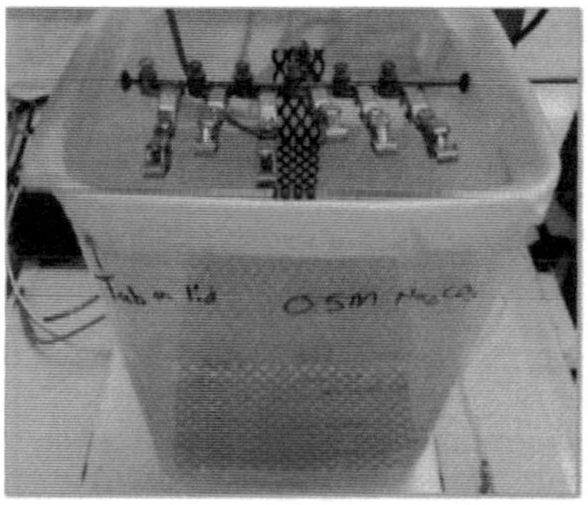

FARADAY TECHNOLOGY, INC.

FARADAYIC® Copper Film CO_2 Electrocatalysts

Electrocatalyst Performance

Electrocatalysis Test Reactor

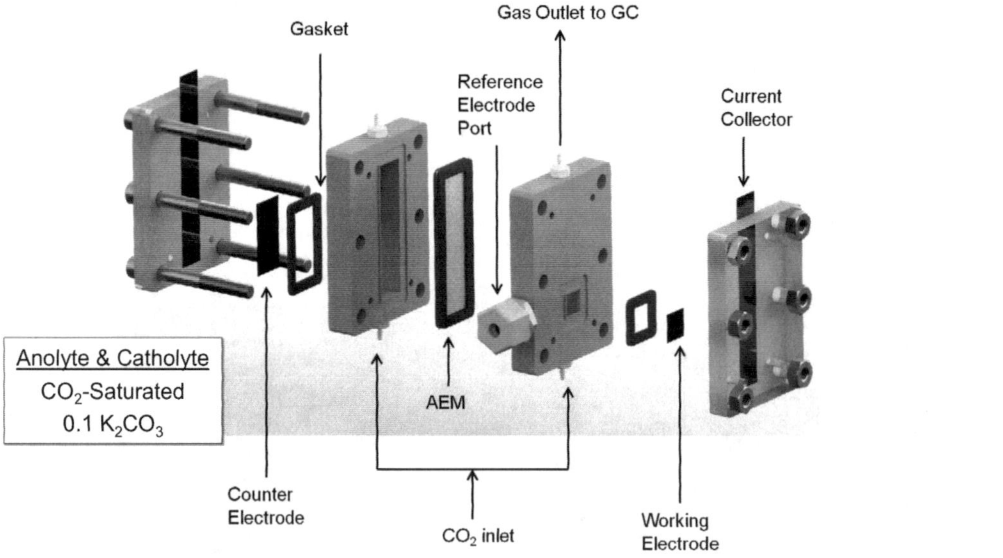

Design adapted from: Kuhl, Jaramillo, et al. Energy & Environmental Science, 5 (2012) 7050-7059.

FARADAYIC® Pulsed ElectroDeposition Waveforms

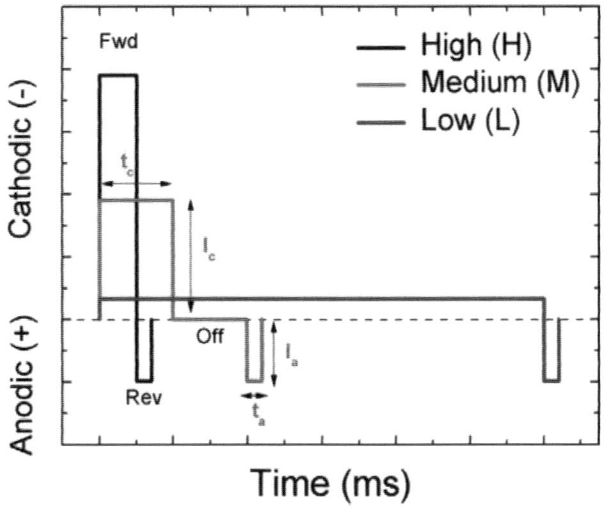

FARADAY TECHNOLOGY, INC.

Electrocatalysis – Effect of Activation Temperature

Pulsed Waveform "Low"

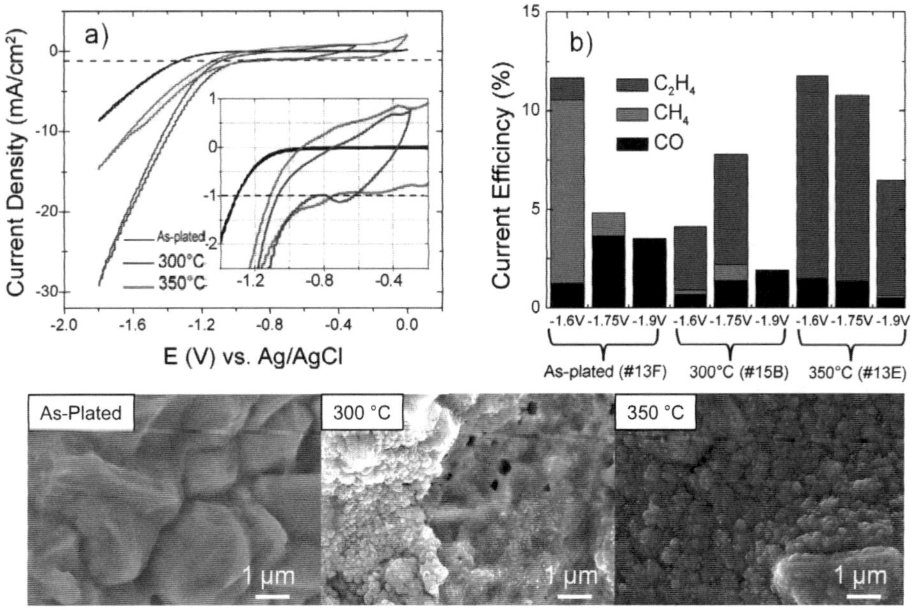

FARADAY TECHNOLOGY, INC.

Electrocatalysis – Effect of Deposition Waveform

Pulse-Deposited Samples Activated at 350 °C

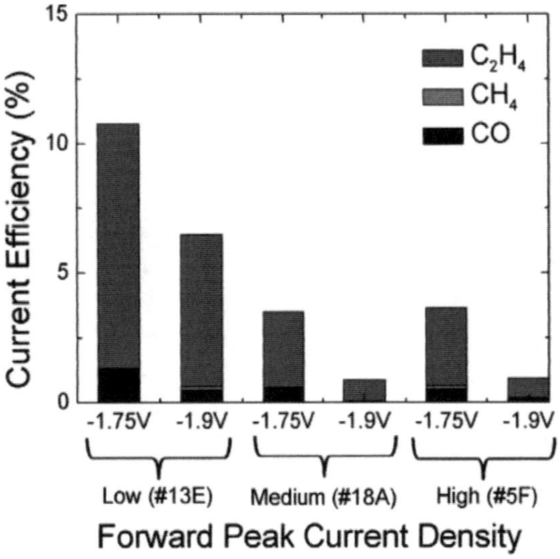

FARADAY TECHNOLOGY, INC.

Electrocatalysis – Effect of Pulsed vs DC Waveform

DC vs Pulsed Waveform "Low"

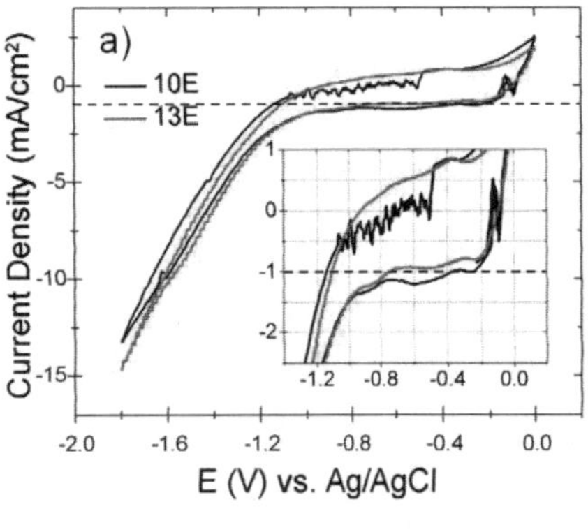

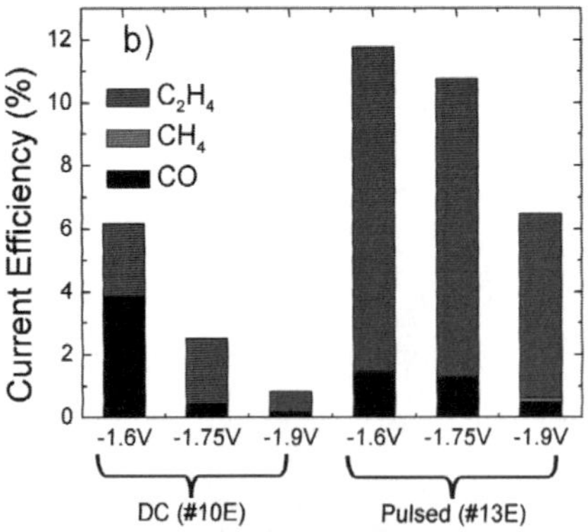

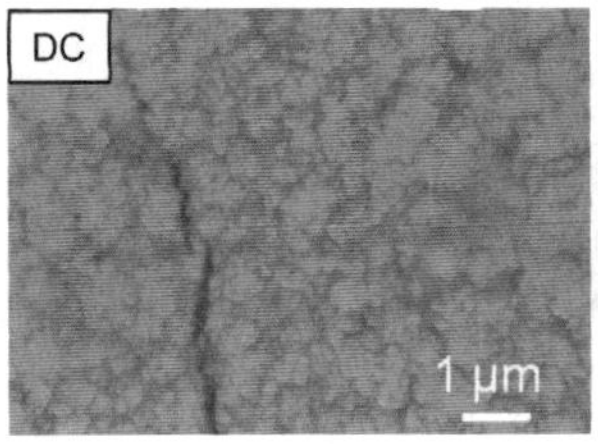

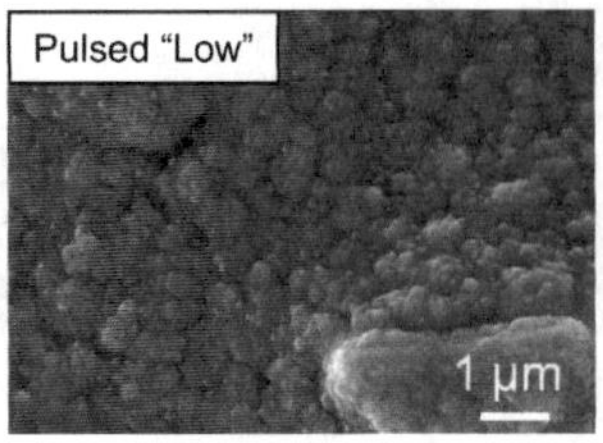

FARADAY TECHNOLOGY, INC.

Amine Electrolyte Additives

Ionic Liquids & 1° Amines <u>Unstable</u>

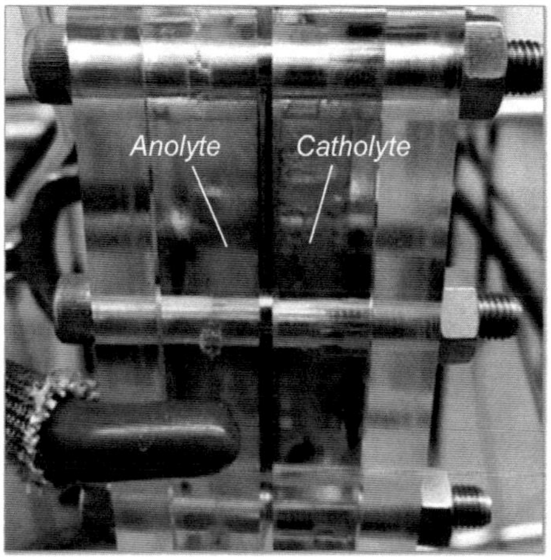

2° & 3° Amines Stable <u>BUT</u> Inhibit CO_2 Conversion @ RT

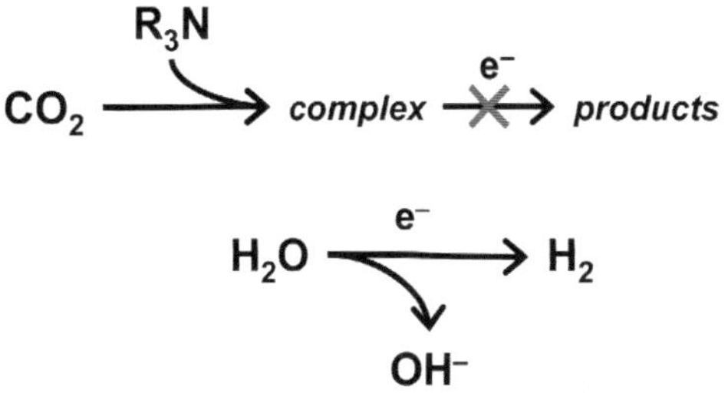

Summary

- Improved CO_2 reduction electrocatalysts via pulsed FARADAYIC® ElectroDeposition
  - Properly tuned deposition waveform superior to DC
    - Greater hydrocarbon production
    - Less cathodic onset potential
  - Activation shifts hydrocarbon selectivity from CH_4 to C_2H_4
    - Activation parameters significant
- Future development
  - Activation/deposition parameters
  - Amine additives at elevated temperature
  - Goals:
    - Increase hydrocarbon (ethylene) selectivity
    - Increase total current density

Acknowledgments

- This work was supported by the United States Department of Energy (contract #DE-SC0015812). Any opinions, findings, conclusions and recommendations expressed in this material are those of the authors and do not necessarily reflect the views of the US DOE.

- Contact Information:

  Brian T. Skinn

  Ph: +1 937 836 7749

  Email: BrianSkinn@FaradayTechnology.com

Thank You!

491354 Pilot Valve CFD Modelling

Abstract:

Pilot valves are utilized in many industrial processes. Increasing their performance efficiency as well as reducing cavitation are main objectives of the designers. We have numerically studied the motion of a 3D pilot valve to investigate the flow field, presence and location of cavitation spots as well as flow forces on the valve components. The motion of the poppet was included through the balance of spring and hydrodynamics forces. A dynamic mesh strategy was utilized to accommodate for the motion of the poppet. History of the volumetric flow rates, flow forces, and the cavitation regions are the main outputs of the simulations. The flow-rate compared favorably with the experiment where the test data was available. The outputs of this study are crucial for optimizing of the design of these valves to enhance the performance and reduce the cavitation.

496398 Multi-Objective Optimization of Solid Sorbent-Based CO_2 Capture Systems

Abstract:

This work presents a multi-objective optimization framework that uses a mixed integer nonlinear programming model (MINLP) for the optimal design and operation of carbon capture plants. A mix of first principle and surrogate-based models have been developed to characterize the process design and operation.

Coal-based power generation is one of the most important sources of energy around the globe. As such, it is important to develop cost-effective carbon capture and storage (CCS) technologies to reduce the environmental impact of power plants. Reducing emissions without affecting the overall cost of the system is difficult. Therefore, to examine the trade-off between cost and emissions, advanced energy analysis tools that consider multiple criteria are needed. The state of the art in technologies for carbon capture include solvents, membrane and solid sorbent systems. This work analyzes the solid sorbent-based adsorption system under different objectives (such as minimizing the cost of electricity and maximizing capture rate, operating vs capital cost, etc.).

In solid sorbent-based capture systems, the process is mainly driven by the interaction of the solid sorbent particles with the flue gas (mainly consisting of N_2, CO_2, and H_2O). The main process units of the solid sorbent systems are multiphase reactors, such as moving bed and fluidized beds. Given the modeling complexity and the difficulty of formulating algebraic models for such processes, previous studies either are generally limited to process-simulation[1] based frameworks or simplified models for mathematical optimization. Process simulators limit the exploration of different process configurations, while in the latest example the validation of the process behavior is limited. Our previous work proposed a mathematical optimization framework, in which both first principles-based and surrogate models[2] were developed to obtain the optimal plant design and operating conditions[3]. Carefully-tuned detailed models implemented in commercial process simulators have been exploited in the preparation and characterization of surrogate models with ALAMO[2]. Model validation and sample generation play a very important role in this mathematical framework.

The proposed mathematical model considers a set of components (CO_2, N_2, H_2O, and Carbamate), streams (utilities, gas, and solids), a set of units (adsorbers and regenerators mixers, splitters, compressors, pumps and heat exchangers), and a set of technologies (two reactor configurations for adsorption and regeneration). **Design decisions** determine the plant layout, by selecting the optimal number of parallel trains, process configuration (number of beds for adsorption and regeneration processes), and equipment design for each unit (diameter, length, area, etc.). At the **operating level**, the decisions involve the optimal flowrates, temperatures, molar fractions of gas and loading of the solids.

Making use of the reduced-order models generated from our previous work, this work uses the normal-constraint method for decision making under multiple objectives. The general idea is to obtain the Pareto frontier generating a set of solutions by solving multiple optimization problems[4]. Finally, one of the Pareto optimal solutions will be selected based on the Utopia-tracking approach[5].

[1] Eslick J. C., Tong C., Lee A., Dowling A. W., Mebane D. S. (2015). Optimization under uncertainty with rigorous process models. 2015 AIChE Annual Meeting.

[2] Cozad A., Sahinidis N. V., Miller D. C. (2014). Automatic Learning of Algebraic Models for Optimization. AIChE Journal, 60, 2211-2227.

[3] Zamarripa M., Eslick J., Lee A., Ajayi O., Wilson Z., Sahinidis N., Miller D. (2016). Optimal Design and Operation of Hybrid CO_2 Capture Systems. 2016 AIChE Annual Meeting.

[4] Messac A., Ismail-Yahaya A., Mattson C. A. (2003). The normalized normal constraint method for generating the Pareto frontier. Structural and Multidisciplinary Optimization, 25, 86-98.

[5] Zavala V. M. (2015). Managing Conflicts among Decision-Makers in Multiobjective Design and Operations. 2015 AIChE Annual Meeting.

497637 Bioelectrosynthesis of Methane from CO_2 for Energy Storage

Abstract:

The use of electricity to drive biologically-mediated reactions (bioelectrosynthesis) is a promising technique for CO2 utilization. Unlike most industrial electrochemical means to convert CO2, bioelectrosynthesis reactions like microbial methanogensis have been shown to have high electron capture efficiency (~90%), and overall energy efficiency (~80%) (Cheng et al., 2009). Bioelectrosynthesis also has the potential to produce a highly specific product stream with limited side reactions.

As we enter an era of abundant and intermittent renewable energy, the conversion of CO2 to fuels at the utility scale is especially desirable. Low-cost electricity during peak production hours can be used to generate fuels for distribution or for storage and later use for off-peak power generation. If the CO2 comes from biological sources or the atmosphere, the resulting fuel can be sold at a premium under Low Carbon Fuel Standards like California's.

Bioelectrosynthesis is being pursued for conversion of CO2 to fuels or chemicals such as hydrogen peroxide or acetate, but the production of methane has several advantages. Methane (as natural gas) is already used as a transportation fuel and has an enormous transportation and distribution network in place. It can be inexpensively stored at large scale in underground reservoirs and can be converted back into electricity with relatively high efficiency at existing natural gas power plants.

An industrially viable scheme for bioelectrosynthesis requires a reactor with high throughput, low overpotential, and a stable biocatalyst. A recent report by Deutzmann et al. (2015) lends significant promise to bioelectrosynthetic methane production. They identified extracellular enzymes that mediate electron transfer from graphite electrodes to methanogenic microbes. The efficiency and stability (> 1 month) of these biocatalysts open a wide range of possibilities for innovative reactor design.

A challenge common among bioelectrosynthetic systems is the need to optimize the "triple junction," where electrons, protons, and substrate are all supplied to the catalyst site while products are removed away before they can further react. The catalyst, in turn, should be supported on a high-surface-area (conductive) material. The biocatalysts reported by Deutzmann et al. have so far only been demonstrated on planar electrodes. However, with extracellular enzymes to mediate electron transfer, a porous electrode can provide radically higher surface area and volumetric throughput.

In this research, we test the activity of Methanococcus maropaludis cells and extracts in a variety of cutting-edge carbon aerogel electrodes. These custom-synthesized, electrically-conductive aerogels have controlled pore network sizes and structures yielding surface areas up to 500x106 m2/m3, 100 times more than what has been used previously in bioelectrosynthesis (Zhu et al., 2015). The aerogels can be 3D-printed into innovate reactor structures that create flow channels for mass transfer while limiting distances at the triple junction.

Here, we report the potential for improved reactor throughput using the combination of enzyme-mediated electron transfer and carbon aerogel electrodes for methane production. The results have implications for reactor design with bioelectrosynthesis and for the viability of CO2 to methane conversion for energy storage.

References

Cheng, S., Xing, D., Call, D. F. & Logan, B. E. Direct Biological Conversion of Electrical Current into Methane by Electromethanogenesis. Environ. Sci. Technol. 43, 3953–3958 (2009).

Deutzmann, J. S., Sahin, M. & Spormann, A. M. Extracellular Enzymes Facilitate Electron Uptake in Biocorrosion and Bioelectrosynthesis. mBio 6, e00496–15 (2015).

Zhu, C. et al. Highly compressible 3D periodic graphene aerogel microlattices. Nat. Commun. 6, 6962 (2015).

Modeling the Sequestration and Transportation of CO_2 in Deformed Coalbed during Enhanced Coalbed Methane Recovery

Quanshu Zeng and Zhiming Wang, State Key Laboratory of Petroleum Resources and Prospecting, College of Petroleum Engineering, China University of Petroleum, Beijing 102249, China
Brian McPherson and John McLennan. Energy & Geoscience Institute at the University of Utah, Salt Lake City, UT 84108, United States

Abstract

Gas adsorption/desorption has a significant effect on coal deformation and permeability evolution during the primary recovery of coalbed methane (CBM) and enhanced coalbed methane recovery (ECBM). The objectives are to: (1) quantify the coal deformation and permeability change caused by methane (CH_4) displacement with carbon dioxide (CO_2), (2) model the sequestration and transportation of CO_2 in deformed coalbed. In this study, the coal deformation during CBM and ECBM recovery was described by substituting simplified local density (SLD) adsorption theory into a theoretical strain model. The strain model was then verified with all CH_4/CO_2 measured coal strain data published in the past 60 years. Next, the permeability evolution was further revealed by rigorously coupling the SLD theory, the strain model, and the cleat permeability model. Finally, the coupled model was further integrated into the Transport of Unsaturated Groundwater and Heat Simulator (TOUGH2) to simulate the ECBM process. The results show that the coupled model can simultaneously describe gas adsorption or desorption, coal deformation, and permeability evolution during CBM and ECBM recovery with only six parameters, including slit length, solid-solid interaction potential energy parameter, surface areas of CH_4 and CO_2, adsorption expansion modulus, and cleat parameter. The coupled model can predict the coal deformation induced by CH_4 or CO_2 adsorption at pressure up to 12 MPa, and the average relative errors between the measured and predicted results are all within 21.48%. The results also show that the adsorption capacity of CO_2 is 2-5 times as large as that of CH_4, and the volumetric strain induced by CO_2 adsorption is 3-5 times as large as that caused by CH_4 adsorption. While the stronger adsorption capacity of CO_2 on coal offers an option for CO_2-ECBM, coal swelling due to CH_4 displacement with CO_2 may narrow down or even close the cleat, significantly reducing the permeability, and thus impacting the injection efficiency. In addition, the original TOUGH2 simulator predicts similar results with several other CBM simulators. However, it's impossible that 90% of CH_4 can be displaced within 90 days. Considering the coal deformation and permeability change due to CH_4 displacement with CO_2, the modified TOUGH2 simulator shows that only 24% of CH_4 is displaced in the first 90 days, and it takes about 1800 days to displace 90% or more. Advances in the understanding of sequestration and transportation of CO_2 help to improve the ECBM technology.

Keywords

Enhanced Coalbed Methane Recovery, Coal Deformation, Gas Adsorption, Permeability Evolution

Introduction

Increasing atmospheric carbon dioxide (CO_2) concentrations are believed to be the major cause of global warming. Several ways to reduce CO_2 emissions include: improve power generation efficiency by upgrading existing plants or building new plants, rely more on renewable energy and nuclear-generated power, and finally, carbon capture and storage (CCS). Of all emission reduction mechanisms, CCS is projected to provide the largest contribution to emission reduction in the coming decade.

Carbon capture and storage is a process of capturing waste CO_2 from large point sources, transporting it to storage sites, and depositing it where it will not enter the atmosphere, normally an underground geological formation. The geological sequestration of CO_2 offers several options, including deep saline reservoirs, oil and gas fields, and unminable coal seams. Of these options, only enhanced oil recovery (EOR) and enhanced coalbed methane recovery (ECBM) provide a value-adding by-product to CO_2 injection, offsetting sequestration cost or, in some cases, making CO_2 injection profitable. This economic benefit is particularly important in the absence of regulatory controls and/or financial incentives for CO_2 sequestration.

CO_2-EOR is a relatively mature technology, however, ECBM is still in its infancy, with most of the CO_2 injection projects in coal seams not to be successful. Advances in the understanding of CO_2 sequestration and transportation may help to improve the ECBM technology. Existing coalbed methane (CBM) simulators, which are developed for primary CBM recovery, have many important features considered, such as: (1) a dual porosity system, (2) a water/gas two-phase fluid, (3) Darcy flow in the cleat system, (4) pure gas desorption in the matrix system, and (5) matrix shrinkage due to gas desorption. However, the recovery process becomes more complex with CO_2 injection. Several additional features have to be considered, such as: (1) methane (CH_4)/CO_2 mixture adsorption, (2) coal swelling due to adsorption difference between CH_4 and CO_2.

This paper is aimed at developing an internally-consistent adsorption-strain-permeability model to simultaneously quantify gas desorption, coal deformation, and permeability change during CBM and ECBM recovery. The remainder of this paper is organized as follows: modeling section presents the model and couple details of gas adsorption, coal deformation, and permeability evolution; results section presents the validation of strain model and coupled model; discussion section presents the ECBM case simulated by modified TOUGH2 simulator.

Modeling

In this section, the coupled adsorption-strain-permeability model is detailed. Gas adsorption, coal deformation, and cleat permeability characteristics are described using simplified local density theory, a theory-based strain model, and a matchstick-based permeability model, respectively.

Simplified Local Density Theory

Rangarajan[7] first proposed simplified local density (SLD) theory by combining mean-field approximation theory and density functional theory, believing that adsorption is the resultant effect of adsorbate-adsorbate and adsorbate-adsorbent molecule interactions. Several assumptions were made for SLD theory[7-10], including

(1) The chemical potential of any point above adsorbent surface is equal to the bulk phase chemical potential;

(2) The chemical potential of any point above adsorbent surface is the sum of adsorbate-adsorbate and adsorbate-adsorbent interactions;

(3) The adsorbate-adsorbent interaction of any point above adsorbent surface is independent of the temperature and the number of molecules;

(4) Coal pores are considered as perfect slits with temperature and pressure distributed uniformly;

(5) All adsorbate and adsorbent molecules are spherical, except for the adsorbate molecules touching slit walls.

Therefore, the chemical potential of any point above adsorbent surface should be the same, or

$$\mu(z) = \mu_{bulk} = \mu_{ff}(z) + \mu_{fs}(z) \tag{1}$$

where, $\mu(z)$ is the chemical potential at z position, J/mol; z is the distance between adsorbate molecule and adsorbent surface, m; μ_{bulk} is the chemical potential of bulk phase, J/mol; $\mu_{ff}(z)$ is the chemical potential caused by the adsorbate-adsorbate interaction at z position, J/mol; and $\mu_{fs}(z)$ is the chemical potential caused by the adsorbate-adsorbent interaction at z position, J/mol.

Since slit shape is assumed for coal pores while adsorbate molecules located between the parallel slit walls, adsorbate molecules have interactions with both walls (Fig. 1), or

$$\mu_{fs}(z) = N_A\left[\Psi_{fs}(z) + \Psi_{fs}(L_s - z)\right] \tag{2}$$

where, interactions between adsorbent and adsorbate molecules can be effectively represented by the Lennard-Jones potential function[11]. Specifically, the interaction between adsorbate molecules and carbon planes after the fourth layer are ignored, or

$$\Psi_{fs}(z) = 4\pi\rho_{atoms}\varepsilon_{fs}d_{fs}^2\left\{\frac{1}{5}\left(\frac{d_{fs}}{z'}\right)^{10} - \frac{1}{2}\sum_{m=1}^{4}\left[\frac{d_{fs}}{z' + (m-1)d_{ss}}\right]^4\right\} \tag{3}$$

where,

$$\varepsilon_{fs} = \sqrt{\varepsilon_{ff} \times \varepsilon_{ss}} \tag{4}$$

$$d_{fs} = \frac{d_{ff} + d_{ss}}{2} \tag{5}$$

$$z' = z + \frac{d_{cc}}{2} \tag{6}$$

and where, N_A is the Avogadro constant, 6.02×10^{23} mol^{-1}; $\Psi_{fs}(z)$ is the potential energy caused by the interaction between adsorbate molecule at z position and the left wall, J; $\Psi_{fs}(L_s - z)$ is the potential energy caused by the interaction between adsorbate molecule at z position and the right wall, J; L_s is the slit width, m; ρ_{atoms} is the carbon atom density, 3.82×10^{19} m^{-2}; ε_{fs} is the potential energy caused by the interaction between adsorbate and adsorbent molecules, J; ε_{ff} is the potential energy caused by the interaction between adsorbate and adsorbate molecules, J; ε_{ss} is the potential energy caused by the interaction between adsorbent and adsorbent molecules, J; d_{fs} is the collision diameter between adsorbate and adsorbent molecule, m; d_{ff} is the diameter of adsorbate molecule, m; d_{ss} is the carbon plane spacing, 3.35×10^{-10} m; z' is the distance between adsorbate molecule and carbon atoms of the first layer, m; and d_{cc} is the diameter of carbon atom, 1.4×10^{-10} m.

The chemistry potentials of bulk and adsorption phases can be described in fugacity forms, or

$$\mu_{bulk} = \mu_0 + RT\ln\left(\frac{f_{bulk}}{f_0}\right) \tag{7}$$

$$\mu_{ff}(z) = \mu_0 + RT\ln\left[\frac{f_{ads}(z)}{f_0}\right] \tag{8}$$

where, interactions among adsorbate molecules can be described by a fluid equation of state. The Patel-Teja-Valderrama equation of state (PTV EOS) can describe the thermodynamic properties of CH_4 and CO_2 with a reasonable accuracy[12], and thus was selected to calculate density and fugacity of the bulk phase and fugacity of the adsorption phase. The physical properties of CH_4 and CO_2 involved in this study are summarized in Table 1 and can be expressed by

$$\frac{p}{RT\rho} = \frac{1}{1-b\rho} - \frac{a\rho}{RT\left[1+(b+c)\rho-bc\rho^2\right]} \tag{9}$$

where,

$$\begin{aligned} a &= (0.66121-0.761057Z_c)\frac{R^2T_c^2}{p_c} \\ &\times\left[1+\left(0.46283+3.58230\omega Z_c+8.19417\omega^2Z_c^2\right)\left(1-\sqrt{T/T_c}\right)\right]^2 \end{aligned} \tag{10}$$

$$b = (0.02207+0.20868Z_c)\frac{RT_c}{p_c} \tag{11}$$

$$c = (0.57765-1.87080Z_c)\frac{RT_c}{p_c} \tag{12}$$

where, μ_0 is the chemical potential at reference state, J/mol; R is the gas constant, 8.314 J/(K·mol); T is the temperature, K; f_{bulk} is the fugacity of bulk phase, Pa; f_0 is the fugacity at reference state, Pa; $f_{ads}(z)$ is the fugacity of adsorption phase at z position, Pa; p is the pressure, Pa; ρ is the density, mol/m^3; a is the attraction term, J·m^3·mol^{-2}; b is the repulsion term, m^3/mol; c is the polarity term, m^3/mol; Z_c is the critical compression factor, dimensionless; T_c is the critical temperature, K; p_c is the critical pressure, Pa; and ω is the eccentric factor, dimensionless.

To calculate the fugacity of the bulk phase and adsorption phase, the PTV EOS can be rearranged as

$$\begin{aligned} \ln\frac{f_{bulk}}{p} &= \frac{p}{RT\rho_{bulk}} - 1 - \ln\left(\frac{p}{RT\rho_{bulk}} - \frac{pb_{bulk}}{RT}\right) + \frac{a_{bulk}/(RT)}{\sqrt{b_{bulk}^2+6b_{bulk}c_{bulk}+c_{bulk}^2}} \\ &\times\ln\left[\frac{2+\rho_{bulk}\left(b_{bulk}+c_{bulk}-\sqrt{b_{bulk}^2+6b_{bulk}c_{bulk}+c_{bulk}^2}\right)}{2+\rho_{bulk}\left(b_{bulk}+c_{bulk}+\sqrt{b_{bulk}^2+6b_{bulk}c_{bulk}+c_{bulk}^2}\right)}\right] \end{aligned} \tag{13}$$

$$\ln\frac{f_{ads}(z)}{p}=\frac{p}{RT\rho_{ads}(z)}-1-\ln\left[\frac{p}{RT\rho_{ads}(z)}-\frac{pb_{ads}}{RT}\right]+\frac{a_{ads}(z)/(RT)}{\sqrt{b_{ads}^2+6b_{ads}c_{ads}+c_{ads}^2}}$$
$$\times\ln\left[\frac{2+\rho_{ads}(z)\left(b_{ads}+c_{ads}-\sqrt{b_{ads}^2+6b_{ads}c_{ads}+c_{ads}^2}\right)}{2+\rho_{ads}(z)\left(b_{ads}+c_{ads}+\sqrt{b_{ads}^2+6b_{ads}c_{ads}+c_{ads}^2}\right)}\right] \tag{14}$$

where, the attraction term of adsorbed phase varies with the relative position of adsorbate molecules. The relation between $a_{ads}(z)$ and L_s/d_{ff} was obtained by integrating the sum of the interactions between an adsorbate molecule and all the adsorbed molecules around it[7].

For $L_s/d_{ff}\geq 3$,

$$\frac{a_{ads}(z)}{a_{bulk}}=\begin{cases}\frac{3}{8}\frac{z}{d_{ff}}+\frac{5}{16}-\frac{1}{8\left(\frac{L_s-z}{d_{ff}}-\frac{1}{2}\right)^3} & \frac{1}{2}\leq\frac{z}{d_{ff}}\leq\frac{3}{2}\\ 1-\frac{1}{8\left(\frac{z}{d_{ff}}-\frac{1}{2}\right)^3}-\frac{1}{8\left(\frac{L_s-z}{d_{ff}}-\frac{1}{2}\right)^3} & \frac{3}{2}\leq\frac{z}{d_{ff}}\leq\frac{L_s}{d_{ff}}-\frac{3}{2}\\ \frac{3(L_s-z)}{8d_{ff}}+\frac{5}{16}-\frac{1}{8\left(\frac{z}{d_{ff}}-\frac{1}{2}\right)^3} & \frac{L_s}{d_{ff}}-\frac{3}{2}\leq\frac{z}{d_{ff}}\leq\frac{L_s}{d_{ff}}-\frac{1}{2}\end{cases} \tag{15}$$

For $3>L_s/d_{ff}\geq 2$,

$$\frac{a_{ads}(z)}{a_{bulk}}=\begin{cases}\frac{3z}{8d_{ff}}+\frac{5}{16}-\frac{1}{8\left(\frac{L_s-z}{d_{ff}}-\frac{1}{2}\right)^3} & \frac{1}{2}\leq\frac{z}{d_{ff}}\leq\frac{L_s}{d_{ff}}-\frac{3}{2}\\ \frac{3L_s}{8d_{ff}}-\frac{3}{8} & \frac{L_s}{d_{ff}}-\frac{3}{2}\leq\frac{z}{d_{ff}}\leq\frac{3}{2}\\ \frac{3(L_s-z)}{8d_{ff}}+\frac{5}{16}-\frac{1}{8\left(\frac{z}{d_{ff}}-\frac{1}{2}\right)^3} & \frac{3}{2}\leq\frac{z}{d_{ff}}\leq\frac{L_s}{d_{ff}}-\frac{1}{2}\end{cases} \tag{16}$$

For $2>L_s/d_{ff}\geq 1.5$,

$$\frac{a_{ads}(z)}{a_{bulk}}=\frac{3}{8}\left(\frac{L_s}{d_{ff}}-1\right) \tag{17}$$

where, a_{bulk} is the attraction term for the bulk phase, J·m^3·mol^{-2}; b_{bulk} is the repulsion term for the bulk phase, m^3/mol; c_{bulk} is the polarity term for the bulk phase, m^3/mol; $a_{ads}(z)$ is the attraction term for the adsorption phase, J·m^3·mol^{-2}; b_{ads} is the repulsion term for the adsorption phase, m^3/mol; c_{ads} is the polarity term for the adsorption phase, m^3/mol; and ρ_{bulk} is the density of bulk phase, mol/m^3; and $\rho_{ads}(z)$ is the density of adsorption phase at *z* position, mol/m^3.

Equilibrium criteria can be eventually obtained by substituting Eqs. (2), (7), and (8) into Eq. (1), or

$$f_{ads}(z) = f_{bulk} \exp\left[-\frac{\Psi_{fs}(z) + \Psi_{fs}(L_s - z)}{k_B T} \right] \tag{18}$$

where, k_B is the Boltzmann constant, 1.38×10^{-23}J/K.

According to SLD theory, the Gibbs excess adsorption can be obtained by integrating the density difference between adsorption and bulk phases along the slit length. Since slit pore includes two walls (Fig. 1), half the surface area is considered in Eq. (19), or

$$n^{Gibbs} = \frac{A_s}{2} \int_{\frac{3d_{ff}}{8}}^{L_s - \frac{3d_{ff}}{8}} \left[\rho_{ads}(z) - \rho_{bulk} \right] \mathrm{d}z \tag{19}$$

where, n^{Gibss} is the Gibbs excess adsorption amount, mol/kg; and A_s is the surface area of unit mass adsorbent, m^2/kg.

A fluid mixing rule[9] was implemented in the fluid equation of state to describe adsorbate-adsorbate interactions, including CH_4-CH_4, CH_4-CO_2, and CO_2-CO_2 (Fig. 2). Similar to pure component adsorption, the fugacity of CH_4 and CO_2 in both phases can be described by the PTV EOS, revised as

$$\begin{aligned}
\ln\frac{f_i^{bulk}}{x_i p} = {} & \frac{\bar{b}_i^{bulk}\rho_{bulk}}{1 - b_{bulk}\rho_{bulk}} - \ln\left(\frac{p}{RT\rho_{bulk}} - \frac{p b_{bulk}}{RT} \right) + \frac{a_{bulk}}{RT\sqrt{b_{bulk}^2 + 6b_{bulk}c_{bulk} + c_{bulk}^2}} \\
& \times \left[\frac{\bar{a}_i^{bulk}}{a_{bulk}} + 1 - \frac{b_{bulk}\bar{b}_i^{bulk} + 3b_{bulk}\bar{c}_i^{bulk} + 3\bar{b}_i^{bulk}c_{bulk} + c_{bulk}\bar{c}_i^{bulk}}{b_{bulk}^2 + 6b_{bulk}c_{bulk} + c_{bulk}^2} \right] \\
& \times \ln\left[\frac{2 + \rho_{bulk}\left(b_{bulk} + c_{bulk} - \sqrt{b_{bulk}^2 + 6b_{bulk}c_{bulk} + c_{bulk}^2} \right)}{2 + \rho_{bulk}\left(b_{bulk} + c_{bulk} + \sqrt{b_{bulk}^2 + 6b_{bulk}c_{bulk} + c_{bulk}^2} \right)} \right] \\
& + \left[\frac{p}{RT\rho_{bulk}} \frac{b_{bulk}\bar{b}_i^{bulk} + 3b_{bulk}\bar{c}_i^{bulk} + 3\bar{b}_i^{bulk}c_{bulk} + c_{bulk}\bar{c}_i^{bulk}}{b_{bulk}^2 + 6b_{bulk}c_{bulk} + c_{bulk}^2} \right. \\
& \left. + \frac{\left(-b_{bulk} + c_{bulk}\right)\left(\bar{b}_i^{bulk}c_{bulk} - b_{bulk}\bar{c}_i^{bulk}\right)}{b_{bulk}^2 + 6b_{bulk}c_{bulk} + c_{bulk}^2} \right] \left(1 - \frac{1}{\frac{p}{RT\rho_{bulk}} - \frac{p b_{bulk}}{RT}} \right)
\end{aligned} \tag{20}$$

and

$$
\begin{aligned}
\ln\frac{f_i^{ads}(z)}{y_i(z)p} &= \frac{\bar{b}_i^{ads}\rho_{ads}(z)}{1-b_{ads}\rho_{ads}(z)} - \ln\left[\frac{p}{RT\rho_{ads}(z)} - \frac{pb_{ads}}{RT}\right] + \frac{a_{ads}(z)}{RT\sqrt{b_{ads}^2+6b_{ads}c_{ads}+c_{ads}^2}} \\
&\times\left[\frac{\bar{a}_i^{ads}(z)}{a_{ads}(z)} + 1 - \frac{b_{ads}\bar{b}_i^{ads} + 3b_{ads}\bar{c}_i^{ads} + 3\bar{b}_i^{ads}c_{ads} + c_{ads}\bar{c}_i^{ads}}{b_{ads}^2+6b_{ads}c_{ads}+c_{ads}^2}\right] \\
&\times\ln\left[\frac{2+\rho_{ads}(z)\left(b_{ads}+c_{ads}-\sqrt{b_{ads}^2+6b_{ads}c_{ads}+c_{ads}^2}\right)}{2+\rho_{ads}(z)\left(b_{ads}+c_{ads}-\sqrt{b_{ads}^2+6b_{ads}c_{ads}+c_{ads}^2}\right)}\right] \\
&+\left[\frac{p}{RT\rho_{ads}(z)}\frac{b_{ads}\bar{b}_i^{ads} + 3b_{ads}\bar{c}_i^{ads} + 3\bar{b}_i^{ads}c_{ads} + c_{ads}\bar{c}_i^{ads}}{b_{ads}^2+6b_{ads}c_{ads}+c_{ads}^2}\right. \\
&\left.+\frac{\left(-b_{ads}+c_{ads}\right)\left(\bar{b}_i^{ads}c_{ads}-b_{ads}\bar{c}_i^{ads}\right)}{b_{ads}^2+6b_{ads}c_{ads}+c_{ads}^2}\right]\left(1-\frac{1}{\dfrac{p}{RT\rho_{ads}(z)}-\dfrac{pb_{ads}}{RT}}\right)
\end{aligned}
\tag{21}
$$

where,

$$\bar{a}_i^{bulk} = \left(\frac{\partial n a_{bulk}}{\partial n_i}\right)_{\theta,V,n_j} \tag{22}$$

$$\bar{b}_i^{bulk} = \left(\frac{\partial n b_{bulk}}{\partial n_i}\right)_{\theta,V,n_j} \tag{23}$$

$$\bar{c}_i^{bulk} = \left(\frac{\partial n c_{bulk}}{\partial n_i}\right)_{\theta,V,n_j} \tag{24}$$

$$\bar{a}_i^{ads}(z) = \left[\frac{\partial n a_{ads}(z)}{\partial n_i}\right]_{\theta,V,n_j} \tag{25}$$

$$\bar{b}_i^{ads} = \left(\frac{\partial n b_{ads}}{\partial n_i}\right)_{\theta,V,n_j} \tag{26}$$

$$\bar{c}_i^{ads} = \left(\frac{\partial n c_{ads}}{\partial n_i}\right)_{\theta,V,n_j} \tag{27}$$

where, attraction terms are determined by quadratic mixing rules while repulsion and polarity terms are determined by linear mixing rules, or

$$a_{bulk} = \sum_i\sum_j x_i x_j a_{ij}^{bulk} \tag{28}$$

$$a_{ads}(z) = \sum_i\sum_j y_i(z)y_j(z)a_{ij}^{ads}(z) \tag{29}$$

$$b_{bulk} = \sum_i x_i \bar{b}_i^{bulk} \tag{30}$$

$$b_{ads} = \sum_i y_i(z)\bar{b}_i^{ads} \tag{31}$$

$$c_{bulk} = \sum_i x_i \bar{c}_i^{bulk} \tag{32}$$

$$c_{ads}=\sum_i y_i(z)\overline{c}_i^{ads} \tag{33}$$

and

$$a_{ij}^{bulk}=\sqrt{\overline{a}_i^{bulk}\times\overline{a}_j^{bulk}} \tag{34}$$

$$a_{ij}^{ads}(z)=\sqrt{\overline{a}_i^{ads}(z)\times\overline{a}_j^{ads}(z)} \tag{35}$$

where, f_i^{bulk} is the fugacity of component i in the bulk phase, Pa; $f_i^{ads}(z)$ is the fugacity of component i in the adsorbed phase at z position, Pa; x_i is the mole fraction of component i in the bulk phase, dimensionless; x_j is the mole fraction of component j in the bulk phase, dimensionless; $y_i(z)$ is the mole fraction of component i in the adsorbed phase at z position, dimensionless; $y_j(z)$ is the mole fraction of component j in the adsorbed phase at z position, dimensionless; $\overline{a}_i^{bulk}$ is the attraction term of component i in the bulk phase, $J\cdot m^3\cdot mol^{-2}$; $\overline{a}_i^{ads}(z)$ is the attraction term of component i in the adsorbed phase at z position, $J\cdot m^3\cdot mol^{-2}$; $\overline{b}_i^{bulk}$ is the repulsion term of component i in the bulk phase, m^3/mol; $\overline{b}_i^{ads}$ is the repulsion term of component i in the adsorbed phase, m^3/mol; $\overline{c}_i^{bulk}$ is the polarity term of component i in the bulk phase, m^3/mol; $\overline{c}_i^{ads}$ is the polarity term of component i in the adsorption phase, m^3/mol; a_{ij}^{bulk} is the cross coefficient to calculate the attraction term in the bulk phase, $J\cdot m^3\cdot mol^{-2}$; and $a_{ij}^{ads}(z)$ is the cross coefficient to calculate the attraction term in the adsorption phase, $J\cdot m^3\cdot mol^{-2}$.

For adsorbate-adsorbent interactions, both components can be calculated by the Lennard-Jones potential function, or

$$\Psi_i^{fs}(z)=4\pi\rho_{atoms}\varepsilon_i^{fs}\left(d_i^{fs}\right)^2\left\{\frac{1}{5}\left(\frac{d_i^{fs}}{z'}\right)^{10}-\frac{1}{2}\sum_{i=1}^{4}\left[\frac{d_i^{fs}}{z'+(i-1)d_{ss}}\right]^4\right\} \tag{36}$$

where,

$$\varepsilon_i^{fs}=\sqrt{\varepsilon_i^{ff}\times\varepsilon_{ss}} \tag{37}$$

$$d_i^{fs}=\frac{d_i^{ff}+d_{ss}}{2} \tag{38}$$

where, $\Psi_i^{fs}(z)$ is the potential energy caused by the interaction between component i at z position and the left wall (Fig. 1), J; ε_i^{fs} is the potential energy caused by interactions between component i and either wall, J; ε_i^{ff} is the potential energy caused by the interactions among the molecules of component i, J; d_i^{fs} is the collision diameter between component i and carbon atom, m; and d_i^{ff} is the molecule diameter of component i, m.

To implement the CH_4/CO_2 mixture adsorption model, each component of the mixture should satisfy the adsorption equilibrium criterion. In addition, both phases should satisfy the mole fraction constraints, or

$$f_i^{ads}(z)=f_i^{bulk}\exp\left[-\frac{\Psi_i^{fs}(z)+\Psi_i^{fs}(L_s-z)}{k_BT}\right] \tag{39}$$

$$\sum x_i=1 \tag{40}$$

$$\sum y_i=1 \tag{41}$$

Gibbs excess adsorption of the mixture can be calculated as

$$n^{Gibbs} = \sum n_i^{Gibbs} \tag{42}$$

where,

$$n_i^{Gibbs} = \frac{A_i^s}{2} \int_{\frac{3}{8}d_i^{ff}}^{L_s - \frac{3}{8}d_i^{ff}} \left[\rho_{ads}(z) y_i(z) - \rho_{bulk} x_i \right] dz \tag{43}$$

where, n_i^{Gibbs} is the adsorption amount of component i, mol/kg; and A_i^s is the surface area occupied by component i, m^2/kg.

Coal Strain Model

Both reservoir compaction and matrix shrinkage may occur during the primary recovery of CBM, as depicted schematically in Fig. 3. These have opposite effects on coal deformation and permeability evolution. Coal deformation due to reservoir compaction can be described by linear elastic theory, while that due to methane desorption may be predicted on the basis of Bangham and Fakhoury[13]. Both strains are assumed to be independent but superposed, or

$$\Delta\varepsilon = \Delta\varepsilon_m + \Delta\varepsilon_s \tag{44}$$

where, $\Delta\varepsilon$ is the total volumetric strain change, dimensionless; $\Delta\varepsilon_m$ is the volumetric strain change due to reservoir compaction, dimensionless; $\Delta\varepsilon_s$ is the volumetric strain change due to gas adsorption/desorption, dimensionless.

On one hand, the volumetric strain change due to reservoir compaction may be calculated on the basis of linear elastic theory and expressed as

$$\Delta\varepsilon_m = -\frac{3(1-2v)}{E}(p_n - p_0) \tag{45}$$

where, v is the Poisson's ratio, dimensionless; and E is the Young's modulus, Pa; p_0 is the initial pressure, Pa; and p_n is the current pressure, Pa.

On the other hand, Bangham and Fakhoury[13] believed that the adsorption of adsorbate will decrease the surface free energy of adsorbent, and the swell increment induced is proportional to the surface free energy decrease, or

$$\Delta\varepsilon_{sl} = \Delta\Psi \times \gamma \tag{46}$$

where,

$$\gamma = \frac{A_{ss}\rho_c}{K_{ss}} \tag{47}$$

where, $\Delta\varepsilon_{sl}$ is the linear strain change due to gas adsorption/desorption, dimensionless; $\Delta\Psi$ is the surface free energy change due to gas adsorption/desorption per unit mass adsorbent, J/kg; γ is a deformation constant, kg/J; A_{ss} is the specific surface area of adsorbent, m^2/m^2; ρ_c is the coal density, kg/m^3; and K_{ss} is the adsorbent expansion/shrinkage modulus due to gas adsorption/desorption, Pa.

According to the Gibbs adsorption equation, the surface free energy decrease may be characterized as

$$\Delta\Psi = RT\int_{p_0}^{p_n} \frac{n^{Gibbs}}{A_{ss}} \mathrm{d}(\ln p) \tag{48}$$

Combining Eqs. (46) and (48), the linear strain change due to gas adsorption/desorption may be expressed as

$$\Delta\varepsilon_{sl} = \frac{\rho_c RT}{K_{ss}} \int_{p_0}^{p_n} \frac{n^{Gibbs}}{p} \mathrm{d}p \tag{49}$$

To simplify the strain calculation, isotropy is assumed for the coal, and volumetric strain is assumed to be three times as large as linear strain[14-16], or

$$\Delta\varepsilon_s = \Delta 3\varepsilon_{sl} = \frac{3\rho_c RT}{K_{ss}} \int_{p_0}^{p_n} \frac{n^{Gibbs}}{p} \mathrm{d}p \tag{50}$$

Note that n^{Gibss} changes with pressure, and strains can be represented in differential forms and calculated according to Simpson's rule. Assuming that $f(p) = n^{Gibss} / p$, strain change due to gas adsorption/desorption may be rewritten as

$$\Delta\varepsilon_s = \frac{3\rho_c RT}{K_{ss}} \frac{\Delta p}{3} \left[f(p_0) + 4f(p_1) + 2f(p_2) + \cdots + 2f(p_{n-2}) + 4f(p_{n-1}) + f(p_n) \right] \tag{51}$$

Coal Permeability Model

Several cleat permeability models for coal reservoirs have been developed, including the Shi-Durucan (S&D) model[17-20], the Cui-Bustin (C&B) model[21, 22], the Palmer-Mansoori (P&M) model[23, 24] (and its improved version[25]). These models share three assumptions: (1) matchstick geometry, (2) uniaxial strain, and (3) constant overburden stress. Among these models, S&D model was found to have the best performance. The following section provide a brief review of S&D model.

Shi and Durucan[26] derived a permeability model accounting for two competing effects. The former effect is reservoir compaction. The effective stress will increase as the pore pressure decreases, resulting in the closure of cleats and the decrease of cleat permeability. Matrix shrinkage has an opposite effect on the cleat and permeability evolutions as pore pressure reduces and gas desorbs. Thus, an initial permeability decrease due to reservoir compaction is usually compensated by the subsequent increase owing to matrix shrinkage. Whether the permeability is higher than its initial value depends on the net effect of these two competing mechanisms.

The S&D model was derived from the constitutive equation of a linear thermo-elastic porous medium, but replaced the thermal expansion term with an analogous matrix shrinkage term, or

$$\frac{k}{k(0)} = e^{-3C_f \Delta\sigma_h^e} \tag{52}$$

where,

$$\Delta\sigma_h^e = \Delta\sigma_{h,rc}^e + \Delta\sigma_{h,ms}^e \tag{53}$$

and

$$\Delta\sigma_{h,rc}^e = -\frac{v}{1-v} \Delta p_p \tag{54}$$

$$\Delta\sigma_{h,ms}^e = \frac{E}{3(1-v)} \Delta\varepsilon_s \tag{55}$$

where, k is the permeability, mD; $k(0)$ is the initial permeability, mD; C_f is the cleat volume compressibility, Pa^{-1}; $\Delta\sigma_h^e$ is the effective horizontal stress change, Pa; $\Delta\sigma_{h,rc}^e$ is the effective horizontal stress change due to reservoir compaction, Pa; $\Delta\sigma_{h,ms}^e$ is the effective horizontal stress change due to matrix shrinkage, Pa; Δp_p is the pore pressure change, Pa.

In addition to pressure change, the adsorbate component change during ECBM will also contribute to the strain and permeability change, or

$$\frac{k_{CO2}}{k_{CH4}} = e^{-EC_f \Delta\varepsilon_s/(1-v)} \tag{56}$$

where,

$$\Delta\varepsilon_s = \varepsilon_{CO2} - \varepsilon_{CH4} \tag{57}$$

where, k_{CO2} is the permeability of carbon dioxide, mD; k_{CH4} is the permeability of methane, mD; ε_{CO2} is the volumetric strain caused by carbon dioxide adsorption, dimensionless; ε_{CH4} is the volumetric strain caused by methane adsorption, dimensionless.

Model Solution

Since adsorption in the coal strain model here is calculated by simplified local density, while matrix shrinkage in the permeability model is calculated by the strain model, these processes are rigorously coupled. The solution procedure for the coupled adsorption-strain-permeability model is depicted in Fig. 4. Input data include pressure, temperature, physical properties of CH_4 and CO_2, bulk phase mole fraction, physical and elastic properties, and adsorption, strain and cleat parameters of coal. Then, the algorithm calculates density, fugacity, and chemical potential of CH_4 and CO_2 components in both the bulk and adsorption phases, and calculates the adsorbate-adsorbent interaction and induced chemical potential. Next, loop the densities of CH_4 and CO_2 in the adsorption phase until both components satisfy a specific adsorption equilibrium criterion. Finally, adsorption, strain, and cleat parameters of coal are iteratively calculated until predicted results coincide with the adsorption isotherm, strain data, and permeability data.

Results

All CH_4 and CO_2 measured coal strain data[14, 22, 26-28] published during the past 60 years and publicly available were assembled to evaluate the performance of the proposed strain model.

Pressure, temperature, physical properties of CH_4 and CO_2, as well as the physical and elastic properties, adsorption and strain parameters of coal, must be realized prior to the coal strain calculation. The physical properties of CH_4 and CO_2 were extracted from the NIST database[29] while physical and elastic properties of coal were measured directly from coal samples, and adsorption and strain parameters were obtained by fitting the adsorption isotherm and strain data. Specifically, adsorption parameters include slit length, solid-solid interaction potential energy parameter, and surface areas of CH_4 and CO_2. The former two parameters were regressed with the shape of both isotherms while the surface area of each component was regressed with the magnitude of its isotherm. The strain parameter refers to adsorption swelling modulus, which is also the inherent nature of coal. All the input parameters required are summarized in Table 2, and the predicted adsorption and strain results are plotted in Figs. 5 and 6, respectively.

Fig. 5 suggests that CH_4 adsorption increases monotonically with pressure, while CO_2 first increases then decreases with pressure, and the inversion pressure is about 6 MPa. Fig. 5 also suggests that the adsorption capacity of CO_2 is 2-5 times as large as that of CH_4, this indicates coal has a stronger interaction with CO_2. This adsorption difference offers a possibility to deposit CO_2 in unminable coal seams, displace CH_4, thus increase CBM recovery without excessively lowering reservoir pressure.

Fig. 6 suggests that the strain model can predict coal deformation fairly accurately at pressure up to 12 MPa; the average relative errors between measured and predicted results are all within 21.48%, and the largest error occurs is associated with CO_2 adsorption on Wolf Mountain coal[22] (Fig. 6(c)). For all plotted results (Fig. 6), positive expresses expansion, while negative expresses shrinkage. Since volumetric strain is always larger than zero, this means matrix shrinkage is more obvious than reservoir compaction. The volumetric strains induced by both CH_4 and CO_2 adsorption increase monotonically with pressure. In addition, CO_2 adsorption induced volumetric strain is almost 3-5 times as large as that caused by CH_4 adsorption. This may narrow or even close the cleat, significantly reducing the permeability near the injection well, and thus impacting injectivity and limiting methane desorption.

Discussion

In this section, the coupled model was further integrated into the Transport of Unsaturated Groundwater and Heat Simulator (TOUGH2) to simulate the ECBM process.

TOUGH2 are multi-dimensional numerical models for simulating the coupled transport of water, vapor, non-condensible gas, and heat in porous and fractured media. EOS7C-ECBM, an equation of state module for the TOUGH2 program, is suitable for simulating nitrogen (N_2) or CO_2 ECBM process. CH_4/CO_2 mixture adsorption was considered, however, coal swelling due to adsorption difference between CH_4 and CO_2 is ignored. In other words, TOUGH2 fail to quantify the strain and permeability change due to gas adsorption.

A subroutine, of which gas adsorption, coal deformation, and permeability evolution was rigorous coupled, was added to the EOS7C-ECBM module of TOUGH2. The subroutine also offers an option to describe the thermodynamic properties of fluid with PTV EOS, to calculate the adsorption capacity with SLD theory, to calculate the coal deformation with the theoretical model, and to predict the permeability evolution with S&D, C&B, P&M, and improved P&M models.

An ECBM case from Law et al.[30] was selected to test the performance of TOUGH2, and to highlight the coupled adsorption-strain-permeability model. A coal seam with uniform thickness is assumed homogeneous and isotropic. No heat transfer exists between the injected gas and coal seam, the system remains isothermal. The properties of coal seam are summarized in Table 3, and coal properties are summarized in Table 4. In particular, the adsorption parameters were fitted with SLD theory, and the elastic and strain parameters were from the typical values of the San Juan basin[27].

Inverted 5-spot pattern was used in the case, with CO_2 injected to displace CH_4 in the coal reservoir. The injection well has a constant bottom hole flowing pressure of 12.5 MPa while the bottom hole flowing pressure of producing well is constant at 0.275 MPa. Since the entire seepage field is symmetrical, only a quarter was selected, and divided into $11 \times 11 \times 1$ grids, as illustrated in Fig. 7.

TOUGH2 EOC7C-ECBM was first used to simulate the ECBM process, and to compare with other simulators. After 30, 60, and 90 days of gas injection, CO_2 mole fraction distributions in the coal seam are shown in Fig. 8. It can be observed that 90% methane or more was displaced and produced within 90 days. The result is consistent with those obtained by other coalbed methane reservoir simulators, as shown in Figs. 9-12. This indicates TOUGH2 EOC7C-ECBM has a comparable performance with GEM, ECLIPSE, COMET 2, and SIMED II. However, this is not consistent with reality, and may be too optimistic.

With the EOS7C-ECBM module modified with the coupled model, and the ECBM process was re-simulated. After 30, 60, 90, 180, 360, 1800 days of gas injection, CO_2 mole fraction distributions in the coal seam are shown in Fig. 13. Taking into consider the adsorption difference between CH_4 and CO_2, and the induced strain and permeability change, only 24% of the CH_4 is displaced in the first 90 days, and it takes about 1800 days to displace 90% or more. This indicates that ECBM is even worse than primary recovery. This phenomenon was mainly due to coal swelling caused by the adsorption differences between CH_4 and CO_2. Although the matrix swell increment is limited, the cleats are easy to narrow down or even close, thus reduce the permeability near the injection well significantly and impact the injection efficiency. This help to understand the sequestration and transportation of CO_2, and may have a guidance to ECBM technology.

Conclusions

In this paper, a rigorously coupled adsorption-strain-permeability model was developed to quantify the coal deformation and permeability change caused by CH_4 displacement with CO_2, and to reveal the rules of CO_2 sequestration and transportation in deformed coalbed. The following conclusions and recommendations are suggested, based on the results of this study.

(1) With coal deformation due to matrix shrinkage calculated with gas adsorption theory, and permeability evolution calculated with theoretical strain model, these processes were rigorous coupled. The coupled model can simultaneously quantify gas adsorption or desorption, coal deformation, and permeability evolution during CBM and ECBM recovery with only six parameters.

(2) The coupled model can predict the coal deformation induced by CH_4 or CO_2 adsorption at pressure up to 12 MPa, and the average relative errors between the measured and predicted results are all within 21.48%.

(3) The adsorption capacity of CO_2 is 2-5 times as large as that of CH_4, and the volumetric strain induced by CO_2 adsorption is 3-5 times as large as that caused by CH_4 adsorption. While the stronger adsorption capacity of CO_2 on coal offers an option for CO_2-ECBM, this technology was not to be successful due to coal swelling induced by adsorption difference between CH_4 and CO_2.

(4) Considering the coal deformation and permeability change due to CH_4 displacement with CO_2, the modified TOUGH2 simulator shows that only 24% of CH_4 is displaced in first 90 days, and it takes about 1800 days to displace 90% or more.

Acknowledgments

Financial support from the National Science and Technology Major Project of China (2016ZX05044005-001), High School Subject Innovation Engineering Plan of China (111 Project: B12033), and China Scholarships Council Program (201606440099) are gratefully acknowledged.

Reference

(1) Advanced Resources International Inc. Enhanced coalbed methane recovery: Worldwide application and CO_2 sequestration potential. *International Energy Agency*, 1999.

(2) Reeves, S. An overview of CO_2-ECBM and sequestration in coal seams. *AAPG Studies in Geology* **2009**: 17-32.

(3) Godec, M.; Koperna, G.; Gale, J. CO_2-ECBM: A review of its status and global potential. *Energ. Procedia* **2014**, 63(3-4): 5858-5869.

(4) Sloss, D. L. L. Potential for enhanced coalbed methane recovery. *International Energy Agency*, 2015.

(5) Reeves, S.; Taillefert, A.; Pekot, L.; Clarkson, C. The Allison Unit CO_2 - ECBM pilot: A reservoir modeling study. *U.S. Department of Energy*, 2004.

(6) Li, X. C.; Fang, Z. M. Current status and technical challenges of CO_2 storage in coal seams and enhanced coalbed methane recovery: An overview. *Int. J. Coal Sci. Technol.* **2014**, 1(1): 93-102.

(7) Rangarajan, B.; Lira, C. T.; Subramanian, R. Simplified local density model for adsorption over large pressure ranges. *AIChE J.* **1995**, 41(4), 838-845.

(8) Fitzgerald, J. E.; Sudibandriyo, M.; Pan, Z. J.; Robinson, R. L.; Gasem, K. A. M. Modeling the adsorption of pure gases on coals with the SLD model. *Carbon* **2003**, 41(12), 2203-2216.
(9) Fitzgerald, J. E.; Robinson, R. L.; Gasem, K. A. M. Modeling high-pressure adsorption of gas mixtures on activated carbon and coal using a simplified local-density model. *Langmuir* **2006**, 22(23), 9610-9618.
(10) Zeng, Q. S.; Wang, Z. M.; McPherson, B. J.; McLennan, J. D. Theoretical approach to model gas adsorption/desorption and the induced coal deformation and permeability change. *Energ. Fuel* **2017**, 31(8), 7982-7994.
(11) Hasanzadeh, M.; Alavi, F.; Feyzi, F.; Dehghani, M. R. Simplified local density model for adsorption of pure gases on activated carbon using Sutherland and Kihara potentials. *Micropor. Mesopor. Mat.* **2010**, 136(1-3), 1-9.
(12) Valderrama, J. O. A generalized Patel-Teja equation of state for polar and non-polar fluids and their mixtures. *J. Chem. Eng. Jpn.* **1990**, 23(1), 87-91.
(13) Bangham, D. H.; Fakhoury, N. J. The translation motion of molecules in the adsorbed phase on solids. *J. Chem. Soc. Pakistan.* **1931**, 1324-1333.
(14) Levine, J. R. Model study of the influence of matrix shrinkage on absolute permeability of coal bed reservoirs. *Coalbed Methane and Coal Geology* **1996**, 109, 197-212.
(15) Megson, T. H. G. *Structural and stress analysis*; Elsevier Butterworth-Heinemann: Oxford, Britain, 2014 (ISBN No. 9780080999364).
(16) Robertson, E. P. Improvements in measuring sorption-induced strain and permeability in coal. *In Society of Petroleum Engineers - SPE Eastern Regional/AAPG Eastern Section Joint Meeting*; Pittsburgh, PA USA, October 11-15, 2008.
(17) Shi, J. Q.; Durucan, S. Drawdown induced changes in permeability of coalbeds: A new interpretation of the reservoir response to primary recovery. *Transport Porous Med.* **2004**, 56(1), 1-16.
(18) Shi, J. Q.; Durucan, S. A model for changes in coalbed permeability during primary and enhanced methane recovery. *SPE Reserv. Eval. Eng.* **2005**, 8(4), 291-299.
(19) Shi, J. Q.; Durucan, S. Exponential growth in San Juan Basin Fruitland coalbed permeability with reservoir drawdown: Model match and new insights. *SPE Reserv. Eval. Eng.* **2010**, 13(6), 914-925.
(20) Shi, J. Q.; Durucan, S. Modelling laboratory horizontal stress and coal permeability data using S&D permeability model. *Int. J. Coal Geol.* **2014**, 131(1), 172-176.
(21) Cui, X.; Bustin, R. M.; Chikatamarla, L. Adsorption-induced coal swelling and stress: Implications for methane production and acid gas sequestration into coal seams. *J. Geophys. Res.* **2007**, 112(B10202), 1-16.
(22) Bustin, R. M.; Cui, X. J.; Chikatamarla, L. Impacts of volumetric strain on CO_2 sequestration in coals and enhanced CH_4 recovery. *AAPG Bull.* **2008**, 92(1), 15-29.
(23) Palmer, I.; Mansoori, J. How permeability depends on stress and pore pressure in coalbeds: A new model. *In Proceedings - SPE Annual Technical Conference and Exhibition*; Denver, CO USA, October 6-9, 1996.
(24) Palmer, I.; Mansoori, J. How permeability depends on stress and pore pressure in coalbeds: A new model. *SPE Reserv. Eng.* **1998**, 1(6), 539-543.
(25) Palmer, I. Permeability changes in coal: Analytical modeling. *Int. J. Coal Geol.* **2009**, 77(1-2), 119-126.
(26) Ottiger, S.; Pini, R.; Storti, G.; Mazzotti, M. Competitive adsorption equilibria of CO_2 and CH_4

on a dry coal. *Adsorption* **2008**, 14(4), 539-556.
(27) Harpalani, S.; Mitra, A. Impact of CO_2 injection on flow behavior of coalbed methane reservoirs. *Transport Porous Med.* **2010**, 82(1), 141-156.
(28) Pan, Z. J.; Connell, L. D. Modelling of anisotropic coal swelling and its impact on permeability behaviour for primary and enhanced coalbed methane recovery. *Int. J. Coal Geol.* **2011**, 85(3-4), 257-267.
(29) National Institute of Standards and Technology. *Thermophysical properties of fluid systems*; http://webbook.nist.gov/chemistry/fluid/.
(30) Law, D. H. S.; Van der Meer, L. G. H.; Gunter, M. D. Numerical Simulator Comparison study for enhanced coalbed methane recovery processes, part I: Pure carbon dioxide injection. *In Proceedings - SPE Gas Technology Symposium*; Calgary, Alberta, Canada, April 30 - May 2, 2002.

Tables

Table 1 Physical Properties of CH_4 and CO_2

Fluid	T_c	P_c	Z_c	ω_c	d	ε_{ff}/k_B
	K	MPa	-	-	nm	K
CH_4	190.56	4.599	0.288	0.011	0.3758	148.6
CO_2	304.13	7.377	0.274	0.239	0.3941	195.2

Table 2 Model Fitting Results for Coal Strain Induced by CH_4 and CO_2 Adsorption

Database	Coal	T	L_s	ε_{ss}/k_B	A_{CH4}	A_{CO2}	E	v	ρ_c	K_{ss}	δ_{CH4}	δ_{CO2}
		K	nm	K	m^2/g	m^2/g	MPa	-	m^3/kg	MPa	%	%
Levine	Fruitland	308.15	1.50	19.5	104.2	137.3	4400	0.32	1400	1665	5.41	3.50
Bustin	Ardley	298.15	2.59	12.6	19.8	64.8	3000	0.30	1500	559	3.78	4.49
	WM	298.15	1.19	21.5	49.3	53.2	3000	0.30	1340	823	2.07	21.48
Ottiger	Sulcis	318.15	1.80	66.7	83.9	96.7	2410	0.48	1400	1812	8.31	5.06
Harpalani	San Juan	308.15	1.14	22.6	71.6	61.0	3450	0.37	1400	1484	3.37	7.61
	Illinois	308.15	2.08	18.1	51.2	97.2	2119	0.40	1400	816	7.07	9.14

Table 3 Coal Seam Properties

A_c	t_c	$k_{CH4}(0)$	$k_{CO2}(0)$	ϕ_i	T	p_{inj}	p_{pro}	ρ_c	ρ_{water}	S_{water}	S_{CH4}
m^2	m	mD	mD	-	°C	MPa	MPa	kg/m^3	kg/m^3	-	-
647497	9	3.65	0.91	0.001	45	12.5	0.275	1434	990	0.592	0.408

Table 4 Coal Properties

L_s	ε_{ss}/k_B	A_{CH4}	A_{CO2}	E	v	K_{ss}	C_f
nm	K	m^2/g	m^2/g	MPa	-	MPa	MPa^{-1}
2.63	16.7	46.4	70.2	3450	0.37	1295	0.145

Figures

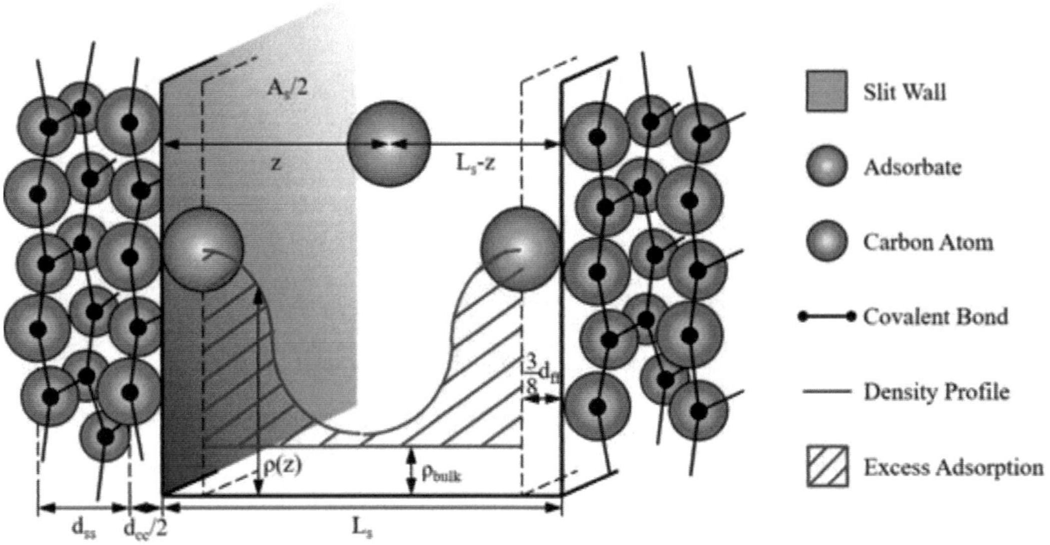

Fig. 1 Slit Pore Characterization in SLD Theory

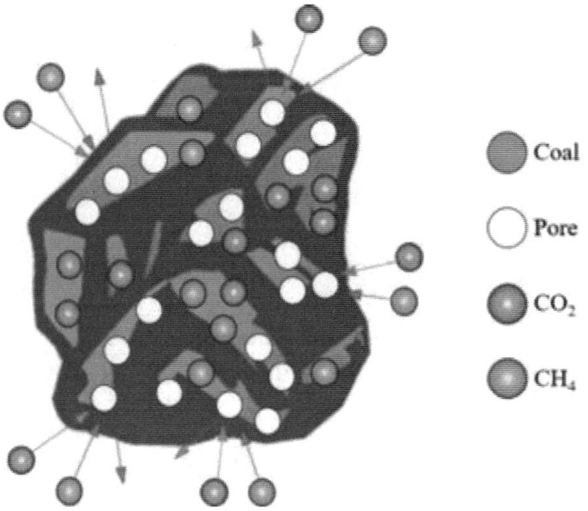

Fig. 2 Schematic Diagram of CH_4/CO_2 Mixture Adsorption on Carbonaceous Adsorbent

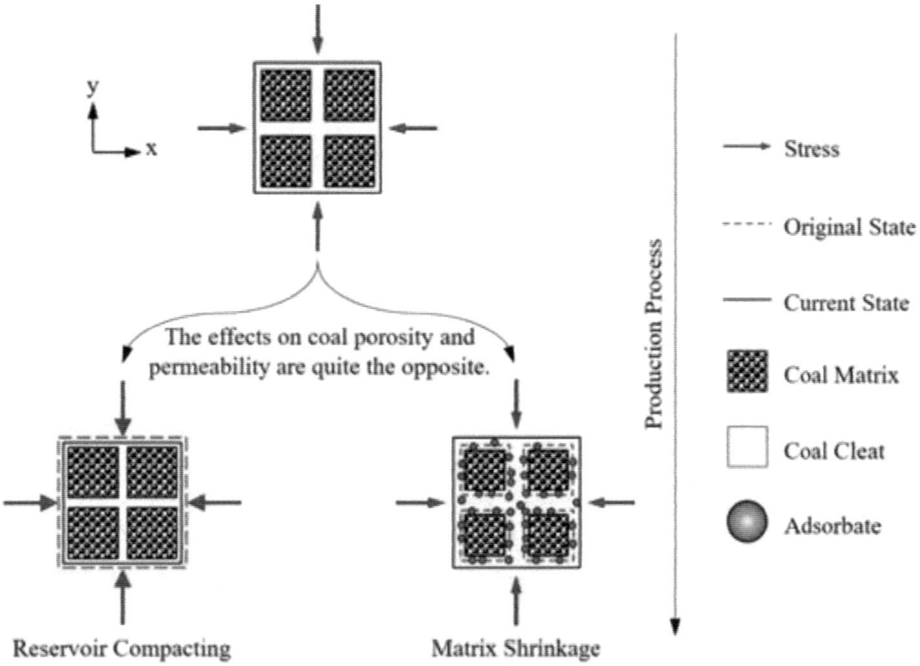

Fig. 3 Reservoir Compaction and Matrix Shrinkage Effects on Coal Permeability

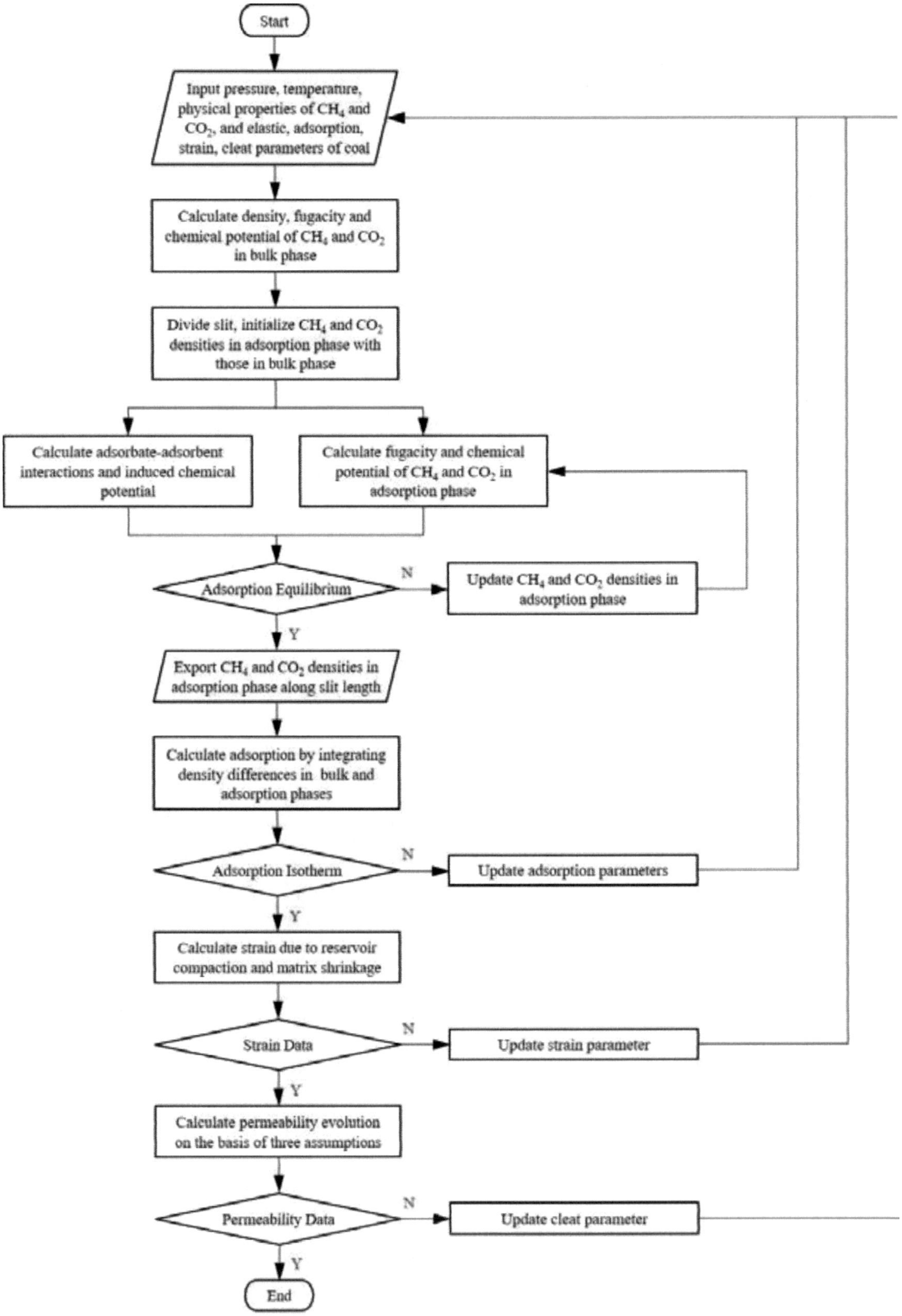

Fig. 4 Solution Procedure for Coupled Adsorption-Strain-Permeability Model

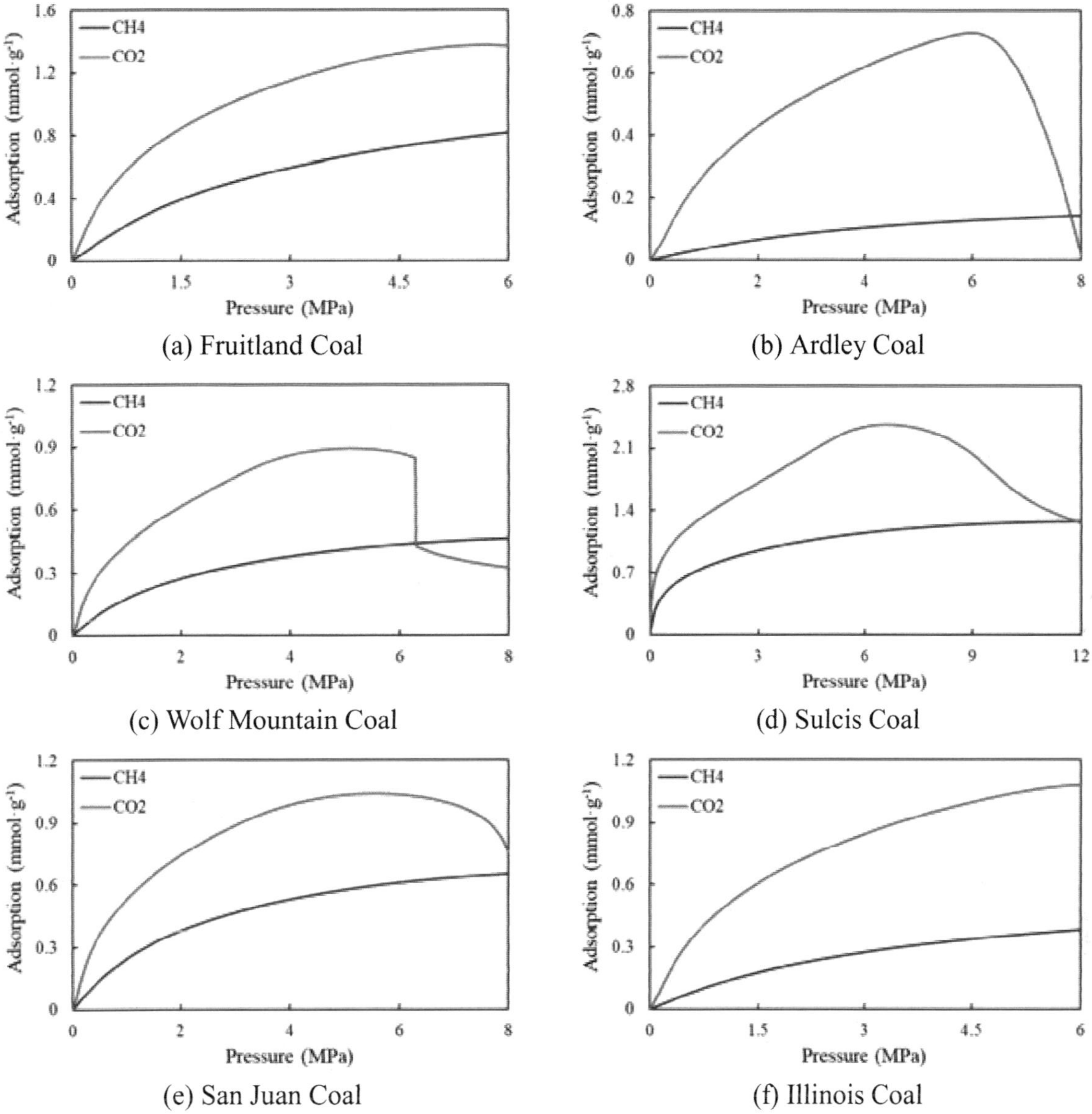

Fig. 5 Adsorption Isotherms of CH_4 and CO_2

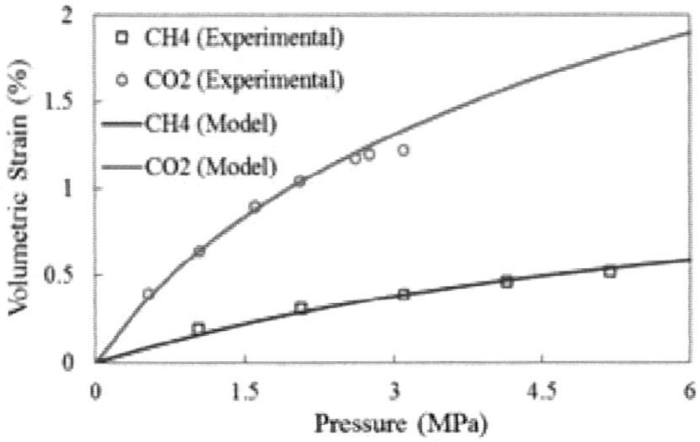

(a) Fruitland Coal

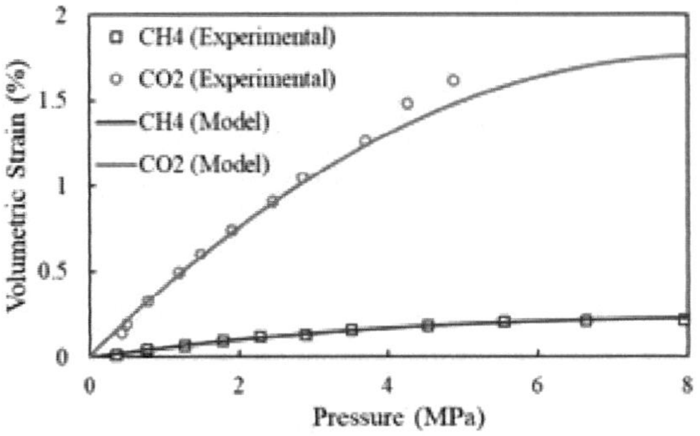

(b) Ardley Coal

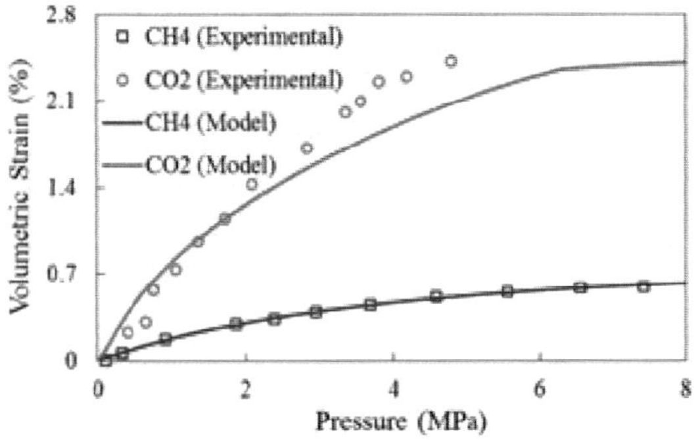

(c) Wolf Mountain Coal

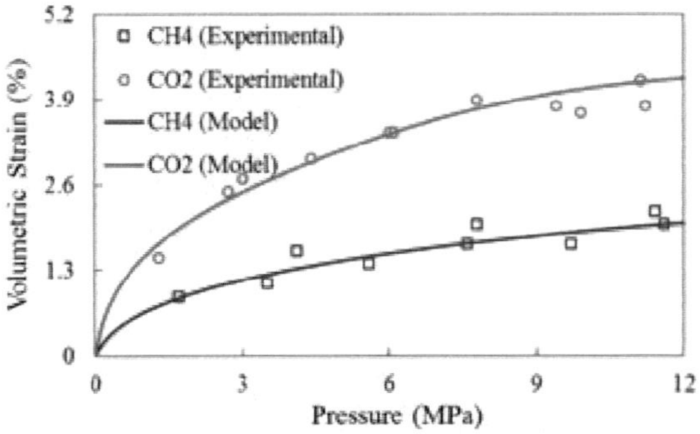

(d) Sulcis Coal

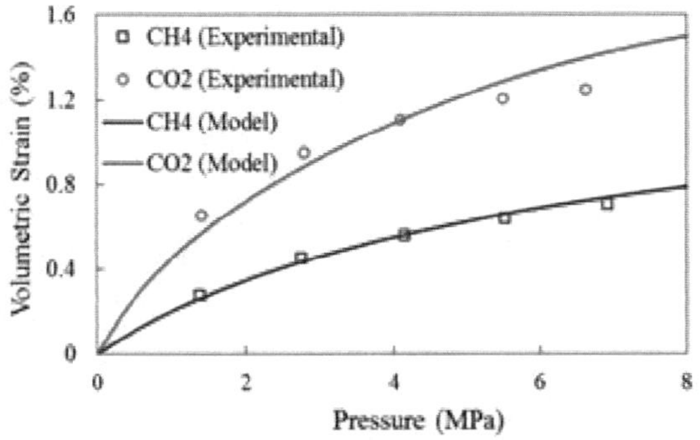

(e) San Juan Coal

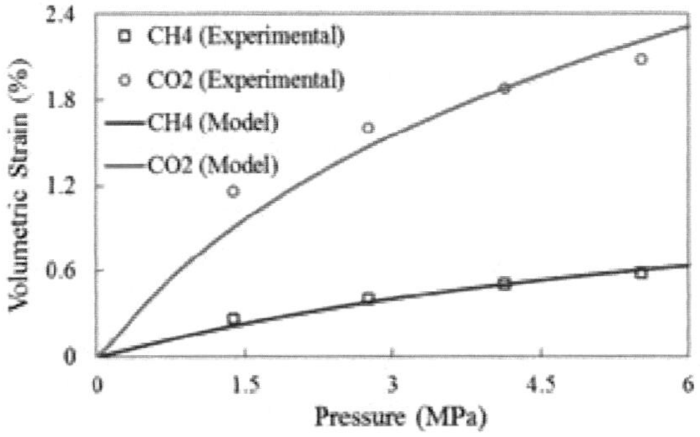

(f) Illinois Coal

Fig. 6 Volumetric Strain with Pressure

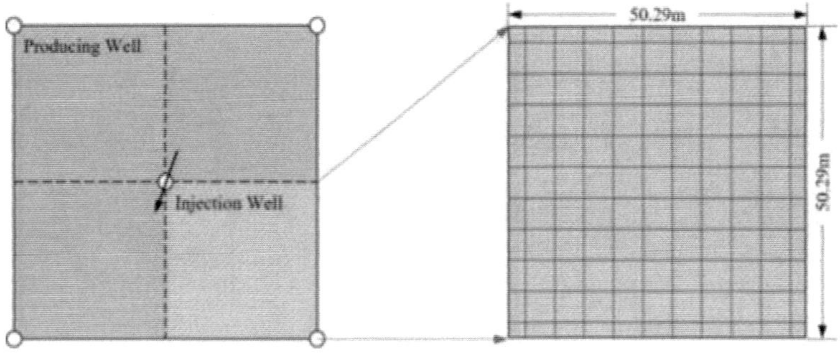

(a) Inverted 5-Spot Pattern (b) 11×11×1 Grid System

Fig. 7 Schematic Diagrams of Well Pattern and Grid System Used

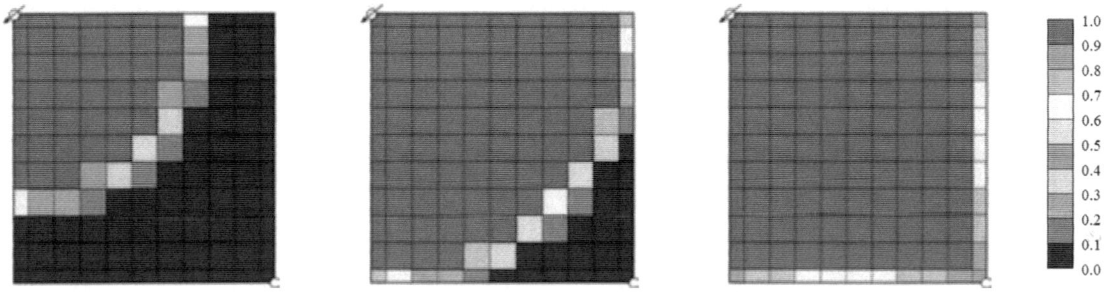

(a) 30 days (b) 60 days (c) 90 days

Fig. 8 CO_2 Mole Fraction Distributions Predicted by TOUGH2

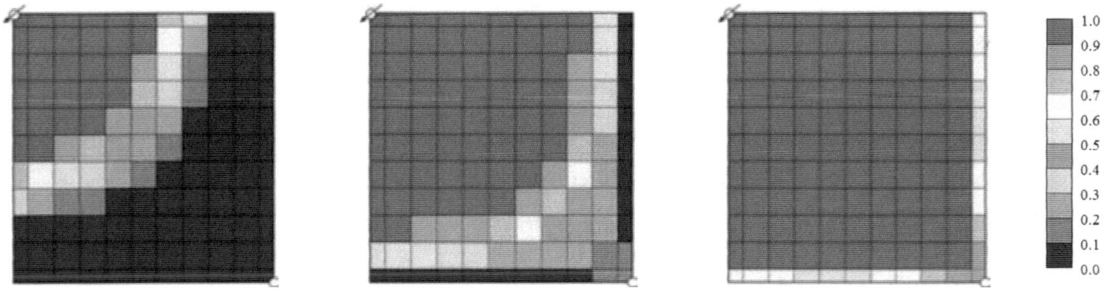

(a) 30 days (b) 60 days (c) 90 days

Fig. 9 CO_2 Mole Fraction Distributions Predicted by GEM

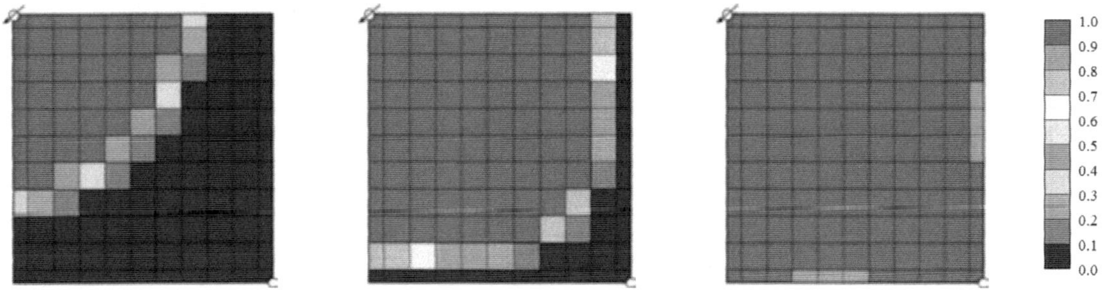

(a) 30 days (b) 60 days (c) 90 days

Fig. 10 CO_2 Mole Fraction Distributions Predicted by ECLIPSE

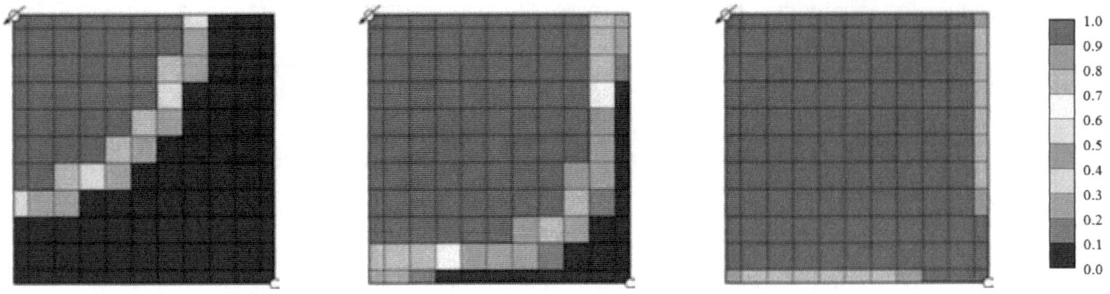

(a) 30 days (b) 60 days (c) 90 days

Fig. 11 CO_2 Mole Fraction Distributions Predicted by COMET2

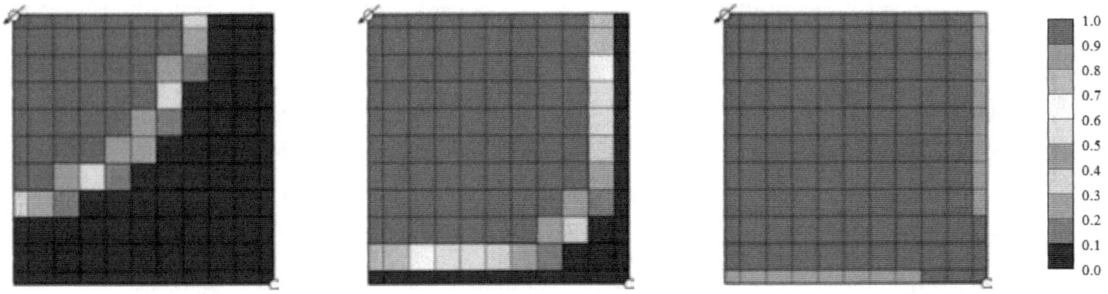

(a) 30 days (b) 60 days (c) 90 days

Fig. 12 CO_2 Mole Fraction Distributions Predicted by SIMED II

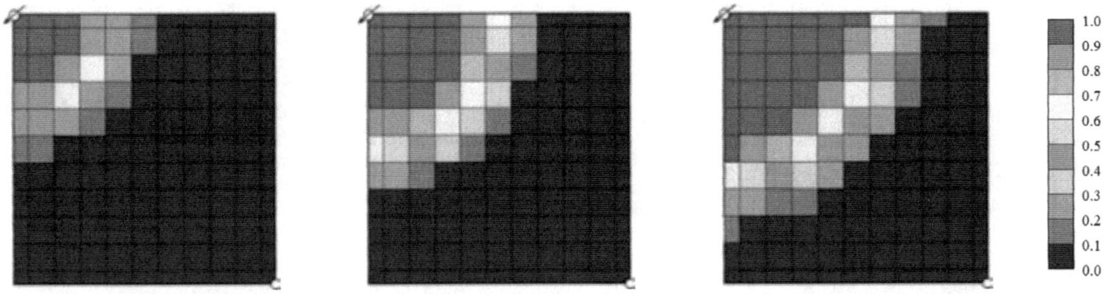

(a) 30 days (b) 60 days (c) 90 days

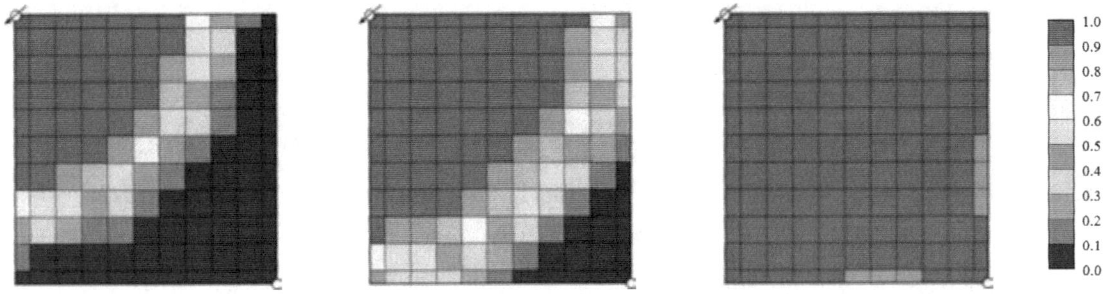

(d) 180 days (e) 360 days (f) 1800 days

Fig. 13 CO_2 Mole Fraction Distributions Predicted by Modified TOUGH2

498318 Integrated CO_2 Capture /Water-Gas Shift with Integrated HEAT Management for IGCC Applications

Abstract:

As one of the 16 transformational carbon capture technologies projects selected by Department of Energy (DOE), Southern Research is developing a combined CO_2 capture water-gas shift (WGS) process for integrated gasification combined cycle (IGCC) applications.Â Here we present a laboratory-scale combined sorbent/catalyst reactor system with advanced integrated heat management to capture 90% of CO_2 from simulated Transport Reactor Integrated Gasifier (TRIG) syngas at warm gas conditions (250 â€" 400 Â° C). A highly efficient regenerable magnesium based sorbent was developed to remove CO_2. It offers high and stable CO_2 capacity with minimum requirement for temperature and pressure swings. Under optimized conditions, the sorbents achieved up to 6 mmol/g of CO_2 working capacity and remained stable over 250 cycles of adsorption and regeneration. The selected WGS catalyst achieved around 95% CO conversion under both continuous feed and combined cycle tests and remained stable during the test duration that lasted over 200 hrs. The approach has potential to reduce costs and increase efficiency for electricity generation from IGCC. Upon successful completion of bench-scale integrated test program, the process will be ready for pilot scale development. This presentation will examine bench-scale results on combined CO_2 capture/WGS process using an integrated heat management reactor.

497907 Solar-Driven Electrochemical Desalination of Seawater

Abstract:

Providing access to clean water is a major global issue, which has been listed as one of the grand challenges of the 21^{st}century. To address this challenge, alongside efforts such as water conservation and recycling, several desalination techniques based on reverse osmosis and electrodialysis have been developed to meet the clean water demands. However, the present techniques either have a low salt rejection or high energy requirement for a desirable salt rejection. Here we propose a solar-powered electrochemical cell utilizing bipolar membranes to produce desalinated water from the seawater. A stack consisting of a bipolar membrane, an anion-exchange membrane, a cation-exchange membrane, and a bipolar membrane separated by seawater is used to electrodialyze seawater to create a desalted stream and two brine streams. The bipolar membranes simultaneously trap salt ions while producing counter-ions to neutralize the brine streams, and prevents salt ions from contacting the electrodes. To analyze the feasibility and robustness of the design, modeling and simulation of the solar-driven electrodialysis were performed using COMSOL Multiphysics. Preliminary results show that at least 99% of salt rejection can be attained from the initial brine concentration of 3.4 M to yield near-neutral pH water containing 0.015 M of salt. The effect of various parameters such as the resistance of seawater, conductivity, and permeability of membrane, flowrate, applied current density, and seawater concentration on the efficiency of desalination will be discussed. The successful implementation of this design could be used in water-stressed areas as well as research facilities on the ocean and in the remote locations. Solar-driven electrodialysis for desalination of water would make the technology portable and easily accessible.

500349 Salt Harvesting of Reverse Osmosis Concentrate By Continuous Chemically-Enhanced Seeded Precipitation

Abstract:

Desalination of inland brackish water via reverse osmosis membrane (RO) technology is often confronted by the need to attain high recovery permeate product recovery in order to reduce the challenge of concentrate (brine) management. Driving toward high recovery is particularly problematic in source water that are of high scaling propensity as is typically the case with agricultural water source and especially agricultural drainage. In order to reach high recovery RO operation while avoiding membrane mineral scaling, a process of chemically-enhanced seeded precipitation (CESP) was proposed as an intermediate process to desupersaturate (with respect to sparingly water soluble salts such as gypsum and calcite) the RO concentrate from a primary RO stage so as to enable subsequent additional desalting in a secondary RO state and hence increased the overall recovery. In the CESP process $CaCO_3$ precipitation (as calcite) is first induced via partial lime-dosed treatment for antiscalants (originating from the primary RO) scavenging, followed by subsequent $CaSO_4$ (as gypsum) seeded concentrate desupersaturation. The CESP process, which was previously demonstrated as a batch process, was explored in the present study to evaluate the feasibility of utilizing the approach as a practical means for achieving high recovery RO under continuous desalting process operation. Accordingly, a novel fluidized bed CESP reactor system, for continuous RO system operation, was developed to systematically investigate the impact of various process conditions (e.g., feed composition and pH, temperature, CESP crystallizer residence time). Continuous RO concentrate demineralization tests were conducted using antiscalant-containing, gypsum-supersaturated RO concentrate solutions. These studies demonstrated that precipitated calcite particles in the initial partial lime treatment step (with minimum dose compared to conventional softening process) enabled significant antiscalants scavenging; this facilitated subsequent mineral salts precipitation via non-chemical (i.e., gypsum seeded) heterogeneous precipitation. Continuous mode operation of the partial-lime and seeded-precipitation fluidized reactor, along with model simulations, demonstrated that steady-state operation with respect to the required level of calcium desupersaturation and a particle size distribution was achievable by optimizing the size and initial mass loading of gypsum seed, as well as the fluidized CESP reactor recycle ratio (i.e., fluidized bed recirculation flow rate relative to the CESP feed flow rate). Results of the current study indicate that enhancement of overall RO water recovery RO from 63% via primary RO up to 90% (or higher) with CESP followed by secondary RO are feasible for raw source water sources that are supersaturated with respect to both gypsum and calcite.

494980 3D Carbon Material and Its Excellent Capacitive Deionization Performance

Abstract:

Capacitive deionization (CDI) as a promising technology for efficient ion removal from brackish water has attracted wide attention. Ion removal mechanism of CDI is based on electric double layer formation on electrode materials at low power supply. We constructed 3D carbon materials with large accessible surface areas and high electrical conductivity which can effectively improve the electrosorption capacity. The excellent capacitive deionization performance, which is better than commercial CDI materials (activated carbon) can attribute to the particular structure of 3D porous frameworks, proper micro/meso/macropore ratio and high electrical conductivity. Thus, 3D carbon materials have great potential for water treatment in industrial application.

499797 Nutrient Recovery from Thermally-Treated Agriculture Waste Using Membrane Distillation

Abstract:

As the modern dairy industry continues to develop, the trend has consistently been a decrease in the number of operations, but an increase in the number cows at each operation. Considering that a single dairy cow produces ~150 pounds of manure each day, the amount of manure that can accumulate at these concentrated dairy operations is significant. Current manure handling methods involve temporary storage of manure and then land application as a low-grade fertilizer; however, this often results in the contamination or eutrophication of local and regional surface waters. When not used locally, the financial burden of transporting the manure to an off-site waste management facility can be overwhelming. Thus, a new management approach that treats the manure on the dairy is sought.

One such management method being developed utilizes an integrated hydrothermal carbonization (HTC)-membrane distillation (MD) process for treating dairy manure on-site. HTC is an energy densification process that converts organic waste material into a valuable solid fuel and a nutrient-rich aqueous product (HAP). The HAP is then treated using MD to concentrate the nutrients for use as a liquid fertilizer and produce a high-quality water for on-site reuse.

In this study, we evaluate the nutrient and water recovery potentials of MD when treating HAP. A bench-scale direct contact MD (DCMD) system was tested with feed temperatures ranging from 35-60°C and distillate (permeate) streams maintained at 25°C, resulting in water fluxes from 6-15 L/m^2-hr. HAP from multiple HTC conditions were evaluated, and recoveries of total nitrogen (TN) and total organic carbon (TOC) were consistently >99%, and overall concentration factors >6 based on total volume were achieved. Because MD rejects nearly all of the nitrogen and carbon in the HAP, the final concentrated solution may be suitable for use as a liquid fertilizer while the high-quality distillate water may be recycled back to the HTC process or used elsewhere on-site. These preliminary results indicate that the combined HTC-MD process is a technically valid alternative to land application of manure or transport for off-site treatment.

496785 Recovery and Reuse of Aluminum in Municipal Wastewater

Abstract:

Coagulant recovery is an attractive option from both economic and environmental sustainability standpoints and has elicited the interest of researchers over the past 10 – 20 years. However, these studies focussed on recovery with minimal investigations of the recycled organics and nutrients. Therefore, in this work, method of recovery of alum coagulant from primary sludge originated from treated raw wastewater and its reuse potential have been investigated and assessed. The sludge was generated using alum as a coagulant in a standard jar test (100 RPM for 1 minute and 30 RPM for 20 minutes) from primary influent obtained from Pottersburg Water Pollution Control Plant. This sludge was later acidified with either HCl or H_2SO_4 to a pH of 0.5 – 2.5 to recover the aluminum (60 minutes of mixing at 170 RPM to solubilize the coagulant followed by 10 minutes of centrifugation at 3700 RPM for separation of the coagulant). Efficiencies increased with pH, varying from 82% - 90% (pH 0.5) to 70% - 78% (pH 1.5) to 39% - 60% (pH 2.5). There was no significant difference in the results using either HCl or H_2SO_4. In addition to that, acid treatment reduced the sludge volume by 80% which would alleviate the cost of sludge disposal. The cost of acid at $0.28/kg and taking into account the incremental aluminum recovery at pH 0.5 vs. 1.5, operating at a pH of 0.5 rather than a pH of 1.5 would not be justified. Given the marginal improvement in going from pH 1.5 to 0.5, on a cost performance basis, the optimum pH for aluminum recovery was chosen as 1.5 and all further experiments were conducted at this pH. The recovered coagulant was recycled twice for coagulation of fresh batches of primary influent and the water quality parameters (total suspended solids, phosphorous, nitrogen, and chemical oxygen demand) were determined. For the fresh alum, the removal efficiencies of total suspended solids, chemical oxygen demand, total phosphorous, and total nitrogen were 85%, 65%, 80% and 33%, respectively. For the recycled aluminum, removals of total suspended solids, chemical oxygen demand and total nitrogen were less affected with coagulant recycle, with removals decreasing by approximately 10 percentage points in each recycle. This efficiency implies that the recovered aluminum's coagulation power is effective, thus reducing oxygen demand and biological sludge volumes. Phosphorous was the most sensitive parameter during coagulant recycling, due to the carryover of o-phosphate with the recovered coagulant and increased by 1 mg/L and 3 mg/L in the first and second cycles respectively. To mitigate the increasing phosphorous load, struvite precipitation was attempted by the addition of magnesium chloride and ammonium hydroxide in a molar ratio of 1:1:1. This proved to be ineffective as it precipitated the aluminum.

Graphical abstract

492626 First Principles of Metal Ion Extraction from Non-Buffered Water with Hydrophobic Deep Eutectic Solvents

Abstract:

Some metals, such as Cd, Pb, and As, are highly toxic at low concentrations.[1,2] A technique often applied for the removal of metal ions from water is liquid-liquid extraction, due to the many advantages it offers over other methods.[3] These advantages include easy to operate equipment, continuous removal of the contaminants and small amounts of extractant needed.[4]

A disadvantage of the commonly used techniques is the need for conventional solvents, which are often toxic, volatile, and flammable. Recently, ionic liquids (ILs) were introduced for the removal of metal ions from water.[5] Due to the electrostatic forces between the cation and the anion, they have a low volatility and are not flammable.[6,7] However, their production is costly and therefore less interesting for industry.

Alternative solvents are deep eutectic solvents (DESs), first introduced in 2003. DESs are composed of two or more solid components interacting via hydrogen bonding and Van der Waals interactions.[8] These interactions transform the two solids into a liquid phase.

A hydrophobic variant of these innovative solvents was presented in 2015 by a number of us.[9] The first hydrophobic DESs consist of decanoic acid and quaternary ammonium salts, which were subsequently used for the extraction of volatile fatty acids from water.[9] In 2015 also a second paper regarding hydrophobic DESs was presented in the literature. In this manuscript biomolecules were removed from an aquatic environment.[10]

Here, metal ion extraction from a non-buffered aquatic environment is shown for the first time.[11] The results show that transition metals can be removed from water with distribution coefficients (*D*) higher than 0.99. When the initial concentration of the model metal ion Co^{2+} is increased from 1 g/L to 4 g/L, a decrease in *D* to approximately 0.83 is observed. When the solvent to feed ratio is varied, as low as 1:10, a decrease of *D* to approximately 0.80 is measured. It was found that the prevailing mechanism is ion exchange between the positively charged Co^{2+} and lidocaine. Furthermore, it is shown that Co^{2+} can be removed within 5 s. Finally, it is shown that it is possible to recover the DES.

References

[1] T. Martin and D. Holdich, *Water Research*, 1986, **20**, 1137–1147.

[2] P. Madoni and M. G. Romeo, *Environ. Pollut.*, 2006, **141**, 1–7.

[3] J. Rydberg, *Solvent extraction principles and practice, revised and expanded*, CRC Press, 2004.

[4] S. Wellens, B. Thijs, and K. Binnemans, *Green Chemistry*, 2012, **14**, 1657–1665.

[5] D. Parmentier, S. J. Metz, and M. C. Kroon, *Green Chemistry*, 2013, **15**, 205–209.

[6] J. P. Hallett and T. Welton, *Chemical reviews*, 2011, **111**, 3508–3576.

[7] P. Wasserscheid and T. Welton, *Ionic liquids in synthesis*, Wiley Online Library, 2008, vol. 1.

[8] M. Francisco, A. van den Bruinhorst, and M. C. Kroon, *Angew. Chem. Int. Ed.*, 2013, **52**, 3074–85.

[9] D. J. G. P. van Osch, L. F. Zubeir, A. van den Bruinhorst, M. A. A. Rocha, and M. C. Kroon, *Green Chemistry*, 2015, **17**, 4518–4521.

[10] B. D. Ribeiro, C. Florindo, L. C. Iff, M. A. Coelho, and I. M. Marrucho, *ACS Sustainable Chemistry & Engineering*, 2015, **3**, 2469–2477.

[11] D. J. G. P. van Osch, D. Parmentier, C. H. J. T. Dietz, A. van den Bruinhorst, R. Tuinier, and M. C. Kroon, *Chemical Communications*, 2016, **52**, 11987–11990.

Savannah River National Laboratory™

OPERATED BY SAVANNAH RIVER NUCLEAR SOLUTIONS

We put science to work.™

Impact of Impurities and Degradation Products During Chemical Processing of High Level Waste in the Defense Waste Processing Facility

Dan Lambert, Wesley H. Woodham, Matthew S. Williams, J. David Newell, Jack Zamecnik

Session 477a, Advances in Nuclear Waste Processing and Management

November 1, 2017

SRNL-STI-2017-00221

Savannah River NUCLEAR SOLUTIONS

FLUOR • NEWPORT NEWS NUCLEAR • HONEYWELL

Outline

- **Background**
- **Flammability Issues**
  - Hydrogen Generation
  - Ammonia Generation
  - Antifoam Degradation
  - Thermolysis
- **Conclusions/Recommendations**

Savannah River Site

- The Savannah River National Lab is located at the Savannah River Site
- Located at the site are forty-three ~1 million gallon tanks filled with high-level radioactive waste (8 have been closed/grouted)
- This is considered the worst environmental hazard in South Carolina

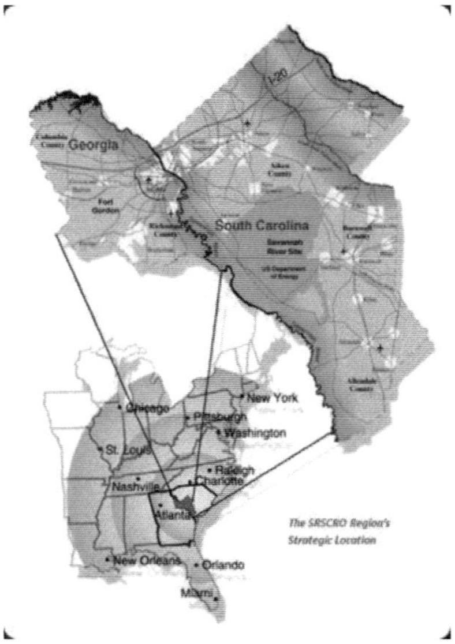

DWPF Process Overview
Formic Acid Flowsheet

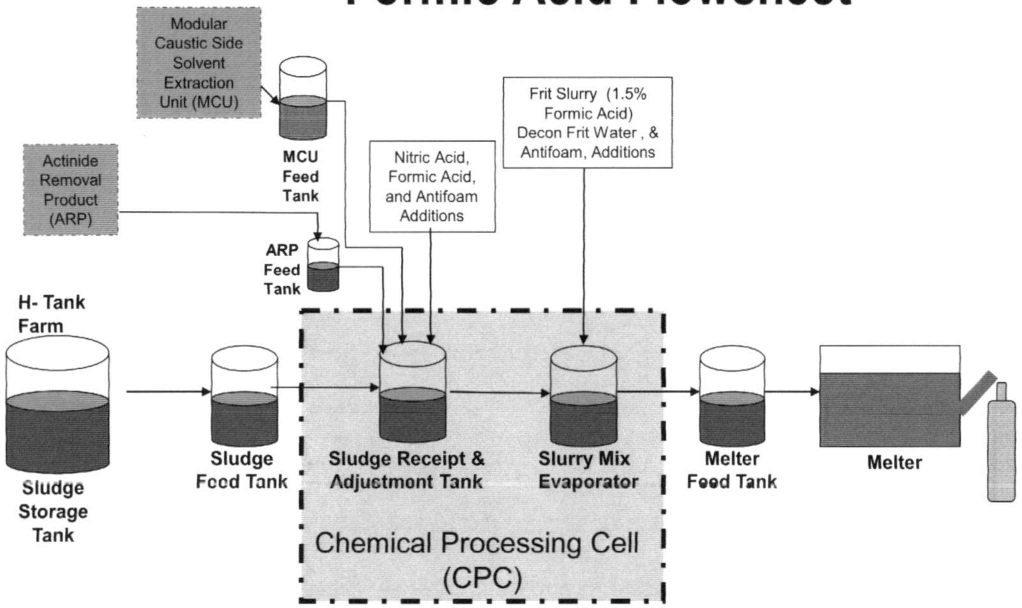

DWPF Process Overview

Glycolic Acid Flowsheet

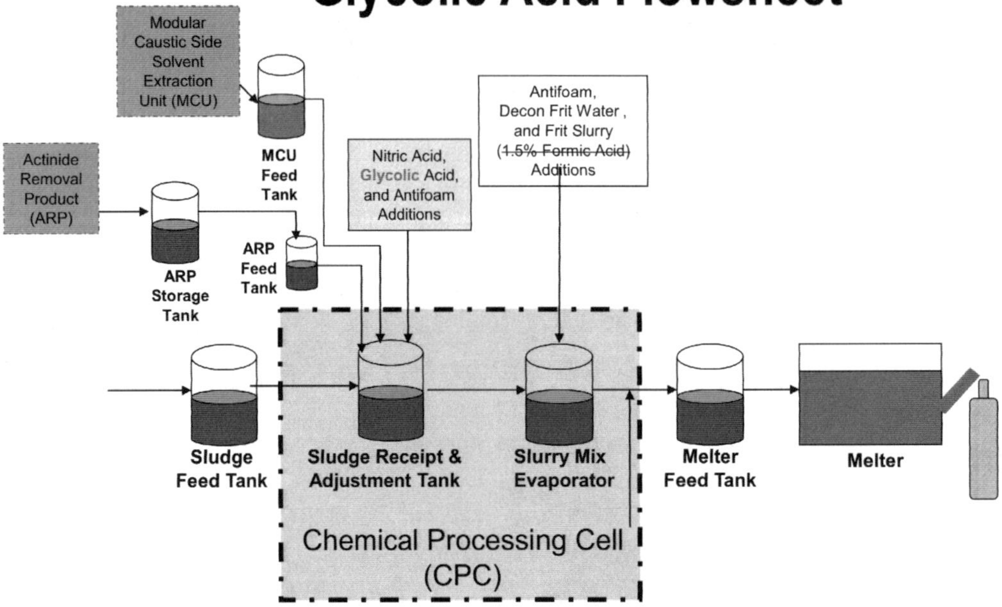

Why Change DWPF CPC Flowsheet?

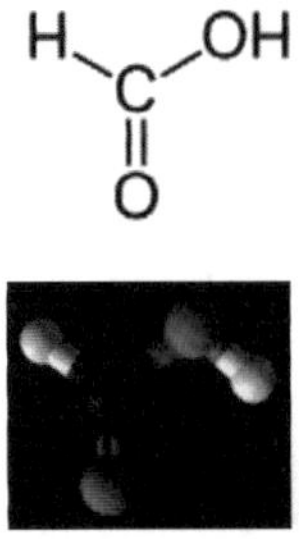

Formic Acid

- **Formic acid not only reduces mercury but also produces hydrogen**
- **If catalytic H_2 can be eliminated**
  - Air purge can be decreased
  - Gas chromatographs not safety significant
- **Without formic acid, rheology is thinner, pH more stable, complexes metals**
- **Glycolic acid does not produce hydrogen**

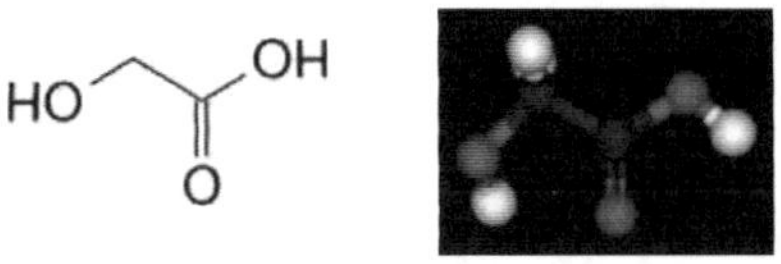

Glycolic Acid

What is Glycolic Acid (hydroxyacetic acid)?

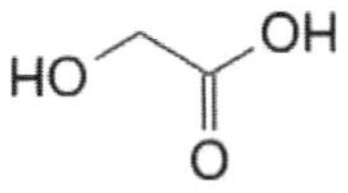

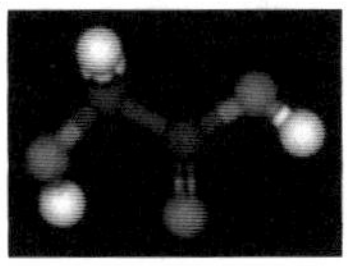

Glycolic Acid

- Smallest α-hydroxy acid
- Colorless, odorless, and hygroscopic crystalline solid
- Highly soluble in water
- A few million kilograms are produced annually

<u>Commercial Uses</u>

Facial Peel

- Physicians 20-70 wt%
- Home use 10-20 wt%

Soap and scum removal

Mineral scale removal

Acidity control

Water treatment

Boiler scale removal

- 2 moles Glycolic:mol Formic

Bioabsorbed polymer sutures

DWPF Flammability Issues

- Noble metals catalyze the decomposition of formic acid to hydrogen

$$HCOOH \rightarrow H_2 + CO_2$$

  - An air purge along with interlocks are used to ensure a flammable mixture of hydrogen and air is never formed
- The strip effluent (contains ^{137}Cs) also contains about 87 ppm entrained solvent (primarily isopar L)
  - The strip effluent is added slowly to minimize the impact
- The antifoam degrades to three flammable degradation products
  - The antifoam addition amount and timing between antifoam additions is controlled
- Ammonia is generated during chemical processing
  - Ammonia scrubbers are used to remove ammonia and prevent ammonium nitrate deposits

Flammable Issue: Hydrogen Generation

- **New flowsheet developed to reduce hydrogen generation**
- **Completed 100 experiments with glycolic acid**
- **Hydrogen generation is extremely low**
  - Very little formate is present (formic acid is impurity in glycolic acid) in runs with noble metals and mercury
  - SRAT purge has been reduced from 230 to 93.7 scfm
    - *Can't be lowered because of strip effluent isopar and antifoam degradation products*
  - SME cycle often has highest H_2 generation, purge remains at 74 scfm
  - In most runs hydrogen is detected only in SME cycle

Typical Hydrogen Profile

- **Very Low H2 generation in SRAT in Scaled Demo Runs GN70-79**

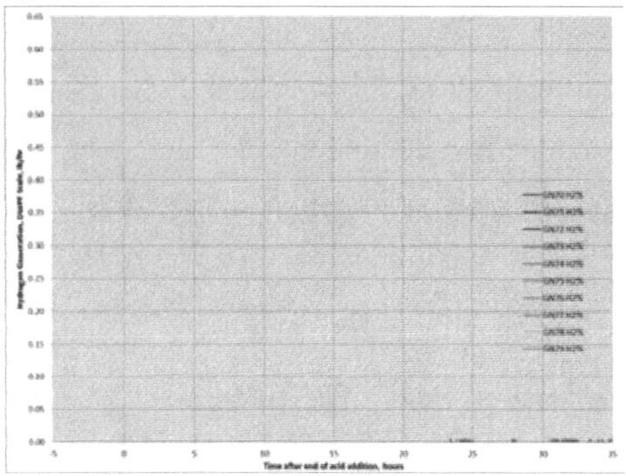

- **Zoom in 65x**

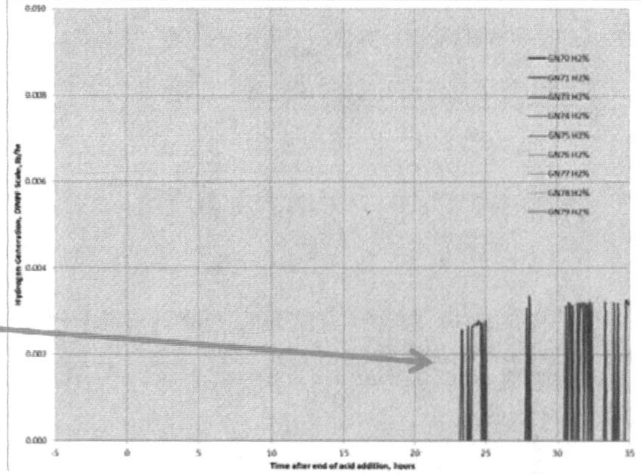

Flammable Issue: Ammonia Generation

- **Ammonia is generated from stepwise reduction of nitrate to ammonia**
  - During chemical processing, pH rises to ~10 in nitric-formic acid flowsheet
  - Equilibrium shifts from NH_4^+ to NH_3
  - Ammonia gas is released from slurry
  - Ammonia reacts with nitric acid to form ammonium nitrate

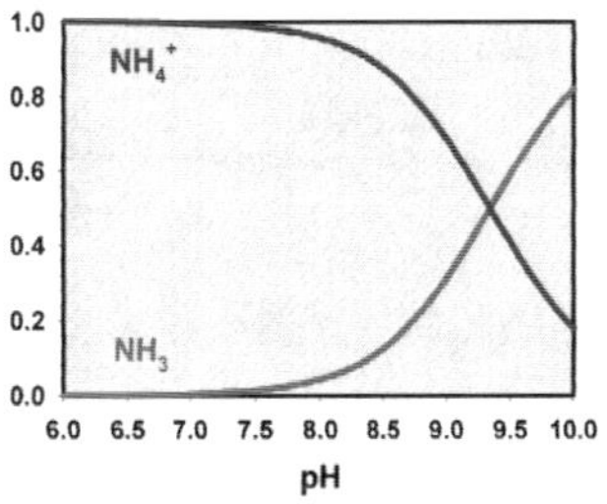

- **NO_2 results from decomposition of nitrite**
- **NO_2 is scrubber by water vapor and produces nitric acid**
  - Ammonia nitrate is an explosion hazard

Flammable Issue: Ammonia Generation

- **Very little ammonia is generated in GN Flowsheet runs**
  - No detected ammonia in offgas
  - Trace NH_4^+ detected in ammonia scrubber solution (35 mg/L vs 18,400 NO_3^-)
- **Low pH of SRAT product pushes equilibrium to ammonium**
  - $NH_3 + H^+ \leftrightarrows NH_4^+$
  - Ammonia would not be released even if produced
- **Glycolic Flowsheet**
  - NH_3 would stay in melter feed and be fed to melter
  - Ammonia scrubbers not needed for NH_3 removal
- **Formic Flowsheet:**
  - NH_3 would be produced in SRAT/SME, released to SME offgas
  - NH_3 would be removed by SRAT/SME scrubbers by SMECT liquid
  - NH_3 would be released in RCT with caustic addition or returned to tank far
  - NH_3 would be removed by RCT scrubber by SMECT liquid
  - In SE to SME runs, 2.5% of nitrate was converted to ammonia

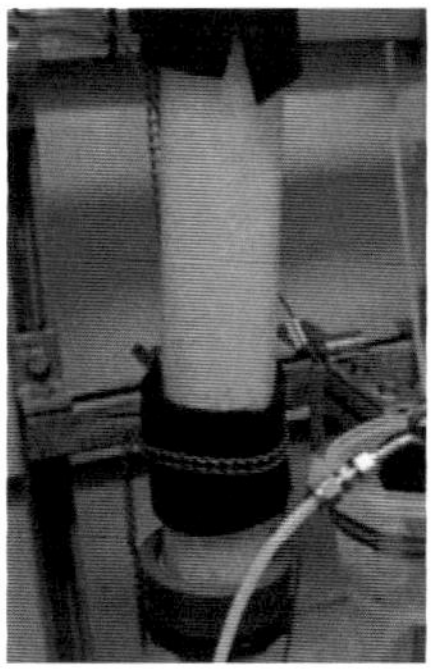

Flammable Issue: Antifoam 747 Degradation

- **Gas is generated during chemical processing**
  - N_xO_x generated due to nitrite destruction
  - CO_2 generate from carbonate destruction
  - Water vapor generated during boiling
- **Small insoluble solids stabilize foam and trap gases**
- **Foam generated during processing could cause a foam-over and lead to contamination, a complete facility shutdown, and the loss of millions of dollars**
- **Antifoam IIT747 implemented in DWPF in 2001 to reduce foaming**
- **Dozens of foam-overs have occurred, stopping processing in DWPF**

Flammable Issue: Antifoam 747 Degradation

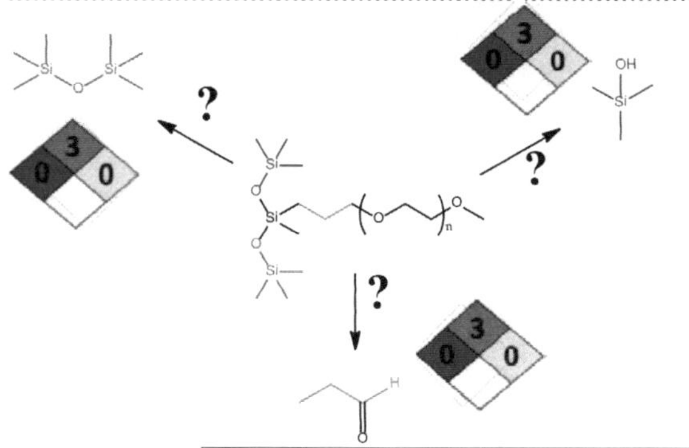

- **May 2015 – A Potential Inadequacy Safety Acquisition (PISA) was issued due to flammable antifoam degradation products which stopped all processing**

Compound	Formula	Structure	Molar Mass, g/mol	Solubility	Lower Flammability Limit, vol %	Boiling Point
Hexamethyl disiloxane (HMDSO)	$C_6H_{18}OSi_2$	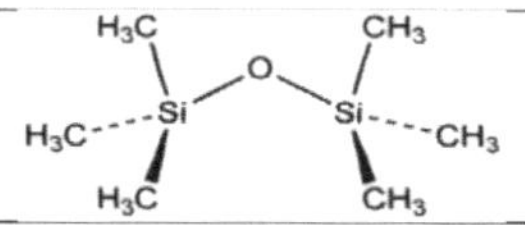	162.38	930.7 ppb	0.5-0.8	100 °C
Trimethyl silanol (TMS)	$C_3H_{10}OSi$	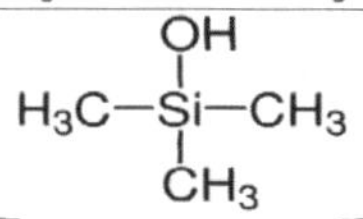	90.20	0.995 g/L	0.5-0.6	99 °C
Propanal	C_3H_6O	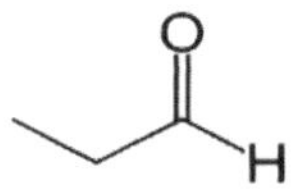	58.08	200 g/L	2.6	49 °C

Why does antifoam degrade so fast?

- **Antifoam 747 is most stable at pH 6.5-7.5**
- **Nitric-formic acid flowsheet**
  - Initial pH is ~12
  - pH drops to 4 by end of acid addition
  - Due to hydrogen generation and consumption of formic acid, pH rebounds to 10
- **Nitric-glycolic acid flowsheet**
  - Initial pH is ~11
  - pH drops to 4 by end of acid addition
  - pH rises slowly to about 5 by completion of processing
- **At initiation of boiling**
  - Gas generation is maximum (foaminess maximum)
  - Liquid boiling surface pH is <0 due to scrubbing of NO_2 as nitric acid

HMDSO Flammability Concern

- **Hexamethyldisiloxane (HMDSO) insoluble antifoam degradation product**
- **Antifoam degrades at low and high pH**
- **Maximum HMDSO is at initiation of boiling**
  - Condensate very acidic
  - Peak temperature
- **Short term fix**
  - Add undiluted antifoam
  - Reduce antifoam addition to 1/6th
  - Add antifoam more frequently

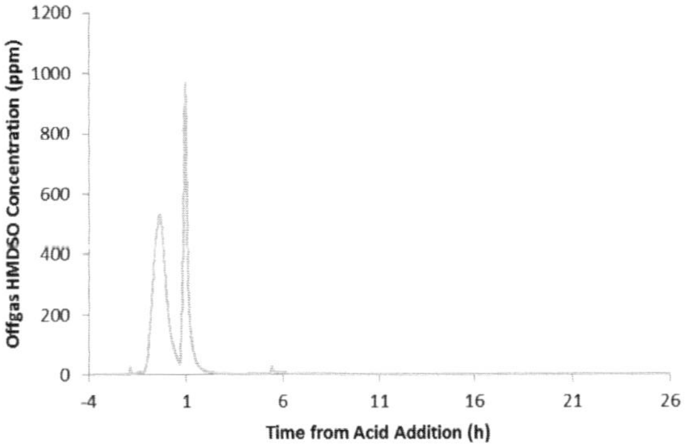

Long Term Fix

- **New antifoam**
  - Hydrolysis resistant antifoam
  - Antifoam that doesn't produce flammable degradation products
- **Change processing to minimize antifoam decomposition**
  - Acidify slurry prior to boiling
  - Complete nitrite destruction prior to boiling
- **Mechanical fix**
  - Spray water to break foam
  - Install bladed in vapor space to break foam
- **Nonantifoam Chemical fix**
  - Flocculate solids
  - Large solids too big to stabilize foam

Flammability Issue: Thermolysis

- **Thermolysis or thermal decomposition of organic impurities leads to hydrogen generation**
- **This can happen in high level waste evaporators, in waste tanks, and in any downstream processing facility**
- **The 3 main organic impurities in the glycolic acid flowsheet are glycolate, formate and oxalate**
- **Formate and oxalate do not produce significant hydrogen via thermolysis, but glycolate does**
- **Although thermolysis is a slow reaction compared to catalyzed formic acid decomposition, it can be an issue in tanks and processing vessels with lower purges and higher temperatures**

Impact of Thermolysis

- **There is virtually no glycolate from chemical processing in DWPF due to glycolate's low vapor pressure**
  - Unless there is a foamover
- **In the melter, entrainment leads to glycolate in the condensate**
  - Higher during bubbled operations
- **Significant hydrogen peak at boiling**
- **How to minimize**
  - Improve antifoam
  - More gentle processing
  - Don't bubble melter unless needed
- **Testing ongoing to determine H_2 generation for glycolate**

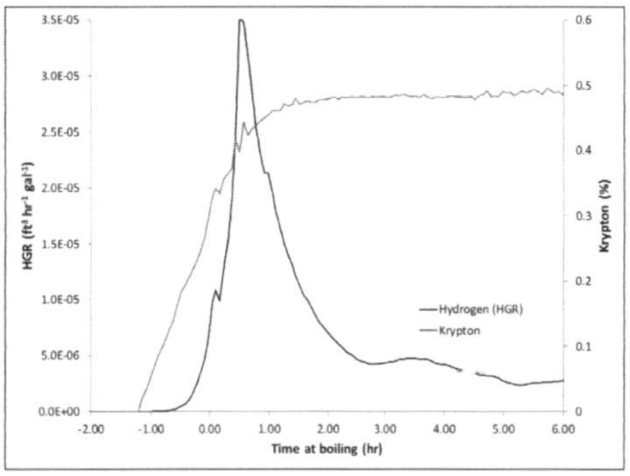

Conclusions

- **The nitric-glycolic acid flowsheet will**
  - Virtually eliminate hydrogen generation
  - Virtually eliminate ammonia generation
- **The current antifoam 747 should be replaced**
- **Determine H2 generation from glycolate thermolysis through testing**

Recommendations

- **Implement nitric-glycolic acid flowsheet in DWPF**
- **Develop hydrolysis resistant antifoam**
  - Processing changes may further reduce antifoam degradation
- **Complete testing to measure H2 generation from glycolate thermolysis**
  - Modify processing to minimize glycolate concentration in condensate

Acknowledgements

- **A large team has been involved in developing this flowsheet**
  - **SRNL researchers:** Mike Stone, Jack Zamecnik, Brad Pickenheim, Chris Martino, David Newell, Wes Woodham, Tom White, Whitney Riley, Frances Williams, Matt Williams, John Pareizs, John Mickalonis, Duane Adamson, et al
  - **SRNL Technicians:** Vickie Williams, Phyllis Workman, Jon Duvall, David Healy, Kim Wyszynski, Courtney Burkhalter, Madison Caldwell, Shielded Cells Technicians, et al
  - **SRR:** Terri Fellinger, Bill Holtzscheiter, Spencer Isom, Mason Clark, et al
  - **Interns:** Clayton Brandenburg, Michelle Luther, et al
- **Work was funded by the Department of Energy/Environmental Mgt**

Questions

Savannah River National Laboratory™
OPERATED BY SAVANNAH RIVER NUCLEAR SOLUTIONS

We put science to work.™

Assessment of Dead End Filtration as a Screening Tool for LAWPS

John Geeting
Reid Peterson
Jarrod Allred
Renee Russell

TOPICS

- Comparison of CFF to DEF
- Apparatus
- AP-105 CUF Testing Results
- AP-105 DEF Testing Results
- Summary of CUF and DEF Simulant Results
- Conclusions

Problem Statement

Millions of gallons of radioactive waste to be treated at Hanford

Hanford tank waste treatment will be initiated by pretreating low activity waste in a facility known as the low activity waste pretreatment system (LAWPS)

The flowsheet integration team proposed that batch filtration tests using small quantities of feed be used to "qualify" each feed campaign prior to delivery to LAWPS.

To that end, testing was conducted to assess this proposal and involved:

- 1. Development of dead end filtration apparatus for use in the hot cell with radioactive feeds.
- 2. Evaluation of dead end filtration relative to cross flow filtration (CFF) with tank waste and simulants

Pacific Northwest
NATIONAL LABORATORY
Proudly Operated by Battelle *Since 1965*

3

Advantages of DEF

- 1. Less feed
- 2. Simpler equipment
  - DEF is commercial off the shelf
  - CUF has many custom parts
  - DEF only requires a balance to calibrate inside the hotcell
  - CUF has Coriolis, magnetic flowmeter and RTDs in cell that have to be replaced on yearly cycle.
  - Other necessary services for CUF include compressed air and chilled water
- 3. Smaller footprint in hot cell
- 4. Lighter

Pacific Northwest
NATIONAL LABORATORY
Proudly Operated by Battelle *Since 1965*

4

Basic Comparison of DEF to CFF

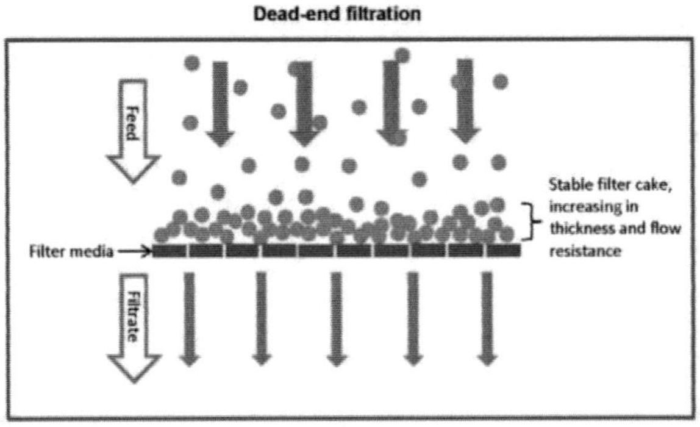

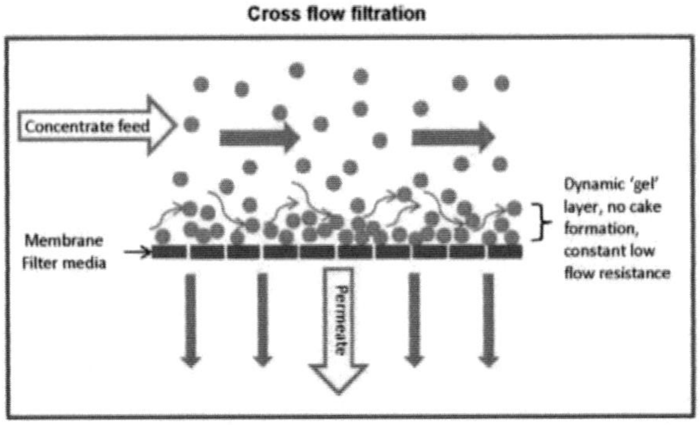

Pacific Northwest NATIONAL LABORATORY

Proudly Operated by Battelle *Since 1965*

Reasoning as to why DEF might be used as a screening tool

- Assumption: LAWPS will only treat supernatant with very low insoluble solids content.
- $R_{total}=R_{media}+R_{fouling}+R_{cake}$
  - R_{media} – approximately the same between DEF and CFF
  - R_{cake} for DEF can grow very large. If small amount of solids then DEF and CFF may have similar resistance.
  - $R_{fouling}$-Are they similar?
- The hope: filtration using CFF and DEF may be sufficiently similar for screening purposes

Pacific Northwest NATIONAL LABORATORY

Proudly Operated by Battelle *Since 1965*

CUF Piping Diagram

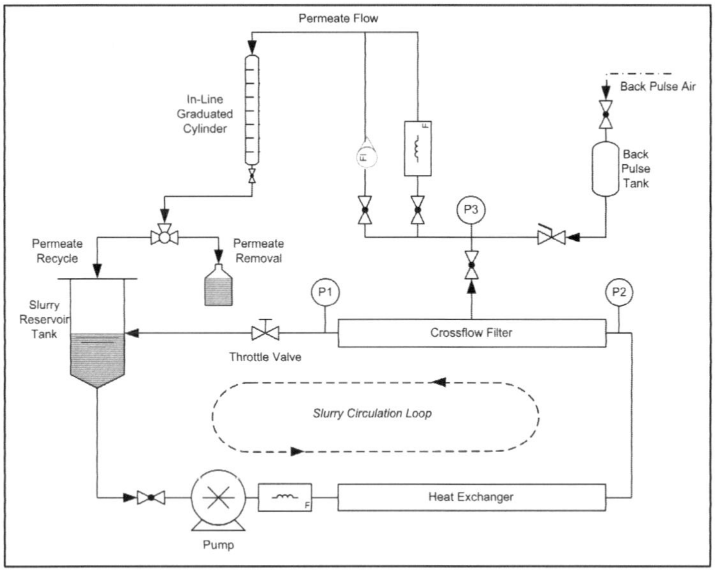

- Closed Loop Recirculation for Slurry
  - Reservoir Tank
  - Pump
  - Heat Exchanger
  - Filter
  - Throttle Valve
  - Magnetic Flow Meter
  - Pressure Gauges
- Permeate Manifold
  - Pressure Gauge
  - Mass Flow Meter
  - In-line Graduated Cylinder
  - Back-Pulse Chamber

The CUF

- Filter element
  - Variable: 0.5 to 2.0 ft
  - 0.1 micron rating Mott element
- System Volume
  - Maximum *4.1 Liters*
  - Operating Minimum *1.2 Liters*
  - Startup Minimum *1.5 Liters*
- Pump
  - Rotary Lobe *Up to 12 GPM*
- Instrumentation
  - Pressures *0 -- 100 psig*
  - Temperature *0 – 100° C*
  - Slurry Flow Rate *Up to 15 GPM*
  - Permeate Flow Rate *Up to 30 lph*

8

Pacific Northwest NATIONAL LABORATORY

Proudly Operated by Battelle *Since 1965*

The DEF

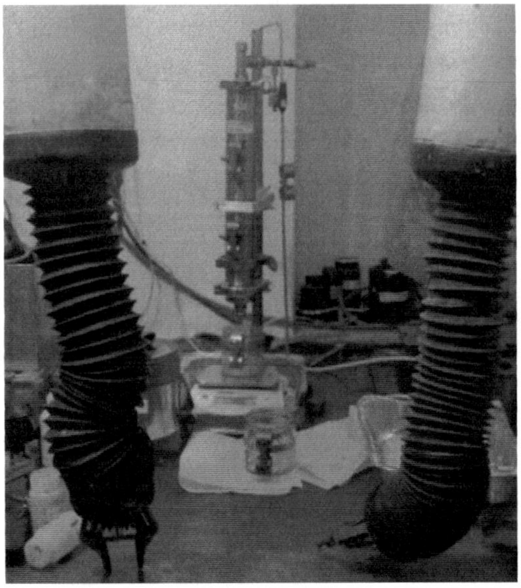

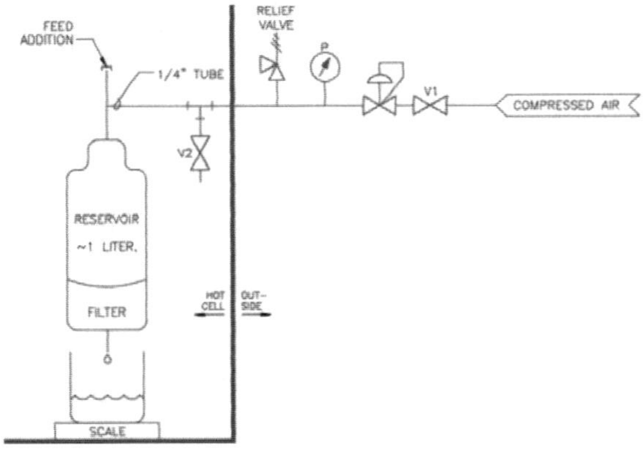

Pacific Northwest
NATIONAL LABORATORY

Proudly Operated by Battelle *Since 1965*

9

Comparison between DEF and CUF

Parameter	CUF	DEF
Filter	0.1 μm sintered SST	0.1 μm sintered SST
Manufacturer	Mott	Mott
Filter thickness (in)	1/16	1/16
Filter shape	Tubular (1/2-in ID)	Flat plate
Filter Area (ft^2)	0.065	0.031
Volume of feed (L) (nominal)	4	0.75
Volume/Area (cm)	66	26
Filtration Method	Cross-flow	Dead-end
Ease of construction	Difficult –many custom parts	Easy- generally COTS
M&TE Replacement	More difficult and expensive	Easy- outside hotcell
Footprint	Uses entire hotcell	Uses small portion of hotcell and can be easily moved

Filtration with AP-105 Tank Waste–completed twice

- Load feed into CUF
  - Diluted to 5.6 M Na
- Operate under recycle for ~ 28 hours
- Dewater to produce product for IX - total of 5 L produced
- Dead end filter balance – approximately 1.5 L (dewatered with varying pressure and constant pressure)
- CUF operated at 14.7 ft/s and various transmembrane pressure
- Temperature at 25C

Tank Waste Composition

	mg/L	Molarity
Na	196,000	8.53
Al	20,750	0.769
K	5525	0.141
S	1810	0.056
P	733.5	0.024
Cr	505	0.010
W	147.5	0.001
Mo	83.5	0.001
B	69.85	0.006
Ni	45.45	0.001
Viscosity	4.67	cP
Density	1.4	kg/liter
137Cs	208	μCi/mL
pH	13	
TOC	0.3	wt%

Pacific Northwest
NATIONAL LABORATORY

CUF- Recycle Test

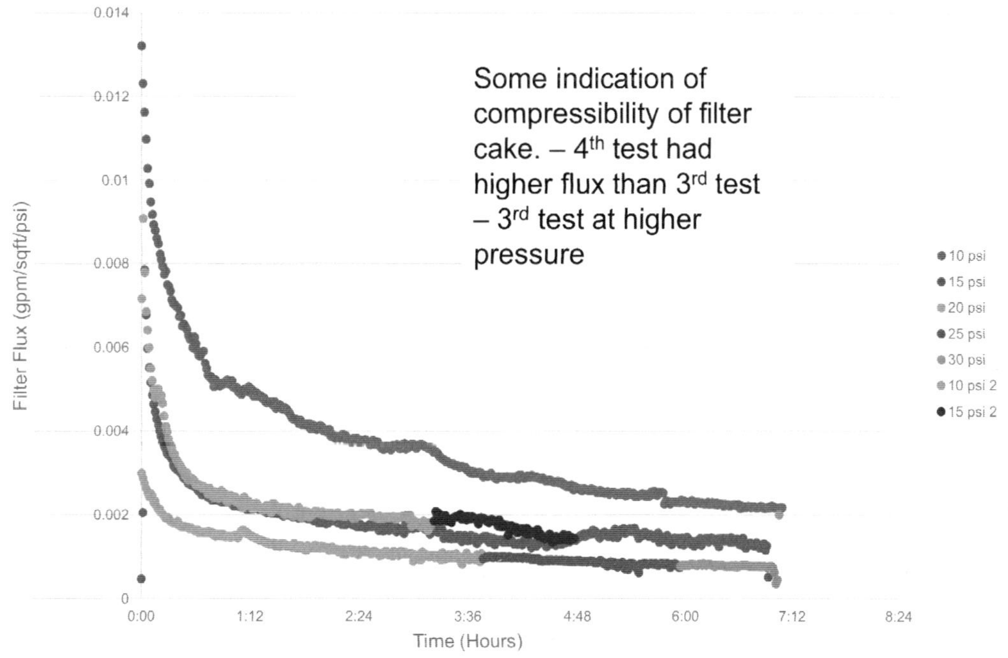

Some indication of compressibility of filter cake. – 4th test had higher flux than 3rd test – 3rd test at higher pressure

CUF- Dewatering Test (prototypic operation)

Compared to backpulsing, increasing TMP had limited impact on flux in terms of magnitude and duration

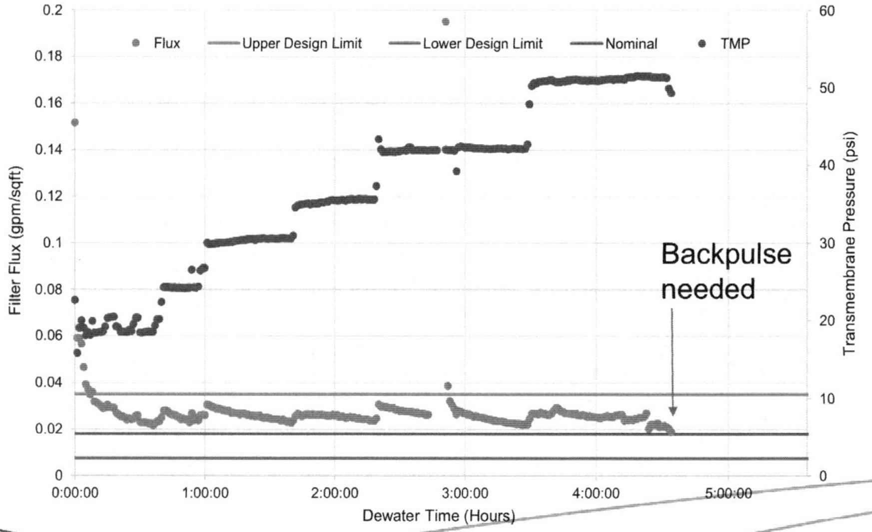

DEF AP-105 Filtration

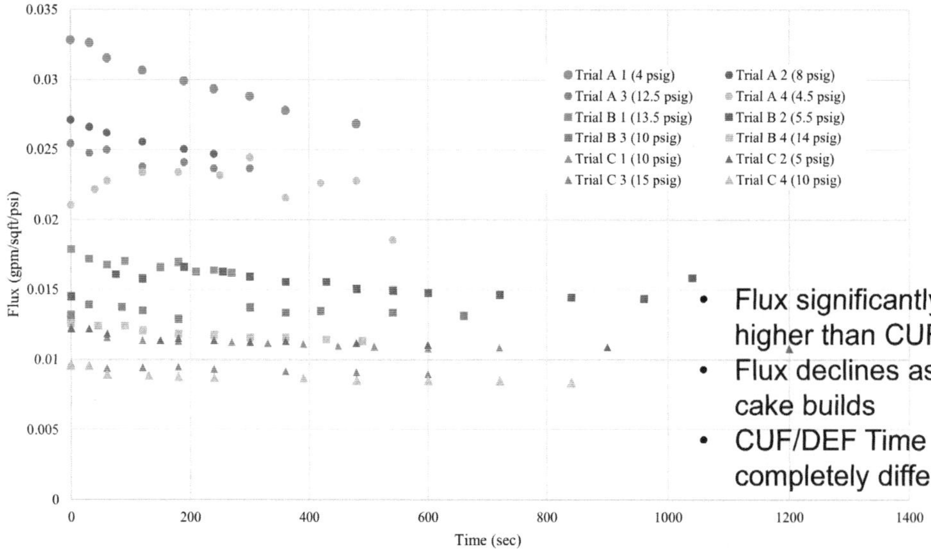

- Flux significantly higher than CUF
- Flux declines as filter cake builds
- CUF/DEF Time scales completely different

DEF Filtration of AP-105 with varying Pressure

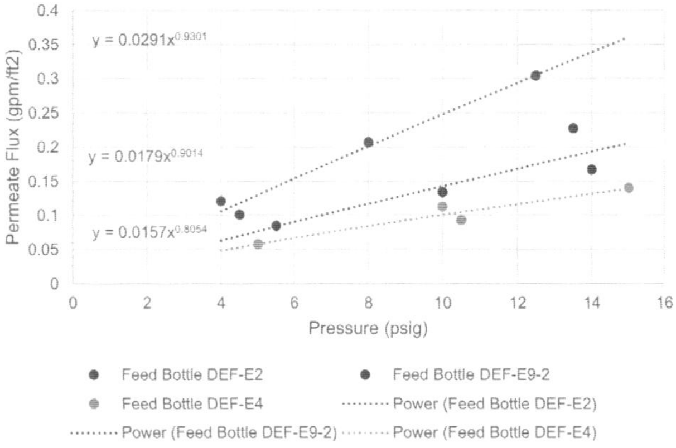

$\delta=0$ incompressible
$0<\delta<1$ moderately compressible
$\delta>1$ highly compressible

Solids are compressible

$J \propto P^{(1-\delta)}$

J- permeate flux (gpm/ft^2)
P-pressure (psig)

Pacific Northwest NATIONAL LABORATORY

Proudly Operated by Battelle *Since 1965*

DEF Filtration of AP-105 at Constant Pressure

K- cake permeability (m^2)
R_m-media resistance (1/m)
C_L – $\phi_s/(\epsilon_s-\phi_s)$
ϕ_s- solid vol fraction in slurry
ϵ_s- solid vol fraction in cake

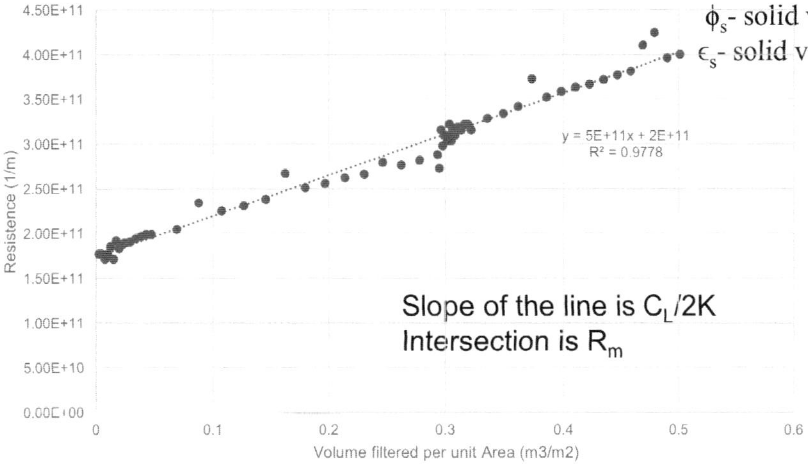

Slope of the line is $C_L/2K$
Intersection is R_m

Resistance of filter cake is NOT negligible

Pacific Northwest NATIONAL LABORATORY

Proudly Operated by Battelle *Since 1965*

DEF Filtration of AP-105

Observed cake thickness ~0.5mm =0.0005m
Volume Filtered per unit Area=0.5m

C_L= L/V~0.001 and K~2x10^{-15} m^2

Ease of Filtration	Permeability (m^2)
Very Easy	10^{-12}
Moderate	10^{-14}
Difficult (AP-105 Solids)	10^{-15}
Very Difficult	10^{-16}

AP-105 feed is supernatant sampled from tank farm with "no apparent" solids.

Small amount of solids present have low permeability and are considered "difficult" to filter

17

Pacific Northwest
NATIONAL LABORATORY

Proudly Operated by Battelle *Since 1965*

Summary of Simulants Available for DEF comparison

LAWPS Simulant	Description	Density (g/mL)	Avg. CUF Long-Term Flux (gpm/ft^2)	DEF Flux (gpm/ft^2)	Ratio
1	5.6 M Na with ~0.5 wt% sodium oxalate solids	1.21	0.04	0.1	2.5
2	5.6 M Na with 0.1 wt% iron oxyhydroxide	1.21	0.03	0.04	1.3
3	Not used				
4	8 M Na with ~0.25 wt% precipitated solids	1.35	0.03	0.07	2.3
5	4 M Na with ~0.1 wt% boehmite solids	1.18	0.1	0.2	2.0
6	6 M Na with ~1 wt% precipitated sodium phosphate solids	1.28	0.08	0.05	0.7

Comparison is difficult (different time scales, compressible cake, growing filter cake on DEF, and filter fouling on CUF. (Flux ratio for AP-105 is ~6.)

Comparison of DEF to CUF Flux for Simulants

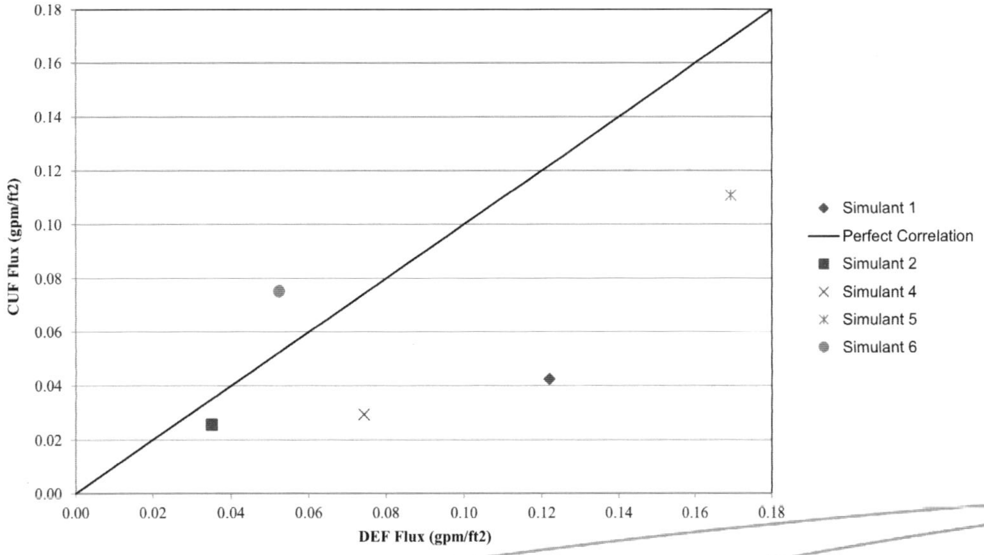

Pacific Northwest
NATIONAL LABORATORY

Proudly Operated by Battelle *Since 1965*

Conclusions

- Nature of the data make comparisons between DEF and CFF difficult
  - Limited correlation between DEF and CFF flux is due to fundamental differences
    - CFF method minimizes cake build up, and therefore cake resistance, allowing for more filter fouling
    - Different time scales of testing
- DEF is not a good substitute for qualifying feeds at lab scale, especially with small volume of feed
- DEF and CFF methods are complementary
  - DEF better suited for collecting and analyzing solids
  - DEF provides information on cake properties
  - CFF is prototypic of plant operations and provides better flux data and can better demonstrate backpulsing and filter cleaning methods
- AP-105 Filter cake is compressible. Demonstrated with both CFF and DEF.
- AP-105 Filter cake permeability measured. Demonstrated with DEF.
- Increasing TMP was shown not to be effective for long. Demonstrated with CUF.
- Bench scale testing shows filter cleaning will need to be done every 7 to 14 hours. Demonstrated with CUF.

ORY

Proudly Operated by Battelle *Since 1965*

BACK UP Information

Solids Residue - DEF

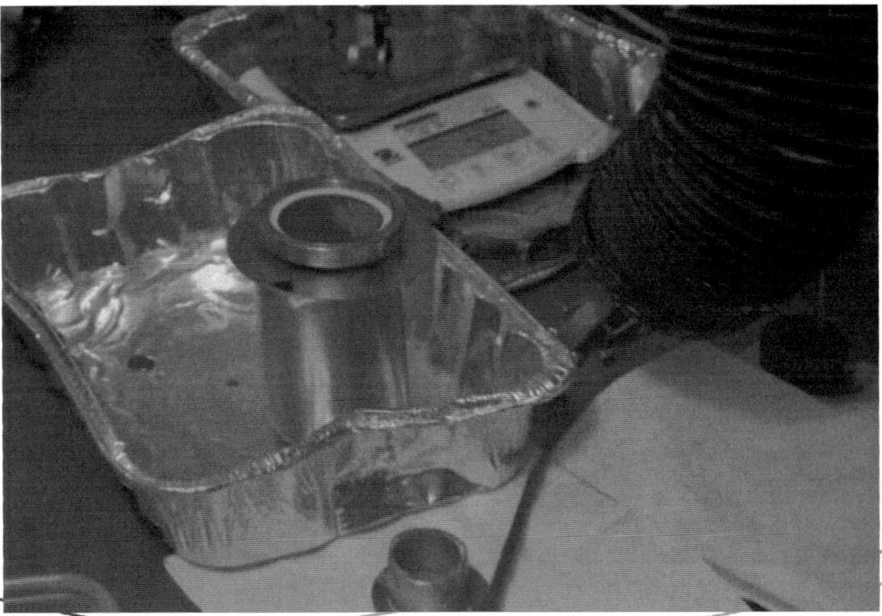

Solids characterization – top 10 components

(Analyte)	ppm
Na	199,500
C	92,000
Al	35,550
Cr	33,700
Fe	18,000
Mg	9,645
Ni	6,150
B	4,445
K	4,220
Pb	2,235

Na:C ratio in line with sodium oxalate

Pending additional microscope analysis

Preliminary data, still pending review

Investigation of High-Level Waste Glass Melting Using X-Ray Computed Tomography

Alexander S. Choi, Donald H. Miller, David M. Immel, and Frank G. Smith

2017 AIChE Annual Meeting, Minneapolis, MN
October 30 – November 3, 2017

High-Level Waste (HLW) Processing at Savannah River Site (SRS)

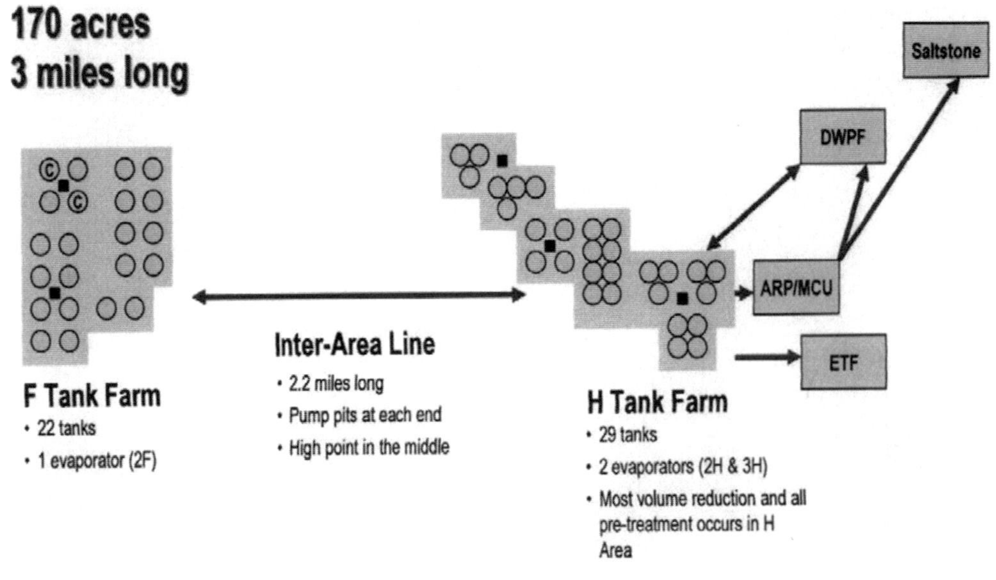

Defense Waste Processing Facility (DWPF)

- Components of HLW Glasses:
  - *Waste – Sludge (Fe, Al, Na, Mn, U, Tc, Am) and Salts (Cs, Sr)*
  - *Glass Forming Frits (Si, B, Li, Na)*
- Goals:
  - *Reduce life-cycle cost by maximizing waste throughput.*
  - *Reduce canister production by maximizing waste loading.*
  - *Optimize frit composition for maximum melt rate.*
- Current Strategy:
  - *Determine candidate frits using property/composition models.*
  - *Bench-scale melt rate testing of candidate frits to select one that yields highest melt flux – expensive and time-consuming process.*
  - *Semi-empirical modeling of melting rate.*

Melt Rate Furnace (MRF)

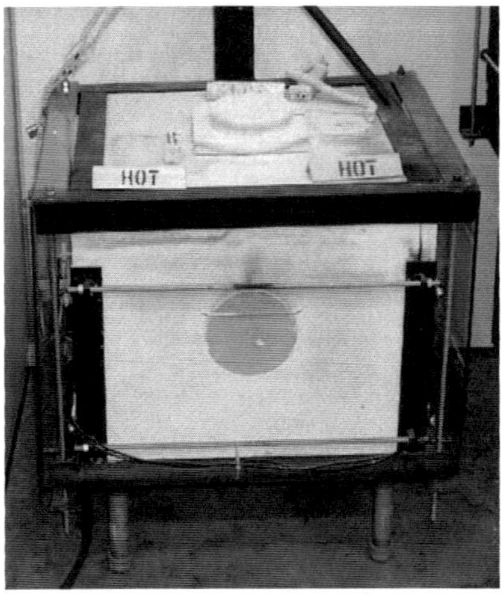

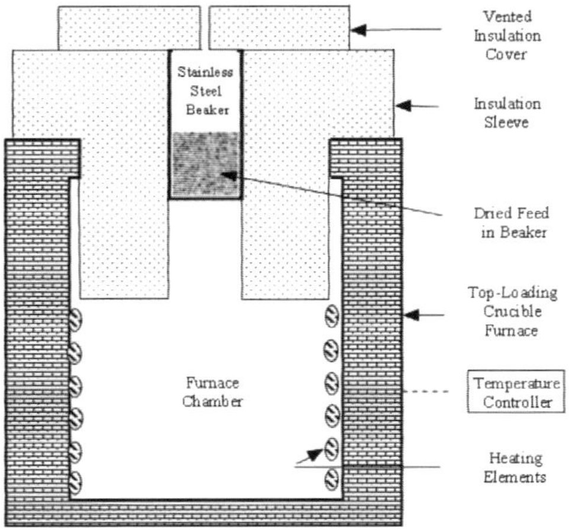

- Lindberg/Blue M crucible furnace (Model CF56822C) by Thermo Fisher Scientific.
- 19 cm diameter inner chamber with bottom 20.3 cm heated.
- 5 cm and 10 cm diameter stainless steel beakers
- Batch size = dried waste simulant/frit blend to produce ~50 g of glass when fully melted.
- One-dimensional vertical heating in the beaker to simulate the cold cap melting.
- Batch heating time = 20 - 22 min.

Visual Determination of Melting Rate

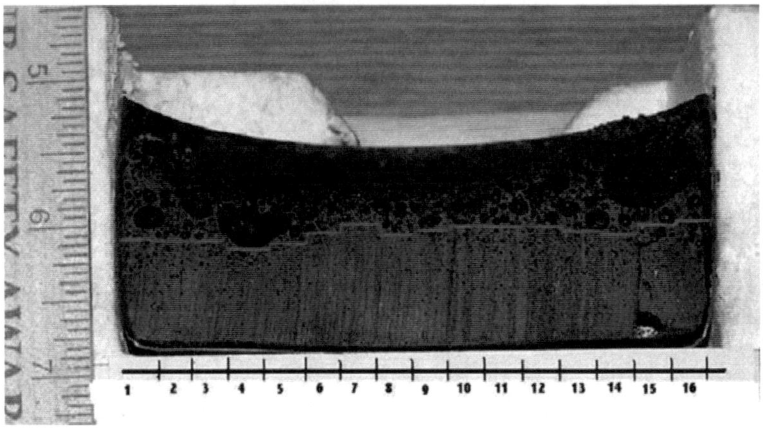

- Linear Melt Rate (LMR) is determined as the average of measured glass heights in each of the 16 horizontal segments divided by the time each sample was fired.
- LMR was useful in identifying melting accelerants and ranking relative melting rates of candidate frits.
- However, LMR becomes increasingly subjective as glass height (red line) becomes more difficult for feeds with increasing concentration of gas generators (i.e., in the presence of more numerous bubbles or voids).
- Bubble distribution and thus glass height of each cross-section will not remain the same when rotated around 360° along the center axis.

X-Ray Scan

- 5 cm-diameter beakers containing heat-treated samples were X-ray scanned using a lens-coupled digital radiography system consisting of:
  - GE/Seifert 420 Isovolt X-ray generator
  - Hamamatsu CsI scintillation screen
  - Roper Scientific Quantix CCD camera fitted with a Nikon Nikkor 105 mm lens
- A four-axis motion control system was used for sample positioning.
- Control and processing software was written using the Labview programming language.
- For each sample, 360 digital radiographs were taken at one degree angular increments.
- X-ray images were taken along a cross-section of the sample through the center axis perpendicular to the source.
- The potential at the X-ray source was set at 300 kV.
- A 1.25 mm stainless steel filter was used at the source to harden the source beam and a 0.13 mm lead filter was used at the scintillator to attenuate X-ray scatter.
- Feldkamp cone-beam reconstruction was performed on the sample data using software developed at Lawrence Livermore National Laboratory to produce 3-D volumetric data at a resolution of 0.0033 mm^3 representing a 0.15 mm sided voxel.

Computed Tomography (CT)

- A segmentation approach was used to build 3-D volumetric data of heat-treated samples at a resolution of 0.0033 mm^3 representing a 0.15 mm-sided voxel.
- Radiographic attenuation in a homogeneous material follows Lambert-Beer's Law:

$$I = I_0 e^{-\mu L} \quad (1)$$

where I_0 = intensity of the incident X-ray beam
I = intensity of the attenuated X-ray beam
μ = attenuation factor of the material in mm^{-1}
L = material dimension or length of a voxel side in mm.

- The dimensionless quantity μL = CT density (ρ_{CT}), which is correlated to material density.

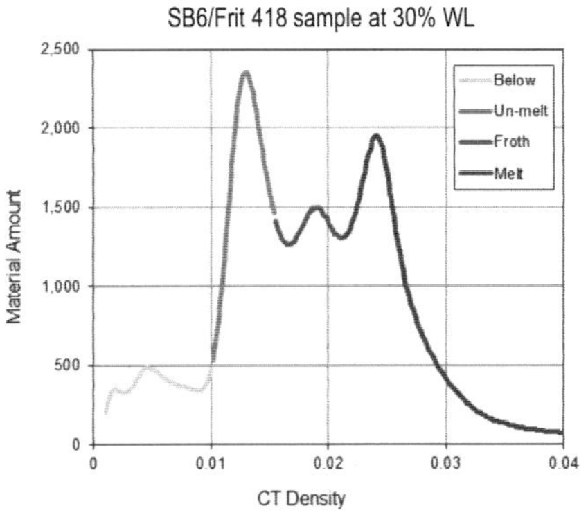

3-D volumetric data for typical samples shows four morphologically-distinct layers:

- Melt: Fully melted glass with little or no gas bubbles.
- Froth: Melted material interspersed among gas bubbles.
- Unmelt: Feed material that may have begun to soften but maintains its original granular structure.
- Below: Material of very low but non-zero density indicating gas-filled voids or radiographic artifacts.

Threshold CT Densities

- Threshold boundaries between adjacent layers were determined by differentiating and smoothing the histogram point-by-point and finding significant negative-to-positive zero crossings of the 3rd derivative of the histogram.

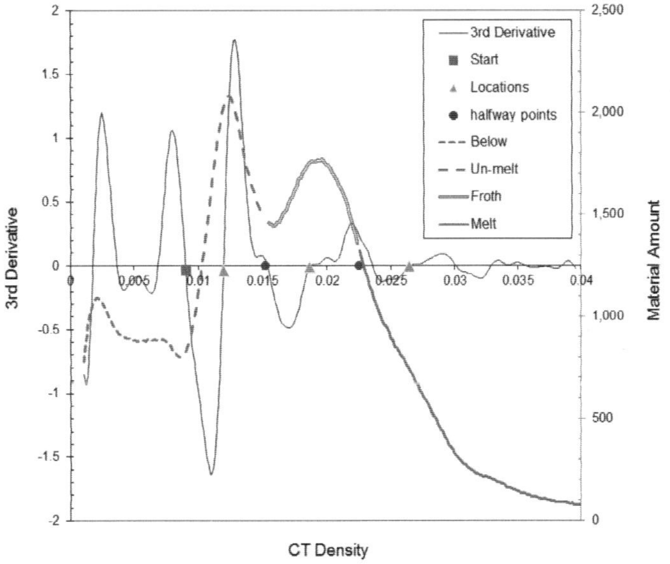

Compositions of HLW Simulants and Frit 418

Simulant	SMEM-1	SMEM-2	SMEM-3	SMEM-4	SB6
Elements (>0.1 wt%)	wt% calcined	wt% calcined	wt% calcined	wt% calcined	wt% calcined
Fe	32.00	31.60	11.15	12.30	18.90
Al	8.99	9.14	23.15	23.75	15.90
Mn (IV)	3.03	0.52	1.54	0.30	5.92
Mn (II)	1.01	0.18	1.96	0.40	
Ca	3.88	2.05	3.27	1.81	0.95
Mg	0.38	2.64	0.31	2.50	0.45
Ni	0.23	2.61	0.20	2.28	2.54
Cr	0.02	0.30	0.02	0.27	-
Cu	0.07	0.06	0.05	0.06	0.18
Si	1.50	1.51	1.42	1.32	0.14
Na	13.80	14.80	16.80	15.90	15.40
K	<0.10	<0.10	0.62	<0.10	0.07
Ce	0.11	0.11	0.10	0.10	<0.01
Ba	0.08	0.10	0.06	0.06	<0.01
Pb	0.11	0.08	0.09	0.09	<0.01
Zr	0.06	0.12	0.07	0.06	0.22
S	0.31	0.31	0.37	0.41	0.37
Total	65.58	66.13	61.18	61.61	61.04
Anions	mg/kg slurry	mg/kg slurry	mg/kg slurry	mg/kg slurry	mg/kg slurry
Cl	125	130	1,125	133	701
NO_3	22,100	27,200	22,700	25,400	45,600
SO_4	204	<100	772	<100	931
COOH	51,400	48,900	51,100	47,300	89,000
C_2O_4	<100	<100	<100	<100	399

Target Composition of Frit 418

Oxide	wt %
B_2O_3	8
Li_2O	8
Na_2O	8
SiO_2	76
Total	100

X-Ray Images of Bounding Sludge/Frit 418 Feeds at 36% Waste Loading (WL)

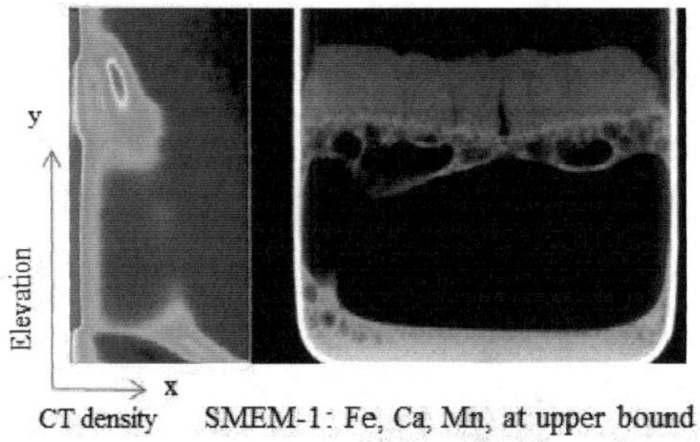

SMEM-1: Fe, Ca, Mn, at upper bound
Al, Mg, Ni at lower bound

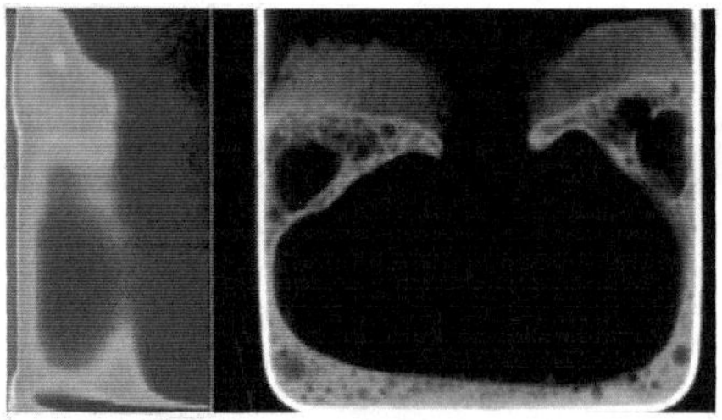

SMEM-2: Fe, Mg, Ni at upper bound
Al, Ca, Mn at lower bound

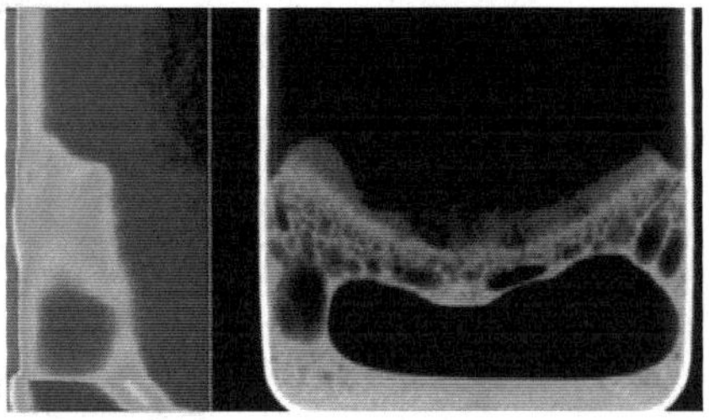

SMEM-3: Al, Ca, Mn at upper bound
Fe, Mg, Ni at lower bound

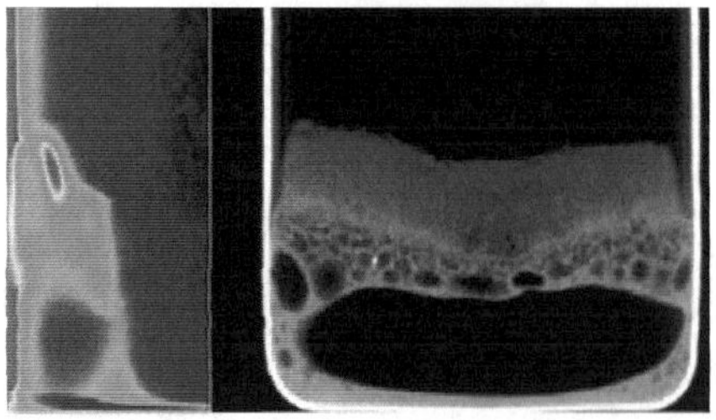

SMEM-4: Al, Mg, Ni at upper bound
Fe, Ca, Mn at lower bound

Histograms of Bounding Sludge/Frit 418 Feeds

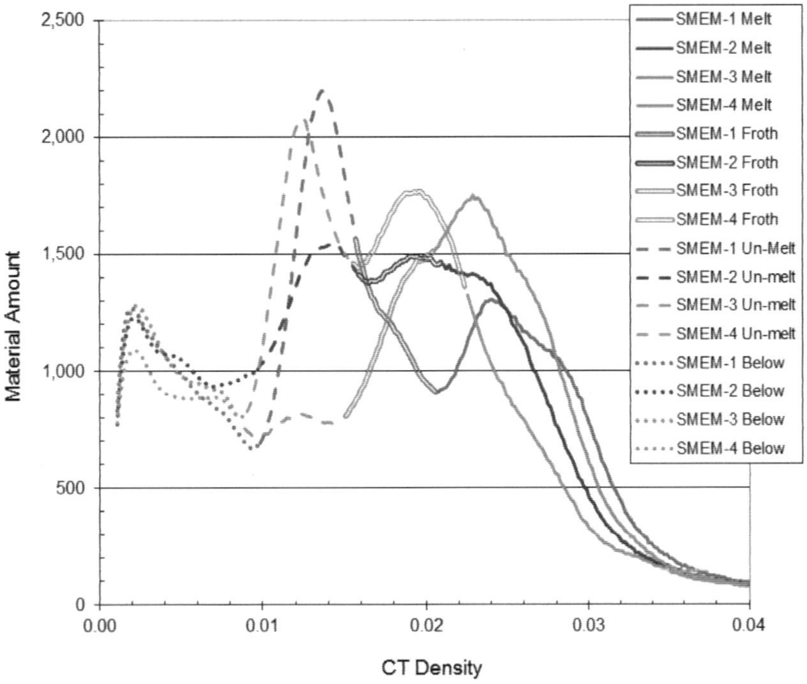

Material Amount and Density of Heat-Treated Bounding-Sludge/Frit 418 Feeds

Sample ID	Material Amount					Material Density (mm^{-3})				
	Total	Un-melt	Froth	Melt	Melt + Froth	Overall	Un-melt	Froth	Melt	Melt + Froth
SMEM-3	161,430	27,064	40,805	93,562	134,366	5.784	3.417	5.347	7.572	6.722
SMEM-2	165,923	47,278	46,189	72,456	118,645	5.414	3.716	5.462	7.656	6.620
SMEM-1	174,012	57,499	36,410	80,103	116,513	5.453	3.866	5.314	7.863	6.838
SMEM-4	160,654	55,857	58,867	45,930	104,797	4.967	3.551	5.423	7.978	6.308
std dev/ mean	3.7%	29.8%	21.3%	27.5%	10.3%	6.2%	5.4%	1.3%	2.4%	3.4%

Projected Composition of SMEM Glasses and Calculated Properties

Oxides:	SMEM-1 (mol%)	SMEM-2 (mol%)	SMEM-3 (mol%)	SMEM-4 (mol%)
Fe_2O_3	6.88	6.68	2.34	2.52
Al_2O_3	4.00	4.00	10.05	10.07
MnO	1.76	0.30	1.49	0.29
CaO	2.32	1.21	1.91	1.03
MgO	0.37	2.57	0.30	2.36
NiO	0.09	1.05	0.08	0.89
Cr_2O_3	0.01	0.07	0.00	0.06
CuO	0.02	0.02	0.02	0.02
SiO_2	54.80	54.39	53.30	52.84
Na_2O	13.28	13.41	14.22	13.71
K_2O	-	-	0.18	-
Ce_2O_3	0.01	0.01	0.01	0.01
B_2O_3	4.86	4.83	4.74	4.70
Li_2O	11.33	11.25	11.03	10.96
BaO	0.07	0.08	0.05	0.05
PbO	0.01	0.01	0.01	0.01
ZrO_2	0.02	0.03	0.02	0.01
SO_3	0.14	0.08	0.22	0.44
Total	99.97	99.99	99.97	99.97
net alkali ion field strength* (1/Å^2)	69.997	70.203	67.341	65.254
non alkali ion field strength (1/Å^2)	15.807	28.337	13.168	25.328
total modifier field strength (1/Å^2)	85.804	98.540	80.510	90.582
calculated viscosity at 1,150 °C (Pa·sec)	3.445	3.338	13.086	13.879
calculated liquidus, T_l (°C)	937.6	1,068.3	822.0	946.9

Factors Affecting HLW Glass Melting Rate

- Total field strength (*FS*) of modifier ions is calculated as the weighted sum of individual modifier field strengths:

$$FS = \sum_{k=1}^{n} C_k \left(z_k / a_k^2 \right) \qquad (2)$$

where C_k = concentration of modifier k (as oxide) in mole fraction
z_k = valence charge of modifier k
a_k = effective ionic radius of modifier k in Å

The lower the *FS*, the less rigid the network structure (hence, lower melt density), allowing more large radionuclides into the network.

- The higher the rate of gas evolution from salt decomposition, the lower the rate of heat transfer.

As the nitrate decomposes over a wider temperature range than the formate, it has a greater adverse impact on the melting rate.

- The higher the concentration of *Al*, the more modifier ions are tied up with AlO_4^- tetrahedra, making the network structure more rigid and the melt viscosity higher.

- However, the higher the melt viscosity, the more sluggish the melt in closing the channels through which evolving gases escape.

X-Ray Images of Heat-Treated SB6/Frit 418 Feeds at Varying WL

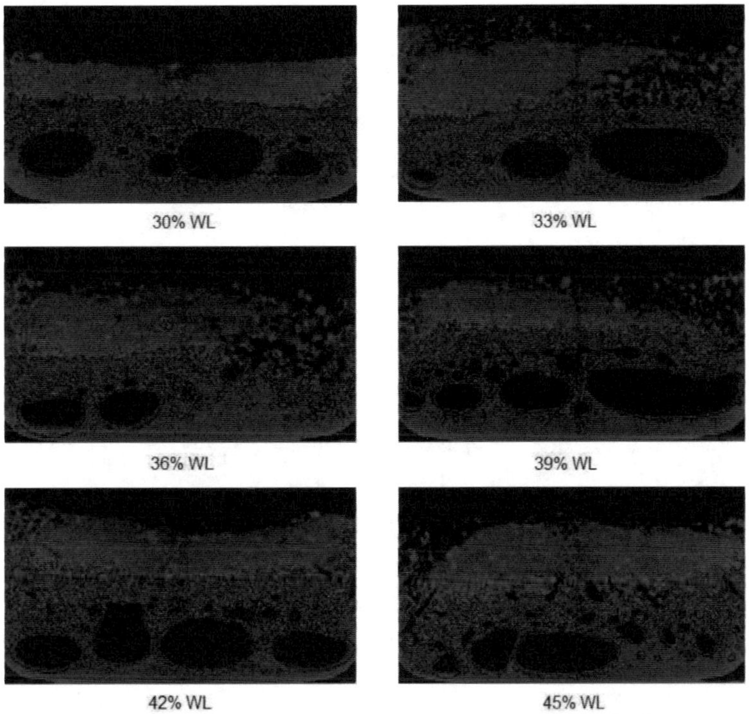

Histograms of SB6/Frit 418 Feeds at Varying WL

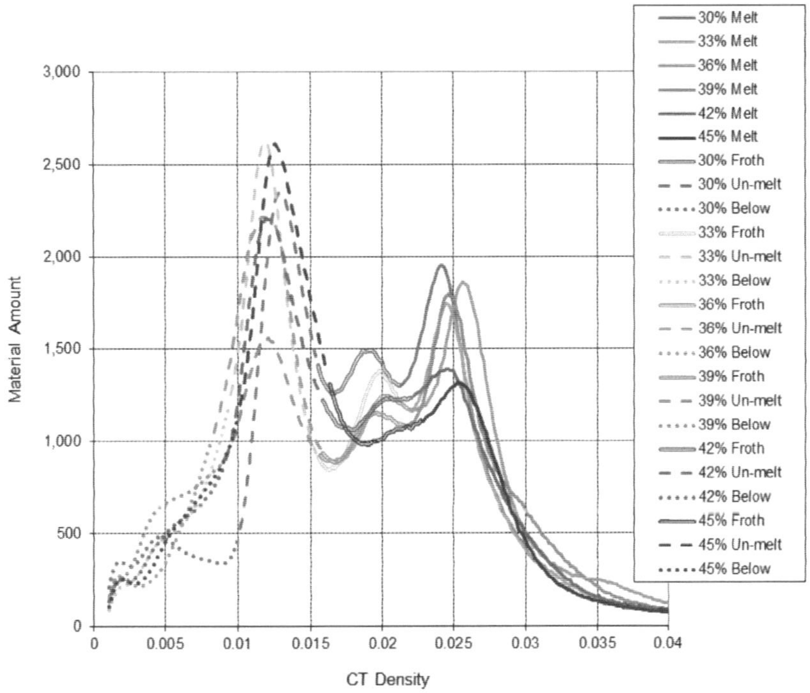

Material Amount and Density of Heat-Treated SB6/Frit 418 Feeds

Waste Loading (wt%)	Material Amount					Material Density (mm-3)				
	Total	Un-melt	Froth	Melt	Melt + Froth	Overall	Un-melt	Froth	Melt	Melt + Froth
30	180,710	55,996	49,417	75,297	124,714	5.399	3.854	5.451	7.623	6.584
33	176,536	66,460	43,512	66,563	110,076	5.070	3.595	5.589	7.789	6.740
36	195,145	75,999	43,380	75,766	119,146	5.015	3.474	5.690	8.048	6.993
39	166,307	50,761	38,222	77,324	115,546	5.396	3.582	5.557	7.910	6.938
42	177,440	68,310	45,327	63,803	109,130	5.106	3.662	5.624	7.935	6.778
45	174,004	74,998	42,602	56,405	99,007	5.150	3.868	5.760	8.054	6.876
std dev/ mean	3.2%	25.6%	61.0%	39.9%	32.1%	3.2%	4.3%	1.9%	2.1%	2.2%

Relative Melting Rates vs. Waste Loading of SB6/Frit 418 Feeds

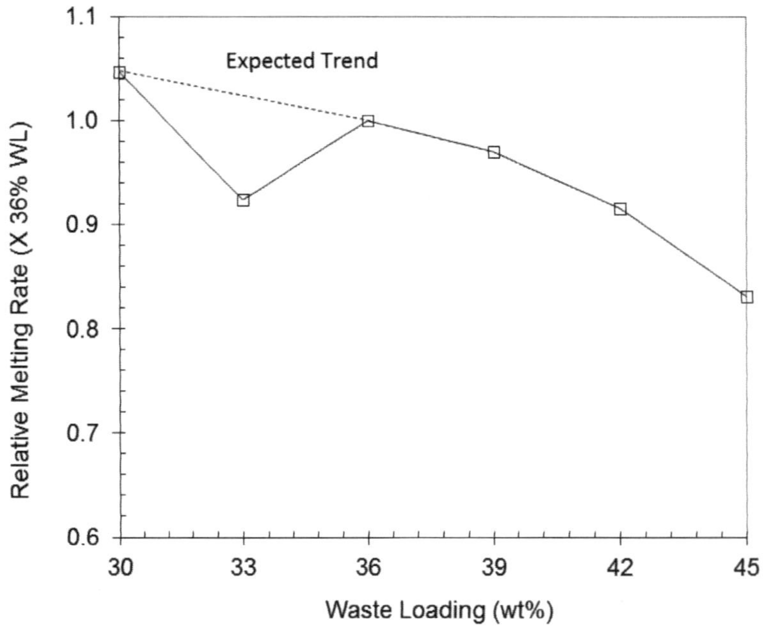

Reproducibility of Data

- The reproducibility of CT results was checked using duplicate samples of:
  - Four different simulant compositions.
  - Each simulant spiked with high and low concentration of gas generators (or salts).
- The concentration of salts showed no appreciable effect on the scatter of data.
  - The standard deviation ranged from 3.9 to 13.7% of their respective means for feeds with higher concentrations of salts and from 0.02 to 15.7% for feeds with lower concentrations of salts.
- Due to small sample size, the root-mean squared error (RMSE) was used as a generalized standard deviation encompassing all four feeds:

Relative RMSE of All Samples

Unmelt	Froth	Melt	Melt+Froth
7.2%	4.5%	8.2%	1.7%

Relative RMSE of Two Samples Re-Scanned One-Year Later

Unmelt	Froth	Melt	Melt+Froth
1.7%	1.1%	1.1%	1.0%

Conclusions

- X-ray CT is capable of providing:
  - Visual images of batch-to-glass conversion, including gas evolution and resulting bed expansion.
  - A quantitative measure of relative melting rates of various waste/frit blends in a consistent and objective manner.
  - A framework for optimizing frit design to accommodate ever-changing waste batch chemistry.
- X-ray images of heat-treated samples revealed the presence of four morphologically distinct layers: Melt, Froth, Unmelt and Below, as defined earlier.
- Quantifying the amount of material in each layer by CT and combining the material amounts in the Melt and Froth layers would give a viable measure of relative melting rate rankings that is representative of continuously-fed melters.
- The HLW glass melting rate is at least affected by:
  - Total field strength of modifier ions
  - Rate of calcine gas evolution
  - Concentration of Al
  - Melt viscosity

Acknowledgements

- Funding was provided by:
  - U.S. Department of Energy EM-31 Technology Development & Deployment (TDD) Program
  - U.S. Department of Energy Office of River Protection (DOE-ORP) at Hanford, WA
- Dr. Pavel Hrma of Pacific Northwest National Laboratory (PNNL) and Dr. Albert Kruger of DOE-ORP for technical review of the paper.
- This work has been published in *International Journal of Applied Glass Science (2017).*

493564 Greenhouse Gas Intensities and Energetic Productivity Dynamics of Giant Global Oilfields: A Life Cycle Approach

Abstract:

The steady increase in world energy consumption, coupled with record-breaking earth surface temperatures have induced governments to implement targets for reducing future greenhouse gas (GHG) emissions. Use of oil products contributes ~35% of global GHG emissions, and the oil industry itself consumes 3-4% of global primary energy. Most GHG analysis of the oil sector has focused on comparing the impacts of oil-derived transport fuels to other transport fuel options. This is a reasonable focus as the largest source of emissions from oil use is mostly the combustion of finished fuels (e.g., gasoline burned in an automobile). However, emissions from oil and gas extraction and processing are sometimes significant, particularly for unconventional resources or when energy-intensive enhanced oil recovery (EOR) strategies are used. Also from a national viewpoint, oil and gas extraction can cause a significant fraction of domestic emissions in fossil fuel exporting countries like Canada (~20% of national emissions), Russia (~20%), and Norway (~28%). Since oil resources are becoming increasingly heterogeneous -- requiring different extraction and processing methods -- GHG and energy studies should evaluate oil sources using detailed project-specific data.

Unfortunately, prior oil-sector GHG and energetic productivity analyses rely on "snapshot" data of varying quality and vintage, and have largely neglected the fact that the energy intensity of producing oil can change significantly over the life of a particular oil project. These changes result from age-related changes in engineering practice (e.g. water/gas/steam injection) and processing requirements (e.g., fluid separation). Neglecting temporal trends in oil-sector emissions is problematic for long-term climate and integrated assessment modeling, where emissions trends over decades are of interest.

Here, via a detail engineering-based life cycle approach, we use decades-long historical data from twenty-five globally significant oil fields (> 1 billion barrels ultimate recovery) to model GHG intensities and energy return on investment (EROI) from oil production as a function of time. The net energy ratio (NER) and external energy ratio (EER) are used as two measures to estimate oil fields EROI. We find that volumetric oil production declines with depletion, but this depletion is accompanied by significant growth – in some cases over tenfold – in per-MJ GHG emissions. At the same time while energetic productivity decline significantly, with some fields seeing declines in excess of 90%. Depletion and reservoir exhaustion requires increased energy expenditures in drilling, oil recovery, and oil processing.

Probabilistic Monte Carlo simulation is used to draw generalized historical trends from the aggregated data of onshore/offshore, and light/medium/heavy oil fields: over 25 years, this model predicts about two-fold increase in GHG intensities, and ~ 30% and ~ 20% declines in NER and EER, respectively. We also derive a general relationship for projecting evolving dynamics of GHG emissions, NER, and EER trends over time.

In summary, the per-MJ GHG emissions and EROI from oil fields change from "oil-to-oil", but also differ significantly "time-to-time" over a field's productive life. These trends have implications for long-term climate and energy system modeling, as well as for climate policy, due to potential large increases in carbon intensity and extraction energy as global oilfields age. These effects may result in significant underestimation of future energy demand in the global oil sector in long-run models like integrated assessment models.

Economic and Environmental Evaluation of Olive Mill Wastewater Treatment Methods for a Self-Supplied American Olive Oil Mill (AIChE Abstract #493715)

Chae Jeong and Benjamin Davis
The Cooper Union, New York, NY

Introduction

Olive oil is a popular consumer product that is highly versatile. It is sold as a food product but also used in other products such as cosmetics and pharmaceuticals.[1] Consequently, there is a large market for olive oil with a production estimate of 3.225 million metric tons in the 2015/2016 season.[2] Spain, Italy, and Greece are the largest producers and contribute about 97% of European olive oil production, with Europe contributing 60% of the world production.

Olive oil extraction methods date back as far as 3000 B.C. – since then, these methods have changed significantly, requiring new machinery and heavy utilities,[3] making it possible to create more olive oil in a shorter amount of time. This has allowed for greater revenue generation, but also presents additional processing difficulties. One main difficulty is the cost to implement and maintain the new technology – modern production methods require water and electricity, investment in capital, and additional costs to manage waste.

In the United States (U.S.), olive oil production occurs at a much smaller scale, with about 14,000 metric tons produced yearly (against 1.3 million metric tons produced in Spain which is almost 100 times more).[4] However, the industry is not absent; Americans consume an average of 80 million gallons of olive oil annually, the second largest non-European consumer in the world.[5] Major olive oil producers in the U.S. have recently started operating at larger scale, mainly in California. Most U. S. olive oil mills cultivate their own olives, including those operating at the larger industrial scale. Smaller mills require much more human labor to process their olives and thus it is economically and socially interesting to consider the benefits of implementing a continuous milling operation. The question is: how profitable can industrial scale production be given a mill that only processes their own olives, i.e., no outside suppliers? Especially for smaller mills, it is essential that the system is operating as efficiently as possible. There have been various studies done to optimize the production of olive oil, mostly with regards to maximizing yield. Much of the work done has been experimental, such as studies of effects of coadjuvant addition on oil yield and effects of malaxing time on the yield as well as quality of oil.[6,7] The work we present uses experimental work found in literature to build a model to optimize the production process which will maximize the operating revenue of a self-supplied mill, but economics do not give the whole picture.

In addition to concerns of profitability are concerns of the environmental impacts of the new technology. Processing equipment for continuous olive oil milling operations requires much more water input than rudimentary oil pressing methods. With this comes the increase of waste called olive mill wastewater (OMW), which is comprised of process water used to wash olives and aid separation of the oil from the fruit plus the vegetable water that is separated from the fruit. The majority of this OMW is vegetable water (which is about 50% wt. of the olive fruit[8]) and water from decanting. Both streams contribute to the high phenolic content of OMW, which makes it highly toxic to plants and bacteria. OMW produced by the olive oil production process can have a phenolic content from 500 mg/L up to 24000 mg/L, which is extremely high.[9] The allowable limit of phenolic industrial wastewater for release into freshwater aquatic systems in the U.S. is 3.4 mg/L and therefore, it is not recommended that OMW be released into the environment without treatment.[10]

Some treatment methods include:

1. *Evaporation Ponds*: The most common way to dispose of OMW is collection in an evaporation pond.[8] Extra processing is not required and the capital cost of a pond is fairly low. However, the pond requires a large amount of land for installation, which may not be available on an olive orchard, and OMW has an unpleasant odor, causing nuisance for the surrounding community. In addition, potential spills from the pond can be harmful for the olive orchard and the environment.[8]
2. *Spreading OMW as Fertilizer*: The most cost efficient method of OMW disposal is direct spreading onto the olive orchard; this does not require any extra processing and therefore has low capital cost. This is a viable method for self-supplied olive oil mills where the olives are grown on site and there is land to spread the OMW on. Magdich et al. suggest that using OMW as fertilizer can potentially be agriculturally beneficial because of the minerals and nutrients such as potassium, nitrogen, and magnesium present.[11] This should induce microbial growth in the soil and in turn positively impact the growth of olives. However, if the recommended dose of OMW (100 m^3/ha) is used, it would only account for a small percentage of the OMW; if all of the OMW is spread, the dosage would be too high and therefore harmful to the olive orchard.
3. *Adsorption Using Activated Carbon*: Adsorption using activated carbon (AC) is a generally popular method of water treatment. AC is proven to have high efficiency of separation, high adsorption capacity, and low operating costs. It is used in a variety of applications such as groundwater remediation, drinking water filtration and air purification. [12] Adsorption via AC is effective in removing certain organics, including phenols making it a potentially viable treatment method for OMW. [12] However, commercial AC is too expensive to employ and not enough AC would be generated from the biomass found on the olive orchard (pomace, tree branches, leaves, etc.) to fully treat the OMW.
4. *Biodegradation by Activated Sludge*: Biodegradation has advantages to other separation methods, including adsorption, because a concentrated stream of phenols would not be generated. The phenols would be broken down and consumed by biomass and further processing would not be necessary. OMW can create an environment that is favorable for microbial growth; in some cases, growth is not effected by the polyphenolic content of OMW and microorganisms will partially degrade the phenolic compounds. Thus, provided that a robust microorganism was cultivated or engineered, biological treatment may prove to be effective in reducing the phenolic content of OMW. However, it is difficult to account for the large range of phenolic concentrations expected for OMW. In cases of high concentrations (above 3000 mg/L)[13], the OMW would need to be diluted (with about 7 times as much dilution water as existing wastewater in the worst case), which would require significantly increase water usage.
5. *Membrane Separation and Valorization*: Processes like reverse osmosis, microfiltration, and ultrafiltration are commonly used for large-scale removal of impurities from water streams.[12] A viable membrane separation process needs to yield a valuable product because relative to the revenue generated by selling olive oil, a membrane separation train would have high capital costs and would have to make a profitable product to justify the investment. Literature suggests that the biophenols found in OMW offer an array of health benefits. Compounds like hydroxytyrosol, tyrosol, and oleuropein have shown antioxidant, anticancer, anti-inflammatory, and cardioprotective properties, making them desirable food, cosmetic, and pharmaceutical ingredients.[14, 15, 16] These phenols can potentially be separated from OMW using membrane technology to both treat the wastewater and isolate a product of value. Because this treatment method would address both the environmental concern of OMW and potentially prove economically viable, it will be explored further.

Our model of an olive oil production process was set up to reflect five main steps of the olive oil production process[17]:

1. *Olive cultivation*: Olive harvest begins in September; olives are harvested by shaking trees and collecting the fruit that falls. This can be done mechanically or manually. Once harvested, the olives are sent to the processing facility where they are sorted by color and condition. The olives that are in the best condition (ripe, fresh, unbruised, green, etc.) are processed first. The remaining olives are put in storage until they are ready to be processed.
2. *Olive washing*: The sorted olives are deleafed and washed with water in the cleaning and washer section. The olives are cleaned of any dirt or pesticides.
3. *Olive milling:* The washed olives are transferred to be crushed by a mill. The mill expels oil from the olive flesh and pits, creating an olive paste (pomace).
4. *Olive malaxing:* The paste created in the milling step is mixed in the malaxing step. This helps create a continuous oil phase. Water or coadjuvants are often added to this step to aid the process.
5. *Olive oil decanting:* The paste and oil enter a separation step that involves centrifugation and decantation to separate the oil from the pomace. This can occur in a two-phase or three-phase decanter. A two-phase decanter generates an OMW/pomace slurry waste while the three-phase decanter generates OMW and an olive press cake waste.[18]

A schematic of equipment corresponding to the process steps listed above is shown below in Figure 1.

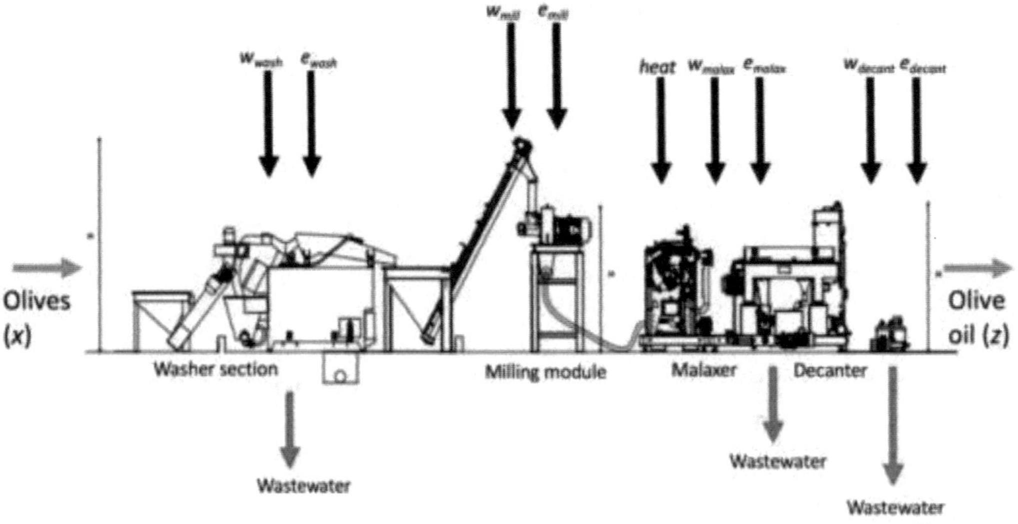

Figure 1: Process equipment schematic for a commercial olive oil production process.[19]

The process begins with the washer, which requires the input of clean water and electricity to operate (the water and electricity requirements for the washing step is labelled w_{wash} *and* e_{wash}, respectively). The washed olives are then transferred to a mill to be crushed, which requires electricity (e_{mill}). The malaxing step requires electricity (e_{malax}) and heated water (q, w_{malax}), which aids the separation process. Once malaxed, the paste and oil enter a decanter, where the oil is separated from the pomace (solids). This is assumed to occur in a two-phase decanter for this model, which uses less process water than a three-phase decanter. The decanter requires electricity (e_{decant}) and process water (w_{decant}) to operate.

Process Model and Economic Optimization

These five steps are typical of a continuous industrial-scale olive oil processing mill and are all considered in our model. The problem posed in this optimization model seeks to maximize the operating revenue of an American olive orchard given the implementation of continuous processing methods. Decision variables include the amount of oil that is produced and the time of the process. The problem takes into account the cost of capital, the cost of operations, variations in malaxing time, and the final yield of olive oil from the process. The optimization problem can be written as:

$$
\begin{aligned}
\text{maximize} \quad & \text{gross operating revenue} = (\text{revenue from oil}) - (\text{production costs}) \\
\text{subject to} \quad & \text{conditions affecting oil yield} \\
& \text{capital cost of equipment} \\
& \text{cost of cultivation} \\
& \text{cost of energy use} \\
& \text{cost of water use} \\
& \text{limits on malaxing time} \\
& \text{limits on quantity of olives} \\
& \text{limits on equipment capacity}
\end{aligned}
$$

To calculate the operating revenue generated by an existing olive orchard, the total costs of capital and operation needed to be considered. This is defined as c, the sum of cultivation, equipment, and operating costs, c_{cult}, c_{ops}, and c_{equip}, respectively (all in U.S. dollars).

The cultivation costs are dependent on the total acreage of the orchard and include cost of fertilization, pesticides, irrigation and harvesting, totaling \$300 per acre for mechanical harvesting.[20]

The equipment costs are dependent on capacity ($\dot{m}$) and is expected to have a lifetime of 10 years. A relationship was established using rough cost estimates available on an online resource for olive oil producers/consumers and quotes by a representative of Alfa Laval [19,21]– the cost data is in Table 1 and the resulting capital cost estimation relationship is as follows, annualized over 10 years:

$$10c_{equip} = 361.81\dot{m}^{0.8}$$

Table 1: Equipment capital cost data. [19,21]

Equipment Capacity, $\dot{m}$ (lb/hr)	Estimated Total Cost, c_{equip}
300	\$30,000
500	\$50,000
1100	\$150,000
4000	\$270,000
7000	\$400,000

The operating cost is the total cost of utilities (electricity and water). The cost of electricity ($c_{electricity}$) and cost of water (c_{water}) can be calculated from requirements of each major process unit, which can be found in Table 2. The average cost of electricity and water in California is \$0.15/kWh and \$0.00207/gal, respectively.[22, 23] The unit operations are assumed to be run throughout the production period (t_{total}) and the total electricity and water requirement for each unit is based on the time of operation (apart from the decanter, which has requirements based on the amount of olive oil being processed).

Table 2: Electricity and water requirement for process units. Note that data for 2-phase decanter was converted from kWh/L of oil and m³/L of oil to kWh/lb of oil and gal/lb of oil, respectively.

Process Equipment	Electricity Requirement	Water Requirement
Washing section [9]	e_{wash} = 3.26 kW	w_{wash} = 67 gal/hr.
Hammer crusher (Mill) [9]	e_{mill} = 7.5 kW	--
AlfaOliver 500 (Malaxer) [9]	e_{malax} = 24.5 kW	w_{malax} = 44.91 gal/hr.
2-Phase Decanter [14]	e_{decant} = 0.338 kWh/lb. oil	w_{decant} = 0.613 gal/lb. oil

The revenue calculations deal with the gross sales generated by the olive oil mill (***pz***), where *p* is the selling price of oil in dollars per pound of extra virgin oil (\$/lb) and *z* is the amount of oil produced in t_{total} (lb). The selling price of oil varies from month-to-month; for this model the selling price for September 2016 was used, resulting in a price of \$1.81/lb of extra virgin oil.[24] The amount of oil processed is dependent on the amount of olives that are processed (w_o) and the yield of the process units. The amount of olives that are processed is limited to the amount of olives available from the orchard; an acre can hold 200 trees, which holds about 35 lb of olives/tree.[25]The yield of decanter ($y_{decanter}$) does not vary significantly based on operating conditions and is assumed to be about 90% in this model. However, the yield on a malaxer ($y_{malaxer}$) is dependent on malaxing time (t_m) and temperature (which was experimentally quantified).[26] Yield was shown to generally greatest at a malaxing temperature of 35°C and therefore, the following relation was used to determine the malaxer yield at 35 °C:

$$y_{malaxer} = 0.01 \times (-0.0114{t_m}^2 + 0.989t_m - 4.0942)$$

To run the malaxing process at 35 °C, the energy required to heat the water (q) from 27 °C was determined and included in the total utility requirements.

Changes in malaxing time cause changes in time of operation (τ) and total production time (t_{total}). Time of operation is the sum of the time of operation of each process step in hours. The total production time is calculated from the total amount of olive processed and the capacity of the process, dependent on its basis capacity ($\dot{m}$) and the operation time (τ). The time for washing, milling, and decanting are not accurately quantifiable and were assumed to be 10 minutes each, based on the 1 hour of operation time most common in literature. In cases seen in literature, malaxing time is about 30 minutes, leaving 30 minutes total for washing, milling, and decanting. Limits on malaxing time were determined by the data found in literature to the use the relation formed above; t_m must be between 60 and 30 minutes.[26]

The full optimization model can be found below. As formulated, the problem is a nonlinear programming problem (NLP). The objective is a nonlinear function of two variables: *z* and *c*. The problem consists of 20 variables in total, 17 equality constraints, and 6 inequality constraints. Within the Optimization Toolbox in MATLAB, there exist several functions which are capable of solving NLPs; 'fmincon' was used here, which finds the minimum of constrained multivariable NLP problems and uses the interior-point method as its default setting.[27] The specific problem below assumed a 17-acre orchard.

The model was also adapted to create a function that characterizes the production process of extra virgin oil given the acreage of an olive orchard (a), the malaxing temperature (T), and the selling price of the oil (p).

$$
\begin{aligned}
\underset{x\in\Re^{20}}{\text{maximize}} \quad & pz - c \\
\text{subject to} \quad & z = w_o y_{malaxer} y_{decanter} \\
& y_{malaxer} = 0.01 \times (-0.0114 t_m^2 + 0.9892 t_m - 4.0942) \\
& \tau = (t_m + t_{wash} + t_{mill} + t_{decant}) \\
& \dot{m} t_{total} = \tau w_o \\
& c = c_{equip} + c_{ops} + c_{cult} \\
& 10 \times c_{equip} = 361.81 \dot{m}^{0.8} \\
& c_{ops} = c_{electricity} + c_{water} \\
& c_{electricity} = 0.15(e_{wash} + e_{malax} + e_{mill} + e_{decant} + q) \\
& e_{wash} = 3.26(\frac{t_{wash}}{\tau})(t_{total}) \\
& e_{malax} = 24.5(\frac{t_m}{\tau})(t_{total}) \\
& e_{mill} = 7.5(\frac{t_{mill}}{\tau})(t_{total}) \\
& e_{decant} = 0.338 z \\
& q = \rho w_{malax} C_p \Delta T \\
& c_{water} = 2.07 \times 10^{-3}(w_{malax} + w_{wash} + w_{decant}) \\
& w_{wash} = 67(\frac{t_{wash}}{\tau})(t_{total}) \\
& w_{malax} = 44.91(\frac{t_m}{\tau})(t_{total}) \\
& w_{decant} = 0.613 z \\
& 30 \le t_m \le 60 \\
& 0 \le w_o \le 120785 \\
& \epsilon \le \dot{m} \le 7000
\end{aligned}
$$

where:

$$
x = \begin{bmatrix} w_o \\ t_m \\ z \\ y_{malax} \\ \tau \\ t_{total} \\ c \\ c_{equip} \\ \dot{m} \\ c_{ops} \\ c_{electricity} \\ c_{water} \\ e_{wash} \\ e_{malax} \\ e_{mill} \\ e_{decant} \\ q \\ w_{malax} \\ w_{wash} \\ w_{decant} \end{bmatrix}
$$

and the parameters are:

$$\epsilon = 0.001$$

$$p = \$3980.07/\text{metric ton oil} = \$1.81/\text{lb oil}$$

$$y_{decanter} = 0.9$$

$$t_{wash} = t_{mill} = t_{decant} = 10 \text{ min}$$

$$c_{cult} = (17 \text{ acres})(\$300/\text{acre/year}) = \$5100/\text{year}$$

$$C_p = 4.17 \text{ J/g-°C}$$

$$\rho = 1000 \text{ kg/m}^3 = 0.008345 \text{ lb/gal}$$

$$\Delta T = 35°\text{C} - 27°\text{C} = 8°\text{C}$$

The model was used to characterize the commercialization of the olive oil production for various orchard sizes: 17 acres to reflect the US average olive orchard size, 400 acres to reflect the average acreage for a self-supplied olive oil mill, and 2200 acres to reflect the largest US olive orchard.[28, 29,30] Some key economic findings can be found in Table 3 below.

Table 3: Economics of implementing commercial olive oil production process where selling price is assumed to be $1.81/lb oil.

Quantity	17 acres	400 acres	2200 acres
Amount of Oil Produced (lb)	18,900	444,000	2,440,000
Gross Sales	$34,200	$804,000	$4,420,000
Cultivation Costs	$5,100	$120,000	$660,000
Operating Costs	$2,890	$30,900	$143,000
Total Capital Cost	$23,800	$97,700	$211,000
Total Operating Revenue	$23,800	$643,000	$3,600,000

The model also provides information on environmental implications of commercializing the olive oil production, particularly energy and water usage. Some key resource usage findings are in Table 4.

Table 4: Total electricity and water usage required for commercial olive oil production process.

Quantity	17 acres	400 acres	2200 acres
Total Electricity Usage (kWh)	18,700	201,000	932,000
Total Water Usage (m^3)	78.8	1,460	6,570

With the information from the model, the total amount of OMW that would need to be treated can be determined (Table 5). This is used to cost and size equipment necessary for a membrane separation train.

Table 5: Quantification of total OMW output at various olive orchard sizes.

Quantity	17 acres	400 acres	2200 acres
Washing Water Usage (m^3)	27.2	110	230
Malaxing Water Usage (m^3)	7.75	318	669
Decanting Water Usage (m^3)	43.8	1,030	5,670
Vegetable Water (m^3)	27.4	645	3,550
Total OMW Output (m^3)	106	2,100	10,100

Environmental Impacts and Waste Treatment

We propose a potential separation train for this OMW using information from a study done by Sedej et al.[31] which provided data on a detailed membrane separation process and drying step to refine the phenol rich stream into a powder. Because the phenols extracted from this process will most likely be sold in powder form to be used as food additives, dietary supplements, and/or cosmetic additives, a separation train that includes a drying step is a more complete picture of how a valorization process would look.

The separation process utilizes a 2-step sequential membrane filtration system. The raw waste stream from a two-phase decanter is first sent through a centrifuge to separate the slurry into a solid phase (pomace) and liquid phase (OMW). The OMW is then fed into an ultrafiltration (UF) unit to remove macromolecules and other suspended solids. The permeate is fed into a reverse osmosis (RO) unit to remove the remaining phenols. Because the RO permeate is almost all water, it can be recycled back into the olive oil milling process or used for irrigation. The RO retentate is fed into a spray dryer to refine the phenol-rich stream into a powder. A process flow diagram can be found in Figure 2.

Table 6: Process flow diagram for two-stage membrane separation of polyphenols from OMW, adapted from process described by Sedej et al.[31] *Note:* x_{phenol} *is the mass fraction of phenolic compounds.*

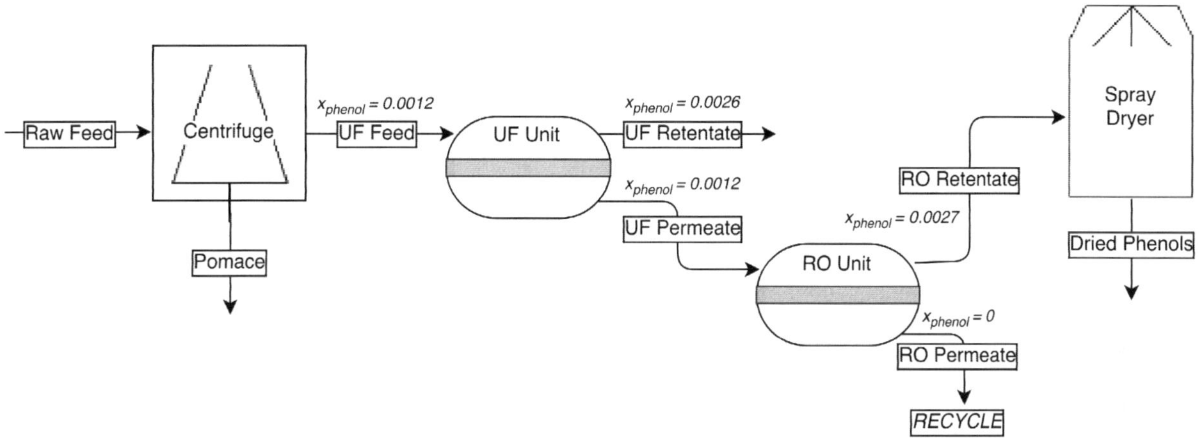

A total annualized cost (TAC) accounting for capital and operation was determined to assess the economic viability of this separation train; capital included a centrifuge, UF membrane unit, RO membrane unit, spray dryer, and two centrifugal pumps. The life time of the capital equipment is expected to be 10 years. The total operating cost includes electricity required to operate the pumps and replacement membranes needed for a year. The TAC can be found in Table 7.

Table 7: Total annualized cost of employing membrane separation train to treat OMW.

Orchard Size	Capital Cost	Operating Cost	TAC
17 acre	$21,400	$1,590	$3,720
400 acre	$47,400	$6,780	$11,500
2200 acre	$93,700	$17,200	$26,500

This annualized cost would need to be offset by selling the spray-dried mixed phenol product generated by the process; this analysis predicts a product sale price to break even of $0.25 per gram for the large (2200 acre) orchard size, which is low enough to be viable. Some key environmental impacts of the membrane separation train are found in Table 8 below.

Table 8: Electricity usage, water recovery, and phenol recovery as a result of implementing separation train.

Quantity	17 acres	400 acres	2200 acres
Electricity Used (kWh)	219	4,340	20,800
Water Recovered (m^3)	57.4	1,140	5,470
Dried Phenols Recovered (kg)	9.42	45.8	105

Future Work

Implementing the membrane separation process seems to be economically feasible for a self-supplied olive oil mill. Future work will focus on a sale price and market for this product as well as separator designs for treating the remaining waste in the UF Retentate. Other factors to consider include choice of treatment method for the UF retentate, potential for heavy metals in the dried product, and cheaper alternative processes for making a similar product. Also, it is important to consider the trade-offs in implementing this separation train; while the process recovers about 54% of the water used to make the olive oil, more electricity is required to operate the separation train. When an evaporation pond is compared to this alternative, it uses less energy but has other sorts of impacts (nuisance/smell, land area, potential for release); for a full picture, a full LCA is required to compare the two alternatives.

Acknowledgements

Ashmera Mohammed contributed to the olive mill economic optimization model.

References

[1] S. Rogers, "20 Unusual Uses for Olive Oil," EcoSalon, 16 Nov. 2011. [Online]. [Accessed 22 Sept. 2016].
[2] World Olive Oil Production Exceeds 3.2M Tons https://www.oliveoiltimes.com/olive-oil-business/italy-leads/51742 (accessed Nov 18, 2016).
[3] History of the Olive https://www.oliveoilsource.com/page/history-olive (accessed Nov 18, 2016).
[4] World Olive Oil Figures - International Olive Council http://www.internationaloliveoil.org/estaticos/view/131-world-olive-oil-figures (accessed Nov 18, 2016).
[5] U.S. Olive Oil Production Facts http://aoopa.org/index.php?n=17&id=17 (accessed Nov 18, 2016).
[6] Moya, M. et al. "Industrial trials on coadjuvants for olive oil extraction." *Journal of Food Engineering* 2010, *97*, 57-63. (accessed Sept 2016)
[7] Clodoveo, M. *Trends in Food Science & Technology* 2012, *25*, 13-23. [Online]. (accessed 11 Nov. 2016)
[8] Fiestas Ros de Ursinos, J.A. and R. Borgja Padilla. "Use and treatment of olive mill wastewater: Current situation and prospects in Spain." Grasas y Aceites 43.2 (1992): 101-106.
[9] Borja, Rafael, Francisco Raposo, and Barbara Rincon. "Treatment Technologies of Liquid and Solid Wastes from Two-Phase Olive Oil Mills." Grasas y Aceites 57.1 (2006)
[10] U.S. Environmental Protection Agency. Phenol: Ambient Water Quality Criteria. Washington, DC. 1978. Print.
[11] Magdich, Salwa et al. "A yearly spraying of olive oil mill wastewater on agricultural soil over six successive years: Impact of different application rates on olive production, phenolic compounds, phytotoxicity, and microbial counts." Science of the Total Environment 430 (2012): 209-216.
[12] Seader, J.D., Ernest J. Henley, and D. Keith Roper. Separation Process Principles: Chemical and Biochemical Operations. 3rd ed. Hoboken, NJ: John Wiley & Sons, 2011. Print.
[13] Marrot, B. et al. "Biogradation of High Phenol Concentration by Activated Sludge in an Immersed Membrane Bioreactor." Biochemical Engineering Journal 30.2 (2006): 174-183.
[14] Obied, Hassan K. et al. "Bioactivity and Analysis of Biophenols Rcovered from Olive Mill Waste." Journal of Agricultural and Food Chemistry 53 (2005): 823-837.
[15] El-Abbassi, Abdelilah, Hajar Kiai, and Abdellatif Hafidi. "Phenolic Profile and Antioxidant Activities of Olive Mill Wastewater." Food Chemistry 132 (2012): 406-412.
[16] Vilaplana-Perez, Cristina et al. "Hydroxytyrosol and Potential Uses in Cardiovascular Diseases, Cancer, and AIDS." Frontiers in Nutrition 1.18 (2014).
[17] M. Wong, L. Eyres and L. Ravetti, "Modern Aqueous Oil Extraction - Centrifugation Systems for Olive and Avocado Oils," in *Green Vegetable Oil Processing*, Urbana, IL: AOCS, 2012, pp. 19-51.

[18] C. Hansen, "Environmental Impact of Olive Oil Processing Waste," EcoMENA, 13 Aug. 2014. [Online]. [Accessed 22 Sept. 2016].
[19] *Compact and continuous olive oil production*; Alfa Laval; pp. 1-4.
[20] Klonsky, K. What Yield, Harvest Cost, and Price are Profitable?. University of California, Davis.
[21] "Sample Costs". *The Olive Oil Source*. N.p., 2016. Web. 12 Dec. 2016.
[22] The Price of Electricity in Your State http://www.npr.org/sections/money/2011/10/27/141766341/the-price-of-electricity-in-your-state (accessed Nov 18, 2016).
[23] McKenzie, R. My California Water Is an Undiluted Bargain http://www.wsj.com/articles/my-california-water-is-an-undiluted-bargain-1430781715 (accessed Nov 18, 2016).
[24] Olive Oil, extra virgin - Monthly Price - Commodity Prices - Price Charts, Data, and News - IndexMundi http://www.indexmundi.com/commodities/?commodity=olive-oil&months=60 (accessed Nov 18, 2016).
[25] Klonsky, K. What Yield, Harvest Cost, and Price are Profitable?. University of California, Davis.
[26] Clodoveo, M. *Trends in Food Science & Technology* 2012, *25*, 13-23. [Online]. [accessed 11 Nov. 2016].
[27] Nonlinear Programming - MATLAB https://www.mathworks.com/discovery/nonlinear- programming.html (accessed Nov 18, 2016).
[28] Taste http://redridgefarms.com/taste (accessed Nov 18, 2016).
[29] EU olive oil farm report. (2012). Belgium: European Commission - Agricultural and Rural Development.
[30] Is American Olive Oil About to Have Its Moment? http://www.bloomberg.com/features/2016-california-olive-oil/ (accessed Nov 18, 2016).
[31] Sedej, Ivana et al. "Membrane-Filtered Olive Mill Wastewater: Quality Assessment of the Dried Phenolic-Rich Fraction." Journal of Food Science 81.4 (2016): 886-896.

498400 Systematic Generation of Hybrid Insulation Materials Via Data Envelopement Analysis

Abstract:

Buildings are responsible of approximately 40% of the EU's energy consumption, and also of the 36% of the CO2 emissions. For that reason, improvements to reduce both, the energy consumption and the energy embodied of the construction materials, are a key strategy to achieve the requirements proposed by the European Commission.

Improvements in new construction can't achieve the goals proposed for 2050, for that reason, interventions in the actual buildings stocks represent an important opportunity. The improvement in this stock can be carried out by multiple ways, but improving the thermal envelope with insulation materials is one of the most extended application, which achieves better performance as thicknesses increases. This represents an important reduction in both, costs and environmental impact associated to the energy consumption. Despite that, such intervention requires initial investments that involve an environmental impact from manufacture, installation, dismantling and disposal of the materials which shouldn't be neglected when suggesting improvements in buildings efficiency due to this embodied energy which can represent an important impact depending on the materials used.

The development of innovative insulation materials is gaining interest in the scientific community in recent years, creating materials with better thermal properties or with low embodied environmental impact, such as PCM or bio-based materials respectively. Despite those improvements, only a marginal fraction of the insulation market is relaying on this materials, for that reason, prefabricated insulation solutions may represent a solution.

A new methodology focused on a Data Envelopment Analysis (DEA) combined with thermal simulations of EnergyPlus has been performed to different insulation materials with the objective to achieve sandwich panels to increase the use of bio-based insulation materials.

The total cost of the intervention and the environmental impact associated to the energy consumption for heating and cooling a building and the life cycle of building materials are evaluated for different insulators and thicknesses, generating a high quantity of inputs to analyse with DEA. To reduce the amount of inputs, an objective reduction has been carried out, identifying those inputs which define the global performance and tanking out those which doesn't gave extra details of the performance.

Once identified the key inputs, DEA finds the efficient thicknesses of some materials, which represent the materials that without combination achieve the better performance, and are also the materials which will be combined to generate the sandwich panels with the projections of the non-efficient solutions to the efficient frontier.

This study has been carried out in a cubicle like building, for 8 different insulation materials and 6 different thicknesses, which varies from 0.01 and 0.26 with increments of 0.05cm. Despite the initial amount of 11 objectives, after the objective reduction only 5 have been analysed, 4 environmental inputs and the total cost of the intervention, achieving 21 efficient solutions and 27 non-efficient ones. The non-efficient solutions have been projected to the efficient frontier, achieving the same amount of new combination of materials, which should be analysed to identify the feasible solutions. Finally those solutions have been implemented in the same model, including the energy simulation and the DEA model, finding 6 new efficient solutions for the 6 feasible projections.

498618 Eco-Efficiency Assessment of EU Manufacturing Sectors Linking Multi-Regional Environmentally-Extended Input-Output Tables and Data Envelopment Analysis

Abstract:

In todayâ€™s globalized market where international trade plays a major role, assessing the environmental footprint of anthropogenic activities and allocating the corresponding environmental responsibilities among the parties involved have become very challenging tasks.

Multi-regional environmental extended input-output models (MREEIO) are gaining wider interest in this context, as they cover a wide variety of economic transactions and environmental loads that provide detailed information about the impact generated worldwide. Previous works have studied the impact of a single region following production and consumption-based assessment approaches, yet they failed to analyze the trade-off between economic wealth created and environmental impact caused (i.e. eco-efficiency).

In this work, we propose an approach that combines MREEIO tables with data envelopment analysis (DEA) for the assessment of the eco-efficiency of the main EU manufacturing sectors considering three environmental impacts - global warming potential (GWP), potential acidifying equivalent (PAE) and tropospheric ozone forming potential (TOFP)-, together with the total economic output. Besides distinguishing between efficient and inefficient sectors, our approach provides improvement targets for those found inefficient that if attained would make them efficient.

We firstly analyze the mismatch between the production and the consumption-based approaches, finding that consumption-based emissions are often higher than production-based ones. DEA is next applied, finding that five out of the 14 EU manufacturing sectors are efficient from a production-based perspective, while four are efficient from a consumption-based one (with only two repeating as efficient in both cases).

We argue that the combination of DEA with MREEIO is particularly appealing for policy making, since the former allows spotting inefficient sectors and determining improvement targets for specific impacts, while the latter informs on the ultimate drivers behind such impacts across international supply chains. Furthermore, the information provided by each accounting approach is complementary. Production-based accounting pinpoints sectors and contaminants that require regulations and/or investment in better technologies. Consumption-based accounting, on the other hand, helps identify ultimate sources of impact, a piece of information that can be used to select alternative cleaner suppliers.

Life Cycle Optimization with Ecosystems As Unit Operations

Tapajyoti Ghosh and Bhavik Bakshi

William G. Lowrie Department of Chemical and Biomolecular Engineering
The Ohio State University, Columbus, Ohio, USA

17AIChE
Annual Meeting, Minneapolis, MN

11/01/17

Motivation For Sustainable Process Design

Shifting of Problems

- Prevent shifting of problems across system boundaries, temporal and spatial scales.
- Design the process/system.
- Account for emissions over the life cycle while optimizing for specific objectives.

Process to Planet framework

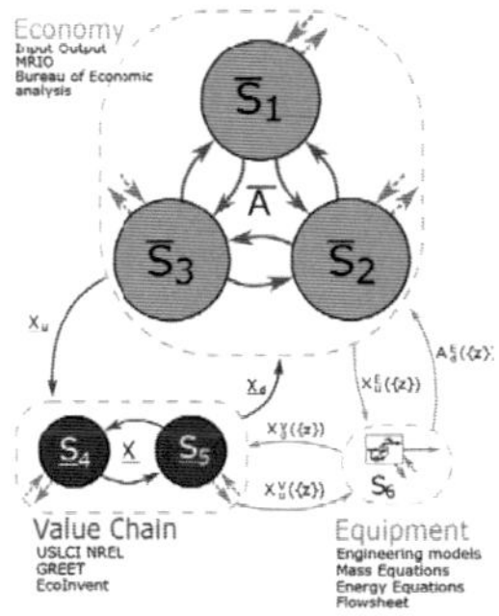

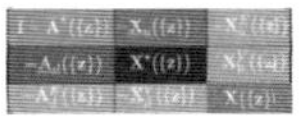

- **Hybrid LCA + Process Design**
- Multiscale framework that combines information at three different scales to perform Life cycle analysis and Sustainable Process Design.
- Expands boundary of Process design by including life cycle through the economy scale along with conventional Process LCA scale.

Hanes and Bakshi. AIChE Journal 2015.

What is Environmental Sustainability?

Ignoring Ecological Capacity!!

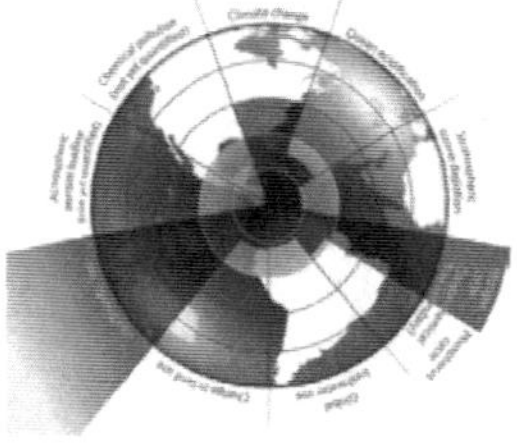

Ecosystems Inclusion in Design

- Why: **A method for defining Sustainability.** Possible method for solving the **"wicked"** problem.
- Reasons: Ecosystems are truly sustainable systems that we know of.
- Needs: Include role of nature in engineering decision making.

Rockstrom et. al, 2009, *Nature*

Techno-Ecological Synergy

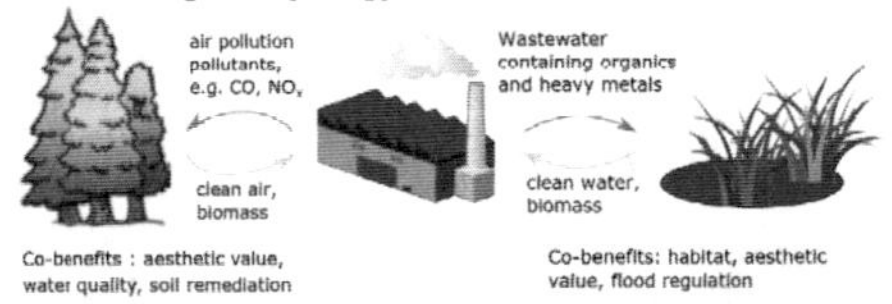

Ecosystems in Design

- The Techno-Ecological Synergy(TES) framework explicitly accounts for the supply of ecosystem services and interconnects technological systems with ecosystems through a symbiotic and synergistic relationship.
- Closing the loop of waste materials, pollutants and resources.

Gopalakrishnan, Bakshi and Ziv, *AIChE Journal*, 2016

Process Design with Ecosystems in Value Chain

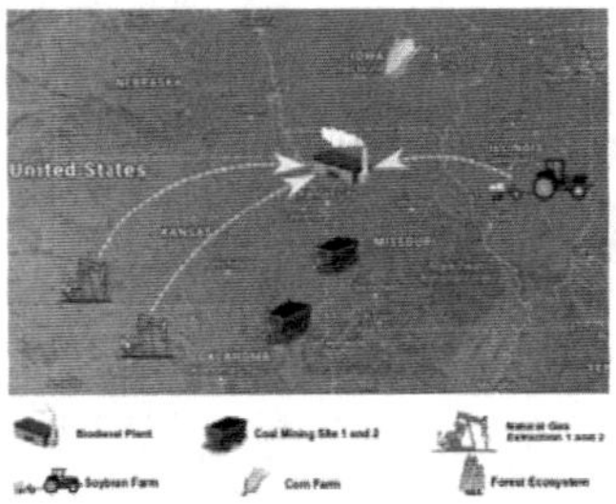

- Coal is available from nearby coal mining sites.
- However, using coal will not comply with their environmental practices due to greater emission quantity.
- Another option is the use of oil from corn over soybean.

Process Design with Ecosystems in Value Chain

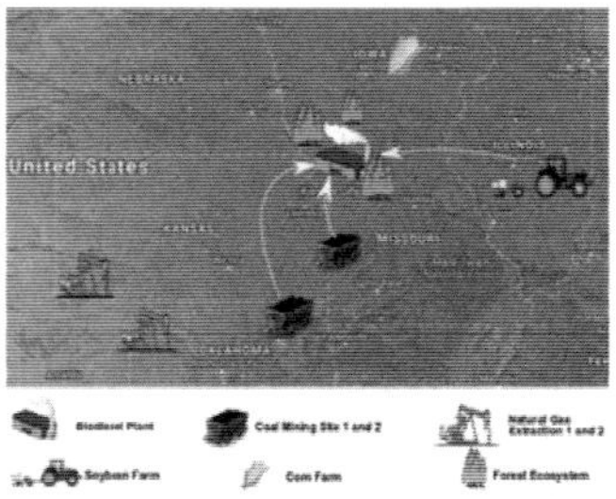

- The biodiesel plant can manage ecosystems at its local site which will result in pollutant capture, leading to emissions offset.
- Additionally the process needs to modify itself for minimum environmental impact.

Challenges Background

Process Design with Ecosystems in Value Chain

- **Emissions trading** if land is unavailable.
- If the coal mining sites have land available, incentives may be provided so that these suppliers maintain ecosystems at their sites.
- Over the entire value chain, ecosystems will result in emission offsets for biodiesel industry.

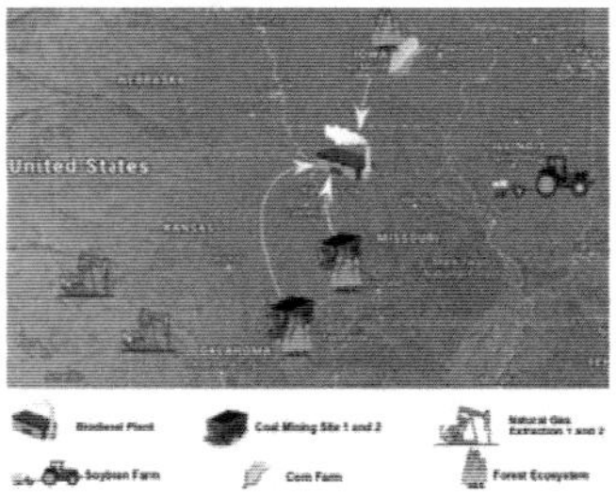

Some other challenges

Product Footprint Improvement

- Major chemical industries are aiming for eliminating toxic chemical use, reducing human risks and improving social conditions not only locally but over the entire supply chain.
- Food packaging industries are shifting towards sourcing produce from farms practicing sustainable agriculture.
- Incustries need **tools and metrics** that are standardized and can calculate product footprints and design sustainable supply chains.
- Can ecosystems provide a better solution to the problem of **determining sustainability**?
- Can considering ecosystems provide a better measure of the supply chain being truly sustainable if it as a whole does not exceed nature's ecological capacity?

Objective

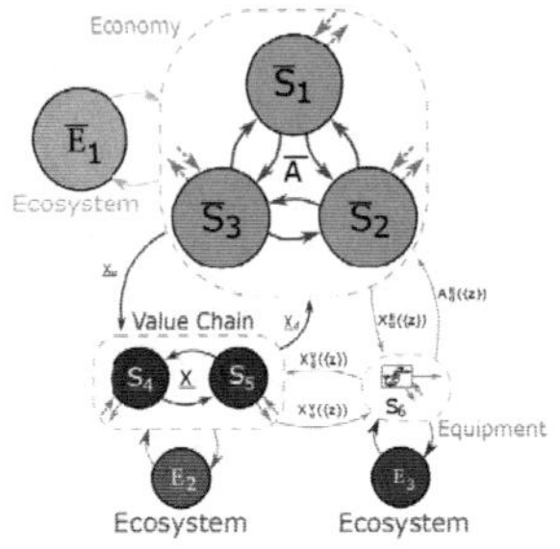

P2P-TES Framework

- We propose a framework that connects **technological systems** and **ecological systems** at multiple spatial scales.
- Triple Design Problem: **Engineering Process**, **Upstream Value Chain** and **Size/Type of ecosystems** at both local and regional supplier sites.

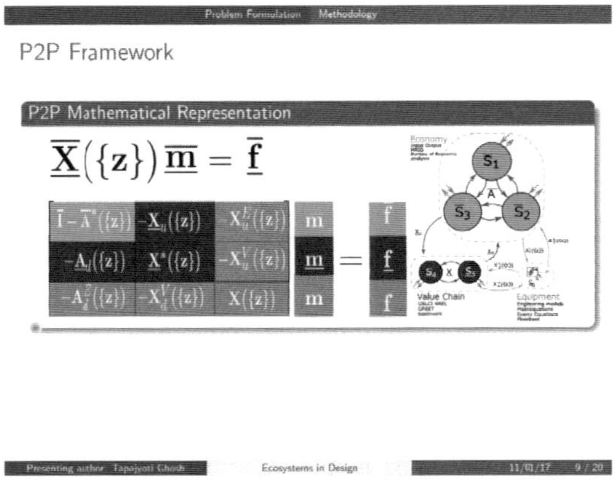
Problem Formulation Methodology
P2P Framework
P2P Mathematical Representation
$\underline{\overline{X}}(\{z\})\,\overline{\underline{m}} = \overline{\underline{f}}$
Economy
Value Chain
Equipment
Presenting author: Tapajyoti Ghosh
Ecosystems in Design
11/01/17 9 / 20

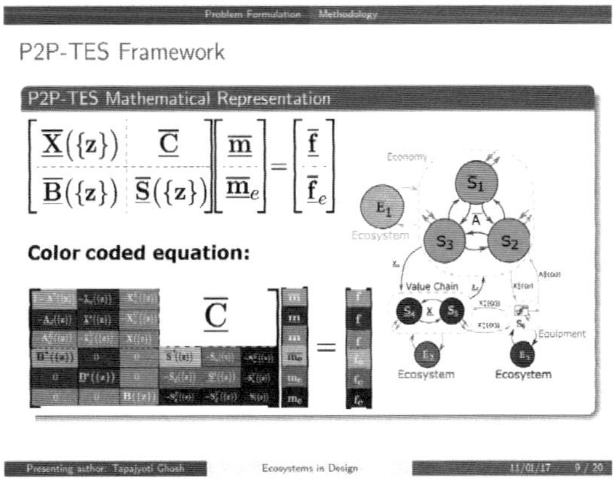
Problem Formulation Methodology
P2P-TES Framework
P2P-TES Mathematical Representation
$\begin{bmatrix} \underline{\overline{X}}(\{z\}) & \overline{\underline{C}} \\ \underline{\overline{B}}(\{z\}) & \underline{\overline{S}}(\{z\}) \end{bmatrix} \begin{bmatrix} \overline{\underline{m}} \\ \overline{\underline{m}}_e \end{bmatrix} = \begin{bmatrix} \overline{\underline{f}} \\ \overline{\underline{f}}_e \end{bmatrix}$
Color coded equation:
Economy
Ecosystem
Value Chain
Equipment
Ecosystem
Ecosystem
Presenting author: Tapajyoti Ghosh
Ecosystems in Design
11/01/17 9 / 20

Optimization Formulation

Objectives and Constraints

- *Minimize* $\begin{cases} \underline{\overline{\mathbf{Z}}} = \sum_i^N \underline{\overline{\mathbf{f}}}_e^i \\ \mathbf{Z}^{\$} = -\mathbf{NPV}(\{\mathbf{z}\}, \underline{\overline{\mathbf{m}}}_e) \end{cases}$
- subject to constraints,
- $\left[\begin{array}{c:c} \underline{\overline{\mathbf{X}}}(\{\mathbf{z}\}) & \underline{\overline{\mathbf{C}}} \\ \hdashline \underline{\overline{\mathbf{B}}}(\{\mathbf{z}\}) & \underline{\overline{\mathbf{S}}}(\{\mathbf{z}\}) \end{array}\right] \left[\begin{array}{c} \underline{\overline{\mathbf{m}}} \\ \hdashline \underline{\overline{\mathbf{m}}}_e \end{array}\right] = \left[\begin{array}{c} \underline{\overline{\mathbf{f}}} \\ \hdashline \underline{\overline{\mathbf{f}}}_e \end{array}\right]$
- $\mathbf{H}(\{\mathbf{z}\}) \geq 0$ Energy, mass balances, process parameters in process engineering model
- $\underline{\overline{\mathbf{m}}}_e \geq 0$ Constraint for ecosystem sizes to be non-negative
- $\underline{\overline{\mathbf{m}}} \geq 0$ Constraint for system unit sizes to be non-negative
- $\mathbf{S}(\{\mathbf{z}\}) \geq 0$ Ecosystem modeled as unit operations

Visualization of Biodiesel Network

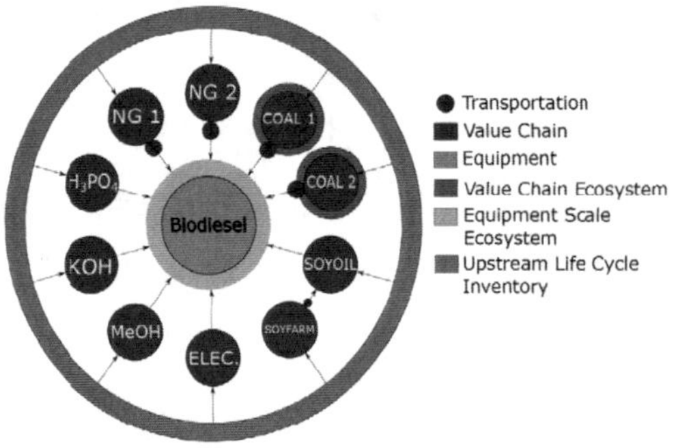

Visualization of Biodiesel Network

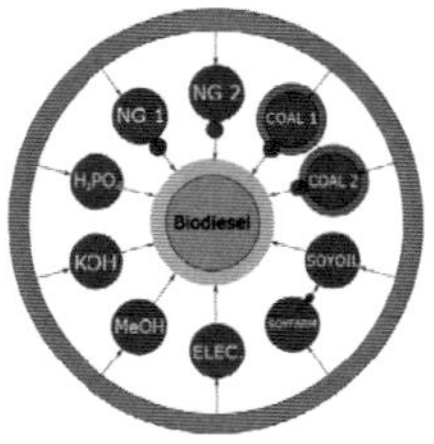

Case Study

- **Process Design** of a **biodiesel** manufacturing process.
- Design decision for choosing between two technology choices for meeting energy requirements - **Coal** and Natural Gas.
- Choice between **two** coal mining sites and **two** natural gas sites.

Visualization of Biodiesel Network

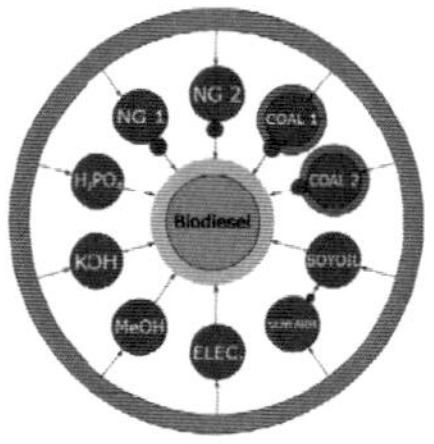

Case Study

- Design of ecosystem sizes at the supply chain and at the local manufacturing site.
- Choice between ecosystem types - Oak and Elm at the mining sites.
- Minimization of SO_2 emissions.
- Maximization of **Net Present Value** of Biodiesel process.

Scenario 1: Energy Source - Natural Gas

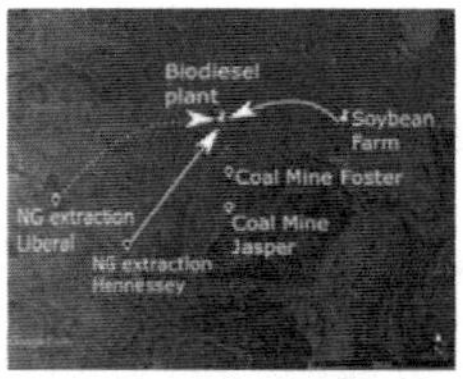

Objective

- Decision is made on the basis of **maximizing** the Net Present Value while **minimizing** SO_2 emissions.

Objective Function Values

- NG Supplier at Hennessey chosen.
- NPV: **$2.98 Million**
- SO_2: **106 Kg/Year**

	SO2 (kg/year)	NPV (Million $)
Scenario 1	106	2.98
Scenario 2		
Scenario 3		
Scenario 4		

Scenario 2: Energy Source - Coal

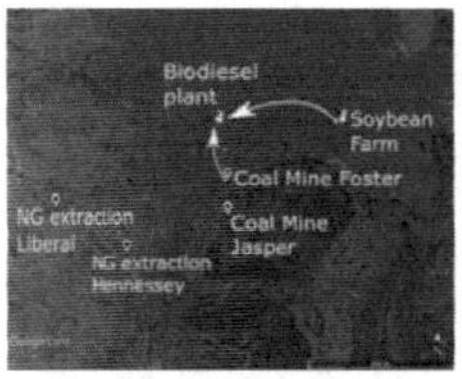

Observation

- Decision is made on the basis of **maximizing** the Net Present Value while **minimizing** SO_2 emissions.
- Consumption of coal results in increase of emissions.

Objective Function Values

- Coal Supplier at Foster chosen.
- NPV: **$3.53 Million**
- SO_2: **5482.53 Kg/Year**

	SO2 (kg/year)	NPV (Million $)
Scenario 1	106	2.98
Scenario 2	5482.53	3.53
Scenario 3		
Scenario 4		

Scenario 3: Ecosystem at Biodiesel Site

Scenario 3

- In this scenario, we consider that the biodiesel industry has land available for maintaining a forest ecosystem.
- The forest ecosystem results in SO_2 deposition.

Objective Function Values

- Coal Supplier at Foster chosen
- NPV: **\$3.519 Million**
- SO_2: **0 Kg/Year**
- Area of Oak ecosystem: 241 acres

	SO2 (kg/year)	NPV (Million \$)
Scenario 1	106	2.98
Scenario 2	5482.53	3.53
Scenario 3	0	3.519
Scenario 4		

Scenario 4: Ecosystem at Supply Chain Sites

Scenario 4

- In this scenario, Biodiesel industry does not have land available.
- Incentives for managing ecosystems at the supply chain.
- Coal mining sites have land available.

Scenario 4

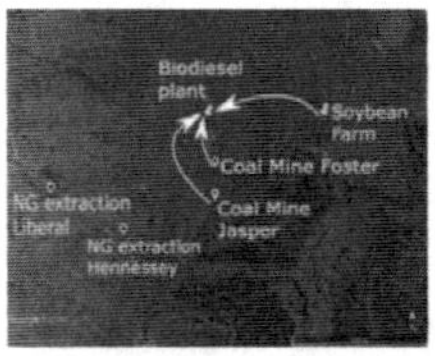

Objective Function Value

- Oak forestry Ecosystem
- NPV: **$3.524 Million**
- SO_2: **0 Kg/Year**

Results

- Land area available for the ecosystems at each mining site was not enough to offset entire life cycle emissions.
- Coal is sourced from both mining sites. Allocation of ecosystem service credits have been achieved through monetary burden.
- Disaggregation of supply chain results in increased robustness.

Observations

	SO2 (kg/year)	NPV (Million $)
Scenario 1	106	2.98
Scenario 2	5482.53	3.53
Scenario 3	0	3.519
Scenario 4	0	3.524

- This framework can be used for performing LCA of supply chain networks, design of supply chains, process design with ecosystems at multiple scales simultaneously.
- This case study only considered **SO_2 emissions.**
- There are real life scenarios where this actually can be applied. Example - **Industrial Symbiosis**.
- Demonstrates how **emissions trading** can be applied through a design framework. Introduces concept of **Payment for Ecosystem services.**

Temporal Variation of SO_2 Deposition

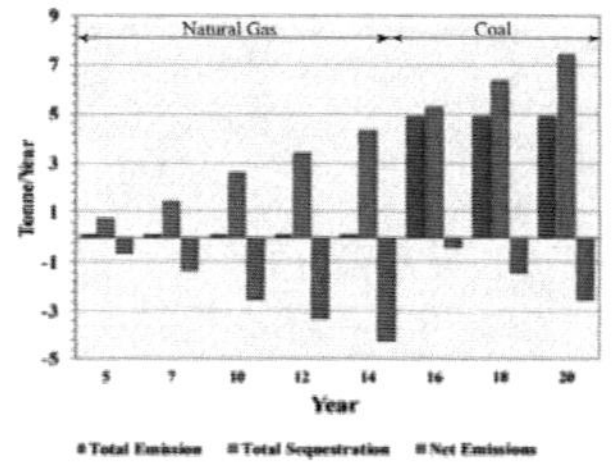

Ecosystem Dynamics

- SO_2 dry deposition by trees is dynamic and increases over time as trees mature.
- Coal use becomes viable only after year 14 when plantation is viable enough to remove added emissions due to coal use.

Multiobjective Pareto Front

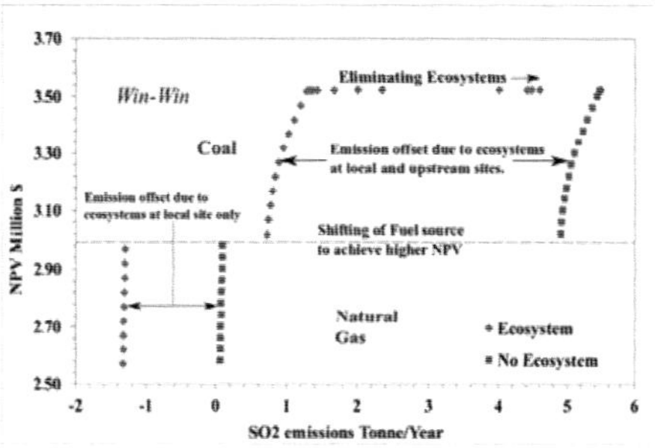

Ecosystems at Biodiesel and Coal extraction sites

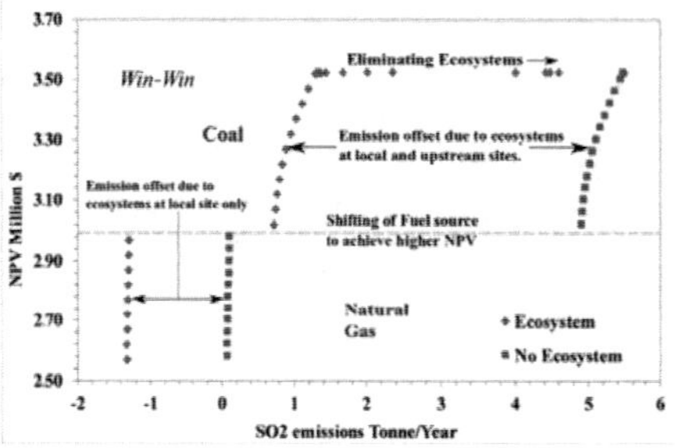

Conclusion

- **P2P-TES** framework can prove to be a valuable tool for industries to gauge environmental sustainability criteria, not just at the local scale but over the entire supply chain.
- This framework allows designing of upstream supply chains to achieve desired objectives.
- Including Ecosystems in the sustainable process design problem expands the solution space and results in "win-win" solutions, affects supply chain design and might prove to be a measurement of sustainability.

Future Work

- Incorporation of the economy scale and ecosystems at that scale.
- Incorporate yearly variation of ecosystems, service shed definition, transportation emission and ecosystems allocation.

Acknowledgement

Project Fund - Sustainable and Resilient Economy Program at Ohio State

Process Systems Engineering Group

Thank You!

Designing **Techno-Ecological Synergies** while accounting for Ecosystem Dynamics. *(681c), Thursday 8:50 AM.*
Design of CO_2 conversion to Dimethyl Carbonate by the **Process-to-Planet** Multiscale Modelling Framework. *(658), Thursday 8:00 AM.*

ghosh.117@osu.edu, bakshi.2@osu.edu

499162 Techno-Economic Assessment of Dry Reforming of Methane Process

Abstract:

The reaction of methane and carbon-dioxide to produce syngas (mixture of CO and H_2) is called Dry Reforming of Methane (DRM). DRM has received considerable interest in academia and industry due to CO_2 being one of the reactant feed. Most of the work done in DRM research involves novel catalysts to reduce the coking problem and thermodynamic studies to estimate product composition at different operating conditions. Very few studies have addressed the issue of understanding the CO_2 balance of the whole process and one recent study has used an optimization-based approach to study CO_2 fixation rates for different syngas ratios. If the motivation for DRM is to reduce the carbon footprint of producing syngas, then it is crucial to track CO_2 flows in and out of the system. As DRM has not been tried on an industrial scale yet, quantifying the environmental benefits and plant economics of the process is necessary in the path of commercialization.

In this work, our focus is on understanding the maximum potential of the DRM reaction to reduce CO_2 emissions of a process. Unlike previous studies which mainly considered reformer duty as the energy penalty, in our work, all major sources of emissions in each process pathway have been accounted. LINGO software has been used for equilibrium calculations and optimization. For sake of brevity, we have presented the analysis in two parts. In the first study, DRM is studied with other available reforming technologies in parallel. In the second study, we have investigated the potential of producing syngas of high H_2/CO ratio from a stand-alone DRM unit operating at low temperatures (500-700 °C).

From the reaction stoichiometry of DRM, it can be inferred that normally at industrial reforming conditions (800-1200 °C), DRM alone will produce a syngas of H_2/CO ratio of 1 and below. Hence, to produce syngas of higher syngas ratios, we need another existing syngas production technology (Partial Oxidation – POx or Steam Methane Reforming – SMR) in parallel to make up the desired syngas ratio. These 3 processes (DRM, POx and SMR) are in parallel and the optimization model selects the optimum choices to minimize carbon emissions of the overall unit while producing syngas of the desired ratio. The results indicate that DRM has the potential of having a net CO_2 fixation (net reduction in CO_2 across the process boundary) only for low syngas ratio of below 1. At syngas ratio of 2 and above, the optimization model does not select DRM. This work signifies the importance of understanding the CO_2 balance of the DRM process before proposing it as an emissions reduction strategy. The operating costs of producing syngas in the optimized syngas production networks are also presented.

In the second part of the study, we have only looked at a stand-alone DRM unit to study different scenarios and its potential in producing syngas of high H_2/CO ratio. To avoid the carbon formation regime of 600-750 °C, industrial reformers are operated at very high operating temperatures. However, most of the catalysis studies of DRM seem to focus on the 500-700 °C range, and the reason is to reduce the reformer duty which will help enhance the CO_2 fixation rates. These studies entirely ignore that coke is removed as CO_2 and this negatively impacts the CO_2 balance. Hence, it is essential to quantify the carbon emissions before claiming that DRM has lesser carbon footprint. We have analyzed the performance of different DRM catalysts and compared the CO_2 balance for different ways of catalyst regeneration. These results have been compared with the industry benchmarks of SMR and POx. The results show that coke management plays a major role in deciding the CO_2 balance of DRM process. If the coke formed can be removed physically (which is still an experimental challenge), then DRM alone can produce syngas of high H_2/CO ratio as it produces the additional H_2 by methane decomposition. And such a unit will have significantly lower carbon footprint than existing technologies (SMR, POx). However, if coke is removed by burning in air, it will result in CO_2, which will negatively affect the CO_2 balance. The results presented in this section will provide directions to researchers in the DRM area to realign focus to those operating conditions where we have better CO_2 fixation rates. The operating costs of such a novel unit have been estimated and compared with industry standards of SMR and POx.

This study will help understand the impact of DRM unit on carbon emissions and the operating costs of a syngas generation unit. The results can then be compared with other CO_2 reduction strategies in the plant to prioritize projects aiming towards a low-carbon emissions future.

498504 Prediction of Cesium Loading on Crystalline Silicotitanate Ion Exchange Resin

Abstract:

At the Savannah River Site (SRS) waste saltcake stored in the SRS Tank Farms is dissolved in water and processed through the Tank Closure Cesium Removal (TCCR) system to remove radioactive cesium and strontium from the high-activity waste solution. The TCCR system is a crystalline silicotitanate (CST) ion exchange (IX) system comprised of a carousel of two small lead-lag IX columns. Concerns have been raised that cesium could accumulate on the CST resin at concentrations that exceed the safety threshold of 170 Ci/L, potentially necessitating a costly redesign of the TCCR system.

A computer model was developed by Professor Rayford G. Anthony of Texas A&M University (TAMU) to simulate IX equilibria between electrolytic solutions and CST solid resin. The TAMU model was used to estimate cesium loading on CST as a function of initial cesium concentration in the waste solution.

This paper presents the computational results for a wide range of waste solution and discusses the effects of key ions (i.e., Na^+, OH^-, K^+) on cesium loading.

Computational Optimization of Difficult Physics-Based Project Designs Using Massively Parallel Solution Methods

2017 AAEES Excellence in Environmental Engineering and Science ™ Grand Prize Competition Winner - Research Category

Presenter:
Larry M. Deschaine, Ph.D., P.E.
AICHE Annual Meeting
November 1, 2017

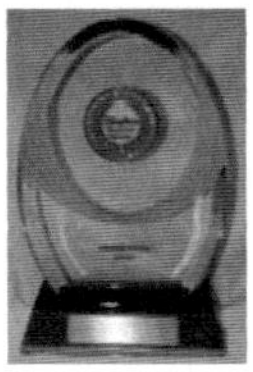

Why PBMO™ Subject Matter Expert Decision Support

- ▼ National Research Council (2013) findings
  - Large number of complex groundwater sites requiring cost effective long term management
- ▼ Distributed computing platform
  - Rapid & automated analysis of large number of simulations
- ▼ Link optimization algorithms with physics based model
  - Optimal pump and treat (today)
  - Source delineation
  - Plume delineation/Long term management
  - Treatment trains

How can we bridge from where we are, to where we need to go?

2

Current Problems/Challenges

▼Time and Uncertainty:

- Pump-and-treat (P&T) remedies have been operating continuously for many years with little progress toward remedial goals
- Most P&T remedies lack exit strategies

▼Cost:

- Hundreds of millions of dollars are being spent annually on P&T
- Emerging contaminants such as the group of chemicals commonly referred to as "PFAS" (i.e., per- and polyfluoroalkyl substances), which are associated with foams used in firefighting, will lead to more extensive use of P&T remedies with increased costs.

▼Deficiency:

- Traditional approaches for remedial optimization are inefficient:
  - Expert opinion with little quantitative foundation
  - Trial-and-error "Brute Force" model simulations, limited number or over-simplified model simulations to achieve computational efficiency

HGL

3

Design Optimization Goals

Site Investigation

- ▼ Source location and identification (SLI)
- ▼ Plume delineation and tracking (PDT)
- ▼ Optimal long–term monitoring (LTMO)
- ▼ Minimize life-cycle cost

Remedial Optimization

- ▼ Control plume migration
- ▼ Optimize groundwater drawdown
- ▼ Minimize or cap cost
- ▼ Minimize or cap time
- ▼ Maximize or surpass contaminant removal or destruction goal

Honor property access restrictions and engineering constraints

HGL

4

Physics-Based Management Optimization ™ (PBMO ™)

▼ Cloud-based PBMO™ Tool:

- Uses *computational optimization algorithms* for improving cost, time and certainty of remedial projects
- Solves highly complex optimization problems *rapidly* (minutes to hours)
- Provides optimal solutions for *remedy design, implementation, maintenance and exit strategies*

▼ PBMO™:

- Accommodates public-domain MODFLOW/MT3DMS, and HGL's MODFLOW-SURFACT™ cutting-edge software for highly robust and efficient parallel simulations on the Cloud.

5

PBMO™ Medallion Conceptualization

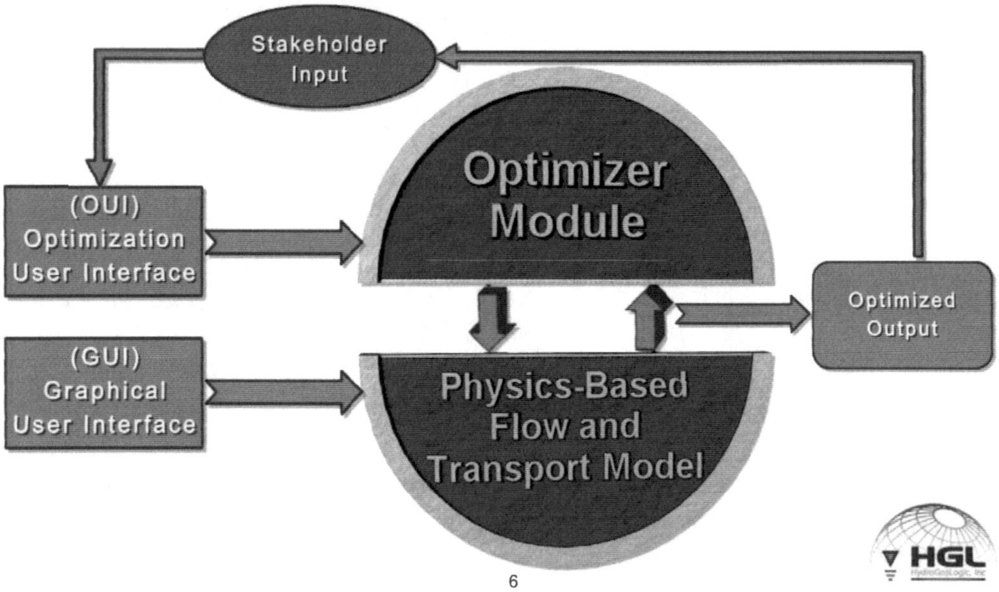

6

PBMO^{TM} Optimization Tools

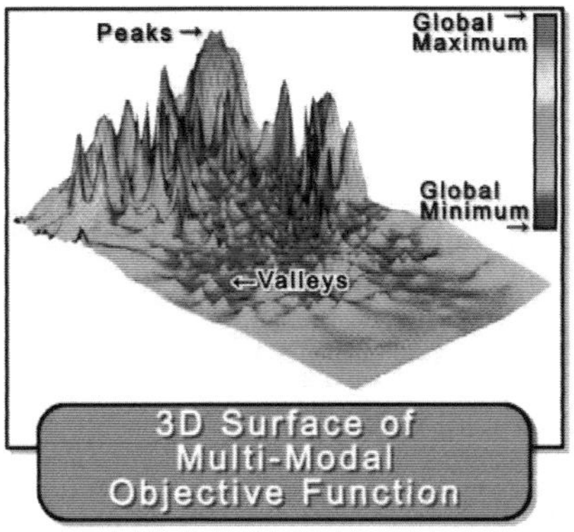

▼Key elements of computational optimization include:
- Objective Functions
- Decision Variables
- Constraints

Optimizer Module

PBMO^{TM} Optimization Algorithms

Single Objective Deterministic Guided Search Algorithms:

Sequential Linear Approximation (SLA)
Sequential Quadratic Programming (SQP)
Generalized Reduced Gradient (GRG)
Branch and Bound (B&B)
Multi-Start with local search (MS+LS)
Lipschitz Global Optimization (LGO ©)
Constraint Programming (Logic-based optimizer)
-Semi-definite programming (SDP)
-Interior-point method (IPM)
-Multi-level coordinate search (MLCS)

Multiple Objective Algorithm:

Analytical Hierarchy Process (AHP)

Parallel Search Methods

Global Adaptive Random Search (GARS)
Adaptive Global Random Search (AGRS)
Adaptive Design of Experiment (ADOE)
Monte-Carlo / Latin Hypercube
Multi-start Adaptive Partition (MSAP)

Machine Learning/Data Mining Algorithms:

WEKA (Machine Learning)
RapidMiner (Machine Learning)
Fuzzy Logic
Reinforcement Learning
Information Theoretics

Built with modern Fortran and Python

Optimizer Module

Physics-Based Models in PBMO™

- ▼ MODFLOW-based groundwater flow and contaminant transport simulation codes
  - MODFLOW/MT3DMS
  - HGL's MODFLOW-SURFACT™
  - Future: MODFLOW-USG (unstructured gridding)
- ▼ Can be customized to accept codes for surface water and integrated surface water/groundwater modeling
  - HEC-RAS
  - HEC-HMS
  - MODHMS®

9

PBMO™ Execution on the Cloud

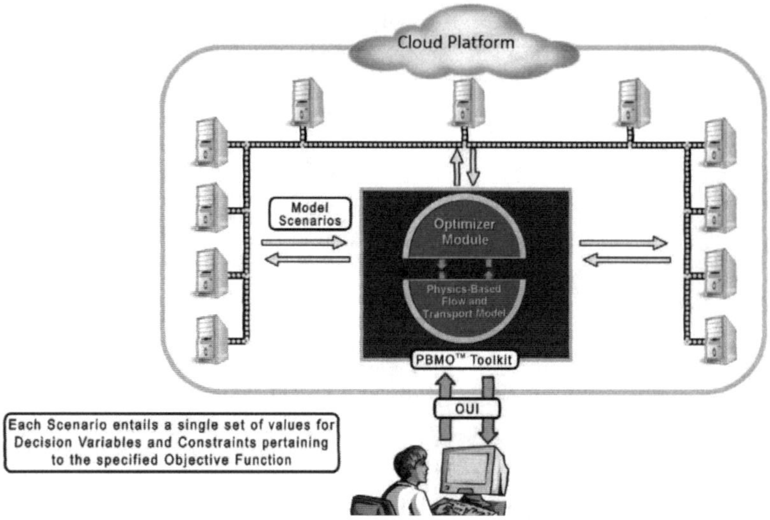

10

PBMO^{TM} Optimization Computational Engine (OCE)

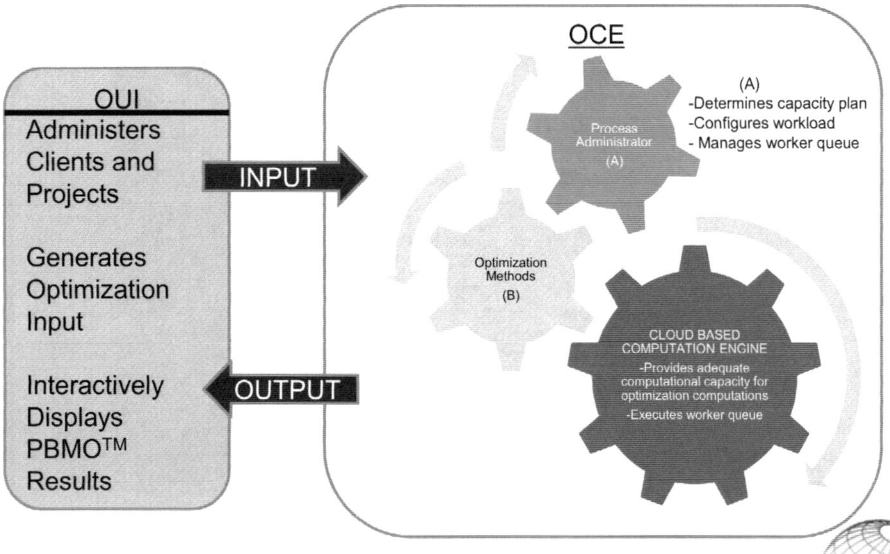

11

PBMO^{TM} Optimization Computational Engine (OCE) – In Action

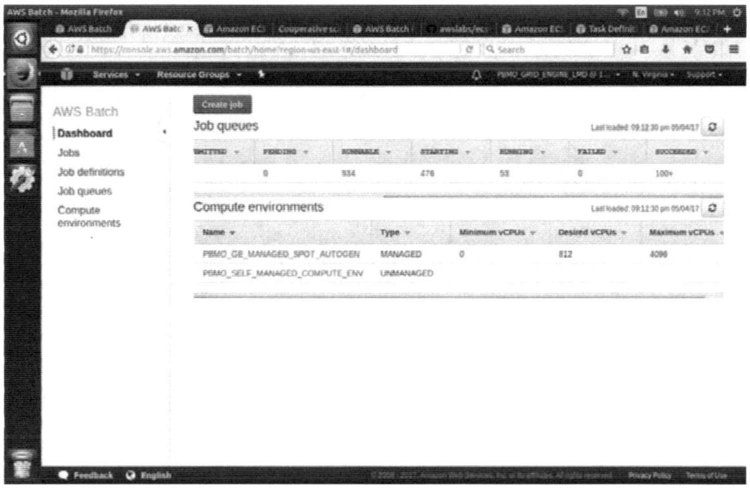

12

Example: Garvey Elevator Superfund Site, Hastings, NE
Site Location and Operational History

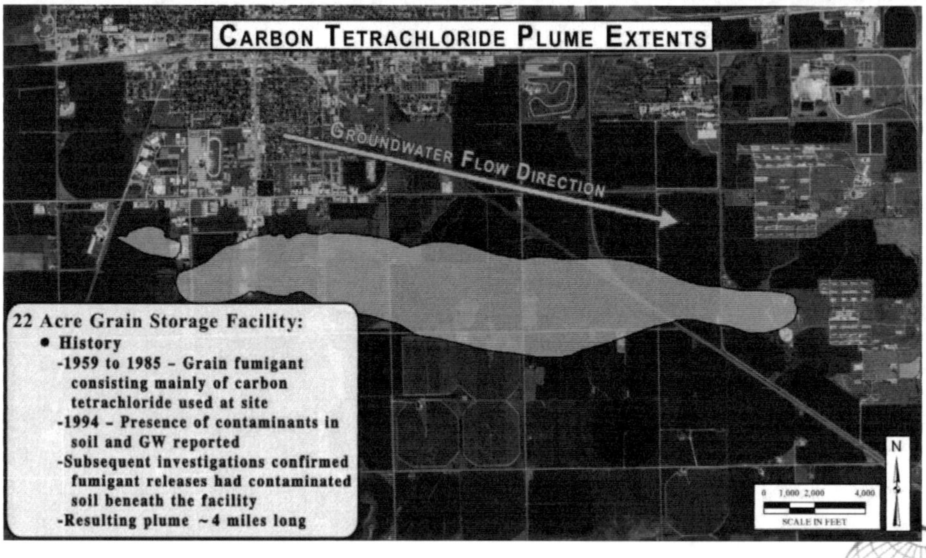

HGL

13

Example: Garvey Elevator Superfund Site
PBMO^{TM} Design Criteria

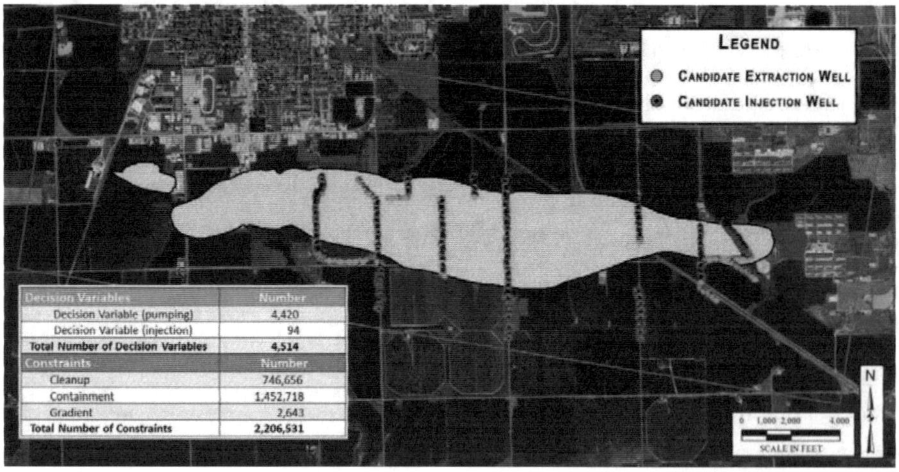

Decision Variables	Number
Decision Variable (pumping)	4,420
Decision Variable (injection)	94
Total Number of Decision Variables	**4,514**
Constraints	Number
Cleanup	746,656
Containment	1,452,718
Gradient	2,643
Total Number of Constraints	**2,206,531**

HGL

14

Example: Garvey Elevator Superfund Site PBMO™ Optimal Solution

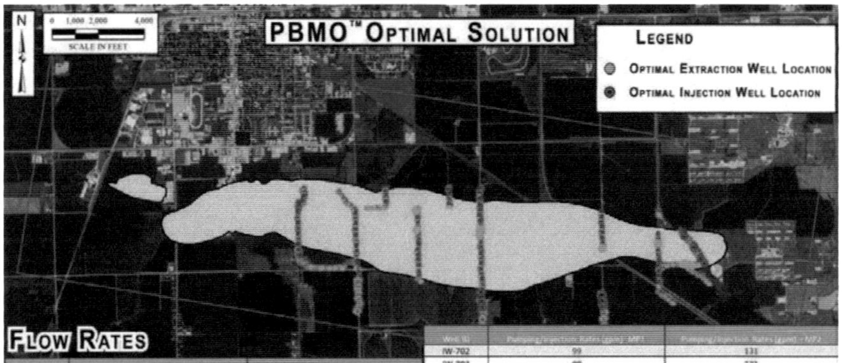

Well ID	Pumping/Injection Rates (gpm) - MP1	Pumping/Injection Rates (gpm) - MP2
EW-103	69	69
EW-102	188	188
EW-201	394	394
EW-202	397	397
EW-303	235	300
EW-304	88	150
EW-302	326	326
EW-301	73	73
Total Extraction	1,770 gpm	1,897 gpm

Well ID	Pumping/Injection Rates (gpm) - MP1	Pumping/Injection Rates (gpm) - MP2
IW-702	99	131
IW-703	99	131
IW-704	99	131
IW-701	99	131
IW-501	229	229
IW-502	229	229
IW-503	229	229
IW-601	229	229
IW-602	229	229
IW-603	229	229
Total Injection	1,770 gpm	1,897 gpm

15

Example: Garvey Elevator Superfund Site Remedy Performance Snapshots

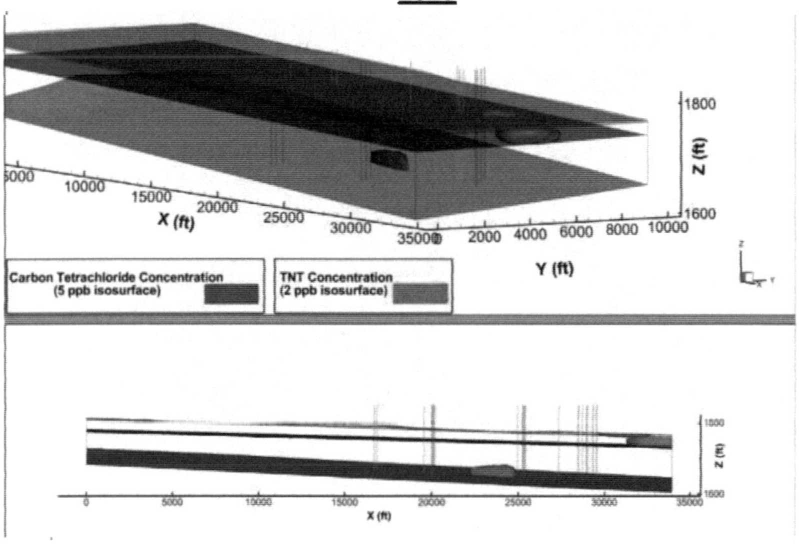

16

PBMO™ Future Enhancements

Problems and Challenges

▼ Optimal management of complex groundwater restoration sites

- Limited characterization data
- Multiple plausible conceptualizations
- Incorporating treatment trains
- Changing regulatory framework
- Stakeholder acceptance

PBMO™ Future Enhancements

▼ Source location identification (SLI)

▼ Plume delineation and tracking (PDT)

▼ Optimal long–term monitoring (LTMO)

17

PBMO™ Future Enhancements: Source Location Identification (SLI)

▼ Optimal identification of source location is crucial for determining the nature and extent of groundwater contamination

▼ Sources consist of DNAPL, LNAPL and emerging contaminants (PFAS/PFOA)

▼ Sources of contaminants can be emanating from an individual source or from multiple sources

18

Source Delineation Concept

▼ Begin with Multiple Potential Source Locations with Various Likelihoods of being Actual

Subject matter experts assess potential source locations, and likelihoods

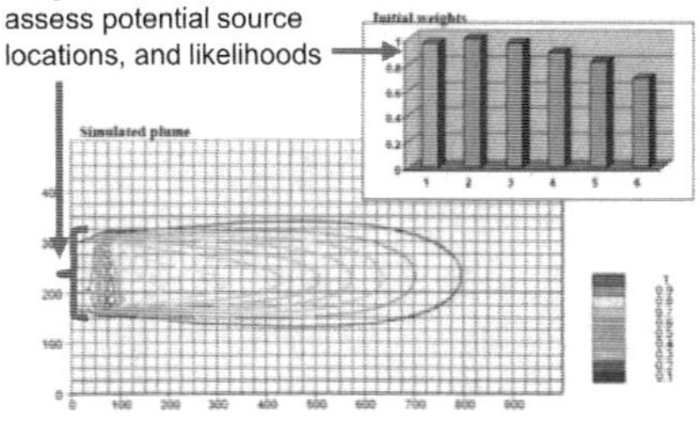

▼ End with the Source Location(s) Identified

Algorithm automatically develops sampling strategy to identify the true source area

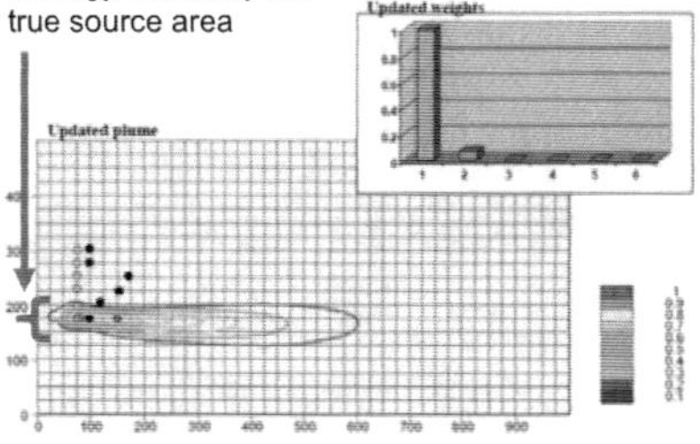

Iterative Sampling locations (●)

Ref: Pinder, C. F., Ross, J., Dokou, Z. Optimal Search Strategy of the Definition of DNAPL Source Location, SERDP, 2009

19

Source Delineation Site Demonstration

▼ Demonstration Report

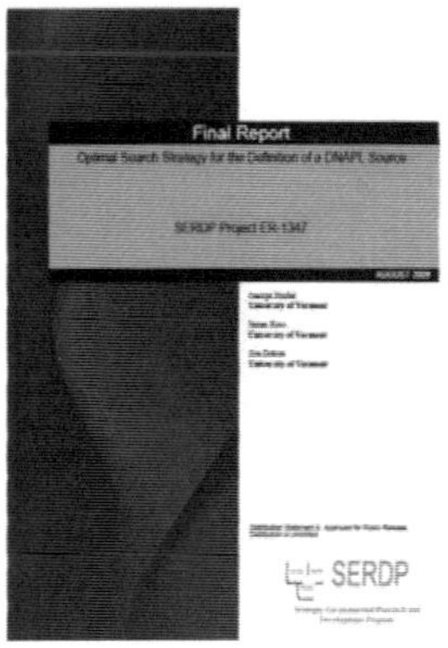

▼ Technology demonstrated at the Anniston Army Depot in Anniston, AL

▼ Multi-layer model (residuum overlying weathered bedrock)

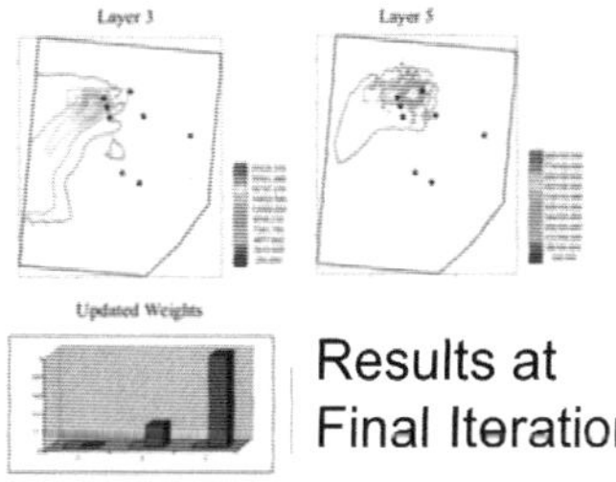

Results at Final Iteration

HGL is extending this technology to incorporate it into PBMO™

20

PBMO™ Future Enhancements: Plume Delineation and Tracking (PDT)

- ▼ Enables stakeholders and decision makers to base plume migration mitigation actions on a solid informational foundation
- ▼ Tools being incorporated in PBMO™ include:
  - Optimal monitoring well network design
    - ➢ Locate monitoring wells to maximize information content at a specified point in time
  - Long-term monitoring optimization
    - ➢ Locate monitoring wells to maximize information content at a specified time horizon

21

Case Study: TCE Monitoring Well Assessment for the Ogallala Aquifer

- ▼ Stakeholder Reviews:
  - Independent Review Team Recommends Formation of Technical Advisory Group (TAG)
  - TAG Unanimously Recommends Customized Tool for Plant Evaluation (Risk Assessment & Corrective Measures Study)
  - Application of Plume Delineation Module Initiated
- ▼ Results: The Plume Delineation Technology Assessed That:
  - The Existing MW Network Reduced Uncertainty by 93.4%. Optimal by 98%.
  - Only 1-2 Additional Optimally Located Wells Required Versus the Initial 6-12 Deep (~600 Feet) Wells Envisioned by the Stakeholders
- ▼ Acceptance: All Analyses and Conclusions were Accepted by Stakeholders
  - High Confidence in Analyses
- ▼ ROI Cost Savings:
  - Cost of Analysis: $96K
  - Well Avoidance Costs: $2M
  - ROI: 20:1

22

Presentation Summary and Conclusions

- ▼ New paradigm for decision support pertaining to environmental restoration and water resource management
  - Management optimization linked with physics-based models on distributed computing platform to provide real-time solutions
- ▼ Proven applications on P&T systems
  - Optimization of existing systems
  - Optimal design for new systems
- ▼ Expanding capabilities to provide decision support on all aspects of large, complex environmental restoration and water resource management challenges

23

Contact for Additional Information

Larry M. Deschaine, Ph.D., P.E.

HydroGeoLogic, Inc.

ldeschaine@hgl.com

24

504253 An Evaluation of Organically Bound Tritium (OBT) Research

Abstract:

Tritium (^3H) is radioactive isotope of hydrogen which decays via beta emission. Tritium incorporated into the tissues of plants and animals is referred to as organically bound tritium (OBT). Tritium ingested as tritiated water or water vapor (HTO) provides a limited dose that is subject to rapid turnover in animals. OBT, largely formed through photosynthesis, represents a more sustained dose when ingested. Elevated radiation exposure to DNA in cells can cause mutation that alter cell function and a serious may result, e.g., cancer. That is, just as water containing ordinary hydrogen and oxygen is a component of all living cells, HTO can also be incorporated into the cells of the body. Recent evidence demonstrates incorporation of OBT into plant cells so consumption by animals through the food chain is also possible. Research indicates that a fraction of ingested tritium can remain in the human body for more than ten years. However, dose standards for OBT do not exist and the literature highlights this gap.

The objective of this study is to assess organically bound tritium (OBT) research with a focus on impacts on organisms, including plants and animals. A thorough literature review of OBT was performed using various databases and reports from federal agencies. Research findings offer measures and influences of OBT however, standard protocols for assessing OBT is suggested.

489606 3D-Interconnected PVA-Citric Acid Porous Hydrogels Embedded with Potassium Copper Hexacyanoferrate Nanoparticles for Enhanced Removal of Radionuclide Cesium

Abstract:

A potassium copper hexacyanoferrate-embedded poly(vinyl alcohol)-citric acid hydrogel film (HPC) was prepared via a facile two-step method for the effective Cs^+ removal. The synthetic procedure consisted of the simultaneous Cu dispersion and incorporation, followed by the diffusion of potassium hexacyanoferrate accelerated by acetone evaporation. The diffusion-derived KCuHCF formation in the preformed hydrogel facilitated not only the preservation of the 3D-interconnected hydrogel structure, but high dispersion of the nanostructured KCuHCF. Using acetone as a non-solvent, the reverse diffusion rate of the incorporated Cu in the hydrogel matrix was greatly reduced; hence a large amount of KCuHCF was loaded in the matrix.

The as-obtained HPC demonstrates the superior Cs^+ adsorption capacity and selectivity with significantly fast removal kinetics. In the adsorption isotherm, the HPC showed a very high Cs^+ uptake of 667 mg/g KCuHCF at the saturated concentration. Moreover, the uptake process was exceptionally fast with 99.5 % of Cs^+ ions adsorbed within 30 min, and the overall Cs^+ removal was 99.9 % from a dilute Cs^+ solution (9.18 ppm). Even when tested in seawater, the HPC exhibited almost unaltered Cs^+ removal efficiency above 99.5 %, and the high distribution coefficient K_d value of 7.7×10^5 mL/g at an extremely low Cs^+ concentration (0.67 ppm, V/m = 1000 ml/g), which highlighted the exceptional Cs^+ affinity over all competing cations. In addition, this novel adsorbent represented the stable and high Cs^+ removal efficiency of 99.9 % over a broad pH range from 2 to 10.

These enhanced removal properties are closely related to the synergistic effects of rich K^+ contents in KCuHCF, well dispersion of the nanoparticles, and the fast swelling kinetics of the hydrogel matrix. Therefore, combined with the simple synthetic procedure and low production cost, the HPC is one of the most promising adsorbents for the rapid and selective removal of Cs^+ from contaminated environments.

487948 Metal-Organic Frameworks Uio-66 and MIL-125 Nanoparticles Enhance the Performance of Thin Film Nanocomposite Membrane for Water Desalination

Abstract:

Two metal-organic frameworks (MOFs) UIO-66 and MIL-125 nanoparticles (NPs) were filled into thin film nanocomposite (TFN) membranes to increase their performance in reverse osmosis application. Membranes were prepared by the interfacial polymerization of 2% *m*-phenylenediamine (MPD) in aqueous solution and 0.15% trimesoyl chloride (TMC)-isooctane [1] solution. Polysulfone (PSU) sheets were prepared by phase inversion and employed to support the synthesized membranes. Different NPs weight ratios, 0.05, 0.1, 0.15, 0.2, and 0.3 %, were used to study the impacts of these additives on membrane efficiency. The as-prepared membranes and NPs were characterized using different techniques for better understanding of the observations. At the optimum addition, the water flux was enhanced from 62.5 L/m^2 h to 74.9 L/m^2 h by adding 0.15% UIO-66 and to 85.0 L/m^2 h by adding 0.3% MIL-125. The NaCl rejection was not significantly affected by embedding UIO-66 and even slightly improved by MIL-125, always at â‰¥ 98.5% as tested at 2000 ppm salt concentration and 300 psi transmembrane pressure.

References

1- Mohammed Kadhom, Jun Yin, Baolin Deng, â€œA Thin Film Nanocomposite Membrane with MCM-41 Silica Nanoparticles for Brackish Water Purificationâ€, *Membranes*, 6, 50, (2016) 1-12.

492605 The Time Evolution of Fouling Development upon Design Aspects of Direct-Flow Filtration

Abstract:

Key words: hollow fibre membrane, direct-flow filtration, packing density, membrane fouling, filtration resistance

Current membrane manufactures are prone to particle removal and membrane cleaning technologies because of the very challenge fouling issue that they are trying to address. It is that fouling decreases the filtration flux over time during constant pressure filtration and contrastively increases the transmembrane pressure difference (TMP) during constant flux filtration, which both restrain the productivity of the filtration process. In literature, most fouling studies have been directed to the constant pressure cross-flow filtration performances, however filtration operations with constant fluxes are more preferred in practice. This study explores the time evolution of fouling development upon design parameters of constant flux direct-flow filtration (DFF), which permits the use of smaller diameter fibres and generates less tangential flow compared with classical cross-flow filtration (CFF). DFF is usually achieved by capping one end of a CFF hollow fibre module to maintain no outlet flow from the module.

Fouling formation and its effect on filtration are influenced by various parameters, such as filtration resistance, backwash cleaning frequencies and module design parameters, out of which resistance is mostly emphasised. As reported in the literature, the rate of change of resistance is proportional to the instantaneous filtration resistance raised to a power, which varies from filtration to filtration. Hence, our investigation started with a DFF model developed by dividing the module into thousands of sections along the axial distance, which allows us to adapt a positional-changing resistance for each filtration circle. The model was validated by comparing the results with the mathematical model resulting from solution of the Navier-Stokes equations, and one group of results are given in Fig.1, Fig.2 and Fig.3. Whilst there is little evolution in the TMP profiles (Fig.1), the evolution of the resistance profile (Fig.2) results in the changing profiles of local fluxes (Fig.3). These results reflect a particular fouling model and will be contrasted with two other fouling models in the presentation. The models will be manipulated with module design parameters aiming to explain the fouling mechanism and approach for optimization of DFF.

Fig. 1. The transmembrane pressure differences (TMP) upon the operating circles versus axial distance at constant flux 70 LMH. The module had a length of 1.6m and 55% packing density of 1.3mm outer diameter fibres.

Fig. 2. The local resistance in log scale upon the operating circles versus axial distance at constant flux 70 LMH. The module had a length of 1.6m and 55% packing density of 1.3mm outer diameter fibres.

Fig. 3. The local flux upon the operating circles versus axial distance at constant flux 70 LMH. The module had a length of 1.6m and 55% packing density of 1.3mm outer diameter fibres.

495089 Overcoming Operating Pressure Barrier in High Recovery Membrane Desalination Via Hybrid RO-NF Processes

Abstract:

In the application of reverse osmosis (RO) membrane technologies for inland brackish water desalination, achieving high product water recovery (up to 90 % or higher) is critical. Even a small enhancement in water recovery has significant impacts on concentrate volume reduction, which in turn directly affect the capacity requirements and thus overall costs of concentrate management and disposal solutions. RO water recovery in inland water desalination is often limited due to two major challenges: a) membrane mineral scaling, and b) mechanical limit of operating pressure. The challenge of membrane mineral scaling can be overcome via antiscalant treatment, as well as the integration of intermediate concentrate demineralization technologies to remove mineral scale precursors from primary RO concentrate. Given that osmotic pressure rises rapidly with water recovery, mechanical limit of operating pressure in membrane systems imposes an upper limit to product water recovery, in addition to the high capital cost of high pressure pumps. New methods are needed for addressing such pressure-imposed water recovery limit in extending the operating range (and thus applicability) of pressure-driven membrane technologies. The present study addresses the issue of concentrate volume reduction in inland brackish water desalination via high recovery hybrid RO-NF process integration. The hybrid RO-NF processes can operate at higher water recovery levels relative to conventional RO processes, while keeping the required operating pressures within the limits imposed by process components (e.g., pumps, membrane elements, and pressure vessels). For example, process analysis indicates that a two-stage hybrid RO-NF process for desalting inland brackish water (e.g., subsurface agricultural drainage water, mine drainage water, produced water, etc.) at up to 90% water recovery can have up to 54% lower operating pressure (e.g., 700 psig vs. 1,500 psig for 15,000 mg/L feed TDS) relative to a conventional two-stage RO process. However, in order to maintain high product water quality, the reduction in the required applied pressure necessitates recycling of NF permeate to RO feed, which increases the required volumetric feed capacity of pumping (and thus energy consumption). Notwithstanding, the above increase in energy requirement remains modest (i.e., 17% higher than a two-stage RO process). Based on process modeling, analysis, and proof-of-concept experimental data, a set of fundamental principles for process design and optimization of hybrid RO-NF processes will be presented with the aim of demonstrating the promising applicability of pressure-driven membrane technologies for desalination of highly saline waters.

495868 Development and Performance of a Silver Nanoparticle Impregnated Reusable Anti-Biofouling Membrane in a Continuous Cross Flow Membrane Module

Abstract:

It is well-known that biofouling decreases membrane-life significantly with simultaneous decrease in permeate flux. To resolve this issue, silver nanoparticle (Ag-NP) - as a biocidal material - has been impregnated in a polyethersulfone (PES) membrane; the latter being typically used in water treatment applications. The main challenges in using such a membrane would be to: (i) prevent membrane biofouling, (ii) achieve zero bacterial cell (*E. coli, MTCC 1302*) count in the permeate water, (iii) maintain silver (Ag) concentration in the permeate stream within the permissible limit of drinking water and (iv) obtain a high tensile strength of the membrane for continuous performance. Addressing these factors would ensure a long and productive service-life of the membrane.

Therefore, the goal of the present research was to develop and test the anti-biofouling efficacy of an Ag-NP incorporated, sulfonated, polyethersulfone (SPES) membrane and reuse it over multiple cycles. We call it as the Ag-SPES membrane and tested it in a laboratory-scale cross-flow membrane module. To this end, the PES membrane having an overall thickness of 150 μm was first sulfonated and then Ag-NPs were selectively impregnated only on the upper, external surface (having a thickness of 1.74 μm) of the membrane by UV-reduction. This leaves the bulk of the membrane and its lower surface free of any Ag-NP. Cross-section FEG-SEM images of the Ag-SPES membrane confirm the selective presence of Ag-NPs only on the upper surface of the otherwise thicker (150 μm) membrane. This ensures better anti-biofouling performance with lower Ag loading, since most of the Ag on the upper surface can come into contact with the *E. coli* in the inlet, contaminated water; whereas there is no extra Ag in the bulk or lower surface of the membrane, which otherwise does not come into contact with *E. coli*, and is hence of no consequence. Such selective Ag-NP impregnation as a thin layer in the post-membrane formation stage, ensures superior anti-biofouling performance with a much lower amount of Ag itself.

Now, in actual operation, contaminated, inlet water contains a low concentration of microorganisms (like *E. coli*), which overt time attaches to the membrane surface and slowly leads to biofouling after a somewhat long duration. However in current experiments, a very high *E. coli* cell concentration (10^4 CFU/ml of *E. coli*) was purposefully used for reducing the timescale of biofouling. This facilitates an easy check on anti-biofouling efficacy of the Ag-SPES membrane.

To this end, *E. coli* cell suspension at 10^4 CFU/ml of *E. coli* concentration was dispersed in PBS (phosphate saline buffer) and was passed through both PES (as control) and Ag-SPES (as test) membranes. However, *E. coli* cells having diameter 0.8-1 μm and length 1.5-2.5 μm are much larger than the 200 nm pore diameter of the membrane. Therefore, cells do not pass through the pores and instead attaches on the membrane surface. This was confirmed by both confocal microscopy and cryo-FEG-SEM images of PES membrane used in the membrane module, showing presence of only live bacterial cells on the outer surface of the membrane. This leads to biofouling and hence as expected, PES membrane showed a continuous and significant decrease in permeate water flow rate, from 64.5 to 12.4 ml/min, after the passage of 240 L of water (in 4 h). In contrast, due to the presence of Ag-NPs on the upper surface of Ag-SPES membranes, cells attached on this membrane are completely killed, showing only a marginal decrease in permeate flow rate (from 58.3 to 50.2 ml/min in 4 h).

In addition, a very low Ag concentration (0.01 mg/L) was also maintained in the permeate stream, which is well below the recommended limit of 0.1 mg/L (WHO) for use as drinking water. Finally, due to sulfonation and impregnation of Ag-NPs, the PES membrane showed an increase in maximum tensile strength from 2.78 MPa to 3.92 MPa, which is beneficial from a mechanical standpoint of membrane integrity in a water treatment process.

Therefore, we could achieve all the four targets set out by us in the beginning. Furthermore, we also found that the Ag-SPES membrane can be synthesized with different, controlled amounts of Ag loading (e. g. 4, 8.8 and 15 wt.% Ag), by tuning the degree of sulfonation of the Ag-SPES membrane. On capturing time-dependent confocal microscopy (using SYTO-9 and propidium iodide) images it turns out that the Ag-SPES membrane with 8.8 wt.% Ag loading (used in the previous results of this study) shows maximum permeability and best anti-biofouling performance. In contrast, the membrane with 4 wt.% Ag is sub-optimum and hence takes a longer time to kill the attached *E. coli* cells; whereas the membrane with 15 wt.% Ag shows a significant decrease in permeability due to pore blockage from excess Ag.

During long usage over repeated cycles, eventually the Ag content in the membrane may decrease, which might affect the anti-biofouling performance of the membrane. To simulate this condition, the Ag-NPs were purposefully and completely removed from the membrane by acid wash. Subsequently, the membranes were again reloaded with the same amount of Ag loading. It was observed that this does not affect the membrane performance. For example, Ag-SPES membrane reloaded with the optimum 8.8 wt.% Ag (after 4 cycles of operation and reloading), shows an almost identical permeate flow rate of 49.71 ml/min after 4 h, compared to 50.2 ml/min (after 4 h) obtained for the fresh, new membrane.

Thus this work demonstrates the viability of the reusable, anti-biofouling, Ag-SPES membrane, to produce decontaminated water over a long period. This was achieved without compromising the mechanical strength of the membrane and maintaining a permissible Ag concentration in the treated water, which is safe for potable purposes. Additionally, after certain periods of operation, Ag-NP could be selectively impregnated as a thin layer on the upper surface of the membrane repeatedly, gaining back the same degree of performance in continued operation.

497918 Performance of a Cost-Effective Micro-Scale Biochar/Psf Mixed Matrix Hollow Fiber Membrane in the Treatment of Methylene Blue Simulated Wastewater

Abstract:

Colorful dye containing water has increasingly become an important issue with the increasing public concerns on environmental protection. A micro-scale biochar particles/PSF mixed matrix hollow fiber membrane (MMM) was successfully prepared for the removal of methylene blue from water. The static and dynamic adsorption of methylene blue by MMM was investigated. The results revealed that the MMM exhibited a high removal efficiency of methylene blue under a wide pH range of 4 to 10. The adsorption process on biochar and MMM obeyed the intraparticle surface diffusion model and Langmuir isotherm model. At neutral pH, the maximum adsorption capacity was for 544.459 mg/g for biochar and 165.808 mg/g for MMM, respectively. Regeneration could be achieved by using 1 M NaCl in 90 % ethanol aqueous solution, with the desorption rate above 92 %.Furthermore, the MMM displayed good performance in treating simulated methylene blue wastewater through continuous filtration. More importantly, the MMM showed an excellent reusability for methylene blue removal, achieving81 % of the permeate yield of the fresh MMM after 3 regeneration cycles. Finally, the adsorption mechanism studies indicated that the adsorption of methylene blue was associated with electrostatic interaction, hydrogen bonding as well as hydrophobic interaction.

499081 Treating Poultry Processing Wastewaters By Ultrafiltration

Abstract:

Commercial poultry processors use an average of 26.5 L of water per bird during processing of live birds to meat. It has been estimated that in 2007, US poultry processors slaughtered 8.9×10^9 broilers, producing 16.0×10^6 metric tons of ready-to-eat meat. Poultry processing water (PPW) is an example of an oily wastewater. These wastewaters contain large amounts of proteins, fats and carbohydrates from meat blood, skin and feathers. Before discharge, the majority of the soluble and organic particulate matter must be removed.

Use of membrane filtration to treat PPW is attractive as one can avoid the use of chemical additives as are often used in processes such as dissolved air flotation. In addition, appropriately selected membrane pore sizes can provide an absolute barrier to pathogens. In this study different ultrafiltration membranes were evaluated for treatment of PPW. Pretreatment of the feed wastewater is critical in order to suppress membrane fouling. Here pretreatment using electrocoagulation has been investigated.

Our results indicate that electrocoagulation followed by sedimentation could be a viable pretreatment technology prior to ultrafiltration. The evolution of the floc size distribution after sedimentation has been determined using laser diffraction particle sizing. Optimizing the sedimentation time is critical when developing a viable process. In this work we have investigated different sedimentation times. Flux decline, and rejection of a number of different ultrafiltration membranes in tangential flow filtration mode have been investigated. The BOD, COD, TDS in the feed, retentate and permeate have been determined. In addition membrane surface properties have been characterized using scanning electron microscopy, contact angle and zeta potential measurements. Membrane performance has been related to membrane surface properties. Finally the membranes were systematically cleaned and regenerated between runs. Our results indicate that while ultrafiltration is feasible for treating PPW.

500862 Ceramic Hollow Fiber Supported Metal-Organic Framework Uio-66 for Enhanced Adsorptive Separations

Abstract:

Ceramic Hollow Fiber Supported Metal-Organic Framework Uio-66 for Enhanced Adsorptive Separations

Chenghong Wang,[a, b] Dr. Melanie Lee,[a] Dr. Xinlei Liu,[a] Dr. Bo Wang,[a]

Professor J. Paul Chen,[b] and Professor Kang Li*[a]

[a] The Barrer Center, Department of Chemical Engineering, Imperial College London, London SW7 2AZ, United Kingdom

[b] NUS Graduate School for Integrative Sciences & Engineering (NGS), Department of Civil and Environmental Engineering, National University of Singapore, Singapore, 117456, Singapore

ABSTRACT

A novel ceramic hollow fiber was prepared through controllable spinning and sintering to form a two-distinct-layer geometry: one very thin barrier layer containing 3D-pore network structure at the lumen, and one unique layer containing a plurality of conical micro-channels towards the shell side.[1] These micro-channels not only reduce the mass transfer resistance as a transport network, giving rise to competitive permeation fluxes, but also form pockets within which functional materials can be readily deposited.

Recently, metal-organic frameworks (MOFs), a new class of porous materials, has exhibited a great potential in various applications, owing to their unique properties like exceptionally high porosity, large surface area and customizable chemical functionality.[2] One of the representative examples is UiO-66, a zirconium-based MOF, which has shown the best adsorption capacity for arsenic removal from wastewater.[3] However, the current discussion regarding MOF adsorbent is majorly based on the particle samples. In order to be industrially applicable, the powder materials have to be specifically shaped or supported, since in practical dispersed particles would easily leak through the application compartment, resulting in challenging spent-particles-separation issues and severe safety concerns.[4]

Herein, we developed and optimized a functional adsorptive membrane by incorporating UiO-66 into the ceramic hollow fiber. The fast adsorption kinetics provided by UiO-66 allows an efficient adsorptive separation process to be realized within the micro-channels, and the thin barrier layer can prevent any loss of the active adsorbents from leaking out. As a proof of concept, this adsorptive membrane was found to be capable of effectively remediating arsenic contaminated water and producing potable water recovery (less than 10 ppb – required by World Health Organization). In comparison, to achieve a similar performance, the conventional adsorption packing column required eight times amount of active UiO-66 adsorbents. Hence, this adsorptive membrane has exhibited a promising potential to be applied for industrial wastewater decontamination. Looking forward, different functional membranes can be prepared based on specific adsorptive applications, as a wide range of adsorbents can be loaded into this novel ceramic hollow fiber, of which the micro-channel size and barrier layer pore size can be easily manipulated during fabrication.

Figure 1. Schematic diagram of arsenic contaminated water remediation by the adsorptive membrane. The inset demonstrates an enlarged cross-sectional view of the adsorptive membrane. Blue molecule: water; green molecule: arsenic pollutant.

KEYWORDS

MOF, ceramic hollow fiber, adsorption, arsenic, water remediation

REFERENCES

[1] M. Lee, B. Wang and K. Li, *J. Membr. Sci.*, 2016, **503**, 48.

[2] H. Furukawa, K. E. Cordova, M. O'Keeffe and O. M. Yaghi, *Science*, 2013, **341**, 1230444.

[3] C. Wang, X. Liu, J. P. Chen and K. Li, *Sci Rep*, 2015, **5**, 16613.

[4] J. He, T. S. Siah and J. Paul Chen, *Water Res.*, 2014, **56**, 88.

ACKNOWLEDGEMENTS

This work was supported by the research funding provided by Engineering and Physical Sciences Research Council in the United Kingdom (Grant no. EPSRC, EP/J014974/1). C. Wang gratefully acknowledges the National University of Singapore Graduate School for Integrative Sciences and Engineering scholarship.

Adsorption and Kinetic Studies of using Entrapped Sewage Sludge Ash in the Removal of Chemical Oxygen Demand from Domestic Wastewater, with Artificial Intelligence Approach

Rasha A. SaryEl-deen[1]*, Ahmed S. Mahmoud*[1]*, M.S. Mahmoud*[1]*and Mohamed K. Mostafa*[2]

[1]*Sanitary and Environmental Institute (SEI), Housing and Building National Research Center (HBRC), Egypt*

[2]*Environmental Engineering Program, Zewail City of Science and Technology, Giza, Egypt*

Abstract

Huge amounts of sludge are produced daily from municipal wastewater treatment plants. These solid organic wastes produce active ash after complete combustion which can be used in many applications. This study explores different adsorption and kinetic models that can describe the adsorption process of chemical oxygen demand (COD) into entrapped sewage sludge ash in alginate biopolymer (AG/SSA) and implementation of the obtained result in original domestic wastewater sample. The AG/SSA and stander COD solution were prepared in the laboratory. X-Ray Diffraction (XRD) was used for SSA characterization. The effect of AG/SSA on COD removal was studied at different absorbent dose, contact time, stirring rate, pH, and initial COD concentration. The results indicated that AG/SSA is effective in the removal of COD from aqueous solution, where removal efficiencies of 61 and 91% were achieved for 800 and 100 mg/L initial COD concentration, respectively, after 60 min of contact time using dose 8 g/L at pH 5 with fixed stirring rate 200 rpm. The equilibrium isotherm of COD was determined using Freundlich, Langmuir, Koble–Corrigan, Toth, Hill, Khan, Redlich and Elovich models. The kinetic adsorption of COD onto AG/SSA was determined by Avramin, pseudo first order, Intraparticulate, and pseudo second order models. The adsorption data of COD fitted well to Redlich-Peterson model with lowest summation of error of 1.671 and Avramin kinetic model with lowest summation of error of 1.669. Artificial neural network (ANN) with a structure of 5 – 10 – 1 was used to predict the COD removal efficiency. The proposed ANN was found to be effective in simulating the performance of AG/SSA for COD removal, where a high r^2-value was recorded to be 0.9925, 0.9996, and 0.9997 for training, validation and test plots, respectively.

Introduction and Background

Chemical Oxygen Demand (COD) is an important water quality indicator which indicates the expected environmental effects from discharging the wastewater to the nearest water body. Higher COD levels lead to reducing dissolved oxygen (DO) levels in the effluent due to the presence of high level of oxidizable organic materials (Perason and Frangipane, 1975). A reduction in DO level in the receiving water may increase risk to aquatic life due to the existence of anaerobic conditions (Mathew et al., 2013). COD concentration in the effluent should be reduced to a level that allow the treated wastewater to be reused in agricultural purposes.

Sewage Sludge (SS) is a byproduct introduced during the treatment processes in municipal wastewater treatment plants (WWTPs). About 2 million tons dry solids were produced in Egypt in 2005 from WWTPs (Hassan 2015). Such waste should be well managed to reduce health and environmental impacts from spreading SS on land (Karnitz et al., 2007; Mattenberger et al., 2008). Sewage Sludge reuse applications may include organic fertilizer, road surfacing, and land filling (Yu and Zhong 2006). Sewage sludge ash (SSA) is a by-product produced during the incineration of SS after dewatering in incinerators (Cheremisinoff, 2003). SS after incarnation can also be used in wastewater treatment applications. Entrapping SSA in alginate biopolymer has an advantage of reducing the retention time of wastewater treatment process through reduce settling time and to use the specific character of alginate in the same process.

This work attempts to examine the removal efficiency of COD from domestic wastewater by entrapped SSA in alginate polymer. The relation between COD removal efficiency and the variation in initial COD concentration, adsorbent dose, pH, contact time, and stirring rate were studied. Isotherm study was employed by applying different isotherm models in order to accurately model the adsorption data. Diffeternt kinetic models were also conducted to analyze the kinetic data. Artificial neural network (ANN) were conducted to detect the correlation between the experimental factors and the COD removal efficiency.

Materials and Methods

Preparation of sewage sludge ash

The sewage sludge ash (SSA) produced from dried of sewage sludge cake (DSSC) sample delivered from El-Berka wastewater treatment plant by burned (DSSC) in oven at 600°C for 2 hr, the ground SSA sample was further screened. The finer part of SSA sample passing through the sieve number 200 (75μm) was collected and used in the following experiments.

Preparation of encapsulated SSA in alginate biopolymer AG/SSA

Sodium alginate (2 g) was dissolved in 100 ml deionized water (DW) at room temperature (22 ± 2°C). The alginate–water mixture was stirred for 20 minutes using *magnetic* stirrer at 600 rpm until complete dissolution was achieved. The alginate–water mixture was then left for another 30 minutes at room temperature to ensure that the alginate did not float in the solution and to allow air/gas bubbles that are generated during the mixing and stirring processes to escape. The alginate solution was mixed with 1 g of SSA and mixed using magnetic stirrer at 350 rpm for 20 min. The mixture was promptly dropped into a 2 gram $CaCl_2$ solution/100ml distilled water (2% w/v) $CaCl_2$ solution at room temperature using syringe (drop by drop 1drop/ 3 sec). The Ca-alginate gel beads started to form once the sodium alginate/SSA mixture came in contact with $CaCl_2$ solution.

Chemical composition of sewage sludge ash

Table 1 lists the chemical composition of SSA which is mainly consist of silicon dioxide (SiO_2), calcium oxide (Cao), ferric oxide (Fe_2O_3), and sulfur trioxide (SO_3) with a percentage of 33.83, 18.45, 9.56 and 8.01%, respectively.

Table 1. Chemical composition of SSA

Element primary composition %	SiO_2	Al_2O_3	Fe_2O_3	Cao	P_2O_5	MgO	SO_3	K_2O	Na_2O	ZnO	PbO	CuO	TiO_2
	33.83	5.44	9.56	18.45	5.65	2.08	8.01	1.63	0.62	0.55	7.45	0.17	0.95
Heavy metals mg\kg	**Cu**	**Zn**	**Ni**	**Mn**	**Fe**								
	22.77	10.72	4.45	10.46	55								
Moisture content %	12												
Bulk density g\cm³	0.386												

Batch adsorption studies

The adsorption of COD onto encapsulated Sewage Sludge Ash in alginate biopolymer (AG/SSA) was studied by batch technique. The COD removal efficiency was studied at pH: 3, 5, 7, and 9, adsorbent dose: 2, 4, 6, 8, and 10 g/L, stirring rate: 150, 200, 250, and 300 rpm, and contact time: 15, 30, 45, 60, 75, and 90 min. A known weight of adsorbent 8g for AG/SSA was equilibrated with 1000 mL of aqueous COD solution of known concentrations (100, 200, 400, 500, 600, 700, and 800 mg/L) in 1000 mL of Erlenmeyer flasks and shacked at 200 rpm at fixed temperature of 25±2°C for a known period of time 60 min. After equilibration, the suspension of the adsorbent was separated from solution by filtration using Whatman No. 42 filter paper and the concentration of COD remaining in solution was measured using UV spectrophotometer after digestion using COD digester according to ASTM 2005 stander method for water and wastewater (Federation and Association, 2005). The percentage of removal efficiency was calculated using Equation 1. The amount of COD sorbed by weight of AG/SSA nanoparticles was calculated using Equation 2.

$$Sorption\ (\%) = \left(\frac{C_0 - C_e}{C_0}\right) \times 100 \qquad \text{Eq. 1}$$

where C_o is the initial concentration of COD in solution (mg/L) and C_e is the equilibrium concentration of COD in solution (mg/L)

$$Q_e = \frac{(C_0 - C_e)V}{m} \qquad \text{Eq. 2}$$

where Q_e is the equilibrium adsorption capacity (mg/g), V is the volume of aqueous solution (L), and m is the dry weight of the adsorbent (g)

Isotherm studies

Different isotherm studies were conducted for monolayer and multilayer adsorption process using nonlinear equations including Freundlich, Langmuir, Redlich–Peterson, Hill, Sips, Khan, Toth, Koble–Corrigan, Jovanovich, and Hossein, as shown in Table 2.

Table 2. Nonlinear equations of isotherm models

Isotherm model name	nonlinear equations	Constants		References
Langmuir	$Q_e = \frac{q_m K_a C_e}{1 + K_a C_e}$	K_a	Langmuir constant (L/mg)	(Armagan and Toprak, 2013)
		q_m	maximum monolayer adsorption capacity (mg/g)	

Freundlich	$Q_e = K_F C_e^{1/n}$	n	Freundlich constant related to adsorption intensity (dimensionless)	(Armagan and Toprak, 2013)
		K_F	Freundlich constant related to adsorption capacity ((mg/g)(mg/L)-1/n)	
Redlich–Peterson	$Q_e = \frac{K_R C_e}{1 + a_R C_e^{b_R}}$	K_R	Redlich–Peterson isotherm constant (L/g)	(Redlich and Peterson, 1959)
		a_R	Redlich–Peterson isotherm constant (1/mg)	
		b_R	Redlich–Peterson isotherm exponent	
Hill	$Q_e = \frac{Q_H C_e^{n_H}}{K_D + C_e^{n_H}}$	Q_H	Hill isotherm maximum uptake saturation (mg/L)	(Hill, 1910)
		K_D	Hill constant	
		n_H	Hill cooperativity coefficient of the binding interaction	
Sips	$Q_e = \frac{Q_s (K_s C_e)^{B_e}}{1 + (K_s C_e)^{B_e}}$	K_s	Sips isotherm model constant (L/g)	(Koble and Corrigan, 1952)
		B_s	Sips isotherm model exponent	
Khan	$Q_e = \frac{Q_K B_K C_e}{(1 + B_K C_e)^{A_K}}$	Q_k	theoretical isotherm saturation capacity (mg/g)	(Khan et al., 1997)
		B_k	Khan isotherm model constant	
		A_k	Khan isotherm model exponent	
Toth	$Q_e = \frac{K_t C_e}{(a_t + C_e)^{\frac{1}{t}}}$	K_t	Toth isotherm constant (mg/g)	(Toth, 1971)
		a_t	Toth isotherm constant (L/mg)	
		T	Toth isotherm constant	
Koble–Corrigan	$Q_e = \frac{A C_e^D}{1 + B C_e^D}$	A	Koble–Corrigan isotherm constant ($L_n mg_{1-n}$/g)	(Koble and Corrigan, 1952)
		B	Koble–Corrigan isotherm constant $(L/mg)_n$	
		D	adsorption intensity	
Jovanovich	$Q_e = q_{max}(1 - e^{K_j C_e})$	qmax	(mg/g) is the maximum uptake	(Jaroniec, 1975)
		KJ	Javanovic isotherm constant (L/g)	

Kinetic study

To determine the suitable time for reaching the equilibrium state, COD solutions were placed in contact with AG/SSA at different times, at room temperature. The amount of COD removed at time t, Qt (mg/g), was calculated using equation 3.

$$Q_t = \frac{(C_o - C_t)V}{W} \qquad \text{Eq. 3}$$

where C_o is the initial concentration (mg/L), C_t is the concentration of COD at time t (mg/L), V is the volume of the solution (L), and W is the AG/SSA dosages

To investigate the mechanisms of the kinetic process the following kinetics techniques were applied: pseudo-first and second-order, Avrami, Elovich, and Intraparticle. The equation of each model is listed in Table 3 (Ho and McKay, 1998, Ho and McKay, 1999, Avrami, 1940, Zeldowitsch, 1934b, Fola et al., 2016).

Table 3. shows nonlinear equations of kinetic models

Kinetic model name	Equations
Pseudo First Order	$Q_t = Q_e(1 - e^{(-K_1 t})$
Pseudo Second Order	$Q_t = \frac{Q_e^2 K_2 t}{(1 + Q_e K_2 t)}$
Elovich	$Q_t = \frac{1}{\beta} Ln(\alpha \beta_t)$
Avrami	$Q_t = Q_e(1 - e^{(-K_{av} t^{n_{av}}})$
Intraparticle model	$Q_t = K_{id} t^{0.5} + C_i$

Error functions calculations

Different error parameters were calculated in order to identify the most suitable isotherm and kinetic models that describe the treatment process. Due to the ingrained bias that resulting from linear form of isotherm and kinetic models, the nonlinear form was good alternative parameter which provide an accurate technique to determine isotherm and kinetic parameters from the original equation forms (Malek and Farooq, 1996, Khan et al., 1996). To evaluate the better fit of nonlinear models the error function is required (Ho et al., 2002). The five error functions were used for all models and the ideal model was selected from the summation of these errors. The error parameter equations show in Table 4 (Marquardt, 1963).

Table 4. shows errors mathematical equation.

Error name	Equations
The Sum of the Squares of the Errors (ERRSQ)	$\sum_{i=1}^{p}(q_{e,meas}-q_{e,calc})_i^2$
A Composite Fractional Error Function (HYBRD)	$\sum_{i=1}^{p}\left\|\frac{(q_{e,meas}-q_{e,calc})^2}{q_{e,meas}}\right\|_i$
A Derivative of Marquardt's Percent Standard Deviation (MPSD)	$\sum_{i=1}^{p}\left(\frac{q_{e,meas}-q_{e,calc}}{q_{e,meas}}\right)_i^2$
The Average Relative Error (ARE)	$\sum_{i=1}^{p}\left\|\frac{q_{e,meas}-q_{e,calc}}{q_{e,meas}}\right\|_i$
The Sum of the Absolute Errors (EABS)	$\sum_{i=1}^{p}\left\|q_{e,meas}-q_{e,calc}\right\|_i$
Chi error	$\sum_{i=1}^{p}\frac{(q_{e,calc}-q_{e,meas})^2}{q_{e,calc}}$

Neural network structure

An artificial neural network was developed to predict COD removal efficiency, which consists of input, hidden, and output layers. The neural network structure was expressed as 5 – 10 – 1. The data from the five independent variables are transferred to the input layer. All the available data are normally divided into 60, 20, and 20% for training, validation, and testing procedures, respectively (Nasr et al., 2017). The network type is "Feed-forward backprop", while the training and the adaption learning functions are "Trainlm" and "Learngdm", respectively. The "tansig" and "purlin" functions were selected for hidden and output layers, respectively.

Results and Discussions

Characterization of SSA

Powder X-ray diffraction analysis (XRD)

The diffraction angle (2θ) ranged from 5° to 50° and runs at 0.0167° steps. The X-ray diffraction for SSA sample was conducted and the results showed that the major mineral phases in the specimens were silicon dioxide (SiO_2), calcite ($CaCO_3$), and anhydrite ($CaSO_4$), as shown Figure 1. The sample consists mainly of quartz (SiO_2) with a percentage of 84%.

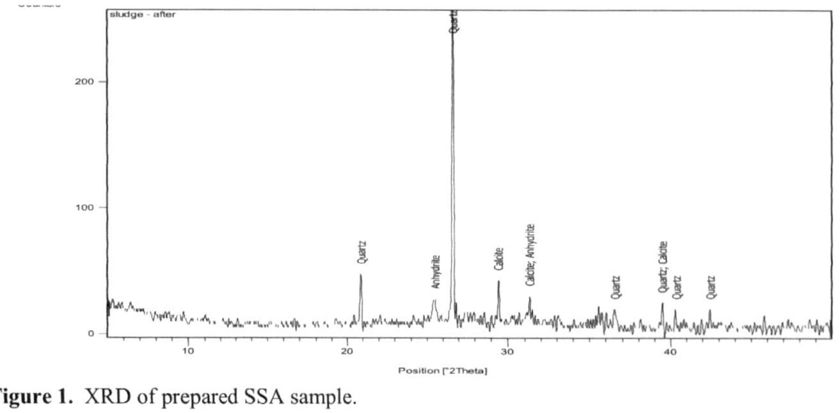

Figure 1. XRD of prepared SSA sample.

Effect of operating parameters

Effect of pH

The effect pH on COD removal efficiency was studied at AG/SSA dosage: 8 g/L, stirring rate: 200 rpm, initial COD concentration: 400 mg/L, and contact time: 60 min (Fig. 2a). A decrease in COD removal efficiency from 78% to 77.9, 66, and 32% was associated with an increase in pH from 3 to 5, 7, and 9, respectively. This result indicated that the most efficient media to remove COD at pH ranges from 3 - 5. In the acidic media, organic compounds are broken down into smaller compounds which makes it easier to be absorbed.

Effect of contact time

The effect of contact time on COD removal efficiency was studied at AG/SSA dosage: 8 g/L, stirring rate: 200 rpm, initial COD concentration: 400 mg/L, and pH: 5 (Fig. 2b). An increase in COD removal efficiency from 63 to 69, 74, 78, 79, and 81% was associated with an increase in contact time from 15 to 30, 45, 60, 75, and 90 min, respectively. At the beginning of adsorption, a large number of abundant sites on the AG/SSA surface were available to adsorb greater amounts of the organic materials. The minimum effective time was recorded as 60 min. The effective time was selected as 60 min due to the small variation of the removal efficiencies between times 60, 75, and 90 min.

Effect of adsorbent dose

The effect of AG/SSA dosage on COD removal efficiency was studied at stirring rate: 200 rpm, initial COD concentration: 400 mg/L, pH: 5, and contact time: 60 min (Fig. 2c). An increase in COD removal efficiency from 45 to 56, 72, 78, and 79% was associated with an increase in adsorbent doses from 2 to 4, 6, 8 and 10 g/L, respectively. A large number of vacant sites on AG/SSA surface were available as a result of increasing the adsorbent dosage, which adsorbed large amounts of organic materials.

Effect of stirring rate

The effect of stirring rate on COD removal efficiency was studied at AG/SSA dosage: 8 g/L, initial COD concentration: 400 mg/L, pH: 5, and contact time: 60 min (Fig. 2d). The removal efficiency of COD was 76% at a stirring rate of 150 rpm, and increased to 78% at a stirring rate of 200 rpm. This is mainly due to improving the opportunity of AG/SSA to come in contact with organic compounds. The removal efficiency remains constant with further increase in stirring rate to 250 and 300 rpm. This results indicated that the suitable stirring rate is 200 rpm.

Effect of the initial concentration

The effect of COD initial concentration on its removal efficiency was studied at AG/SSA dosage: 8 g/L, stirring rate: 200 rpm, pH: 5, and contact time: 60 min (Fig. 2e). A decrease in COD removal efficiency from 91% to 88, 83, 78, 72, 68, 65, and 61% was associated with an increase in COD concentration from 100 mg/L to 200, 300, 400, 500, 600, 700, and 800 mg/L, respectively. At low COD concentrations, the ratio of abundant sites to the total organic compounds was high, and thus, all organic compounds were bound to the AG/SSA surface.

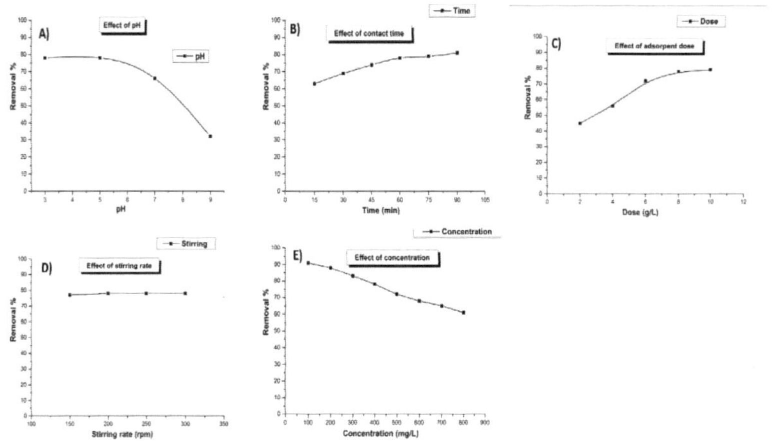

Figure 2. Effect of operating parameter on COD removal using AG/SSA (a) pH, (b) contact time, (C) AG/SSA dosage, and (e) initial COD concentration

Adsorption study

Figure 5 describe nonlinear relations between different adsorption isotherm models. The resulted data from adsorption study using nonlinear models indicating that Redlich-Peterson is the most suitable model that describe the adsorption process with lowest summation of error of 1.671, as shown Table 5. This result indicated that the removal of COD by GA/SSA could undergo both mono and multilayer adsorption, and the Redlich-Peterson model constants were appropriate for representing the adsorption process.

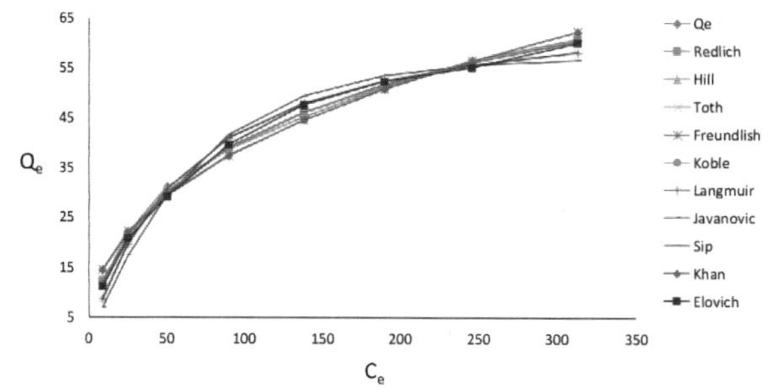

Figure 3. Isotherm studies for nonlinear models

Table 5. shows the results of nonlinear adsorption models

	Redlich- Peterson		Hill		Sips		Khan		Toth	
Constants	Kr	5.700	QH	112.1	Qs	105.3	Qk	4.080	Kt	8.378
	Br	0.043	nH	0.635	Ks	0.005	Bk	2.590	at	63.94
	G	0.470	KD	32.73	Bs	0.660	Ak	0.593	t	8.378
			ERRORS							
Chi error		0.124		0.154		0.134		0.821		1.538
ERRSQ		3.042		3.759		3.916		16.592		34.174
HYBRD		0.128		0.158		0.135		0.975		1.324
MPSD		0.007		0.009		0.006		0.074		0.077
ARE		0.166		0.187		0.180		0.412		0.556
EABS		4.262		4.846		5.227		8.981		15.213
Error sum		**7.729**		**9.114**		**9.599**		**27.853**		**52.883**

	Koble –Corrigan		Javanovic		Freundlich		Langmuir	
Constants	A	3.305	Qm	57.36	Kf	5.949	Qo	70.183
	B	0.030	Kj	0.014	n	2.444	b	0.016
	D	0.646						
			ERRORS					
Chi error		0.143		5.097		0.922		1.538
ERRSQ		3.826		93.773		18.255		34.174
HYBRD		0.145		3.790		1.120		1.324
MPSD		0.007		0.219		0.087		0.077
ARE		0.182		0.956		0.437		0.556
EABS		4.972		25.501		9.256		15.213
Error sum		**9.275**		**129.335**		**30.077**		**52.883**

Kinetics Studies

The results from the adsorption data indicated that Elovich model is the preferred model with lowest error 1.669, as shown in table 6. This result also indicated that the desorption of organic compounds on GA/SSA surface took place at initial times of the sorption process, and the Elovich model constants were appropriate for representing the adsorption kinetics

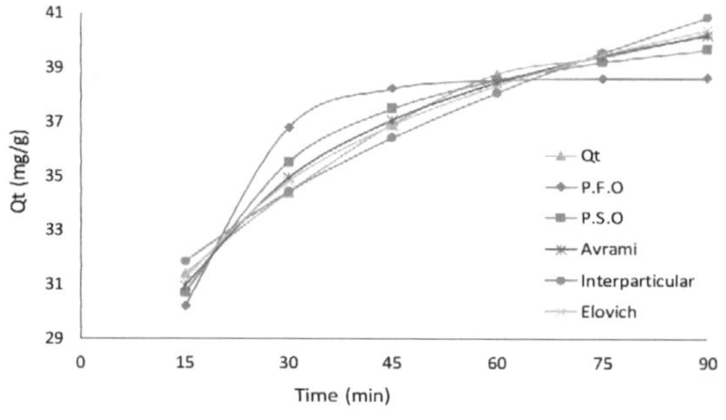

Figure 4. Kinetic isotherm models

Table 6. shows the results of different Kinetic models.

	P.F.O	**P.S.O**	**Elovich**	**Avrami**	**Intraparticle**
Constants	$Q_e = 38.61$ $K_1 = 0.101$	$Q_e = 42.13$ $K_2 = 0.004$	$\alpha = 154.29$ $\beta = 0.1958$	$Q_e = 45.279$ $K_{av} = 0.435$ $N_{av} = 0.358$	$K_{id} = 1.611$ $C_i = 25.562$
			ERRORS		
Chi error	0.337	0.072	0.011	0.018	0.035
ERRSQ	12.308	2.531	0.394	0.611	1.314
HYBRD	0.345	0.073	0.011	0.018	0.035
MPSD	0.010	0.002	0.000	0.001	0.001
ARE	0.211	0.096	0.033	0.045	0.065
EABS	7.582	3.412	1.219	1.573	2.400
Error sum	**20.793**	**6.186**	**1.669**	**2.265**	**3.850**

Artificial neural network

Fig. 5a shows the training process which is characterized by 6 validation checks, 6 epochs, and gradient value equal 0.41434. The training step was terminated because the gradient value has exceeded the value of the least error (1e-5). Fig. 5b shows the mean squared error performance versus epoch number. The MSE of training process decreased with the increase in the number of iterations, while the MSE of the validation and test processes increased with the increase in the number of iterations. An increase in the MSE of the validation process may refer to overfitting, but in this case the overfitting has not occurred because the test and validation curves were comparable. These results revealed that the best validation performance was 0.15397 at epoch 0. Fig. 5c presents the regression plot between targets and outputs. High r^2-value for training, validation and test plots indicated that the proposed ANN has fitted the data well, which is recorded as 0.9925, 0.9996, and 0.9997, respectively. The overall r^2-value was recorded as 0.9951, which indicated that about 99.51% of the variations in COD removal efficiency were described by the studied five factors, while the remaining 0.49% was not described by the proposed model.

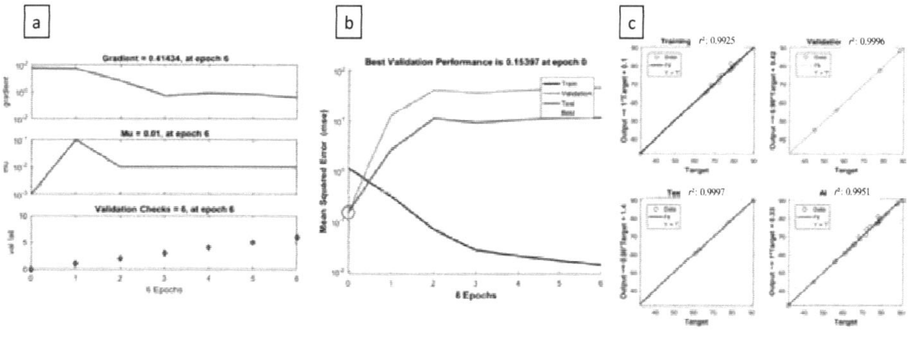

Figure 5. ANN results: a) Training, b) Best validation performances, and c) Regression plot

Conclusions

This study has indicated that CA/SSA is a strong adsorbent material for COD removal from domestic wastewater. The adsorption data of COD fitted well to Redlich-Peterson model and Avramin kinetic model. The highest removal efficiency of 91% was achieved at initial COD concentration 100 mg/L, contact time 60 min, Ca/SSA dose 8 g/L, pH 5, and stirring rate 200 rpm. The proposed ANN was found to be effective in simulating the performance of AG/SSA for COD removal, as indicated by the high overall r^2-value of 0.9951. The created ANN can then be used to predict COD removal efficiencies for different sets of input data.

References

AVRAMI, M. 1940. Kinetics of phase change. II transformation-time relations for random distribution of nuclei. *The Journal of Chemical Physics,* 8, 212-224.

CHAPAGAIN, A. K., HOEKSTRA, A., SAVENIJE, H. & GAUTAM, R. 2006. The water footprint of cotton consumption: An assessment of the impact of worldwide consumption of cotton products on the water resources in the cotton producing countries. *Ecological economics,* 60, 186-203.

FEDERATION, W. E. & ASSOCIATION, A. P. H. 2005. Standard methods for the examination of water and wastewater. *American Public Health Association (APHA): Washington, DC, USA.*

FOLA, A. T., IDOWU, A. A. & ADETUTU, A. 2016. Removal of Cu2+ from aqueous solution by adsorption onto quail eggshell: Kinetic and isothermal studies. *Journal of Environment & Biotechnology Research,* 5, 1-9.

H, F. 1906. Adsorption in solutions. *Phys Chemie,* 57, 384.

HILL, A. V. 1910. The possible effects of the aggregation of the molecules of haemoglobin on its dissociation curves. *J Physiol (Lond),* 40, 4-7.

HO, Y.-S. & MCKAY, G. 1998. Kinetic models for the sorption of dye from aqueous solution by wood. *Process Safety and Environmental Protection,* 76, 183-191.

HO, Y.-S. & MCKAY, G. 1999. Pseudo-second order model for sorption processes. *Process biochemistry,* 34, 451-465.

HO, Y., PORTER, J. & MCKAY, G. 2002. Equilibrium isotherm studies for the sorption of divalent metal ions onto peat: copper, nickel and lead single component systems. *Water, Air, and Soil Pollution,* 141, 1-33.

HOEKSTRA, A. Y. & HUNG, P. Q. 2005. Globalisation of water resources: international virtual water flows in relation to crop trade. *Global environmental change,* 15, 45-56.

JARONIEC, M. 1975. Adsorption of gas mixtures on homogeneous surfaces Extension of Jovanovic equation on adsorption from gaseous mixtures. *Chem. zvesti,* 29, 512.

KHAN, A., AL-WAHEAB, I. & AL-HADDAD, A. 1996. A generalized equation for adsorption isotherms for multi-component organic pollutants in dilute aqueous solution. *Environmental technology,* 17, 13-23.

KHAN, A., ATAULLAH, R. & AL-HADDAD, A. 1997. Equilibrium adsorption studies of some aromatic pollutants from dilute aqueous solutions on activated carbon at different temperatures. *Journal of colloid and interface science,* 194, 154-165.

KOBLE, R. A. & CORRIGAN, T. E. 1952. Adsorption isotherms for pure hydrocarbons. *Industrial & Engineering Chemistry,* 44, 383-387.

LANGMUIR, I. 1918. The adsorption of gases on plane surfaces of glass, mica and platinum. *Journal of the American Chemical society,* 40, 1361-1403.

MALEK, A. & FAROOQ, S. 1996. Comparison of isotherm models for hydrocarbon adsorption on activated carbon. *AIChE Journal,* 42, 3191-3201.

MARQUARDT, D. W. 1963. An algorithm for least-squares estimation of nonlinear parameters. *Journal of the society for Industrial and Applied Mathematics,* 11, 431-441.

NASR, M., MAHMOUD A., FAWZY M. & RADWAN A. 2017 Artificial intelligence modeling of cadmium(II) biosorption using rice straw. *Appl water sci.,* 7(2), 823–831.

PARRIS, K. 2011. Impact of agriculture on water pollution in OECD countries: Recent trends and future prospects. *International Journal of Water Resources Development,* 27, 33-52.

REDLICH, O. & PETERSON, D. L. 1959. A useful adsorption isotherm. *Journal of Physical Chemistry,* 63, 1024-1024.

TEMPKIN, M. & PYZHEV, V. 1940. Kinetics of ammonia synthesis on promoted iron catalyst. *Acta Phys. Chim. USSR,* 12, 327.

TOTH, J. 1971. State equations of the solid-gas interface layers. *Acta Chim Acad Sci Hungar,* 69, 311-328.

WANG, H., DU, S. & LIU, S. 2013. *The Investigation and Assessment on Groundwater Organic Pollution*, INTECH Open Access Publisher.

WHO, J. & CONSULTATION, F. E. 2003. Diet, nutrition and the prevention of chronic diseases. *World Health Organ Tech Rep Ser,* 916.

ZELDOWITSCH, J. 1934a. Adsorption site energy distribution. *Acta phys. chim. URSS,* 1, 961-973.

ZELDOWITSCH, J. 1934b. Über den mechanismus der katalytischen oxydation von CO an MnO2. *Acta Physicochim. URSS,* 1, 364-449.

489883 The CO_2, NO_x, SO_x Adsorbing Capacity of Polyaniline-Based Composite Materials

Abstract:

Polyaniline (PANI)/ attapulgite, vermiculite and diatomite nanocomposites co-doped with hydrochloric acid (HCI) and ammonium persulfate (APS) were prepared by in-situ polymerization method. The existing of additive materials acted as the skeleton of PANI. The samples were characterized by FT-IR, XRD, SEM, BET, UV spectrophotometer and TGA. The results showed that the composites have the structure of attapulgite and the nature of the polyaniline. The composites doped with attapulgite, vermiculite and diatomite had even better adsorption capacity for acid gases such as CO_2, SO_2 and NO_2 under the experimental conditions than that of PANI itself. Those composite materials have great potential on dealing with the fuel gas in thermal power industry and protecting environment.

491323 Performance of Ion Selective Electrodes (ISE) on Wastewaters from Power Plants

Abstract:

The latest U.S. Environmental Protection Agency (EPA) effluent limitations guidelines and standards for the steam electric power generating point source categoryâ€™s final rule published in November 2015 mandates the limitations of toxic metals and other chemical compounds from coal-fired power plants. On an annual basis, the rule reduces the amount of toxic metals, nutrients, and other pollutants that steam electric power plants are allowed to discharge (no more than 1.4 billion pounds per year) and it also reduces water withdrawal by 57 billion gallons per year. Currently, a Flue Gas Desulphurization (FGD) system is used to remove SO_2 from flue gas produced from the coal-fired power plant boiler. In the FGD system, limestone reacts with SO_2, producing gypsum. The system also uses various chemicals, like organosulfides, to remove trace heavy metals (e.g., selenium, mercury, arsenic etc.) from FGD wastewater. Chemicals required for the FGD system are calculated based on coal sulfur and heavy metal content, and the numerical discharge limit regulated by EPA. The plant operators manually collect samples of the discharge water as well as the FGD wastewater periodically, and measure the chemical quantity by spectroscopy in the laboratory, which is time consuming, costly, and tedious. Often, response to a parameter tuning takes days and the plant may discharge over the limit in the meantime.

As of 2015, more than 66 companies are manufacturing water related sensors in the world but only a handful of them market Ion Selective Electrodes (ISE). An ISE (with its own internal reference electrode) is immersed in an aqueous solution containing the ions to be measured, together with a separate, external reference electrode. The external reference can be completely separate or incorporated in the body of the ISE to form a Combination Electrode). The electrochemical circuit is completed by connecting the electrodes to a sensitive (milli)volt meter using special low-noise cables and connectors. A potential difference is developed across the ISE membrane when the target ions diffuse through from the high concentration side to the lower concentration side.

In this project, we have investigated the feasibility of using inexpensive sensors for FGD water measurements and ISE sensors for chemical composition. FGD wastewaters were collected from AEPâ€™s coal-fired power plant. Before testing the ISE electrodes in FGD wastewaters, it is important to quantify the targeted compounds in the FGD wastewater. Therefore, we have used high accuracy gas chromatography, ion chromatography, high-performance liquid chromatography, and induced coupled plasma available for organic and inorganic compounds detection and quantification. Targeted compounds for this project were mercury, selenium, arsenic, nitrate, nitrites, calcium, and chloride. ISEs purchased from Thermo-scientific and NIKO (London, UK) were tested on the known FGD wastewaters individually to determine the performance of each electrodes.

In future, we will select the working electrodes and make a sensor package for wastewater quality monitoring. The package will be designed as a self-powered, wireless, water-resistant, robust, and cheap. The sensor package is expected to demonstrate at the AEPâ€™s site for the real-time data monitoring.

492119 Cesium Removal By Immobilization of Potassium Copper Hexacyanoferrate in a Cellulose-Hydrogel Network

Abstract:

Treatment of radioactive wastewater has become important due to an enormous amount of nuclear waste and the potential uncontrolled release. For adsorption of radioactive cesium, potassium copper hexacyanoferrate (KCuHCF) was immobilized in a functionalized cellulose hydrogel. The encapsulation by the cellulose-based hydrogel can solve the applicability problem of KCuHCF caused by sub-micron sized particles, and the superabsorbent property improved the adsorption kinetics. KCuHCF-cellulose hydrogel composites (HCF-gels) exhibited rapid Cs^{+} recovery with 90% removal of 20 ppm Cs^{+} in 1 h. They also exhibited high cesium capacities of 274-308 mg/g, which was attributed to the abundant sorption sites in the carboxymethyl cellulose. Moreover, the HCF-gels were stable and could adsorb cesium ions sufficiently in a wide range of pH (3-10). Remarkably, the HCF-gels could adsorb Cs^{+} ions selectively (>90%) even in seawater condition which contains highly concentrated other cations such as Na^{+} and Mg^{2+}. The excellent cesium removal performance comes from the high accessibility of hydrogel property and the well-dispersed ion-exchanger particles in the composites.

Degradation behavior of palm oil mill effluent in Fenton

Disni Gamaralalage*[,1], Osamu Sawai[2] and Teppei Nunoura[1,2]
[1]Department of Environment Systems, The University of Tokyo, Japan.
[2]Environment Science Center, The University of Tokyo, Japan.
ERSITY OF TOKYO
*2254185932@edu.k.u-tokyo.ac.jp

1. Introduction

- Palm oil mill effluent (POME) is odorous-brownish liquid, identified as the most harmful waste in palm oil production.

Characteristics of POME in mg/L except pH

pH	COD	TOC	TN	TP	TS	TSS	TVSS
4.7	50000	19000	520	510	66300	44000	38300

- Applied biological treatments discharge poor quality effluents.
- Fenton oxidation,
  - ✓ degrades hardly decomposable
  - ✓ requires a least energy
- But, behavior of POME in Fenton is yet to study.

Fenton oxidation: $Fe^{2+} + H_2O_2 \rightarrow OH\cdot + Fe^{3+} + OH^-$ (1)

2. Objective

Study of organic degradation behavior of POME in Fenton process

3. Materials and Methods

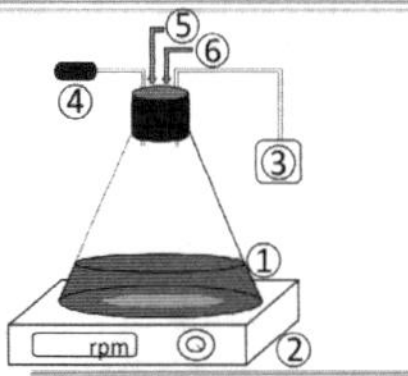

① POME
② Magnetic stirrer
③ Pressure meter
④ Gas sampling point
⑤ Reagents
⑥ pH control

Reaction conditions
- Reagents, $[H_2O_2]_0$: 0.5-3 g/L $[Fe^{2+}]_0$: 0.5 g/L
- pH: 2-5
- Stirring: 500 rpm
- Specimen: 0.2 L
- Head space: 0.39 L
- pH control: H_2SO_4 (18M), NaOH (2M)

Analysis
- TOC and TN: TOC and TN analyzer (TOC-V CPN with TNM-1, SHIMADZU)
- $[H_2O_2]$, $[Fe^{2+}]$ and [Fe]: UV-VIS (UV-1650PC, SHIMADZU)
- Ion analysis: Ion chromatograph (Dx-120, DIONEX) and TN/TP meter (TNP-10, TOA-DKK)
- Compound analysis: HPLC (JASCO) and GC-MS (GCMS-QP 2010, SHIMADZU)
- Gas analysis: GC-TCD (GC-2014, SHIMADZU)

4. Results and Discussion

4.1 Behavior of organics, Fe^{2+} and H_2O_2

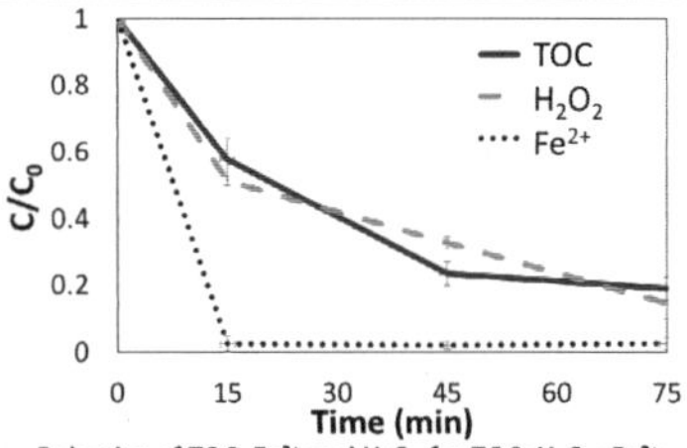

Behavior of TOC, Fe^{2+} and H_2O_2 for TOC: H_2O_2: Fe^{2+} molar ratio of 1:4:0.6 at pH 2-3.

$Fe^{3+} + H_2O_2 \rightarrow HO_2\cdot + Fe^{2+} + H^+$ (2)

$Fe^{2+} + OH\cdot \rightarrow Fe^{3+} + OH^-$ (3)

$OH\cdot + H_2O_2 \rightarrow HO_2\cdot + H_2O$ (4)

$2H_2O_2 \rightarrow O_2 + 2H_2O$ (5)

$OH\cdot + OH\cdot \rightarrow H_2O_2$ (6)

- ➢ TOC degraded faster at the beginning (by $OH\cdot$) and slower afterwards (by $HO_2\cdot$). $OH\cdot$ availability in the system became less by consuming in Eq. 3 and 4.
- ➢ Fe^{2+} decreased faster than its regeneration, due to fast Fe^{2+}-initiated reaction (Eq. 1) and slower Fe^{3+}-initiated reaction (Eq. 2).
- ➢ H_2O_2 reduced gradually compared to Fe^{2+} decrease, due to its regeneration in the system by Eq. 6, despite its participation in Eq. 4 and self-degradation in Eq. 5.

Carbon balance in the system for TOC: H_2O_2: Fe^{2+} molar ratio of 1:1:0.6 with no pH control. (Carbon concentrations are given in mg-C/L)

	TOC	Acetic	Propanoic	Butanoic	Pentanoic	Hexanoic	Phenol	Acetamide	Phthalic	Formic	Others	Inorganic C	CO_2 gas	Total C
0 min	16100	5880	1820	1750	708	319	479	11.3	0	0	5130	1030	0	17100
15 min	10800	4290	0	0	0	0	0	344	5450	718	0	5420	1.44	16200
45 min	8940	3250	0	0	0	0	0	265	4670	222	533	4940	12.1	13900
75 min	8640	3160	0	0	0	0	0	218	3770	146	1350	4840	24.6	13500

- o All the carboxylic acids present in POME (except acetic) were converted into formic and phthalic acids, gaseous CO_2 and inorganic C.
- o Decrease in total C in the liquid fraction indicated formation of solids in the system.

4.2 Behavior of N and P

N and P in the system for TOC: H_2O_2: Fe^{2+} molar ratio of 1:1:0.6 without controlling pH. (N concentrations are in mg-N/L and TP is in mg-P/L)

	TN	NO_2^--N	NO_3^--N	NH_4^+-N	Acetamide	Other N	N_2 gas	TP
0 min	656	81	119	210	6.6	239	0	509
15 min	519	51.9	34.8	9.25	201	222	118	0.5
45 min	460	50.1	36.7	47	155	171	96	0.12
75 min	473	42.1	34.8	135	127	134	30.7	2.67

- N was dropped as 36% in NO_2^--N, 71% in NO_3^--N and 96% in NH_4^+-N at 15 min with forming N_2 gas and acetamide.
- P removed extensively forming insoluble complexes with Fe.

4.4 Effect of $[H_2O_2]$ and optimum conditions

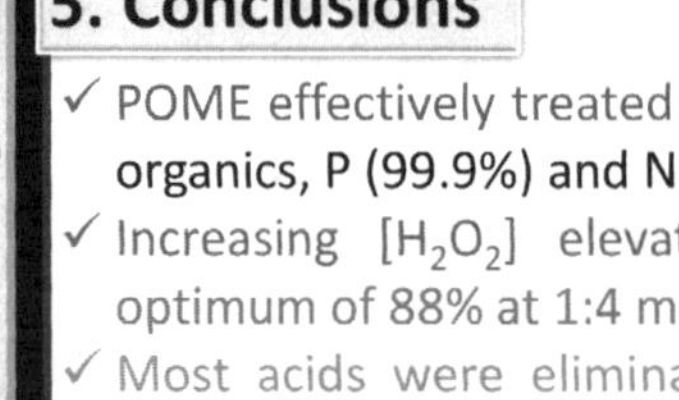

TOC reduction % regarding $[H_2O_2]$ for pH 2-3. (TOC: 190 ppm, $[Fe^{2+}]$: 500 ppm at 0 min)

- ❖ Increasing H_2O_2 increased TOC reduction up to 1:4.
- ❖ Excess H_2O_2, scavenges $OH\cdot$ and less reactive $HO_2\cdot$ produced (Eq. 4). Thus, TOC reduction dropped at higher $[H_2O_2]$.

4.3 Effect of pH

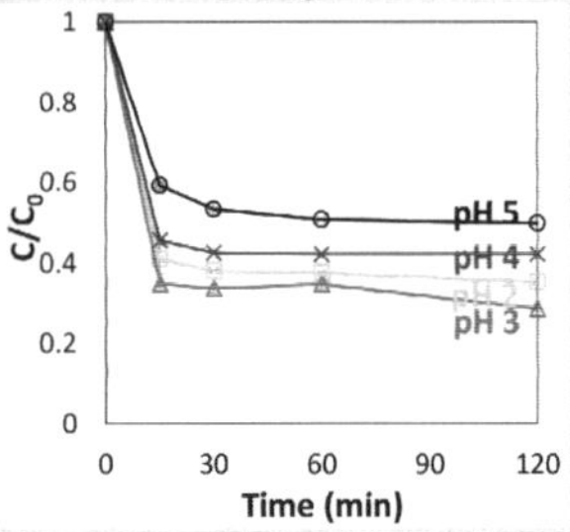

TOC behavior at TOC: H_2O_2: Fe^{2+} molar ratio of 1:2:0.6 at different pH.

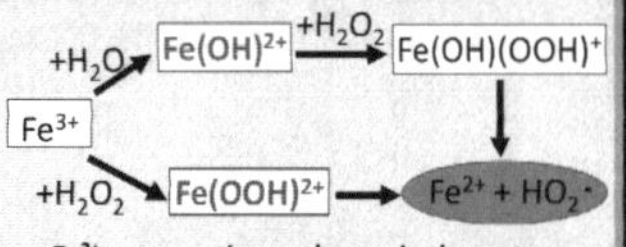

Fe^{2+} regeneration pathways in the system.

- ❑ At pH 2: All Fe are in liquid fraction, but Fe^{2+} forming is not at optimum.
- ❑ At pH 4 and 5: Less Fe in liquid fraction due to precipitating as $Fe(OH)_3$.
- ❑ At pH 3: $Fe(OH)^{2+}$ and $Fe(OOH)^{2+}$ form in optimum concentrations and they contribute for regenerating Fe^{2+}.

5. Conclusions

- ✓ POME effectively treated by Fenton with degrading organics, P (99.9%) and N (21%).
- ✓ Increasing $[H_2O_2]$ elevated TOC reduction with optimum of 88% at 1:4 molar ratio of TOC:H_2O_2.
- ✓ Most acids were eliminated in a short time and highly biodegradable formic, acetic and phthalic acids were formed.
- ✓ Low pH showed high oxidation, optimum-pH 3.
- ✓ Fenton suits as stand-along or pretreatment before biological treatment for POME.

NUNOURA LABORATORY

492267 Removal of Bromine from Tetrabromobisphenol a in an Amine Aqueous Solution Under Hydrothermal Conditions

Abstract:

Tetrabromobisphenol A (TBBPA) is widely employed as one of the common flame-retardants, and the large amounts contained in consumer electronical products have been discharged year by year. The disposal by incineration or landfill results in emission or evolution of toxic compounds produced. Supercritical water was subjected to electrical products containing brominated flame-retardants. While the debromination was effectively occurred, the reactor corrosion may often be observed. Thus, a new effective debromination process is demanded.

In this study, a new debromination process is demonstrated such that a hot compressed water in the presence of amine as a nucleophilic reagent is subjected to a brominated flame-retardant compound such as TBBPA under hydrothermal conditions. Using a small bomb-type batch reactor, TBBPA and a certain amount of an amine aqueous solution were loaded in the reactor, and the reactor was heated in a molten salt bath. After the intended time elapsed, the reactor was removed from the bath, and quickly cooled by water. The products recovered were analyzed by a total organic carbon analyzer, an ion chromatograph, a GC-MS and an HPLC.

Various amine species were found to be effective for the debromination, and methylamine (MA) was the most effective. For, example, 92 % bromine in the ion form in TBBPA with MA was recovered in the product solution at 260 deg C for 25 min while 48% bromine was removed in an aqueous solution of sodium hydroxide at the same condition. According to the analysis of the products the reaction can be considered to proceed via the scission of C-C bonds to form a single aromatic compound along with the debromination. In the presentation debromination of actual electric substrate was also demonstrated in a MA aqueous solution.

494186 Iron/Palladium Nanoparticle Functionalized Membrane System forÂ Chlorinated Contaminates Treatment

Abstract:

Functionalized polyvinylidene fluoride (PVDF) membrane platforms were developed for environmentally benign in-situ nanostructured Fe/Pd synthesis and remediation of chlorinated contaminates (PCB and TCE). To prevent leaching and aggregation, nanoparticle catalysts were integrated into membrane domains functionalized with poly (acrylic acid). Furthermore, nanoparticles of 16-19 nm were observed inside the membrane pores. The quantification of nanoparticles properties (size, distribution and composition) versus depth underneath the membrane surface was also investigated by cross-sectioning the membrane samples with focused ion beam (FIB). The Fe/Pd nanoparticles immobilized membrane showed excellent performance in the degradation of chlorinated organics and also the repeated degradation experiments were conducted to demonstrate the reusability of Fe/Pd immobilized membranes.

To increase the accuracy of kinetic analysis and further simulation of PCB dechlorination, the correlation between nanoparticle properties and depth inside membrane pores was studied by using FIB. Particle size was uniform inside membrane pores at different depths (particle size: 16.7±0.7 nm) but slightly smaller than those nanoparticles located on the surface (19.4±3.2 nm). Furthermore, the fairly similar surface coverage values (0.169±0.011) demonstrated the evenly distributed nanoparticles inside the membrane pores. Besides, the element composition of particles as well as the functionalized membrane domain were analyzed in line-scan mode by using energy dispersive spectroscopy (EDS).

Over 96% degradation of 3,3',4,4',5-pentachlorobiphenyl was achieved at a residence time of 14.7 seconds in the membrane pores. The ksa was calculated (using LFR model) to be 0.171 L/(m^2h) in convective flow mode, which is 2.5 times the rate obtained in batch mode. The effects of temperature and pH values were also investigated.

In XRD analysis, the Fe/Pd particle samples (which were deliberately oxidized and then reduced) exhibited the same crystalline patterns as the original samples. The membrane was tested for reactivity after four degradation cycles with regeneration between each cycle. The increase of surface particle size of 22% resulted in a decrease of 9.7% PCB conversion for the 4 hr reaction time. [1]

This research is supported by the NIEHS-SRP grant P42ES007380.

Reference:

[1] Wan, Hongyi, L. Ormsbee, and D. Bhattacharyya. "Pore functionalized PVDF membranes with in-situ synthesized metal nanoparticles: Material characterization, and toxic organic degradation." Journal of Membrane Science 530 (2017): 147-157.

495498 Development of Challenging Technology on Combustion Hindrances in Commercial Solid Refuse Fuel Combustion Facility

Abstract:

Solid refuse fuel (SRF) is a source of fuel for the generation of steam for power, heating, industrial processing or other uses. To operate SRF combustion plants consistently and reliably, due to harsh conditions, it is essential to solve the various problems such as de-fluidization of mixture of feed and bed materials, fuel-feed failure, reduction of efficiency with temperature drop, degradation of equipment (erosion and corrosion) caused by high temperature and deposit of salt and ash. In this study, we have developed a core technology that can solve various combustion failure phenomena raised in the commercial scale combustion process (waste to energy) having a capacity of more than 50 t/day level. In order to provide stable air flow during steady-state operation and to remove the bottom ash generated during combustion, the design technology of air distributor of the combustion furnace was developed. Chemical additives for the prevention of corrosion caused by clinker formation during SRF burning were applied and the results were analyzed. Furthermore, to investigate the effects of high temperature and corrosive condition on the tube degradation in SRF combustion plants, the tube corrosion test unit was developed and the degree of tube corrosion was examined by varying the operating condition.

495751 Degradation of Organic –Contaminants from Wastewater By Photocatalytic-Methods Via TiO_2 Thin Films and Simultaneous Production of Hydrogen-Preliminary Results

Abstract:

The new focus on the Energy-Water Nexus is the ability to generate energy from wastewater treatment. In particular, contamination from agriculture extensive use of herbicides and pesticides lead to an important contamination of water supply. A four- step technology is being developed to recycle contaminated water and produce hydrogen as potential use in fuel cells. The technology is based on the use of TiO_2 thin films to replace the current costly use of nanoparticles in batch-designed systems that requires elimination of the particles by filtration. The four steps involved in the technology include a- Manufacturing and characterization of the thin films, b-Testing the production of Hydrogen and comparison with other techniques, C-Testing the elimination of agricultural-based contaminants (such as herbicides) and d-Testing the combination of step b and c to simultaneously produce hydrogen and decontaminate water. The production of H_2 via the thin films has been achieve based on pure water. This work is focuses on the use of advanced oxidation techniques, specifically the use of photo-catalytic processes via TiO_2 nanoparticles to degrade herbicides, such as.... (part C). The reactor equipped with UV-lamp and the use the TiO_2 nanoparticles , has been applied to the degradation of contaminants via UV-photocatalytic methods. This research is centered on the reduction kinetics of the pollutants by using UV-photocatalytic processes suitable for the potential *simultaneous* production of H_2 (g) from wastewater. This will be the next effort in the investigation.

496165 $Li_4Ti_5O_{12}$ Pouch Cell Battery System for Selective Lithium Recovery from Aqueous Resources

Abstract:

Lithium shortage in the future has been predicted due to the surge on its demand especially for its use in electric vehicles (EVs). Consequently, recovery of lithium (Li) from other sources such as seawater, brine, and Li containing wastewater has been the center of interest in the energy sector. Discussions are directed towards aqueous sources to potentially alleviate the impending crisis. Ion-exchange using inorganic compounds and electrochemical processes are extensively investigated for Li recovery. However, ion exchange method has significant drawbacks such as slow kinetics and instability of materials. On the other hand, electrochemical Li recovery is challenged with its energy related requirements. However, this method is environmentally benign, has higher production rate and more amenable to scale-up for industrial applications.

Electrochemical Li recovery from aqueous solutions is an attractive alternative to adsorption methods. An electrochemical system that suppresses the high energy requirement and offer high purity Li compounds at a shorter amount of time is necessary. Herein, a modified $Li_4Ti_5O_{12}$ pouch cell battery system for electrochemical lithium recovery is presented. This hybrid reactor is composed of an anode in an organic electrolyte and Li-rich catholyte with Titanium foil or a porous carbon-based electrode as current collector. The two systems were integrated through an anode pouch cell assembly. It contained a lithium superionic conductor (LISICON) solid electrolyte in the open-structured top to separate the organic electrolyte from the aqueous electrolyte. Essentially, there were two reaction cells, one for Li capture and other for Li recovery. In Li capture cell, charging reaction of the system was the driving force to allow lithium ion (Li^+) intercalation into the electrode inside the pouch cell. The Li-rich electrolyte was replaced with a suitable recovery solution to proceed with the Li^+ release in a new cell. In the recovery cell, O_2 (from air) passing through the cathode reacted with Li^+ to form LiOH. On the other hand, if Li_2CO_3 is desired as the end-product, LiOH can be reacted with CO_2 to form the carbonate product.

Repetitive cyclic voltammetry and galvanostatic cycling showed highly selective and stable electrochemical system for aqueous Li^+ uptake and release. Overall, the hybrid electrochemical reactor is capable of selectively capturing Li^+ at a fast rate and has a potential for high energy generation during the recovery process. Hence, this yielded an electrochemical Li recovery system that is environmentally benign and less energy intensive.

This research was supported by NRF funded by the Ministry of Science, ICT and future Planning (2017R1A2B2002109 and 2015R1A2A1A15055407) and the Ministry of Education (2009-0093816).

497036 Sequential Use of UV/H_2O_2â€"(PSF/TiO_2/MWCNT) Mixed Matrix Membranes for Dye Removal in Water Purification: Membrane Permeation, Fouling, Rejection, and Decolorization

Abstract:

Performance of a hybrid ultraviolet/hydrogen peroxide (UV/H_2O_2)â€"mixed matrix membrane system for an azo dye, acid black 1 [AB1], removal in a water purification process was studied. Different mixed matrix membranes embedded with titanium dioxide (TiO_2) nanoparticles, multi-walled carbon nanotubes (MWCNTs), and a mixture of them were fabricated by the phase inversion method. Mixed matrix membranes embedded with MWCNTs resulted in higher pure water flux, and mixed matrix membranes embedded with TiO_2 showed lower flux declines in the presence of AB1. However, all the membranes exhibited very low total organic carbon (TOC) rejection and none of the mixed matrix membranes could decolorize the AB1 solution. UV/H_2O_2 pretreatment of the AB1 solution resulted in enhanced TOC rejection, decolorization, and enhanced antifouling membrane behavior. Combining UV/H_2O_2 with each type of polysulfone (PSF) mixed matrix membranes (PSF/TiO_2, PSF/MWCNT, and PSF/TiO_2/MWCNT) resulted in optimal performance in terms of permeation, flux decline, antifouling, rejection, and decolorization. The hybrid process of UV/H_2O_2-PSF/TiO_2/MWCNT mixed matrix membrane resulted in 270 (L/[m^2.h]) permeation, 29% flux decline, 90% TOC rejection, 99% decolorization, and 99% flux recovery ratio (FRR%).

497769 First-Principles Assessment of Carbon Dioxide (CO_2) Capture Mechanisms in Aqueous Piperazine (PZ) Solution

Abstract:

Energy-efficient and low-cost methods for carbon dioxide (CO_2) capture from flue gases have been sought to curb greenhouse gas emissions while fossil fuels will likely remain a dominant energy source in the next few decades.[1,2] A wet-scrubbing approach using amines has been considered the most promising short-term solution.[3] Its widespread implementation is limited by the high cost associated with the high parasitic energy consumption during solvent regeneration, along with amine degradation and corrosion problems.[4–6] Aqueous piperazine (PZ), a cyclic diamine with two secondary amine groups, has been proposed as an effective solvent for CO_2 capture due to its many advantages such as low regeneration energy, low rates of thermal and oxidative degradation, and low corrosivity relative to other widely used amines including MEA.[7–9] In addition, PZ exhibits a relatively high absorption rate.

Here, we attempt to elucidate molecular mechanisms underlying the reaction of CO_2 with aqueous PZ using combined ab initio and classical force field calculations, including the roles of PZ carbamate ($PZCOO^-$) and protonated PZ (PZH^+).[10] From AIMD simulations combined with RDF analysis, we confirmed that PZ and $PZCOO^-$ may be protonated, while protonation of PZH^+ hardly occurs. As the available basic sites are well-linked via the hydrogen-bonded water network, the relative concentrations of protonated/deprotonated amine species can be predicted from their relative basicities. We also find the probability of protonated PZ carbamate (H^+PZCOO^-) formation to increase with CO_2 loading. This is not only because of the greater availability of PZ relative to $PZCOO^-$, but also because of the thermodynamic favorability of proton transfer from PZH^+ to $PZCOO^-$ (i.e., $PZH^+ + PZCOO^- \rightarrow PZ + H^+PZCOO^-$).

Regarding the tendency of PZH^+ and $PZCOO^-$ to react with CO_2, we found that while $PZCOO^-$ can easily react with CO_2 forming dicarbamate (COO^-PZCOO^-), the direct CO_2 reaction with PZH^+ hardly occurs. Rather, our AIMD simulations consistently demonstrate that the proton is first released from PZH^+, followed by the PZ + CO_2 reaction forming $PZCOO^-$, which can then easily abstract a proton to form H^+PZCOO^-. This behavior has largely been attributed to the low pKa of PZH^+. However, our spatial distribution analysis using classical MD simulations suggests that H_2O tends to more densely pack around PZH^+, which may hinder CO_2 accessibility relative to PZ/$PZCOO^-$, further suppressing the $PZH^+ + CO_2$ reaction.[10]

While CO_2 capture by $PZCOO^-$ appears to be kinetically possible, our free energy calculations predict COO^-PZCOO^- formation to be highly unlikely at thermodynamic equilibrium relative to the monocarbamates.[10] However, the formation of COO⁻PZCOO⁻ at high CO_2 loadings has been reported from nuclear magnetic resonance (NMR) measurements, albeit in small concentrations.[11,12] Analysis of the distributions of PZ/PZCOO⁻/COO⁻PZCOO⁻ near the gas-liquid surface shows that although PZ accumulates at the gas-solvent interface, PZCOO⁻ also remains near the surface.[10] This hints that while PZ may predominantly capture CO_2, the PZCOO⁻ + CO_2 reaction to form COO⁻PZCOO⁻ may be more likely to occur with increasing CO_2 loading as PZ is depleted. Based on our results, we speculate that COO⁻PZCOO⁻ is mostly produced from the PZCOO⁻ + CO_2 reaction near the gas-solvent interface, rather than interconversion with the monocarbamates (PZCOO⁻/H^+PZCOO⁻) in bulk solution.

The results of this study provide important insights into fundamental CO_2 absorption mechanisms in aqueous PZ solution. The improved mechanistic understanding may aid in the development of better kinetic and thermodynamic models to predict and optimize the performance of PZ in concentrated solution and as a rate-promoter in PZ-blends. Furthermore, it may provide valuable guidance on how to mitigate the disadvantages of PZ as well as in the design of new PZ-containing mixed solvents.

References

1. Kenarsari SD, Yang D, Jiang G, et al. Review of recent advances in carbon dioxide separation and capture. *RSC Adv*. 2013;3(45):22739-22773.

2. Figueroa JD, Fout T, Plasynski S, McIlvried H, Srivastava RD. Advances in CO_2 capture technology—The U.S. Department of Energy's Carbon Sequestration Program. *Int J Greenh Gas Control*. 2008;2(1):9-20.

3. Rochelle GT. Amine scrubbing for CO_2 capture. *Science*. 2009;325(5948):1652-4.

4. D'Alessandro DM, Smit B, Long JR. Carbon Dioxide Capture: Prospects for New Materials. *Angew Chem Int Ed Engl*. 2010;49(35):6058-82.

5. Strazisar BR, Anderson RR, White CM. Degradation Pathways for Monoethanolamine in a CO_2 Capture Facility. *Energy and Fuels*. 2003;17(4):1034-1039.

6. Abu-Zahra MRM, Schneiders LHJ, Niederer JPM, Feron PHM, Versteeg GF. CO_2 capture from power plants. Part I. A parametric study of the technical performance based on monoethanolamine. *Int J Greenh Gas Control*. 2007;1(1):37-46.

7. Freeman SA, Dugas R, Van Wagener DH, Nguyen T, Rochelle GT. Carbon dioxide capture with concentrated, aqueous piperazine. *Int J Greenh Gas Control*. 2010;4(2):119-124.

8. Rochelle G, Chen E, Freeman S, Van Wagener D, Xu Q, Voice A. Aqueous piperazine as the new standard for CO2 capture technology. *Chem Eng J*. 2011;171(3):725-733.

9. Zheng L, Landon J, Matin N, Li Z, Qi G, Liu K. Corrosion Behavior of Carbon Steel in Piperazine Solutions for Post-combustion CO_2 Capture. *ECS Trans*. 2014;61(20):81-95.

10. Stowe H, Paek E, Hwang GS. First-principles assessment of CO_2 capture mechanisms in aqueous piperazine solution. *Phys Chem Chem Phys*. 2016;18:25296-25307.

11. Ermatchkov V, Pérez-Salado Kamps Á, Maurer G. Chemical equilibrium constants for the formation of carbamates in (carbon dioxide+piperazine+water) from H-NMR-spectroscopy. *J Chem Thermodyn*. 2003;35(8):1277-1289.

12. Bishnoi S, Rochelle GT. Absorption of carbon dioxide into aqueous piperazine: reaction kinetics, mass transfer and solubility. *Chem Eng Sci*. 2000;55(22):5531-5543.

498104 Nascent Soot Formation By Agglomeration and Surface Growth

Abstract:

Nascent Soot Formation by Agglomeration and Surface Growth

Georgios A. Kelesidis, Eirini Goudeli, Sotiris E. Pratsinis

Particle Technology Laboratory, Institute of Process Engineering,

Department of Mechanical & Process Engineering, ETH Zurich, Switzerland

Nascent soot particles of mobility size 1 â€“ 10 nm are the building blocks of diesel soot, a major environmental pollutant, but also carbon black, a valuable commodity material used in reinforcing tires and other industrial rubber products. Recently, however, major concerns have been raised, since microscopy and mass-mobility measurements have revealed the existence of ultrafine aggregates during nascent soot formation. So, their impact on climate, health and nanomaterials manufacturing needs to be determined accurately.

Here, Discrete Element Model (DEM) simulations are used to investigate the dynamics of nascent soot particle growth after nucleation accounting for simultaneous surface growth and agglomeration in the absence of soot oxidation (Kelesidis et al., 2017). The model is validated with theoretical expressions for pure agglomeration (Goudeli et al., 2015) as well as with surface growth with and without coagulation at full coalescence. The evolution of nascent soot structure is benchmarked against previous numerical studies.

Nascent soot growth by agglomeration with or without acetylene molecule reaction (pyrolysis) is compared to that by full coalescence. Neglecting the non-spherical or fractal-like nature of soot underestimates its aggregate polydispersity up to 40 %. The DEM-derived size distributions of soot growing by agglomeration with surface growth are in good agreement with microscopy (Schenk et al., 2013) and mass-mobility measurements (Camacho et al., 2015) of nascent soot produced in ethylene flames, indicating that surface growth narrows down soot size distributions at low heights above the burner, consistent with literature. The evolution of nascent soot structure from spheres to aggregates quantified by the mass fractal dimension and mass-mobility exponent is also in excellent agreement with experiments (Schenk et al., 2013; Camacho et al., 2015). The effect of soot volume fraction on nascent soot morphology is elucidated. Based on the aggregate projected area, a scaling law is derived for the number and size of the nascent soot primary particles from mass-mobility measurements.

References:

Camacho, J., Liu, C., Gu, C., Lin, H., Huang, Z., Tang, Q., You, X., Saggese, C., Li, Y., Jung, H., Deng, L., Wlokas, I., Wang, H. (2015) *Combust. Flame* **162**, 3810-3822.

Kelesidis, G.A., Goudeli, E., Pratsinis, S.E. (2017) *Proc. Combust. Inst.* **36**, 29-50.

Goudeli, E., Eggersdorfer, M.L. and Pratsinis, S.E. (2015) *Langmuir* **31**, 1320-1327.

Schenk, M., Lieb, S., Vieker, H., Beyer, A., Golzhauser, A., Wang, H., Kohse-Hoinghaus, K. (2013) *PhysChemPhys* **14**, 3248-3254.

499612 N-Isopropylacrylamide (NIPAAm) Based Thermal Responsive Composites for Polychlorinated Biphenyls (PCBs) Removal from Water

Abstract:

Thermally responsive hydrogel based sorbents have gained great attention for environmental remediation, specifically in water treatment, due to their high adsorption capacities and response to external environment change. N-isopropylacrylamide (NIPAAm) is one of the most widely studied thermo-responsive materials, which undergoes reversible phase transition at its lower critical solution temperature (LCST) around 32Â°C. NIPAAm-based thermally responsive materials can be synthesized with various functionalities, which can provide specific interactions with target environment pollutants (e.g., organic dyes, PCBs). Our groupâ€™s recent efforts have focused on applying naturally derived polyphenols to develop materials with binding affinities to such pollutants. In this work, multiple types of NIPAAm-based thermally responsive hydrogels and core-shell magnetic nanoparticles (MNPs) were synthesized and characterized. Their ability to remove specific PCB congeners at both room temperature and elevated temperature was evaluated, and PCB 126 was chosen to serve as a model coplanar congener. We hypothesize that the temperature responsive nature of NIPAAm based composites will facilitate the binding activity through swelling changes around their LCST with contaminants, such as PCBs, captured within their network.

BIODEGRADATIÓN OF CHLORACETANILIDES IN WATER CONTAMINATED USING BACTERIAL MESOPHILIC CONSORTIUMS

17AIChE Annual Meeting, Minneapolis, MN

Boris Guzmán Martínez, Esp; Jose Javier Castro Arellano, PhD;
[1]Laboratorio de Operaciones Unitarias, SEPI-ESIQIE INSTITUTO POLITÉCNICO NACIONAL

Abstract

In this work we present the biodegradation of chloroacetanilides: alachlor (2-chloro-N- (2,6-diethylphenyl) -N- (methoxymethyl) acetamide) and metolachlor (2-chloro-2 ', 6'-diethyl-methoxymethylacetanilide) ; and chlorinated carboxyamides; These herbicides are pre-emergent and the most restricted in current legislation, and unfortunately the most used in the world. The environmental problem with chlorinated xenobiotics is that they accumulate both in soils and in water, especially in lentic waters, and due to their effects on health, being endocrine and mutagenic disruptors, they are normatively located within the priority emerging pollutants to be degraded . For this reason, there is a need to look for alternatives for the degradation of these compounds among which, according to the open bibliography, physicochemical and microbiological degradation processes were obtained. The investigation was carried out on a laboratory scale, using bacterial aggregates to treat polluted oligotrophic lentic water (350 ppm Organochloride). From the results found, possible degradation mechanisms used by these consortiums are proposed and the mineralization of the chemical agent is verified. The selection of the bacterial consortia was obtained (between 60 through PCR RNA 16s / Proteomics). The main participating bacterial groups were identified that showed competitive synergistic properties in biodegradation, and in comparison with those reported in the open literature in similar systems (Degradation greater than 95% after 10 days), Ecotoxicity tests were carried out under international standards (Vibrio fisherii) that validate the remediation achieved, results that will serve as a basis for the development of the solution to the problem posed. Keywords: Consortium, Xenobiotic, Vibrio, COD, TOC

Introduction

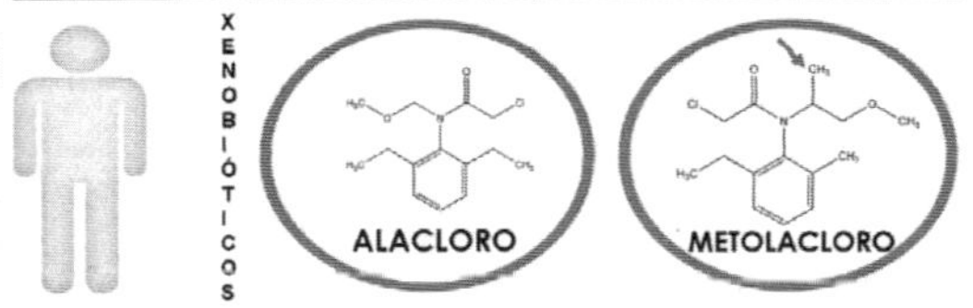

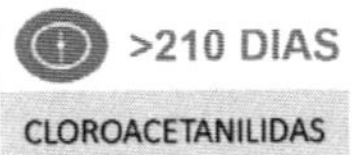

Item	Alachlor	Metolachlor
Numero CAS	15972-60-8	51218-45-2
Peso Molecular	269.77	283.80
Formula Molecular	$C_{14}H_{20}ClNO_2$	$C_{15}H_{22}ClNO_2$
Presión Vapor (mPa) a 25°C	1.86	1.73
Constante de Henry	0.131	3.7 10^-7
Solubilidad (mg/l)	242 at 25°C	530 at 20°C
Kow	2.9	3.13–3.45
Rata de Aplicación	2.8–3.4	1.7–3.4
	3.3–1.6	0.8–1.6

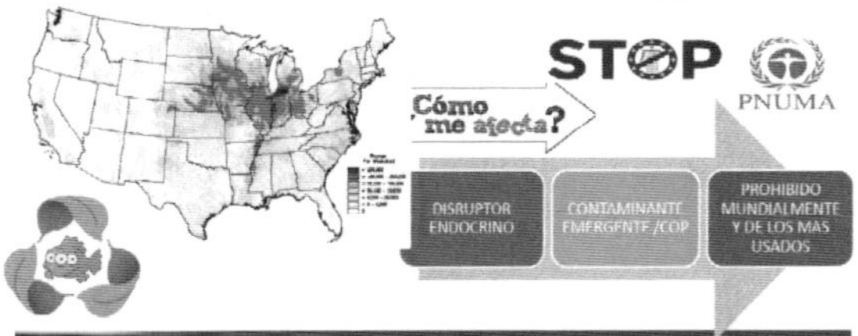

Materiales y métodos

The research was developed under the following methodology:

Selection of the consortium among 60 bacterial aggregates of known remediation, based on the results of adaptation and growth, diversity analysis of the microorganisms present, with McFarland techniques, biochemical tests, MET-cryogenics, and proteomics of the two selected consortiums (MB-PX)

Characterization of initial microbiological, physicochemical, instrumental, environmental and toxicological conditions of water and chlorinated xenobiotic (use of analytical techniques to monitor OD, N, PO3, CO2, pH, COD, TOC, as well as CLAR analysis, GC- MS and UV-vis spectroscopy, FTIR -ATR). From the analysis of results, routes are proposed as degradation mechanisms used by said consortiums and the mineralization of the chemical agent is checked.

Characterization of the end products of degradation and validation of international ecotoxicity of the remediation process (Vibrio fisherii) for endocrine disruption. for the final degradation products under ISO 11348-2.

Tabla 1. Initial Water Characterization

SOLUCION CON ALACLORO (300 ppm)

2-cloro-2',6'-dietil-N-metoximetilacetanilida

No. Lote SZBD163XV

Parámetro	Unidad	Valor
pH	[H+]	7.36
Dissolved oxygen	mg/L	8.2
Chlorides	mg/L	56
Calcium hardness	mg/L	35
Dureza de magnesio	mg/L	15
Hardness	mg/L	50
Conductivity	µS	166
Oxygen Chemical Demand	mg/L	560
Nitrógeno total	mg/L	ND
Total Phosphorus	mg/L	231
Total Solids	mg/L	12.4
Volatile Suspended Solids	mg/L	5.4
Total Organic Carbon	mg/L	260
Temperature	°C	21-22

Tabla 2. . GC-MS Experimental Monitoring of Alachlor

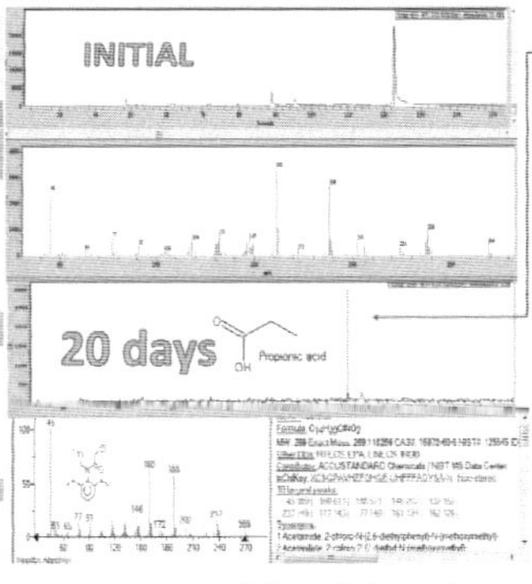

Results

Prueba bioquimica	Resultado
Tinción de Gram	Gram Negativa
Test de hidróxido de potasio	Positivo
Catalasa	Positivo
Relación con el oxígeno	Aeróbica
Metabolismo oxid-ferm de la glucosa	Oxidativo
Hidrólisis de gelatina	Positivo
Hidrólisis de almidón	Positivo
Hidrólisis de Tween	Positivo (lento)
Crecimiento en NaCl 2%	Positivo
Crecimiento en NaCl 5%	Positivo
Fluorescencia en medio King B	Negativo
Crecimiento en medio SX	Positivo

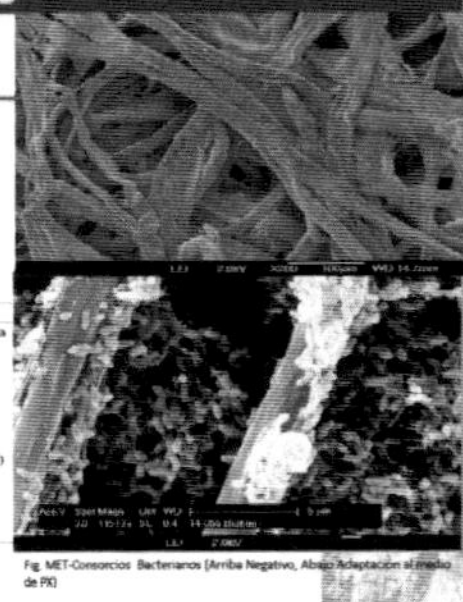

Fig. MET-Consorcios Bacterianos (Arriba Negativo, Abajo Adaptación al medio de PX)

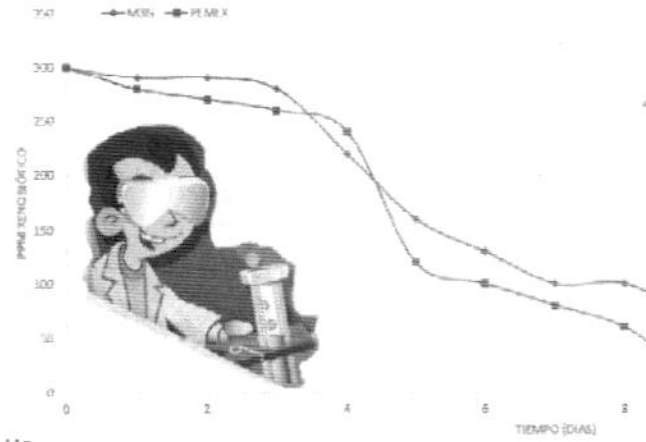

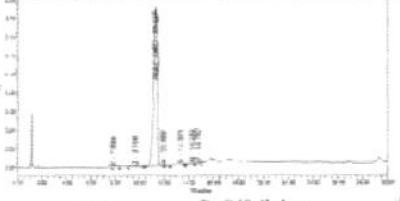

Fig. CLAR Alacloro

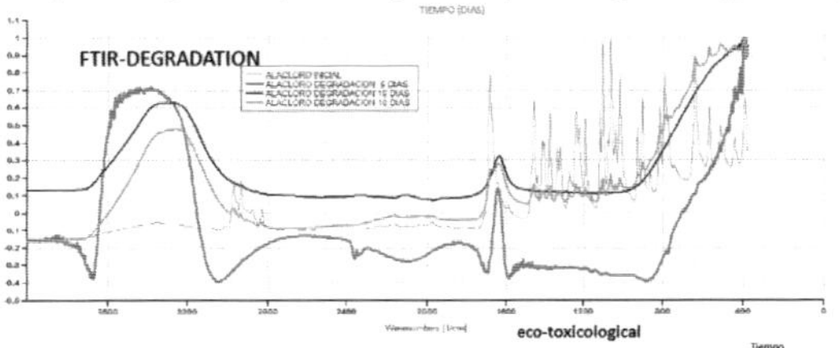

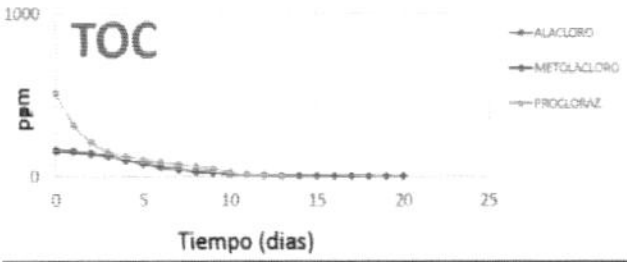

eco-toxicological

ITEM	COMPUESTO ANALIZADO	Tiempo 5'	15'	30
1	ALACHLOR	8	5	4
2	METOLACHLOR	5	3	1
3	PROCLORAZ	11	7	3
4	ALACLORO 20 DIAS	[illegible]	[illegible]	[illegible]
5	METOLACLORO 20 DIAS	[illegible]	[illegible]	[illegible]
6	PROCLORAZ 20 DIAS	115	107	[illegible]
7	MEDIO ACTIVACION BASICO	120	113	108
8	MEDIO ENRIQUECIDO	116	108	88

Degradative / Final Products

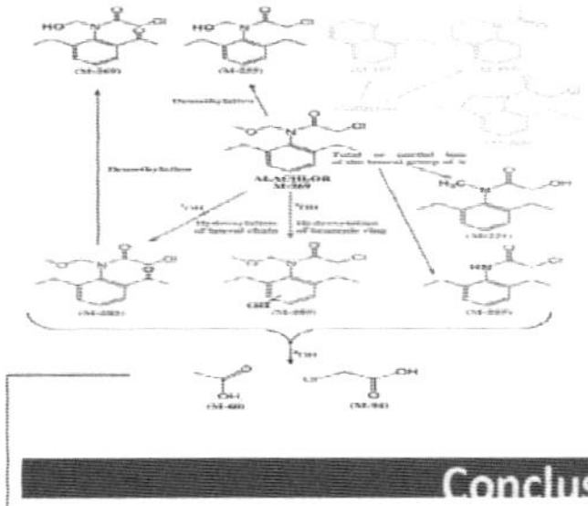

Parámetro	Unidad	Valor
pH	[H+]	6.82
Dissolved oxygen	mg/L	3.0
Chlorides	mg/L	16
Calcium hardness	mg/L	14
Dureza de magnesio	mg/L	12
Hardness	mg/L	28
Conductivity	µS	45
Oxygen Chemical Demand	mg/L	60
Nitrógeno total	mg/L	40
Total Phosphorus	mg/L	5
Total Solids	mg/L	55.9
Volatile Suspended Solids	mg/L	
Total Organic Carbon	mg/L	2
Temperature	°C	23

Conclusions

- The consortiums (XB and PX) were selected from the 60 bacterial aggregates for the degradative development of the chloroacetanilides and chlorinated carboxyamides
- Bacterial species constituting the XB and PX consortium were identified by proteomic and biochemical tests
- The physicochemical characteristics of water for the application of biodegradation protocols are known.
- Removal was achieved> 98% of carboxyamide, 99% of metachlore and 99% of alachlor
- The comparative analysis showed that at 20 days the XB consortium showed a higher percentage of chloroacetanilide removal.
- The uncertainties of the experimental data were determined by the order UC = 0.295 with a confidence level of 95%, K = 2, Expanded uncertainty (U) = 0.59%
- It was possible to study the behavior and the influence of physical-chemical and biological variables that intervene in the process of decontamination and adaptation of the consortium, which will serve as a basis in the design of process processes.
- The concentration of alachlor, metolachlor, prochloraz and metabolites, biomass concentration, was determined throughout the biodegradation process.
- Possible mechanisms of action of the bacterial consortiums and the possible enzymes that participated in the degradative process were found
- Toxicity tests were applied as a mechanism for evaluating the bioremediation process and it was found that for the 3 treated systems, bioremediation is satisfactorily developed

Acknowledgment

IMTA Instituto Mexicano de Tecnología del Agua
CONACYT

References

- Bohuss, I., Rékasi, T., Szikora, S., Barkacs, K., Zaray, G., & Acs, E. (2005). Interaction of acetochlor and atrazine with natural 3 Page 12 of 14 Water Air Soil Pollut (2015) 226: 3 freshwater biofilms grown on polycarbonate substrate in lake Velence (Hungary). Microchemical Journal, 79, 201–205
- Bonet, B., Corcoll, N., Tlili, A., Morin, S., & Guasch, H. (2014).Antioxidant enzyme activities in biofilms as biomarker pollution in a natural system: an active bio-monitoring study.Ecotoxicology Environmental Safety, 103, 82–90
- Dehghani, M., Nasseri, S., & Zamanian, Z. (2013). Biodegradation of alachlor in liquid and soil cultures under variable carbon and nitrogen sources by bacterial consortium isolated from corn field soil. Iran Journal of Environmental Health Science, 10, 21–28.
- Sanchis, S., Polo, A. M., Tobajas, M., Rodriguez, J. J., & Mohedano, A. F. (2013). Strategies to evaluate biodegradability: application to chlorinated herbicides. Environmental Science and Pollution Research. doi:10.1007/s11356-013-2130-y

501344 Colloidal Transport in a Microfluidic Porous Medium with Surface Charge Heterogeneity

Abstract:

Colloidal facilitated transport of low-solubility contaminants in ground water flow has become a major concern recently, since contaminants adsorbed onto natural colloidal particles can travel kilometers over years. Therefore, understanding the behavior of those colloidal particles would provide insights for environmental protection and remediation. In the past most lab-scale experiments focused on studying interactions between particles and homogeneous or geometrically heterogeneous porous media. Hardly any of them tried to decouple the different chemical properties of the porous media grain. In order to shed some light in this aspect, we have fabricated a microfluidic device by packing 10 micron polystyrene (PS) beads with opposite surface charges into a pseudo-three-dimensional porous medium. By injecting 0.5 microns PS particles through the fabricated device, we further study the effect of chemical heterogeneity on the transport and deposition of colloidal particles. The porous medium assembled contains more than 3000 beads and has a size of 500 μm in length, 600 μm in width and 15 μm in height. An on-chip cross channel linked with an off-chip 4-way valve is applied to generate a step input of concentrated colloidal suspension and minimizes the influence of colloidal dispersion before they reach the porous medium. The breakthrough of the particles is recorded at the downstream of the porous medium. The colloidal retention profile is mapped using confocal microscopy. Since beads with different surface charges can be distinguished by fluorescence, we obtain colloidal retention data at the individual grain. With our platform, we aim at developing a relation between the overall collector efficiency and its constitutive collector efficiency under different salt conditions.

501529 Magnetic Nanocomposite Materials As Reusable Adsorbents for Chlorinated Organics in Contaminated Water

Abstract:

The constant growth in population worldwide over the past decades continues to put forward the need to provide access to safe, clean water to meet human needs. There is a need for cost-effective technologies for water and wastewater treatment that can meet the global demands and the rigorous water quality standards and at the same maximizing pollutant efficiency removal. Current remediation technologies have failed in keeping up with these factors without becoming cost-prohibitive. Nanotechnology has recently been sought as a promising option to achieve these goals. The use of iron oxide magnetic nanoparticles as nanoadsorbents has led to a new class of magnetic separation strategies for water treatment. We have developed magnetic nanocomposite systems able to capture polychlorinated biphenyls (PCBs), as model organic pollutants, in aqueous solution, providing a cost-effective water remediation technique. Two distinct methods were employed to develop these polyphenolic nanocomposite materials. The polyphenolic moieties were incorporated to create high affinity binding sites for organic pollutants within the nanocomposites. The first method utilized a surface initiated polymerization of polyphenolic-based crosslinkers and co-monomers on the surface of iron oxide magnetic nanoparticles to create a core-shell nanocomposite. The second method utilized a bulk polymerization method to create macroscale films composed of iron oxide nanoparticles incorporated into a polyphenolic-based polymer matrix, which were then were processed into microparticles. Both methods produce nanocomposite materials that can bind chlorinated organics, can rapidly separate bound organics from contaminated water sources using magnetic decantation, and can use thermal destabilization of the polymer matrix for contaminant release and material regeneration. The polyphenol functionalities used to bind organic pollutants were quercetin multiacrylate (QMA) and curcumin multiacrylate (CMA), which are acrylated forms of the nutrient polyphenols with expected affinity for chlorinated organics. Particles were characterized using transmission electron microscopy (TEM), dynamic light scattering (DLS), Fourier transform infrared spectroscopy (FTIR), thermal gravimetric analysis (TGA) and UV-visible spectroscopy stability analysis. Pollutant binding studies were performed using PCB 126 as a model PCB and chlorinated organic pollutant to determine binding affinity and capacity, and this was quantified using gas chromatography coupled to electron capture detector (GC-ECD). It was demonstrated that the materials effectively bound PCBs in aqueous media.

501552 Colloidal Transport in a Surface Charge Heterogeneous Microfluidic Porous Medium

Abstract:

Colloidal facilitated transport of low-solubility contaminants in ground water flow has become a major concern recently, since contaminants adsorbed onto natural colloidal particles can travel kilometers over years. Therefore, understanding the behavior of those colloidal particles would provide insights for environmental protection and remediation. In the past most lab-scale experiments focused on studying interactions between particles and homogeneous or geometrically heterogeneous porous media. Hardly any of them tried to decouple the different chemical properties of the porous media grain. In order to shed some light in this aspect, we have fabricated a microfluidic device by packing 10 micron polystyrene (PS) beads with opposite surface charges into a pseudo-three-dimensional porous medium. By injecting 0.5 microns PS particles through the fabricated device, we further study the effect of chemical heterogeneity on the transport and deposition of colloidal particles. The porous medium assembled contains more than 3000 beads and has a size of 500 µm in length, 600 µm in width and 15 µm in height. An on-chip cross channel linked with an off-chip 4-way valve is applied to generate a step input of concentrated colloidal suspension and minimizes the influence of colloidal dispersion before they reach the porous medium. The breakthrough of the particles is recorded at the downstream of the porous medium. The colloidal retention profile is mapped using confocal microscopy. Since beads with different surface charges can be distinguished by fluorescence, we obtain colloidal retention data at the individual grain. With our platform, we aim at developing a relation between the overall collector efficiency and its constitutive collector efficiency under different salt conditions.

501714 Elimination of Organic Compounds in Liquid Effluents Using Mexican Natural Zeolite Impregnated with Fe By the Photo-Fenton Process

Abstract:

Abstract

In the present work we present a study of the elimination of the herbicide Glyphosate by the photo-Fenton process using Mexican natural zeolite impregnated with Fe. Because this herbicide is widely used in agriculture for the removal of grasses and shrubs in maize crops. However, because of their possible carcinogenic effect for humans, alternatives should be sought for their disposal of liquid effluents.

The impregnation of Fe on the natural zeolite in acid solution at pH 2.5, using FeCl3 as iron source was performed. The impregnated mineral was analyzed by X-ray photoelectron spectroscopy (XPS) and transmission electron microscopy (TEM), in order to observe the present iron species, because the natural zeolite contains in its structure compounds with iron before being impregnated.

Finally, the impregnated natural zeolite was used in the degradation of the herbicide in solution by means of the photo Fenton process, which searches to improve the elimination of Glyphosate and to obtain byproducts of less environmental impact.

501797 Pretreatment of Solid Wastes from Vegetable Processing for Biofuel Production

Abstract:

Solid vegetable wastes, consisting of mainly onions, were subjected to acid and alkali pretreatment for biofuel production through microbial processes (i.e. aerobic and anaerobic digestion). RSM design was applied to determine optimum conditions (maximum sugar yield, minimum inhibitor yield) for the pretreatments. Preliminary results indicated that optimum yields can be obtained at 3.5 wt.% HCl and 50C, and 7% KOH and 100C for acid and alkali pretreatment, respectively.

496929 Determining the Structure of Hydrothermal Char and Its Effect on Adsorption Capacity

Abstract:

Hydrothermal chars (hydrochars) are a novel carbon material made from the thermal degradation of biomass under liquid water, at moderate temperature and pressure. The moderate conditions and ability to take wet-biomass provide the advantage of allowing hydrothermal carbonization to be a highly tunable process that has produced raw biochars that have been used as fuel, soil amendments, specialty chemicals and adsorbents for heavy metals. To best refine this material and take advantage of how tunable the process is, a better understanding of the structure of the molecule, and its structural kinetics is needed. Studies investigating hydrochar have been conducted using Infrared spectroscopy, NMR spectroscopy, and the corresponding hydrochar model structures are heavily instrument dependent. This points for a need to reconcile hydrochar structures determined from different methods.

A common theme in hydrochar models is the presence of highly oxygenated polycyclic aromatic hydrocarbons (PAHs). Using density functional theory (DFT) to simulate the Raman spectra of PAHs we have investigated the vibrational modes that make hydrothermal chars, such as the characteristic D and G bands often attributed to graphitic content. We have examined the effects molecular size (i.e., number of aromatic rings), presence of functional groups, orientation of PAH rings, and type of aromatic content (arene or furan) on the positions and intensities of different vibrational modes. Using this information we have developed a fitting method for the Raman spectra of hydrothermal chars, which provides information about the aromatic and aliphatic content of the hydrochars.

We then synthesized several glucose based hydrothermal chars at 180 °C and different reaction times. These hydrochars were analyzed using Raman spectroscopy, Infrared spectroscopy, Nuclear Magnetic Resonance, elemental analysis and surface area. The materials were also used to conduct copper adsorption studies, to determine how structural changes in the char influenced its performance as an adsorbate. Chars produced at lower reaction times appeared to also have a highest surface area and capacity for copper(II). On the other hand, the affinity for copper(II) (measured as a function of adsorption per surface area) was shown to increase with reaction time, reaching a maximum for char synthesized with a reaction time of 12 hours. Raman analysis indicated that the char structure changed the most from 2-8 hours. After 8 hours of reaction, the char structure became very stable and changed very little from 8-24 hour reaction time, in parallel with the observed changes in copper surface affinity. We conclude that Raman spectroscopy is a useful tool for the structural determination of hydrochar, and that hydrochar has strong affinity for copper(II), but optimization of the material as a adsorbent depends heavily on optimizing surface areas, which tend to decrease with increasing char yield.

502802 Ceramic Tubular MOF Hybrid Membrane Fabricated Through *in situ* Layer-By-Layer Self-Assembly for Nanofiltration

Abstract:

Nanofiltration has been playing an important role in water purification, in which the developments of novel membrane materials and modules are among significant. Herein, a metal-organic framework (MOFs) hybrid membrane, ZIF-8/PSS
was fabricated on a tubular alumina substrate through a layer-by-layer self-assembly technique. ZIF-8 particles in situ grow into PSS layers to improve their compatibility and dispersion, thereby getting high quality membrane, which was loaded into a steel tubular module for nano-filtrating dyes from water. Under optimized conditions, it shows outstanding
nanofiltration properties toward methyl blue, with the flux of 210 Lm22 h21 MPa21 and the rejection of 98.6%. Furthermore,
the good pressure resistance ability and running stability of the membrane were revealed, which can be attributed to use the ceramic substrate and the inherent stability of ZIF-8. This work thus illustrates a simple approach for fabricating
MOFs hybrid membranes on tubular ceramic substrates, having great potential for industrial applications.

492240 Separation of Oil in Water (O/W) Emulsion Using Commercial Microfiltration Membranes and Sand Filters

Abstract:

Crude oil extraction, refinery as well as frequent offshore oil spillage events unavoidably produced large quantity of oily waste water, which exert heavy burden to environment. The treatment of oily wastewater treatment has become one of the most urgent issues for gas/oil/chemical industries [1].

Two kinds of commercial microfiltration membranes (PES and Nitrocellulose) were characterized and tested for oil in water (O/W) emulsion separation, and compared to the performance of the sand filter with different particle size and bed depth. Wettability of the samples were checked through CA and UCA measurements. Surface properties/morphology were characterized with FTIR/SEM (scanning electron microscope). Separation efficiency were evaluated by checking the turbidity of filtrates and confirmed with TOC values. Experimental results revealed that: 1) The commercial membranes show underwater oleophobicity, with average WCA around 145.8° for PES membrane, with average WCA around 137.1° for Nitrocellulose membrane; 2) Two different membranes present separation efficiency higher than 90% for Dodecanese in water Emulsion, higher than 98% for Hexane in water emulsions [2,3].

Sand filters were studied with different bed depth and particle sizes. Preliminary results showed that the slightly hydrophobic surface can effectively separate the O/W emulsion under gravity, which is superior to the commercial membranes. Under the driving force of 0.1 bar, their performance are equivalent or superior to the commercial membrane, suggesting that the sand filter is a low-cost and effective solution for the separation of O/W emulsions. Table 1 lists the experimental results of the 3 samples for O/W separation.

Table 1. The Separation of O/W Emulsion on 3 media

Medium	Depth/Thickness (mm)	Turbidity (NTU)	Flux (L/m2·h)	TOC (ppm)
Stock solution (Dodecane/water)		**242**		**682.8**
Sands of 75-150μm (gravity)(C_{12})	**40**	**0.6**	**38.4**	**10.9**
Sands of 150-300μm (gravity)(C_{12})	**17**	**31.7**	**188.7**	**54.6**
Sands of 300-500μm (gravity)(C_{12})	**27**	**17.2**	**194.7**	**28.5**
PES membrane (DP= 0.1 bar)	**0.14**	**2.3**	**1698.5**	**56.6**
Nitrocellulose membrane (0.1 bar)	**0.15**	**12.1**	**2239.3**	**58.4**
Stock solution (Dodecane/water)		**185**		**682.8**
Sands of 75-150μm (0.1 bar)(C_{12})	**10**	**73.2**	**887.2**	**18.7**
Sands of 150-300μm (0.1 bar)(C_{12})	**10**	**22.6**	**1596.6**	**48.2**
Stock solution (Hexane)		**264**		**5564.0**
Sands of 75-150μm (0.1 bar)(C_6)	**10**	**45.3**	**702.8**	**12.5**
Sands of 150-300μm (0.1 bar)(C_6)	**10**	**24.8**	**1358.8**	**93.0**

497728 Study of Microplastics in Fresh Water Environment

Abstract:

Abstract

Microplastics, commonly defined as particles or fibers with a diameter < 5 mm consisting of plastic are resistant to natural degradation, leading to high accumulation in environment and this accumulation exposes a global environmental concern for both their physical hazards like entanglement and biological effects like endocrine and reproduction disruption. To date, no standard protocol has been established to collect microplastics and sample treatment methods varies, which makes the results from different studies incomparable. Despite of various approaches developed for MP analysis such as FTIR, micro-R6 Raman and pyrolysis GC, there is still no well established and accepted approach in the measurement of the content of microplastics.

In this study, we will develop a fast and reliable approach for microplastics analysis in water including MP separation and measurement. A new approach is presented for microplastics samples treatment in environmental samples, based on peroxide oxidation and hydrophobic extraction. Our MP particles can be effectively separated from other particles such as natural fibers, dusts and organic tissues by applying our standardized sample treatment protocol. The sample can be directly subjected to filtration and the image-analysis allows particles to be identified and counted. This paper/poster will detail the development of this method and its initial cross-validation by comparison with infrared (IR) microscopy. Microplastics of different sizes could be detected and counted in fresh reservoir water samples.

486368 Towards the Control of Fouling Behavior of Hollow-Fiber Membranes Via High-Throughput Screening

Abstract:

Hollow-fiber (HF) membranes are the preferred configuration for various water and wastewater treatment applications. The fouling of HF membranes is a complex phenomenon that is known to depend on a multitude of parameters and have a severe effect on membrane filtration performance. An ongoing challenge is the significant effort and time that is required to develop and test the effect of different operating parameters. We have recently developed a high-throughput (HT) platform for HF membranes [1] that allows for six filtration tests to be conducted in parallel with each test using a single HF membrane - this design is hereafter referred to as 'high-throughput hollow fiber' (HT-HF) module. As a proof of concept, preliminary experiments were conducted with un-modified polyvinylidene fluoride (PVDF) membranes in constant flux mode to evaluate the fouling tendency for different humic acid (HA) types and concentrations, solution conditions (pH, conductivity), and filtrate flux. It was observed that increasing the pH and decreasing the conductivity results in higher transmembrane pressures and lower apparent sieving coefficients. For example, the final transmembrane pressure (TMP) value decreased from 2.2 to 1.3 psi as the HA solution conditions were changed from 5% phosphate buffered saline (PBS) to 100% PBS while this change in solution conditions has increased the apparent sieving coefficient from 20% to 54%.

Furthermore, we have extended the application of the HT-HF module by performing experiments with a 'library' of potential foulants (humic acids, proteins, polysaccharides), testing different membrane pre-treatment and cleaning strategies, and using surface modified PVDF membranes. Surface modification of PVDF membranes have been studied by different groups in conventional format using different modifying agents such as polydopamine [2–4], polyvinyl alcohol [5], polyvinylpyrrolidone [6] and chitosan [7]. Furthermore, previous studies have shown the usefulness of developing a HT strategy to study the effect of surface modification on polyethersulfone flat sheet membrane performance using a monomer library in 96 well filter plate format towards natural organic matter [8] and protein fouling [9]. However, none of the previous works have studied the surface modification of HF membranes in HT format. In this study, PVDF HF membranes were modified using multiple approaches and modification conditions by using nanoparticles, polyelectrolytes and other polymers deposition in HT format using the HT-HF module. The modified membranes were characterized using contact angle measurements, Fourier transform infrared spectroscopy and scanning electron microscopy. The effect of surface modification condition on membrane hydraulic permeability, TMP profiles and separation performance were studied and the results were compared statistically to each other and to those of un-modified HF membranes. The HT platforms and the results introduced here will be used to develop new strategies for controlling membrane fouling and to better understand membrane fouling behavior.

References

[1] A.S. Kazemi, L. Boivin, S. Mi Yoo, R. Ghosh, D.R. Latulippe, Elucidation of filtration performance of hollow-fiber membranes via a high-throughput screening platform, J. Membr. Sci. 533 (2017) 241–249.

[2] H. Zhu, S. Zhu, A versatile and facile surface modification route based on polydopamine for the growth of MOF films on different substrates, Can. J. Chem. Eng. 93 (2015) 63–67.

[3] L. Shao, Z.X. Wang, Y.L. Zhang, Z.X. Jiang, Y.Y. Liu, A facile strategy to enhance PVDF ultrafiltration membrane performance via self-polymerized polydopamine followed by hydrolysis of ammonium fluotitanate, J. Membr. Sci. 461 (2014) 10–21.

[4] B.D. McCloskey, H.B. Park, H. Ju, B.W. Rowe, D.J. Miller, B.D. Freeman, A bioinspired fouling-resistant surface modification for water purification membranes, J. Membr. Sci. 413–414 (2012) 82–90.

[5] J.R. Du, S. Peldszus, P.M. Huck, X. Feng, Modification of poly(vinylidene fluoride) ultrafiltration membranes with poly(vinyl alcohol) for fouling control in drinking water treatment, Water Res. 43 (2009) 4559–4568.

[6] Z. Xu, L. Li, F. Wu, S. Tan, Z. Zhang, The application of the modified PVDF ultrafiltration membranes in further purification of Ginkgo biloba extraction, J. Membr. Sci. 255 (2005) 125–131.

[7] S. Boributh, A. Chanachai, R. Jiraratananon, Modification of PVDF membrane by chitosan solution for reducing protein fouling, J. Membr. Sci. 342 (2009) 97–104.

[8] M. Zhou, H. Liu, J.E. Kilduff, R. Langer, D.G. Anderson, G. Belfort, High-throughput membrane surface modification to control NOM fouling., Environ. Sci. Technol. 43 (2009) 3865–71.

[9] M. Zhou, H. Liu, J.E. Kilduff, R. Langer, D.G. Anderson, G. Belfort, High Throughput Synthesis and Screening of New Protein Resistant Surfaces for Membrane Filtration, AIChE J. 56 (2010) 1932–1945.

486100 Influence of Backwashing on the Pore Size of Hollow Fiber Ultrafiltration Membranes

Abstract:

Backwashing is a common method for fouling mitigation. However, its impact on the pore-size distribution (PSD) of hollow fiber (HF) membranes has not been studied to date. This study quantitatively assessed the effects of filtration and backwashing cycles on the PSDs of polyacrylonitrile (PAN) and polyvinylidene difluoride (PVDF) HF membranes by evapoporometry (EP) characterization. The membranes were characterized before and after repeated cycles of filtration and backwashing in the absence of any foulants, and for a feed solution of bentonite and humic acid that caused fouling both on and within the membrane pores. Firstly, in the absence of any foulants, backwashing caused the appearance of larger pores, the effect of which was greater for the rubbery PVDF membrane than for the glassy PAN membrane. Secondly, backwashing was more effective in removing the fouling within the larger pores, but could not remove all the deposits within the smaller pores, which provides a mechanistic explanation for the progressive increase in the transmembrane pressure (TMP) with each backwashing cycle. Thirdly, the membranes that did not undergo the 10th backwashing at the end of 10 cycles of filtration and backwashing displayed a marked shift of the PSD towards smaller pores due to the deposition of foulants on and within the largest pores, whereas those that underwent the 10 complete backwashing cycles achieved nearly complete recovery of the larger pores accompanied by an irreversible increase in the diameter of the largest pores. Fourthly, a higher backwashing flux led to similar average pore diameters of the fouled and virgin membranes due to the increased effectiveness in restoring the smallest pores, but the corresponding higher filtration flux negated the benefits due to a greater fouling extent particularly for the larger pores. Finally, in order to achieve the desired permeation and rejection properties, possible enlargement of the pores needs to be taken into consideration when choosing a UF membrane and when specifying the backwashing intensity and protocol.

494413 A Simple, Reactive Approach to Mitigate Fouling and Concentration Polarization in Reverse Osmosis Systems

Abstract:

A Simple, Reactive Approach to Mitigate Fouling and Concentration Polarization in Reverse Osmosis systems

Rajarshi Guha[a], Boya Xiong[b], Michael Geitner,[a] Tevin Moore[a], Darrell Velegol[a], **Manish Kumar[a]**

[a]Department of Chemical Engineering, The Pennsylvania State University, University Park, PA 16802, United States

[b]Department of Civil and Environmental Engineering, The Pennsylvania State University, University Park, PA 16802, United States

Abstract

Fouling of high pressure membranes is a recalcitratnt problem, particularly in reverse osmosis (RO) and nanofiltration (NF) systems. Colloidal, organic and biological fouling are the most commonly encountered challenges in industrial desalination and wastewater treatment processes and there is no convenient way to clean the membrane without interrupting operation or resorting to expensive fouling mitigation infrastructure. Therefore, significant energy savings can be achieved by real time fouling elimination and concentration polarization mitigation. Although several important technological improvements were achieved in membrane design and fouling control in past few years, for example- TiO_2 embedded composite antifouling membranes, photothermal nanoheater membranes or carbon nanotube containing composite electrolytic membranes, which are promising in reducing different types of fouling. However, all the reported methodologies suffer from requiring extensive new infrastructure, lack of scalability and most importantly, efficiency without loss in productivity.

Here, we have demonstrated a simple scalable approach of coating RO/ NF membrane with bioinspired adhesive polymer polydopamine and copper oxide nanoparticles, which decomposed H_2O_2 (0.006% - 0.024%) to oxygen bubbles and thereby, swept foulants away from the surface and at the same time reduced concentration polarization by enhancing mass transfer co-efficient. The polydopamine layer embedded the nanoparticles and acted as the reactive catalytic surface. Additionally, polydopamine layer helped to protect the underlying polyamide layer from hydroxyl radicals by acting as free radical scavenger. Such composite membrane eliminated flux decline with colloidal foulants in repeated runs and even increased the flux level beyond the baseline value with organic humic acid foulants. Enhanced catalytic activity of humic acids in presence of H_2O_2 and reduction of concentration polarization were found to be responsible for such flux enhancements. Embedded copper oxide nanoparticles also retarded *E. Coli* attachment to the membrane surface in stirred cell experiments. On the other hand, such composite membrane was found to eliminate concentration polarization upon H_2O_2 addition by bubble induced mixing. Therefore, such scalable facile approach holds promise to mitigate fouling in industrial membrane systems, reduce concentration polarization and thereby, saving considerable energy.

Figure. Schematic of catalytic/ reactive membrane assembly and realization of mechanism of action.

495199 Graphene Oxide Membranes in Extreme Operating Environments: Concentration of Kraft Black Liquor

Abstract:

Kraft weak black liquor (WBL) is a hot (80-95°C), corrosive (pH ~12), high total solids (15-18 wt%), and high-volume (~500 gal/min in a typical pulp and paper mill) by-product of biomass pretreatment in papermaking and other processes. WBL contains lignin, hemicellulose, organic acids, inorganic salts, and water. Dewatering of WBL is currently performed by a six-stage evaporator train to produce strong black liquor (SBL, up to 80 wt% solids). More than 500 million tons/year of WBL are generated worldwide, and the concentration of WBL to SBL by evaporation is highly energy-intensive (about 0.2 Quads/year in the US alone). Membrane-based concentration of black liquor is attractive as an energy-efficient alternative, but challenging due to the harsh operating conditions and high fouling potential of BL.[1]

This talk will discuss our pursuit of chemically robust, low-cost, and scalable membranes for black liquor concentration, focusing in particular on the development of graphene oxide (GO) nanofiltration membranes for lignin rejection. Thin (< 250 nm), uniform, and defect-free GO membranes were fabricated on macroporous poly(ethersulfone) (PES) supports and shown to possess excellent stability in real mill-derived BL during prolonged operation (1 month) at realistic temperature and feed conditions. Furthermore, high lignin rejection (~99%) and flux (~20 $kg/m^2/hr$) was obtained when concentrating BL feeds at transmembrane pressures (TMPs) up to 50 bar and temperatures up to 85°C. We discuss in the detail the characterization of the GO membrane structure, morphology, and permeation properties.[2] A preliminary discussion of the economics and techniques for scale-up of the present membranes will be included.

1. Kevlich, N. S.; Shofner, M. L.; Nair, S., Membranes for Kraft Black Liquor Concentration and Chemical Recovery: Current Progress, Challenges, and Opportunities. *Separation Science and Technology* **2017**, *52* (6), 1070-1094.

2. Rashidi, F.; Kevlich, N. S.; Sinquefield, S. A.; Shofner, M. L.; Nair, S., Graphene Oxide Membranes in Extreme Operating Environments: Concentration of Kraft Black Liquor by Lignin Retention. *ACS Sustainable Chemistry & Engineering* **2017,** *5* (1), 1002-1009.

500398 Water Permeation through Structural Defects of Single-Layered Graphene Oxide Membranes

Abstract:

Comparing with traditional methods, membrane-based water treatment technologies are essential due to the low energy consumption and high efficiency in removing contaminants or salts. As the flux cross the membrane decreases with membrane thickness, to balance the pay off between permeability and selectivity, membrane should possess pores large enough to allow the passage of water molecules but small enough to block the large solutes, and the membrane thickness should be as thin as possible to provide high flux, which makes the "ideal" membrane might be in the form of one atomic thickness and with suitable pores. Materials like graphene and its derivatives just meet this requirement and have been considered as a promising candidate to prepare ultrathin membranes which mostly closing to the "ideal" state. However, the harsh conditions to prepare macroscopic single-layer graphene with sufficient densities and size controllable pores have hindered its practical applications. Herein, by clarifying two distinct water transportation mechanisms for membranes with sub-monolayer and multilayer GO coverage, a methodology to fabricate nominal single-layered GO membrane was proposed for the first time. The Water flux exhibited a transition from linear to exponential decrease as the GO coverage increase, the condition to prepare nominal single-layered GO membrane could be extrapolated from the turning point of the transition, the as-fabricated membrane which with thickness closing to one-carbon-atom exhibits high water permeance around 64 $L \cdot m^{-2} \cdot h^{-1} \cdot bar^{-1}$, and defects on GO flakes provide major contribution for the membrane sieving property. By evaluating the separation performance of this membrane with rigid molecules, the effective defect size of GO was determined to be ~1.2-1.7 nm. This membrane with nominal single-layer GO cover also show great potential in protein separation.

502718 Braided-Reinforced Thin Film Composite (TFC) Nanofiltration Hollow Fiber Membranes

Abstract:

Hollow fiber membranes are attractive for use in water treatment processes because of their high packing density relative to flat sheet and spiral wound configurations. They are also easily submersed in tanks for bioreactor applications. For water purification in particular, they are commonly use in microfiltration (MF) an ultrafiltration (UF) applications. To enable low pressure drop in fiber lumens, inner diameters of these fibers can be rather large, thus limiting the pressure tolerance of these membranes. Low pressure tolerance of larger capillary fibers has largely excluded the use of hollow fiber from higher pressure membrane processes reverse osmosis (RO) and nanofiltration (NF) except for cases where integrated asymmetric membranes are considered. Use of the thin-film composite (TFC) platform with polyamide based selective layers, while common with flat sheet membranes for RO an NF, have not found commercial success and have largely been relegated to the academic community for studies on forward osmosis (FO). In this work, we show that hollow fiber TFC membranes can be made for higher pressure applications. To address the pressure tolerance issues, we build our TFC using a commercial hollow fiber PVDF UF membrane that is reinforced with an embedded braid as a support. Onto this support is formed the selective barrier layer *in-situ* from piperazine (PIP) an trimesoyl chloride (TMC) monomers. Through careful optimization the monomer concentration and reaction conditions, we have demonstrated softening with a 89 % $MgSO_4$ rejection and a permeance of 8 LMH/bar. This work represents the first time such a reinforce hollow fiber membrane has been used as a support for a TFC membrane and could broaden opportunities for hollow fiber membranes in processes where salt selection (RO, pressure retarded osmosis) is required.

497318 Naphthenic Acids Removal from High Tds Produced Water By Persulfate Mediated Iron Oxide Functionalized Catalytic Membrane, and By Nanofiltration

Abstract:

Increasing stress on the existing water resources necessitates advancement in water treatment technology. During oil extraction process, oil industries generate large amounts of produced water containing organic contaminants, such as naphthenic acids (NA) and very high concentrations of inorganic salts. Recovery of potable water from such water is highly energy intensive due to its high salt concentration. Here, we explored catalytic properties of iron oxide (Fe_xO_y) functionalized membranes in oxidizing NA from water containing high concentrations of total dissolved solids (TDS) using persulfate as an oxidizing agent. Catalytic decomposition of persulfate by Fe_xO_y functionalized membranes followed pseudo-first order kinetics with an apparent activation energy of 18 Kcal/mol. Fe_xO_y functionalized membranes were capable of lowering the NA concentrations to less than discharge limits of 10 ppm at 40 ^{O}C. Oxidation state of iron during reaction was quantified. Membrane performance was evaluated for extended operating period. Coupled nanofiltrationâ€"functionalized membrane was evaluated. Commercially available nanofiltration membranes were investigated to selectively retain NA from water containing high concentrations of dissolved salts. Commercial NF membranes, NF270 (Dow) and NF8 (Nanostone) had NA rejection of 79% and 82%, respectively. The developed Fe_xO_yÂ functionalized membrane can then treat the lower volume concentrated NF stream (retentate) for efficient removal of NA.

492248 Effect of MnO_2 Catalyst and Electrode Geometry on Nonthermal Plasma Reactor Combined with Ceramic Filter for Trichloroethylene Decomposition

Abstract:

Volatile organic compounds (VOCs) emissions have been controlled by Air Pollution Control Act in Japan. There are two major types of VOC processing methods: incineration and adsorption. However, these methods are suitable for treating VOCs with high concentration levels and flow rates. On the other hand, a third method—nonthermal plasma decomposition—can effectively eliminate VOCs with low concentration levels and flow rates. Therefore, the nonthermal plasma decomposition method has been drawing more attention these days as an alternative approach because of its advantages over conventional methods. In our previous work, we developed a new type electrode system for gas treatment in which nonthermal plasma and a ceramic filter were integrated and applied it for VOCs treatment. In this study, to improve the energy efficiency and to reduce by-products of the reactor system, the effect of electrode geometry and the addition of MnO2 catalyst in the system were investigated in the application for trichloroethylene (TCE) decomposition. In the experiments, three electrodes such as rod, bolt, and plate geometries were employed and two locations of setting the MnO2 catalyst as "inside the reactor" and "downstream of the reactor" were tried. The results demonstrated that energy efficiency to decompose TCE was improved by combining with the MnO2 catalyst for both location conditions of "inside" and "downstream" of the reactor. As to by-product reduction, catalyst location of "inside" the reactor showed significant effect. Although small amount of phosgene, dichloroacetylchloride (DCAC), pentachloroethane (PCE) and tetrachloroethane generation was observed for the case of the reactor system without catalyst, phosgene, DCAC, tetrachloroethane were under the detection limit for the case of "inside" the reactor. Among three different electrode geometries, the plate electrode demonstrated highest energy efficiency for TCE decomposition regardless of catalyst location.

492739 Sustainability Assessment of Cement Manufacturing Processes Based on Emergy and Ecological Footprint Analysis

Abstract:

Due to its high environmental impact and energy intensive production, the cement industry needs to adopt more energy efficient technologies to reduce its demand for fossil fuels and impact on the environment. Bearing in mind that cement is the most widely used material for housing and modern infrastructure, the aim of this paper is to analyze the Emergy and Ecological Footprint of different cement manufacturing processes for a particular cement plant. There are several mitigation measures that can be incorporated in the cement manufacturing process to reduce the demand for fossil fuels and consequently reduce the CO2 emissions. The mitigation measures considered in this paper were the use of alternative fuels and a more energy efficient kiln process. In order to estimate the sustainability effect of the aforementioned measures, Emergy and Ecological Footprint were calculated for four different scenarios. The results show that Emergy, due to the high input mass of raw material needed for clinker production, stays at about the same level. However, for the Ecological Footprint, the results show that by combining the use of alternative fuels together with a more energy efficient kiln process, the environmental impact of the cement manufacturing process can be lowered.

499562 A Bi-Dimensional Experimental Cell and Image Analysis: An Effective Device to Validate Model Predictions Applied to Electro-Transport of Heavy Metals

Abstract:

Empirical and theoretical studies have demonstrated that applying an electrical potential to solutes or particles, helps to improve their transport efficiency better than other known methods. Predictions based on an analytical continuum-based model, developed by Tíjaro-Rojas (2015), are being validated by means of a bi-dimensional cell filled with an electrolyte solution containing a charged dye. The dye is used to facilitate visualization of transport; the radial spreading of the dye is captured by means of pictures taken at different times and processed with an image analysis software. The initial concentration of the dye is known and the mass transport of the charged particles is monitored by means of light intensity obtained from the pictures. The light intensity is calibrated to concentration profiles using the Beer-Lambert law. Experimental and numerical results (from the model predictions) are compared in order to validate the model performance for this specific case. The validation of this model fosters understanding related to heavy metal remediation techniques currently being applied around the world to clean contaminated sites.

Reference: Tijaro-Rojas, R. (2015). *Role of micro and macro-heterogeneities in electro-kinetic soil cleaning: An area-averaging approach with dynamic simulations* (Doctoral dissertation, Tennessee Technological University).

498075 Application of Semi-Batch Reverse Osmosis (RO) Operation for Water Purification in Small Communities

Abstract:

Various remote communities around the world are confronted with impaired local water supplies due to both increased salinity and rise in nitrate level. In such communities establishing and maintaining reliable distributed water treatment systems for the production of safe drinking water is a major challenge. Water treatment systems for deployment in remote communities must possess the following attributes: a) provide a physical barrier for reliable contaminant removal, b) high operational flexibility to handle geographical and temporal variability of feed water quality and other operational requirements (e.g., water recovery), c) amenable to process automation, d) compact foot-print capable for fixed or mobile deployment, and e) standardized system configuration (for ease of maintenance and operator training). In this regard, membrane-based (nanofiltration (NF) and low-pressure reverse osmosis (RO)) water treatment is most suitable for meeting the above needs. Accordingly, in the present work, a novel desalination pilot system was developed that can operate in either semi-batch or continuous steady-state mode. This system enabled the generation of fundamental data needed to compare the performance of semi-batch and continuous RO operation with respect to energy consumption as well as attainable recovery. Experimental data were then compared with predictions of RO model for continuous and semi-batch RO. Analysis of the semi-batch process was accomplished considering the influence of axial salt dispersion during the concentrate flushing period. The present study suggests that dispersion within the RO system affects the efficiency of flushing (i.e., removal of excess salt buildup in the RO system during the filtration period), which in turn has a profound impact on the specific energy consumption. It is also shown that, for a given product water recovery, there exists an optimal system flushing rate at which energy optimal performance is attained. Moreover field studies have demonstrated effective field development of the concept of compact semi-batch RO for both salinity reduction and nitrate removal to provide safe drinking water for remote communities.

Contaminant Removal from Effluents through the Use of *Moringa oleifera* Lam. Seeds for Application in Drinking and Water Resources Recovery Facilities and in Ecuadorian Rural Communities

Andrés S. Lagos[1], Marlon M. Pico[1], Esteban R. Nuñez[1], Alina L. Trávez[1], M. Fatme Troya[1], Israel Sornoza[1], Juan S. Villarreal[1], José C. Andrade[1], Aldo J. Cunalata[2], Felipe D. Bastidas[2], Mario Caviedes[3], Michel E. Vargas-Vallejo[2] and Andrea C. Landázuri[1]*

[1]Universidad San Francisco de Quito USFQ, Colegio de Ciencias e Ingenierías – El Politécnico, Departamento de Ingeniería Química, GICAS, Calle Diego de Robles y Vía Interoceánica, Campus Cumbayá. Casilla Postal 17-1200-841, Quito, Ecuador.

[2]Universidad San Francisco de Quito USFQ, Colegio de Ciencias e Ingenierías – El Politécnico, Departamento de Ingeniería Mecánica, GICAS, Quito, Ecuador.

[3]Universidad San Francisco de Quito USFQ, Colegio de Ciencias e Ingenierías – El Politécnico, Departamento de Ingeniería en Agroempresa, GICAS, Quito, Ecuador.

Abstract

The proposed project is part of the multidisciplinary *Moringa Project in Ecuador*, where the main areas of investigation are based on science and engineering fields for application in drinking and water resources recovery facilities as well as in rural communities. This water project is led by Chemical and Mechanical Engineering members part of the Engineers Without Borders initiative from Universidad San Francisco de Quito, Ecuador. *Moringa oleifera* Lam. leafs, seeds and pods are characterized by having adsorptive properties and can be used as natural coagulants and flocculants, as well as filtration medium. Moringa varieties have become an environmental alternative to commonly used reagents used in drinking and wastewater treatment which have been proven to be more expensive than natural sorbents. Aware of the excessive cost of chemicals used in water treatment it is worth looking for substitutes and to propose new environmentally friendly technologies that could be adapted in drinking and water resources plants. Technically, results will help to gain insights about metal and emerging contaminant transport and fate, identify tracers, establish heavy metal removal methods, investigate and propose economically feasible removal methods, develop and implement unit operations at a pilot plant and larger scale, produce MO based products and technologies, and promote the strengthening of water quality regulations in Ecuador.

Introduction

Moringa oleifera Lam. (MO) (Figure 1), is a crop native to South Asia that grows in the Himalayan foothills and is widely cultivated across the tropics (FAO, 2016). In Ecuador, moringa plantations exist in regions such as Santa Elena Peninsula and Pedernales, Manabí. In Ecuador, MO is currently used for medicinal and dietary purposes (Landázuri, Cahuasqui, Egas, & Lagos, 2016).

The leafs, seeds and pods are characterized by having adsorptive properties (Refer to Figure 2 where the irregular morphology of the seeds that could aid MO's capability of retaining contaminants in water) and can be used as natural coagulants and flocculants, as well as filtration medium (Martín, C., Martín, G., García, A., Teresa Fernández, 2013), (Ravikumar & K, 2013). Moringa varieties have become an environmental alternative to commonly used reagents in drinking and wastewater treatment (García-Fayos, B.; Arnal, J.;Alandia, 2012; Matouq, Jildeh, Qtaishat, Hindiyeh, & Al Syouf, 2015) which have been proven to be more expensive than natural sorbents (Araújo et al., 2010). MO based materials that can provide solutions to specific purposes such as: metal removal, emerging contaminant removal (e.g. caffeine, irgasan), turbidity removal, bacteria and larvae removal, among others.

Wastewater treatment coverage in Quito is less than 1%, meanwhile, sewer and drinking water supply is 92.7 and 98.45%, respectively (EPMAPS, 2016). Wastewater is usually discharged directly into open bodies, instead of being used for other purposes such as irrigation, recreation or power generation (Lonergan, 2013). It is an important milestone that by the first term of 2017 the first water resource recovery facility, located in Quitumbe (southern Quito-Ecuador) started operation. Therefore, several emerging issues have arisen that needs to be investigated in this subject new to Ecuadorian reality. This plant will be a great source of information for these studies. Industrial discharges and anthropogenic activities are important sources of metal contamination in water bodies (Araújo et al., 2010). This contamination tends to persist and accumulate in the food chain, therefore constitute a threat to human beings, animals and the surroundings. In order for a metal to be removed from a given solution, it has to be adequately sequestered (Volesky & Holan, 1995). The search of new technologies for metal removal from wastewater has directed the attention to biosorption techniques. Moreover due to MO's morphological characteristics (Figure 2), it has been found that MO has the potential to remove heavy metals from aqueous effluents (Araújo et al., 2010), (Gopalakrishnan, Doriya, & Kumar, 2016).

The study by Voloshenko-Rossin et al. (2015) (Voloshenko-Rossin et al., 2015) where several water samples in specific points along the San Pedro-Guayllabamba-Esmeraldas Rivers were collected showed that components such as carbamazepine (anticonvulsive and mood stabilizer), acesulfame (sweetener) persisted throughout the water path, making them ideal to be used as wastewater tracers. Caffeine, for example, was one of the most abundant organic compounds present in one of the most important waste water discharge. Some of these components persist even after the affluent to a drinking water plant in Esmeraldas was treated.

Research results will help gain insights about metal and emerging contaminant transport, identify tracers, investigate and propose economically feasible removal methods, develop and implement unit operations at pilot scale, produce MO based products and technologies, and promote the strengthening of water quality regulations.

Research Objectives

The objectives pursued by the Engineering, Applied Sciences & Simulation Group (GICAS, in Spanish) from USFQ focus on heavy metal, organic emerging (persistent) contaminants, and bacterial removal from effluents. The first phase of the project consisted on performing batch experiments to explore: metal and caffeine removal, and coagulation-flocculation processes where different pretreatments were applied to MO seeds. Artificial water and real effluents (river water and waste water effluent) were used.

From a rigorous and scientific point of view, on a second phase, the dynamic water filtration system team is exploring the effects of other parameters that may influence the adsorption capacity of this biological material, such as particle size, pH, and the use of specific sections of MO seeds, including inner and outer layers and different combinations of those, the use of multicomponent and real water, etc.

From a process engineering perspective, on a third phase, the goal is to propose laboratory and pilot scale technologies that can help mitigate or observe a desire effect during a process. For this goal, the characterization of MO is necessary to complement and/or to contribute during the decision making process before, during and after the application of a process. Processes (mechanisms) that are currently being studied are: adsorption, coagulation & flocculation, and sedimentation.

Methodology

Coagulation-flocculation experiments used crushed MO without husk in a dose range between 50-1500 mg MO per liter of water sample and were submitted to jar tests to simulate the mixing and sedimentation processes in water treatment facilities. Doses of 10-50 mL of aluminum sulfate per liter of water sample were investigated. In addition, other set of experiments explored MO particle size ranges between 355-2000 μm using a fixed amount of MO equal to 50 mg/L. Tests consisted on the addition of 900 mL of contaminated water samples and 100 mL of the heterogeneous MO aqueous solutions, configured under an agitation of 100 rpm during one minute, and a subsequent 25-minute agitation under 20 rpm. After this time, all jars were allowed to rest for 2 hours while monitoring pH. *E. Coli*, COD and turbidity were the selected response variables.

Metal removal adsorption experiments used artificial metal-containing water (Cd, Pb and Fe). Firstly, seeds were treated by the extraction of their oil by Soxhlet method. Metal solutions were prepared and subjected to contact with the adsorbent while varying different parameters: contact time, initial ion concentration in water, pH, adsorbent dose and multicomponent water. The amount of metals in water was measured by atomic absorption spectrophotometry. Several levels of the following parameters were tested: pH 4.0-6.0; contact time: 0-60 min; dose of MO 0.0-2.0 g/L; concentration of Cd: 0.2-1.0 mg/L, Fe: 1.0-5.0 mg/L, Pb: 2.0-10 mg/L, and the conditions of greatest removal of these ions were identified based on statistical significance tests. A dose of 1 g of crushed and oil free MO seeds were added per liter of artificial solution while keeping a constant stirring during 1 hour and 100 rpm.

Caffeine experiments used a 100 ppm stock solution which was treated under a Fenton process that involved 300 ppm of hydrogen peroxide with 1 to 10 mg Fe (2+) per liter of solution in acidic media at a pH of 2.8. Photo-fenton processes include a UV source with three 20 W blacklight tubes (peak wave length of 352 nm). Caffeine removal using MO involved washing the seeds with distilled water, followed by an overnight oven dried procedure at 60 ºC, prior to the extraction of the proteic fraction using NaCl and subsequent filtration. Atomic absorption spectroscopy (AAS) for metal removal and high performance liquid chromatography (HPLC) for caffeine quantification were used.

Research Findings

Metal Removal Project

This research studies the removal of heavy metals present in water (cadmium, iron and lead) through *Moringa oleifera* Lam. seeds (MO), defining the adsorption kinetic variables and parameters that affect it. The best conditions identified were: contact time of 1 h, pH 6.0, MO dose of 2.0 g/L and initial metal concentrations of 0.2 mg/L, 1.0 mg/L and 2.0 mg/L for Cd, Fe and Pb, respectively. With these conditions, the obtained ion load at equilibrium, q_e, was 0.106 mg Cd/g MO (95.6% removal), 1.850 mg Fe/g MO (90.4% removal) and 1.156 mg Pb / g MO (97.0% removal). Regarding multicomponent tests, it was found that MO removal capacity decreases to 16.3% (Cd), 8.73% (Fe) and 10.8% (Pb), due to ion competition for available active adsorption sites.

The model that best represent adsorption was Langmuir for all three cases (Figure 1); and, all metals followed a pseudo-second order kinetics as well, supported by previous studies (Madzvamuse, Kugara, & Shumba, 2015). This is a pioneer study in Ecuador to future projects that can test new technologies applied to industrial discharges.

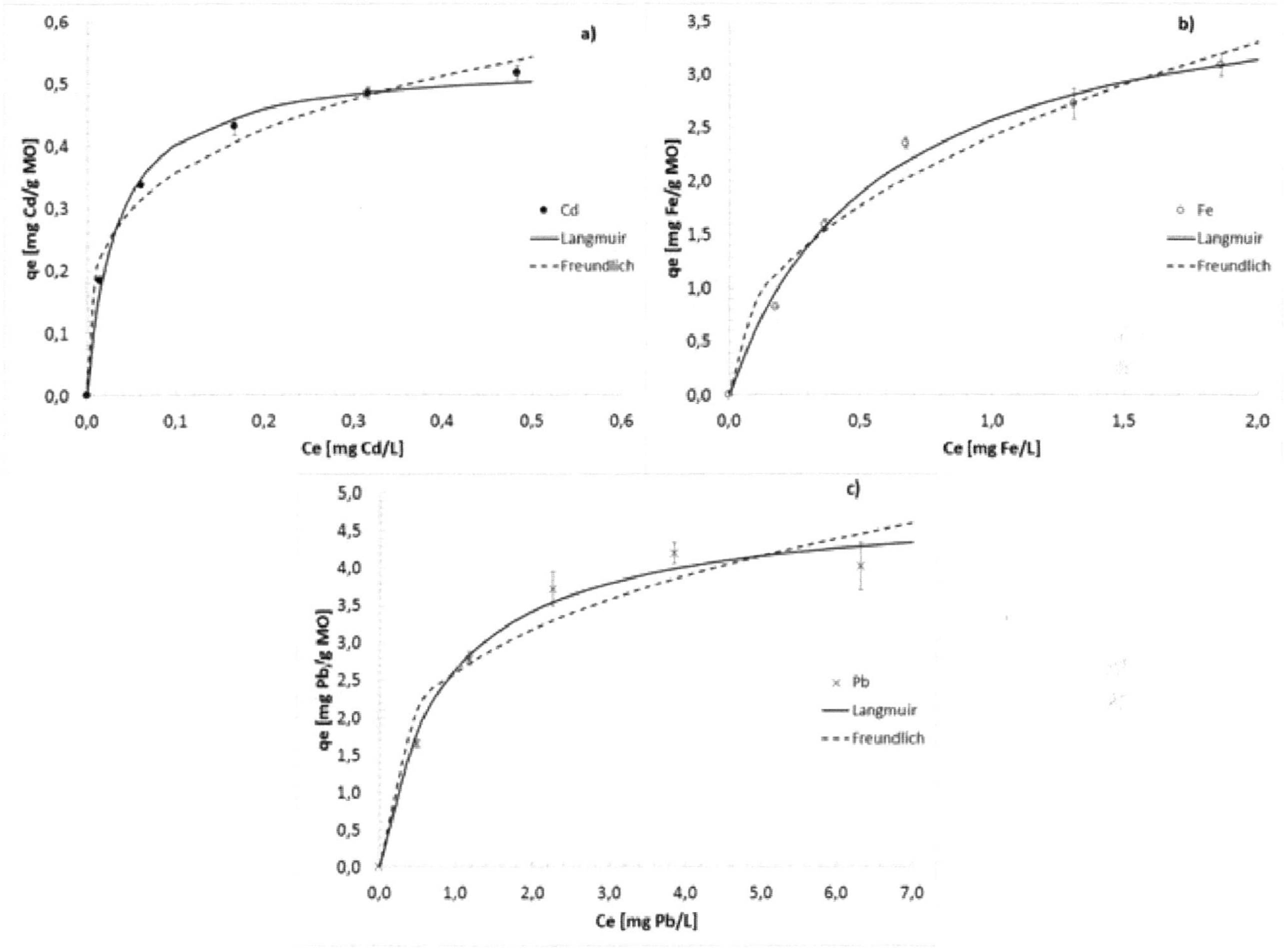

Figure 1. Adsorption isotherms for a) cadmium, b) iron, and c) lead.

Caffeine Removal Project

For perspective, caffeine is one of the most abundant organic compound present in one of the more representative wastewater discharges from the Metropolitan District of Quito and shows some degradation throughout the Esmeraldas watershed (Voloshenko-Rossin et al., 2015).

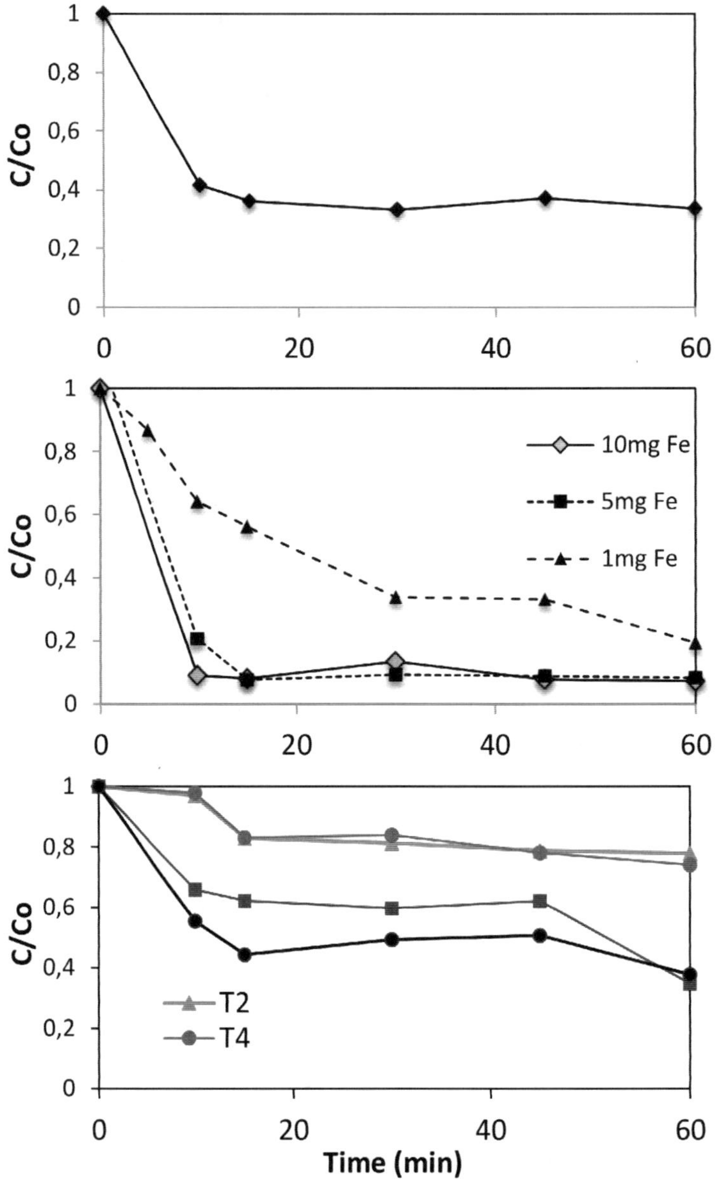

Figure 2. Remaining fraction of caffeine as a function of time using a) Fenton process with 10 mg of Fe, b) photo-fenton process for 1-10 mg Fe(2+) and pH value of 2.8, c) MO with different treatments, being TD1 and TD2 the ones that considers the proteic fraction of MO. Initial concentration of caffeine equals 100 ppm. Fenton process was carried out with 300 ppm hydrogen peroxide concentration. Photo-fenton involved the addition of UV light with a 20W black light tubes (352 nm peak wavelenght).

Caffeine removal from artificial water using the proteic fraction of MO was explored. The best MO treatments are those where the whole seed was used for the extraction of protein with salt (TD1 and TD2). Best removal efficiency obtained is that with photo-fenton (92%), followed by, fenton, and MO with 65% for both treatments (Figure 2).

Coagulation and flocculation Project

Water treatment consists on various successive steps and two of these processes are known as clarification and flocculation-coagulation. In Ecuador, the use of aluminum sulfate is very common, but its impact within the environment is a cause of concern because of the potentially toxic effects under certain pH conditions. For this reason, a natural flocculant-coagulant agent, *Moringa oleifera* Lam. seeds, were investigated. In this research project, the optimum range size of crushed moringa seeds was evaluated during coagulation and flocculation processes. Natural coagulant-flocculant action was explored in three different water sources: Monjas River, San Pedro River and the effluent from the Pilot Water Resource Recovery Facility from the Water and Sanitation Company in Quito, Ecuador. It has been found that for all three water sources, a particle size of 1000 μm of crushed provided the best *E. Coli*, turbidity and COD removal, for instance, being able to reach values below current Ecuadorian normative. Globally, COD, turbidity and E. Coli, were reduced in 54%, 80% and 90%, respectively.

In addition to this, the yield of *Moringa oleifera* Lam. seeds as a natural coagulant agent; and aluminum sulfate, as a synthetic coagulant, was evaluated during coagulation-flocculation processes. The variables monitored were turbidity, COD, *E.Coli* and pH in order to demonstrate that a natural coagulating agent can have a yield equal to or greater than a synthetic agent. By means of experimental design applied to different water sources of the Metropolitan District of Quito, it is concluded that a concentration of 250 mg of *Moringa oleifera* Lam. per liter of water (M1), and a concentration of 10 mL of aluminum sulfate per liter of water (S1), are the treatments that showed, on average, greater removal of turbidity (90%) and COD (60%). Regarding *E. Coli*, a concentration of 10 mL of aluminum sulfate per liter of water is the best treatment. In addition, treatments that included *Moringa oleifera* Lam. maintained pH values between 7.4 and 7.7; showing buffering capacity. This first study shows a possible partial or complete substitution of aluminum sulfate to reduce the disadvantages with respect to health and economic viability in the tertiary treatments for drinking and or water resource recovery facilities in Ecuador.

Conclusions and Recommendations

It is important to emphasize that this is a long term project and that wastewater treatment is fairly new to Ecuadorian reality and it will help us gain fundamental understanding, explore additional effects, and find correlations between contaminants. Research keeps on going to understand the effects of other parameters that may influence the adsorption capacity of this biological material available in Ecuador, such as particle size, pH, and the use of specific sections of MO seeds, including inner and outer layers and different combinations of those. Research interests rely on specific objectives where a collaborative approach within fields is desired. From a process engineering perspective an ultimate goal is to propose laboratory and pilot scale technologies that can help mitigate or observe a desire effect during a process. For this goal, the characterization of MO is necessary to complement and/or to contribute during the decision making process before, during and after the application of a process. Therefore, it is imperative to have an integrative approach between the principles explained through "Basic Science", "Experimentation", "Modeling & Simulation", and "Applications & Manufacture"; where the importance of characterizing MO through analytical/instrumental methods (which is related to a phenomenon), along with the validation through modeling, to obtain an expected effect is necessary.

Current projects involve the optimization of batch experiments and the investigation of dynamic and continuous filtration systems which is an important step towards proposing new and greener technologies of metal and emerging contaminant removal. This Project has the potential to be implemented as part of the various Engineer Without Borders projects which are being carried out by the Student Chapter at Universidad San Francisco de Quito, and through other type of communitarian project. Results will not only involve technical aspects but also social impact. Evidently, the inclusion of green methodologies in wastewater plants, and also the possibility of planting moringa oleifera in affected areas, can benefit the environment and improve the quality of life of inhabitants, Quito's ecosystem and downstream. It is worth notice that moringa oleifera exist in Pedernales, which was tragically devastated after april's 2016 earthquake.

References

1. Araújo, C. S. T., Alves, V. N., Rezende, H. C., Almeida, I. L. S., De Assunção, R. M. N., Tarley, C. R. T., … Coelho, N. M. M. (2010). Characterization and use of Moringa oleifera seeds as biosorbent for removing metal ions from aqueous effluents. *Water Science and Technology*, *62*(9), 2198–2203. https://doi.org/10.2166/wst.2010.419
2. EPMAPS. (2016). Programa para la descontaminación de los ríos de Quito.
3. FAO. (2016). Traditional Crop of the Month: Moringa. Retrieved from http://www.fao.org/traditional-crops/moringa/en/?amp%3Butm_medium=web&%3Butm_source=faohomepage
4. García-Fayos, B.; Arnal, J.;Alandia, S. (2012). Estudio de la descontaminación de efluentes líquidos con elevadas concentraciones de metales pesados (pp. 1098–1108). Valencia: XVI Congreso Internacional de Ingeniería de Proyectos.
5. Gopalakrishnan, L., Doriya, K., & Kumar, D. S. (2016). Moringa Oleifera: A Review on Nutritive Importance and its Medicinal Application. *Food Science and Human Wellness*, *5*(2), 1–8. https://doi.org/10.1016/j.fshw.2016.04.001
6. Landázuri, A. C., Cahuasqui, J., Egas, D., & Lagos, A. (2016). Metal adsorption in aqueous media using moringa oleifera seeds as an alternative method for water treatment in Ecuador. *In Progress*.
7. Lonergan, K. (2013). No Title.
8. Madzvamuse, A., Kugara, J., & Shumba, T. (2015). A Remediation Study Using Moringa oliefera Seed (Uncarbonised and Carbonised) for the Adsorption of Chromium , Cadmium and Lead from Synthetic Water. *International Research Journal of Pure and Applied Chemistry*, *7*(3), 110–121. https://doi.org/10.9734/IRJPAC/2015/16195
9. Martín, C., Martín, G., García, A., Teresa Fernández, E. H. & J. P. (2013). Potenciales aplicaciones de Moringa oleifera. Una revisión crítica. *Pastos Y Forrajes*, *36,*(2), 137–149.
10. Matouq, M., Jildeh, N., Qtaishat, M., Hindiyeh, M., & Al Syouf, M. Q. (2015). The adsorption kinetics and modeling for heavy metals removal from wastewater by Moringa pods. *Journal of Environmental Chemical Engineering*, *3*(2), 775–784. https://doi.org/10.1016/j.jece.2015.03.027
11. Ravikumar, K., & K, P. S. a. (2013). Heavy Metal Removal from Water using Moringa oleifera Seed Coagulant and Double Filtration. *International Journal of Scientific & Engineering Research*, *4*(5), 10–13. Retrieved from http://www.ijser.org
12. Volesky, B., & Holan, Z. R. (1995). Biosorption of heavy metals. *Biotechnology Progress*, *11*(3), 235–250. https://doi.org/10.1021/bp00033a001
13. Voloshenko-Rossin, a., Gasser, G., Cohen, K., Gun, J., Cumbal-Flores, L., Parra-Morales, W., … Lev, O. (2015). Emerging pollutants in the Esmeraldas watershed in Ecuador: discharge and attenuation of emerging organic pollutants along the San Pedro–Guayllabamba–Esmeraldas rivers. *Environ. Sci.: Processes Impacts*, *17*(1), 41–53. https://doi.org/10.1039/C4EM00394B

496850 Performance of a-Stage Process Treating Wastewater Containing Domestic Food Waste from Disposers: An Alternative for Community-Based Sewage Treatment

Abstract:

About one third of the food supply is wasted in the US. Food waste is the second highest component of landfills and is therefore heavily contributing to the production of greenhouse gas emissions. Although nationwide initiatives have been implemented to collect food waste for biological treatment, most household food waste is currently processed with food waste disposers (FWDs) and ends up in the sewage system together with the municipal wastewater. The A-stage of the adsorption/bio-oxidation (A/B) process, is a technology that has recently regained popularity for the primary treatment of domestic wastewater at a decentralized level due to its high efficiency removing pollutants, low energy consumption, small footprint and producing a sludge with high energy content. To our knowledge, there has not been any research undertaken on the potential effects of food waste on the A-stage process. In this study, the impact of food waste on the performance of the A-stage process was studied. In addition, biochemical methane potential (BMP) tests were conducted to evaluate the potential energy generation from the A-sludge with food waste. One lab-scale A-stage reactor was operated with a hydraulic retention time (HRT) of about 60 minutes with a dissolved oxygen (DO) concentration of about 0.7 mg/L and a mixed liquor suspended solids (MLSS) concentration of 2000 mg/L. Overall, the chemical oxygen demand (COD) removal efficiency of the A-stage was about 48% when operated without food waste and 46% with the addition of food waste. BMP tests showed that the biodegradability of the A-sludge treating wastewater with food waste was 20% higher. The implementation of A-stage technology for treating domestic wastewater combined with food waste using FWDs could be an opportunity to increase the energy production from domestic wastewater and decreasing potential GHGâ€™s emissions.

Artificial Intelligence for Organochlorine Pesticides Removal from Aqueous Solutions using Entrapped nZVI in Alginate Biopolymer

Ahmed S. Mahmoud[1], Rasha A. SaryEl-deen[1], and Mohamed K. Mostafa[2]*

[1]Sanitary and Environmental Institute (SEI), Housing and Building National Research Center (HBRC), Giza, Egypt

[2]Environmental Engineering Program, Zewail City of Science and Technology, Giza, Egypt

Abstract

Chlorinated pesticides (ClP) are commonly found in agricultural drainage and represent the most hazardous pollutants that may adversely affect groundwater quality. Many governments are searching for efficient and cost-effective techniques to reduce their negative effects. Nano technologies especially good electron donor nano particles, such as nano zero valent iron (nZVI) has proved to be effective in ClP degradation and adsorption. In this study, entrapped nZVI in calcium alginate biopolymer was used to remove ClP from its aqueous solutions. The effect of entrapped nZVI alginate biopolymer on 20 compounds of ClP was studied at different contact time, absorbent dose, pH, stirring rate, and initial ClP concentration. The results indicated that entrapped nZVI in alginate biopolymer is effective in the removal of ClP from aqueous solutions, where a removal efficiency of 55% was achieved for 60 µg/L extracted blank solution after 45 min of contact time using dose 40 g/L at pH 7 with fixed stirring rate 100 rpm. The removal efficiency of ClP mixture ranged from 41 to 55% for initial ClP mixture concentration of 400 and 60 µg/L, respectively, at the same environmental conditions. Artificial neural network (ANN) with a structure of 5 – 10 – 1 was used to predict the ClP removal efficiency. The proposed ANN was found to be effective in simulating the performance of entrapped nZVI in calcium alginate for ClP removal, where a high r^2-value was recorded to be 0.9194, 1, and 0.9996 for training, validation and test plots, respectively.

Introduction and Background

Volatile organic compounds (VOCs) are class of contaminants and involve liquid or solid organic compounds that can easily evaporate into the gaseous state. In 1997 the Agency for Toxic Substances and Disease Registry (ATSDR) characterized the VOC drinking water pollutants as a past public health hazard and focused its study on VOCs related diseases causing death for men and women. Chlorinated pesticides (ClP) are semi volatile organic compounds that may be present in nature from various sources (Jaward et al., 2004). Also, ClP are of persistent organic pollutants (POPs) (Parween et al., 2014). POPs include nine compounds from organochlorine pesticides (Wurl and Obbard, 2005). Organochlorines have been of great concern due to their occurrence in high concentrations even in remote ecosystems, despite bans on the production and usage (Guruge and Tanabe, 2001). Organochlorine compounds have a negative effect on reproduction cycle of human and animals and lead to hormonic problems (Colborn and Smolen, 1996). Presence of ClP in air or water or food are responsible for related diseases such as Parkinson's disease (PD) (Golbe, 1993) and Alzheimer's disease (AD) (Fleming et al., 1994).

Water saving achieved by reuse of wastewater, which requires low cost effective methods and material to achieve best treatment (Amin et al., 2014). Wastewater dilution can decrease micro pollutant (Verstraeten et al., 2003). Physicochemical treatments, such as coagulation and flocculation are ineffective treatment methods due to high treatment cost (Petrovic et al., 2003, Westerhoff et al., 2005). Biological treatments studies, such as activated sludge and biological trickling filters are also ineffective for the removal of wide range of contaminants (Vieno et al., 2006). Nanotechnologies solve these problems in rapid and effective ways (Braeken et al., 2006).

Due to the different properties of nano particles there are different types of nanomaterials that can be used for wastewater treatments. One of the most effective nanoparticles was magnetic nanosorbents including nano Zero Valent Iron (nZVI). nZVI have effective surface area toward different contaminant removal (El Saliby et al., 2008).

Calcium alginate is a porous biopolymer consisting of linear anionic polysaccharides that allows different contaminants to come in contact with the entrapped material (Bezbaruah et al., 2014). Hence, it has many industrial applications especially in water and wastewater treatment (Brachkova et al., 2010). Alginate beads has been used for organic contaminants removal using different techniques (Shi et al., 2008). By entrapping nZVI into alginate polymer that can reduce the separation cost of nZVI after treatment process.

Materials and Methods

Preparation of encapsulated nZVI

Synthesis of nZVI using drop by drop method; 1.0312 g $FeCl_3.6H_2O$ was dissolved in 100 ml 9/1(v/v) ethanol/water mixture. Also, 0.1 molar from $NaBH_4$ solution was prepared. The $NaBH_4$ solution was poured in a burette and added at a rate of one drop per second into ferric chloride solution and stirring manually. After complete reaction between $NaBH_4$ and $FeCl_3.6H_2O$ the black precipitate appeared. After washing and drying the precipitate using ethanol and dryer with fixed temperature 80 ^{0}C the black nZVI precipitate was ready to entrap into alginate biopolymer. Prepare and dissolve 0.2 g from sodium alginate solution in 100 ml deionized water (DW) at room temperature (22 ± 2 ^{o}C using magnetic stirrer at 600 rpm until complete dissolution was achieved. The alginate solution was mixed with 0.6 g of nZVI and mixed using magnetic stirrer at 300 rpm until all mixture turned to black. The mixture was promptly dropped into a 2% w/v $CaCl_2$ solution using peristaltic pump. Once sodium alginate/nZVI mixture came in contact with $CaCl_2$ solution the encapsulated nZVI in alginate polymer is completely prepared.

Batch adsorption studies

The adsorption of ClP onto encapsulated nZVI in alginate biopolymer (AG/nZVI) was studied by batch technique. The ClP removal efficiency was studied at initial ClP concentration: 60, 200, and 400 µg/L, pH: 3, 5, 7, and 9, adsorbent dose: 3, 4, 5, and 6 g/L, stirring rate: 150, 200, and 300 rpm, and contact time: 15, 30, 45, and 60 min. A known weight of adsorbent dosage (3, 4, 5 and 6 g/L) was equilibrated with 100 ml of the aqueous ClP solution of known concentration (60µg/L) in 250 mL of Erlenmeyer flasks and shacked at 100 rpm at a room temperature of 25 ± 2 °C for a known period of time from 45 min. After equilibration, the suspension of the adsorbent was separated from solution by filtration using Whatman No. 41 filter paper and the concentration of ClP remaining in solution was measured using Master DANI GC (fast gas chromatography) using capillary column DN5 non polar diameter 60m x 0.32mm x 0.25µm, wool linear 2 mm, split/splitless injector and FID detector according to ASTM 2005 stander method for water and wastewater (Federation and Association, 2005). The percentage of ClP compounds removal was calculated using Eq. 1. The amount of ClP sorbed by weight of encapsulated nZVI in alginate was calculated using Eq. 2.

$$\text{Sorption } [\%] = [C_o - C_e/C_o] \times 100 \qquad \text{Eq. 1}$$

where C_o is the initial concentration of ClP solutions, (µg/L) and C_e is the equilibrium concentration of ClP solutions (µg/l)

$$q_e\,[\text{mg/g}] = [[C_o - C_e]V]/\,m \qquad \text{Eq. 2}$$

where qe is the equilibrium adsorption capacity (µg/g), V is the volume of aqueous solution (L), and m is the dry weight of the adsorbent (g)

Isotherm study

Different isotherm studies were conducted for monolayer and multilayer adsorption process using nonlinear equations including Freundlich, Langmuir, Redlich–Peterson, Hill, Sips, Khan, Toth, Koble–Corrigan, Jovanovich, and Hossein, as shown in Table 1.

Table 1. Nonlinear equations of isotherm models

Isotherm model name	nonlinear equations		Constants	References
Langmuir	$Q_e = \frac{q_m K_a C_e}{1+K_a C_e}$	K_a	Langmuir constant (L/mg)	(Redlich and Peterson, 1959)
		q_m	maximum monolayer adsorption capacity (mg/g)	
Freundlich	$Q_e = K_F\, C_e^{1/n}$	n	Freundlich constant related to adsorption intensity (dimensionless)	(Redlich and Peterson, 1959)
		K_F	Freundlich constant related to adsorption capacity ((mg/g) (mg/L)-1/n)	
Redlich–Peterson	$Q_e = \frac{K_R C_e}{1+a_R C_e^{b_R}}$	K_R	Redlich–Peterson isotherm constant (L/g)	(Redlich and Peterson, 1959)
		a_R	Redlich–Peterson isotherm constant (1/mg)	
		b_R	Redlich–Peterson isotherm exponent	
Hill	$Q_e = \frac{Q_H C_e^{n_H}}{K_D + C_e^{n_H}}$	Q_H	Hill isotherm maximum uptake saturation (mg/L)	(Hill, 1910)
		K_D	Hill constant	
		n_H	Hill cooperativity coefficient of the binding interaction	
Sips	$Q_e = \frac{Q_s(K_s C_e)^{B_e}}{1+(K_s C_e)^{B_e}}$	K_s	Sips isotherm model constant (L/g)	(Koble and Corrigan, 1952)
		B_s	Sips isotherm model exponent	
Khan	$Q_e = \frac{Q_K B_K C_e}{(1+B_K C_e)^{A_K}}$	Q_k	theoretical isotherm saturation capacity (mg/g)	(Khan et al., 1997)
		B_k	Khan isotherm model constant	
		A_k	Khan isotherm model exponent	
Toth		K_t	Toth isotherm constant (mg/g)	(Toth, 1971)

	$Q_e = \frac{K_t C_e}{(a_t + C_e)^{\frac{1}{t}}}$	a_t	Toth isotherm constant (L/mg)	
		T	Toth isotherm constant	
Koble–Corrigan	$Q_e = \frac{AC_e^D}{1 + BC_e^D}$	A	Koble–Corrigan isotherm constant (Lnmg1−n/g)	(Koble and Corrigan, 1952)
		B	Koble–Corrigan isotherm constant (L/mg)n	
		D	adsorption intensity	
Jovanovich	$Q_e = q_{max}\,(1 - e^{K_j C_e})$	qmax	(mg/g) is the maximum uptake	(Jaroniec, 1975)
		KJ	Javanovic isotherm constant (L/g)	

Kinetic study

To determine the suitable time for reaching the equilibrium state, ClP solutions were placed in contact with AG/nZVI at different times, at room temperature. The amount of CLP removed at time t, Qt (mg/g), was calculated using equation 3.

$$Q_t = \frac{(C_o - C_t)V}{W} \qquad \text{Eq. 3}$$

where C_o is the initial concentration (mg/L), C_t is the concentration of ClP at time t (mg/L), V is the volume of the solution (L), and W is the AG/SSA dosages

To investigate the mechanisms of the kinetic process the following kinetics techniques were applied: pseudo-first and second-order, Avrami, Elovich, and Intraparticle. The equation of each model is listed in Table 2 (Fola et al., 2016).

Table 2. shows nonlinear equations of kinetic models

Kinetic model name	Equations
Pseudo First Order	$Q_t = Q_e(1 - e^{(-K_1 t)})$
Pseudo Second Order	$Q_t = \frac{Q_e^2 K_2 t}{(1 + Q_e K_2 t)}$
Elovich	$Q_t = \frac{1}{\beta} Ln(\alpha\beta_t)$
Avrami	$Q_t = Q_e(1 - e^{(-K_{av} t^{n_{av}})})$
Intraparticle model	$Q_t = K_{id} t^{0.5} + C_i$

Error functions calculations

Different error parameters were calculated in order to identify the most suitable isotherm and kinetic models that describe the treatment process. Due to the ingrained bias that resulting from linear form of isotherm and kinetic models, the nonlinear form was good alternative parameter which provide an accurate technique to determine isotherm and kinetic parameters from the original equation forms (Malek and Farooq, 1996, Khan et al., 1996). To evaluate the better fit of nonlinear models the error function is required (Ho et al., 2002). The five error functions were used for all models and the ideal model was selected from the summation of these errors. The error parameter equations show in Table 3 (Marquardt, 1963).

Table 3. shows errors mathematical equation

Error name	Equations
The Sum of the Squares of the Errors (ERRSQ)	$\sum_{i=1}^{p} (q_{e,meas} - q_{e,calc})_i^2$
A Composite Fractional Error Function (HYBRD)	$\sum_{i=1}^{p} \left\| \frac{(q_{e,meas} - q_{e,calc})^2}{q_{e,meas}} \right\|_i$
A Derivative of Marquardt's Percent Standard Deviation (MPSD)	$\sum_{i=1}^{p} \left(\frac{q_{e,meas} - q_{e,calc}}{q_{e,meas}} \right)_i^2$
The Average Relative Error (ARE)	$\sum_{i=1}^{p} \left\| \frac{q_{e,meas} - q_{e,calc}}{q_{e,meas}} \right\|_i$
The Sum of the Absolute Errors (EABS)	$\sum_{i=1}^{p} \left\| q_{e,meas} - q_{e,calc} \right\|_i$
Chi error	$\sum_{i=1}^{p} \frac{(q_{e,calc} - q_{e,meas})^2}{q_{e,calc}}$

Characterization of nZVI

X-Ray Diffraction patterns were recorded at a radiation wavelength (Cu K alpha = 1.5418 A°). The X-Ray current and voltage values were 40 kV and 40 mA, respectively. The diffraction angle (2θ) ranged from 5° to 70° at a step size of 0.0167°(Xi et al., 2010). The prepared sample was scanned using scanning electron microscope (SEM) to discover size and morphology of the prepared nZVI particles.

Neural network structure

An artificial neural network was developed to predict CLP removal efficiency, which consists of input, hidden, and output layers. The neural network structure was expressed as 5 – 10 – 1. The data from the five independent variables are transferred to the input layer. All the available data are normally divided into 60, 20, and 20% for training, validation, and testing procedures, respectively (Nasr et al., 2017). The network type is "Feed-forward backprop", while the training and the adaption learning functions are "Trainlm" and "Learngdm", respectively. The "tansig" and "purlin" functions were selected for hidden and output layers, respectively.

Results and Discussion

Characterization of nZVI

XRD analysis of freshly prepared nZVI powder showed presence of two characteristic Iron peak at 2 theta 44.713° and 64.9° with Beta degree 0.384 and0.166, respectively. By applying the XRD results using Bragg's law we can calculate the crystal size was between 23 and 59 nm as shown Fig. 1a. Fig. 1b shows the SEM image which has been taken 1 day after preparation and it is clear that the prepared nZVI particles have formed large nanoclusters due to agglomeration. The nZVI particles also appeared as semi-spherical shape with an average size of 40 nm.

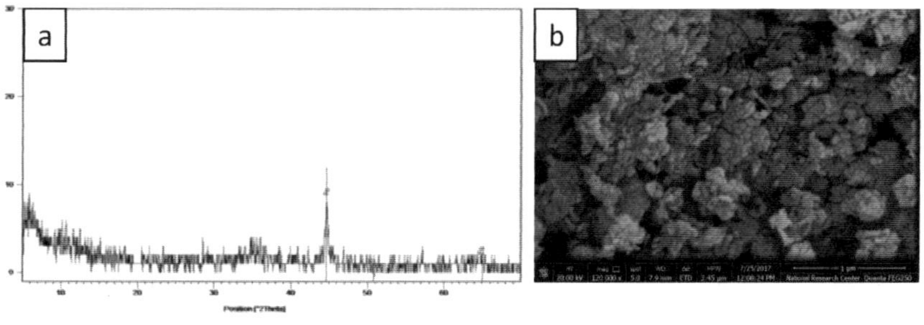

Figure 1. Characterization of nZVI: a) XRD and b) SEM analysis

Effect of operating parameters

Effect of contact time

The effect of contact time on ClP removal efficiency was studied at entrapped nZVI dosage: 40 g/L, temperature: 25 ± 3 °C, pH: 7 ± 0.2, initial ClP concentration: 60 μg/L, and stirring rate: 100 rpm (Fig. 2a). An increase in ClP removal efficiency from 46 to 52, 55, and 56% was associated with an increase in contact time from 15 to 30, 45, and 60 min, respectively. At the beginning of adsorption, a large number of abundant sites on the AG/nZVI surface were available to adsorb greater amounts of the ClP. The minimum effective time was recorded as 45 min. The effective time was selected as 45 min due to the small variation of the removal efficiencies between times 45 and 60 min.

Effect of nZVI dose

The effect of nZVI dosage on ClP removal efficiency was studied at contact time: 45 min, temperature: 25 ± 3 °C, pH: 7 ± 0.2, initial ClP concentration: 60 μg/L, and stirring rate: 100 rpm (Fig. 2b). An increase in ClP removal efficiency from 45 to 55, 64, and 74% was associated with an increase in adsorbent doses from 30 to 40, 50, and 60 g/L, respectively. A large number of vacant sites on AG/nZVI surface were available as a result of increasing the adsorbent dosage, which adsorbed large amounts of ClP.

Effect of pH

The effect of pH on ClP removal efficiency was studied at contact time: 45 min, temperature: 25 ± 3 °C, entrapped nZVI dosage: 40 g/L, initial ClP concentration: 60 μg/L, and stirring rate: 100 rpm (Fig. 2c). An increase in ClP removal efficiency from 23% to 43, 46, and 55% was associated with an increase in pH from 3 to 9, 5, and 7, respectively. This result indicated that the most efficient media to remove ClP at pH range 7.

Effect of stirring rate

The effect of stirring rate on ClP removal efficiency was studied at contact time: 45 min, temperature: 25 ± 3°C, entrapped nZVI dosage: 40 g/L, initial ClP concentration: 60 μg/L, and pH: 7 ± 0.2 (Fig. 2d). The removal efficiency of ClP was 55% at a stirring rate of 100 rpm, and increased to 56 and 57% at a stirring rate of 200 and 300 rpm, respectively. This is mainly due to improving the opportunity of AG/nZVI to come in contact with ClP. Since, no big difference was observed when increasing stirring rate, a stirring rate of 100 rpm was selected for this study.

Effect of the initial concentration

The effect of initial ClP concentration on its removal efficiency was studied at contact time: 45 min, temperature: 25 ± 3 °C, entrapped nZVI dosage: 40 g/L, stirring rate: 100 rpm, and pH: 7 ± 0.2 (Fig. 2e). A decrease in ClP removal efficiency from 55% to 42, and 41% was associated with an increase in ClP concentration from 60 μg/L to 200 and 400 μg/L, respectively. At low ClP concentrations, the ratio of abundant sites to the total ClP in the solution was high, and thus, all ClP were bound to the AG/nZVI surface.

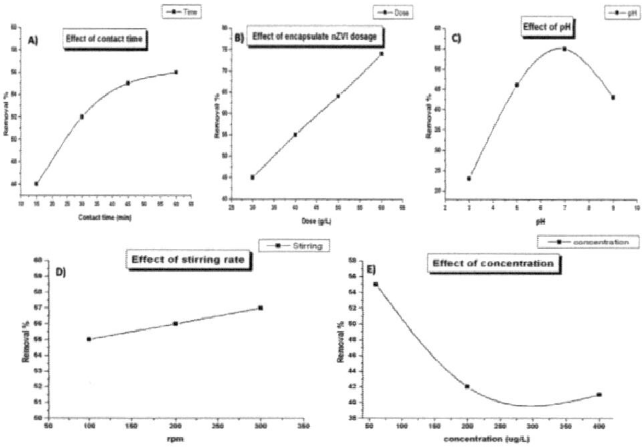

Figure 2. Effect of operating parameter in ClP removal using entrapped nZVI in alginate biopolymer: a) contact time, b) nZVI dosage, c) pH, d) stirring rate, and e) ClP concentration

Isotherm study

Fig. 3 describe nonlinear relations between different adsorption isotherm models. The resulted data from adsorption study using nonlinear models indicating that Sips model is the most suitable model that describe the adsorption process with lowest summation of error of 0.000001, as shown Table 4. Sips isotherm model (Koble and Corrigan, 1952) describe heterogeneous adsorption isotherm process (Günay et al., 2007) and combined form Langmuir and Freundlich model. This model tend to be Freundlich model at low concentration and solve the Freundlich limitation of high concentration during the prediction of monolayer adsorption (Langmuir adsorption model) (Pérez-Marín et al., 2007).

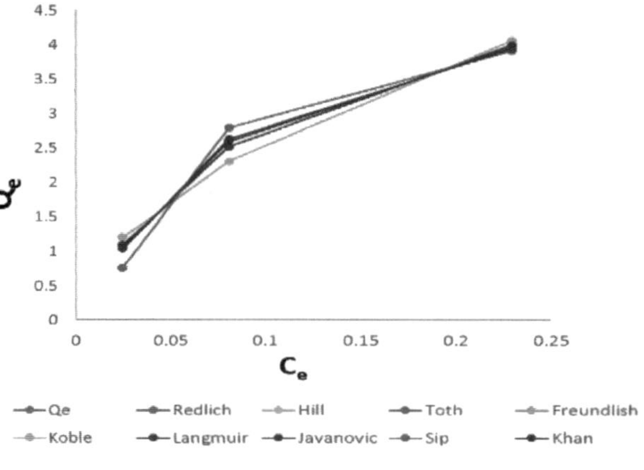

Figure 3. Isotherm studies for nonlinear models

Table 4. shows the results of nonlinear adsorption models

	Redlich- Peterson		Hill		Sips		Khan		Toth	
Constants	Kr	53.940	QH	4.202	Qs	4.202	Qk	4.080	Kt	2.432
	Br	9.114	nH	1.856	Ks	17.85	Bk	2.590	at	0.109
	G	61.03	KD	0.004	Bs	1.856	Ak	0.593	t	2.432
					ERRORS					
Chi error	0.134		0.000		0.000		0.084		0.134	
ERRSQ	0.197		0.000		0.000		0.105		0.197	
HYBRD	0.174		0.000		0.000		0.110		0.174	
MPSD	0.199		0.000		0.000		0.134		0.199	
ARE	0.557		0.000		0.000		0.428		0.557	
EABS	0.700		0.001		0.000		0.472		0.700	
Error sum	1.961		0.001		**0.000**		1.334		1.961	

	Koble –Corrigan		Javanovic		Freundlich		Langmuir	
Constants	A	874.7	Qm	4.27	Kf	9.100	Q_0	5.918
	B	207.9	Kj	11.509	n	1.827	b	9.114
	D	1.854						
				ERRORS				
Chi error	0.000		0.102		0.275		0.134	
ERRSQ	0.000		0.135		0.465		0.197	
HYBRD	0.000		0.135		0.350		0.174	
MPSD	0.000		0.161		0.369		0.199	
ARE	0.006		0.481		0.795		0.557	
EABS	0.010		0.558		1.089		0.700	
Error sum	0.016		1.572		3.344		1.961	

Kinetics Studies

The results from the adsorption data indicated that pseudo second order is the preferred model with lowest summation errors 0.006, as shown in Table 5. This result also indicated that the desorption of ClP on entrapped nZVI in alginate biopolymer depends on two main reactants and alginate nZVI dose, as shown Fig. 4. This result also indicated that the desorption could be chemisorption which occurs through exchange or sharing of electrons between ClP and AG/nZVI, and pseudo second order model constants were appropriate for representing the adsorption kinetics.

Table 5. shows the results of different Kinetic models.

	P.F.O	P.S.O	Elovich	Avrami	Intraparticle
Constants	$Q_e = 0.768$ $K_1 = 0.112$	$Q_e = 0.847$ $K_2 = 0.232$	$\alpha = 0.938$ $\beta = 7.577$	$Q_e = 0.956$ $K_{av} = 0.455$ $N_{av} = 0.325$	$K_{id} = 0.038$ $C_i = 0.497$
			ERRORS		
Chi error	0.001	0.000	0.001	0.000	0.001
ERRSQ	0.000	0.000	0.001	0.000	0.001
HYBRD	0.001	0.000	0.001	0.000	0.001
MPSD	0.001	0.000	0.002	0.000	0.001
ARE	0.049	0.003	0.075	0.030	0.064
EABS	0.035	0.002	0.054	0.022	0.045
Error sum	0.087	**0.006**	0.134	0.053	0.113

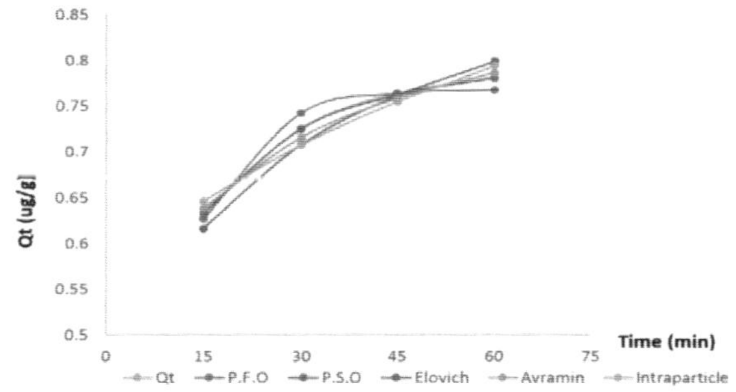

Figure 4. Kinetic isotherm models

Artificial neural network

Fig. 5a shows the training process which is characterized by 6 validation checks, 6 epochs, and gradient value equal 2.8784. The training step was terminated because the gradient value has exceeded the value of the least error (1e-5). Fig. 5b shows the mean squared error performance versus epoch number. The MSE of training process decreased with the increase in the number of iterations, while the MSE of the validation and test processes increased with the increase in the number of iterations. An increase in the MSE of the validation process may refer to overfitting, but in this case the overfitting has not occurred because the test and validation curves were comparable. These results revealed that the best validation performance was 0.0001363 at epoch 0. Fig. 5c presents the regression plot between targets and outputs. High r^2-value for training, validation and test plots indicated that the proposed ANN has fitted the data well, which is recorded as 0.9194, 1, and 0.9996, respectively. The overall r^2-value was recorded as 0.9316, which indicated that about 93.16% of the variations in ClP removal efficiency were described by the studied factors, while the remaining 6.84% was not described by the proposed model.

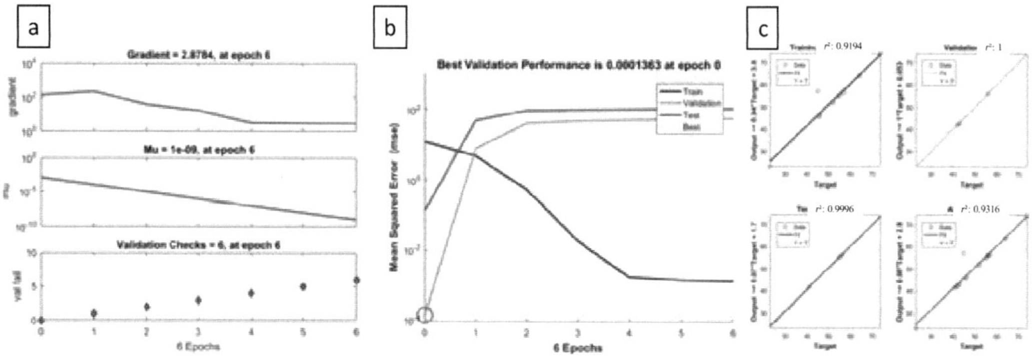

Figure 5. ANN results: a) Training, b) Best validation performances, and c) Regression plot

Conclusions

Results from the batch studies conducted during this research indicated that encapsulated nZVI in alginate was highly efficient on organochlorine pesticides removal from aqueous solutions. The adsorption data of ClP fitted well to Sips isotherm model and pseudo second order kinetic model. The highest removal efficiency of 74% was achieved at initial ClP concentration 60 μg/L, pH 7, stirring rate 100 rpm, contact time 45 min, and AG/nZVI dosage 60 g/L. The proposed ANN was found to be effective in simulating the performance of AG/nZVI for ClP removal, as indicated by the high overall r^2-value of 0.9316. The created ANN can then be used to predict ClP removal efficiencies for different sets of input data.

References

AMIN, M., ALAZBA, A. & MANZOOR, U. 2014. A review of removal of pollutants from water/wastewater using different types of nanomaterials. *Advances in Materials Science and Engineering,* 2014.

BEZBARUAH, A., ALMEELBI, T. B., QUAMME, M. & KHAN, E. 2014. Calcium-alginate entrapped nanoscale zero-valent iron (nzvi). Google Patents.

BRACHKOVA, M. I., DUARTE, M. A. & PINTO, J. F. 2010. Preservation of viability and antibacterial activity of Lactobacillus spp. in calcium alginate beads. *European Journal of Pharmaceutical Sciences,* 41, 589-596.

BRAEKEN, L., BETTENS, B., BOUSSU, K., VAN DER MEEREN, P., COCQUYT, J., VERMANT, J. & VAN DER BRUGGEN, B. 2006. Transport mechanisms of dissolved organic compounds in aqueous solution during nanofiltration. *Journal of membrane science,* 279, 311-319.

COLBORN, T. & SMOLEN, M. 1996. Epidemiological analysis of persistent organochlorine contaminants in cetaceans. *Reviews of environmental contamination and toxicology.* Springer.

EL SALIBY, I., SHON, H., KANDASAMY, J. & VIGNESWARAN, S. 2008. Nanotechnology for Wastewater Treatment: in brief. *Encyclopedia of Life Support System (EOLSS).*

FEDERATION, W. E. & ASSOCIATION, A. P. H. 2005. Standard methods for the examination of water and wastewater. *American Public Health Association (APHA): Washington, DC, USA.*

FLEMING, L., MANN, J. B., BEAN, J., BRIGGLE, T. & SANCHEZ-RAMOS, J. R. 1994. Parkinson's disease and brain levels of organochlorine pesticides. *Annals of neurology,* 36, 100-103.

FOLA, A. T., IDOWU, A. A. & ADETUTU, A. 2016. Removal of Cu2+ from aqueous solution by adsorption onto quail eggshell: Kinetic and isothermal studies. *Journal of Environment & Biotechnology Research,* 5, 1-9.

GOLBE, L. I. 1993. Risk factors in young-onset Parkinson's disease. *Neurology,* 43, 1641-1641.

GÜNAY, A., ARSLANKAYA, E. & TOSUN, I. 2007. Lead removal from aqueous solution by natural and pretreated clinoptilolite: adsorption equilibrium and kinetics. *Journal of Hazardous Materials,* 146, 362-371.

GURUGE, K. & TANABE, S. 2001. Contamination by persistent organochlorines and butyltin compounds in the west coast of Sri Lanka. *Marine Pollution Bulletin,* 42, 179-186.

HILL, A. V. 1910. The possible effects of the aggregation of the molecules of haemoglobin on its dissociation curves. *J Physiol (Lond),* 40, 4-7.

HO, Y., PORTER, J. & MCKAY, G. 2002. Equilibrium isotherm studies for the sorption of divalent metal ions onto peat: copper, nickel and lead single component systems. *Water, Air, and Soil Pollution,* 141, 1-33.

JARONIEC, M. 1975. Adsorption of gas mixtures on homogeneous surfaces Extension of Jovanovic equation on adsorption from gaseous mixtures. *Chem. zvesti,* 29, 512.

JAWARD, F. M., FARRAR, N. J., HARNER, T., SWEETMAN, A. J. & JONES, K. C. 2004. Passive air sampling of PCBs, PBDEs, and organochlorine pesticides across Europe. *Environmental Science & Technology,* 38, 34-41.

KHAN, A., AL-WAHEAB, I. & AL-HADDAD, A. 1996. A generalized equation for adsorption isotherms for multi-component organic pollutants in dilute aqueous solution. *Environmental technology,* 17, 13-23.

KHAN, A., ATAULLAH, R. & AL-HADDAD, A. 1997. Equilibrium adsorption studies of some aromatic pollutants from dilute aqueous solutions on activated carbon at different temperatures. *Journal of colloid and interface science,* 194, 154-165.

KOBLE, R. A. & CORRIGAN, T. E. 1952. Adsorption isotherms for pure hydrocarbons. *Industrial & Engineering Chemistry,* 44, 383-387.

MALEK, A. & FAROOQ, S. 1996. Comparison of isotherm models for hydrocarbon adsorption on activated carbon. *AIChE Journal,* 42, 3191-3201.

MARQUARDT, D. W. 1963. An algorithm for least-squares estimation of nonlinear parameters. *Journal of the society for Industrial and Applied Mathematics,* 11, 431-441.

PARWEEN, M., RAMANATHAN, A., KHILLARE, P. & RAJU, N. 2014. Persistence, variance and toxic levels of organochlorine pesticides in fluvial sediments and the role of black carbon in their retention. *Environmental Science and Pollution Research,* 21, 6525-6546.

PÉREZ-MARÍN, A., ZAPATA, V. M., ORTUNO, J., AGUILAR, M., SÁEZ, J. & LLORÉNS, M. 2007. Removal of cadmium from aqueous solutions by adsorption onto orange waste. *Journal of hazardous materials,* 139, 122-131.

PETROVIC, M., DIAZ, A., VENTURA, F. & BARCELÓ, D. 2003. Occurrence and removal of estrogenic short-chain ethoxy nonylphenolic compounds and their halogenated derivatives during drinking water production. *Environmental science & technology,* 37, 4442-4448.

REDLICH, O. & PETERSON, D. L. 1959. A useful adsorption isotherm. *Journal of Physical Chemistry,* 63, 1024-1024.

SHI, L., HARMS, H. & WICK, L. Y. 2008. Electroosmotic flow stimulates the release of alginate-bound phenanthrene. *Environmental science & technology,* 42, 2105-2110.

TOTH, J. 1971. State equations of the solid-gas interface layers. *Acta Chim Acad Sci Hungar,* 69, 311-328.

VERSTRAETEN, I. M., HEBERER, T., VOGEL, J. R., SPETH, T., ZUEHLKE, S. & DUENNBIER, U. 2003. Occurrence of endocrine-disrupting and other wastewater compounds during water treatment with case studies from Lincoln, Nebraska and Berlin, Germany. *Practice periodical of hazardous, toxic, and radioactive waste management,* 7, 253-263.

VIENO, N., TUHKANEN, T. & KRONBERG, L. 2006. Removal of pharmaceuticals in drinking water treatment: effect of chemical coagulation. *Environmental Technology,* 27, 183-192.

WESTERHOFF, P., YOON, Y., SNYDER, S. & WERT, E. 2005. Fate of endocrine-disruptor, pharmaceutical, and personal care product chemicals during simulated drinking water treatment processes. *Environmental Science & Technology,* 39, 6649-6663.

WURL, O. & OBBARD, J. P. 2005. Organochlorine pesticides, polychlorinated biphenyls and polybrominated diphenyl ethers in Singapore's coastal marine sediments. *Chemosphere,* 58, 925-933.

XI, Y., MALLAVARAPU, M. & NAIDU, R. 2010. Reduction and adsorption of Pb 2+ in aqueous solution by nano-zero-valent iron—a SEM, TEM and XPS study. *Materials Research Bulletin,* 45, 1361-1367.

498387 Bilevel Optimization Applied to the Exhaustive Exploration of Pareto Fronts in Sustainability Studies: Application to the Redesign of the UK Electricity Mix

Abstract:

Sustainability principles have recently emerged as a powerful driver in our society that aims to become more environmentally-friendly and socially responsible. The practical implementation of sustainability principles is deemed challenging, since this task requires the simultaneous satisfaction of economic, environmental and social objectives, where inherent trade-offs naturally arise. In addition, these criteria are difficult to compare and ultimately aggregate into a single indicator (Bhaskar et al., 2000). Therefore, there is still a clear need to devise systematic tools to assist in the design and planning of sustainable systems that could provide an effective and meaningful analysis of sustainability performance.

Multi-objective optimization (MOO) has recently become the preferred approach to assist in the design and planning of sustainable systems (Grossmann and Guillen-Gosalbez, 2010). MOO provides a set of Pareto alternatives rather than a single optimal solution. This set of Pareto optima forms the Pareto frontier, which features non-inferiority (i.e. no objective can be improved without worsening at least another one). Furthermore, MOO requires no articulation of preferences via weights, a controversial step in sustainability problems.

Here, we focus on the systematic exploration of Pareto fronts so as to support decision-making in the post-optimal analysis of the Pareto solutions of MOO models. We mathematically pose the task of exploring the Pareto frontier of an MOO model as a bilevel programming problem. Bilevel programming problems consist of two nested optimization problems where the outer problem is constrained by the optimal solution of the inner one. The outer problem is generally referred to as upper-level problem or leader, while the inner one corresponds to the lower-level problem or follower. The combine use of MOO and bilevel programming was addressed in other works (Calvete and Galé, 2010; Yin, 2002; Ankhili and Mansouri, 2009; Calvete and Galé, 2011), but not in the same manner as we do here. More precisely, in our case the inner problem systematically explores the Pareto front of the MOO model while the outer problem seeks the point in the frontier that best reflects the decision-makers' preferences.

We illustrate the capabilities of this approach through its application to the optimisation of the UK electricity mix according to specific sustainability criteria. The task of redesigning the UK mix is formulated as a linear bi-level programming problem where several economic, environmental and social indicators are considered. This bi-level model is solved to find the closest Pareto point to the current mix considering different distance metrics.

Overall, our approach is an effective technique to support decision-making in the post-optimal analysis of Pareto frontiers. This approach is capable of systematically identifying, from the set of Pareto solutions, the best one according to the decision-makers' preferences, without relying on controversial weighting schemes.

Furthermore, it establishes improvement targets for suboptimal alternatives and provides valuable guidelines on how to improve them. The model also offers the flexibility to predefine the strategy that best matches the users' preferences.

References

Ankhili, Z., Mansouri, A., 2009. An exact penalty on bilevel programs with linear vector optimization lower level. Eur. J. Oper. Res. 197, 36–41. doi:10.1016/j.ejor.2008.06.026

Bhaskar, V., Gupta, S.K., Ray, A.K., 2000. Applications of multiobjective optimization in Chemical Engineering. Rev. Chem. Eng. 16, 1–54. doi:10.1515/REVCE.2000.16.1.1

Calvete, H.I., Galé, C., 2010. Linear bilevel programs with multiple objectives at the upper level. J. Comput. Appl. Math. 234, 950–959. doi:10.1016/j.cam.2008.12.010

Calvete, H.I., Galé, C., 2011. On linear bilevel problems with multiple objectives at the lower level. Omega 39, 33–40. doi:10.1016/j.omega.2010.02.002

Grossmann, I.E., Guillen-Gosalbez, G., 2010. Scope for the application of mathematical programming techniques in the synthesis and planning of sustainable processes. Comput. Chem. Eng. 34, 1365–1376. doi:10.1016/j.compchemeng.2009.11.012

Yin, Y., 2002. Multiobjective bilevel optimization for transportation planning and management problems. J. Adv. Transp. 36, 93–105. doi:10.1002/atr.5670360106

499008 Applying Emission Estimations, Pollution Controls, and Sustainability Evaluations for Generating Chemical Life Cycle Inventories

Abstract:

This presentation will provide a description of an integrated approach to guide, collect, and estimate life cycle inventory (LCI) data for the application of sustainability evaluation and life cycle assessment (LCA) in the design and manufacturing of more sustainable chemicals. Chemical process design and simulation are employed to generate a process flow diagram, energy and material flows, utility demand, and equipment sizing of the chemical manufacturing system. Additional techniques developed by the U.S. Environmental Protection Agency in collaboration with West Virginia University for estimating uncontrolled [1], [2] and controlled [3] emissions from chemical processing equipment are applied to obtain a detailed emission profile for the process.

Moreover, the GREENSCOPE (Gauging Reaction Effectiveness for the Environmental Sustainability of Chemistries with a multi-Objective Process Evaluator) tool is implemented for executing a sustainability assessment when designing and evaluating a chemical manufacturing stage. The GREENSCOPE sustainability evaluation employs environmental, economics, energy, and efficiency indicators. In addition, as one of the main features of GREENSCOPE, a LCI for the chemical synthesis route is obtained. This LCI is an important step in developing life cycle assessments, which determine environmental, economic, and social effects throughout a product's life cycle stages.

The developed methodologies are then applied to a case study of a bio-based chemical production and energy recovery process. The results show improvements in the sustainability evaluation and LCI quantification when compared to commonly used process simulator results. Therefore, the inclusion of inputs, controlled, and uncontrolled emission models enhance the quantitative sustainability evaluation and LCI results, and the subsequent LCAs of chemicals during their manufacturing life cycle stages.

The views expressed in this presentation are those of the authors and do not necessarily represent the views or policies of the U.S. Environmental Protection Agency.

---

[1] Cashman et al. (2016). "Mining Available Data from the United States Environmental Protection Agency to Support Rapid Life Cycle Inventory Modeling of Chemical Manufacturing," *Environ. Sci. Technol.*, 50, 9013-9025.

[2] Smith et al. (2017). "Coupling Computer-Aided Process Simulation and Estimations of Emissions and Land Use for Rapid Life Cycle Inventory Modeling," *ACS Sustainable Chem. Eng.* (Accepted).

[3] Li et al. (2017). "Process and pollution control for sustainability and LCI: application to a bio-based gasification system," *Clean Technol. Envir.* (to be submitted).

498996 A Decision Support Framework for Sustainable Manufacturing

Abstract:

From the sustainability science point of view, industrial sustainability involves a series of system state transitions in an economics-environment-society-based three-dimensional space. The study of system state transition requires methods for conducting continuous sustainability assessment, decision analysis, and decision making in order to derive best possible sustainability actions for implementation to achieve sustainability goals in different decision stages. In this paper, we introduce a new, general decision support framework for achieving short-to-long-term goals for sustainable manufacturing. This framework involves design of a decision-making system that can derive decisions at both the strategic planning and tactical control layers. The strategic planning is designed to assist decision makers in developing optimal short-to-long-term sustainability development (SD) strategies based on system sustainability performance, environmental regulation, etc. The tactical control, on the other hand, is constructed to derive optimal sustainability actions and implement SD strategies in a dynamic domain. Time series forecasting technique will be used to deal with various types of uncertainty information, such as that related to market, material/energy price, etc. This two-layer decision-making framework can provide a structured step-by-step guide to industry in defining SD strategies and deriving sustainability actions systematically. A comprehensive case study on nanocoating manufacturing sustainability is illustrated to demonstrate methodological efficacy.

499160 Techno-Ecological Synergies in Life Cycle Assessment: A General Computational Framework

Abstract:

Life cycle assessment (LCA) is among the most widely used methods to support sustainable engineering. Generalized computational structures have been developed for process model based LCA, input-output LCA, and hybrid LCA [1-3]. However, LCA does not explicitly account for the role of ecosystem services (ES) and the capacity of ecosystems to supply the services demanded throughout the life cycle. Also, the conclusions made by LCA are based on comparisons between alternatives, measuring only relative sustainability performance. This approach encourages technological systems to do "less bad" by reducing environmental impact rather than "more good" by protecting and restoring ecosystems. Therefore, decisions based on relative sustainability and without accounting for the role of ecosystems may, (1) unintentionally increase reliance on scarce ES, and (2) miss out on ecological opportunities for improving the life cycle.

Furthermore, to quantify absolute environmental sustainability, the product's life cycle impacts need to be compared with absolute environmental references such as ecological capacity in the serviceshed of the selected ecosystem service. The framework of techno-ecological synergy (TES) aims to address these challenges [4] by explicitly including the role of ecosystem in mitigating impacts, and referring to ecological carrying capacity as the base for absolute environmental sustainability.

This work develops a rigorous computational framework for incorporating techno-ecological synergies into life cycle assessment, and identifies practical challenges and approaches to enable the proposed extension. The resulting TES-LCA framework extends the computational structure of LCA to explicitly account for ecosystems. Just as technological systems are included in LCA as modules, TES-LCA also includes ecosystems as modules. Interaction between various ecological processes and with technological systems are also captured. Metrics are also defined to quantify ecological overshoot at multiple spatial scales. The basic TES-LCA computational structure is then adapted to account for ES allocation, regional variations and absolute environmental sustainability at the largest scale at which the ES operates (serviceshed). This framework can thus provide unique insights about absolute environmental sustainability for both individual activities and product's life cycles.

Applications demonstrate that compared to conventional LCA, TES-LCA can capture interaction between ES in an explicit manner, account for absolute sustainability, and identify novel improvement strategies through ecosystem restoration. The framework will be applied to a biofuel supply chain to demonstrate the robustness of TES-LCA while identifying the associated challenges, which point out the directions for future research work.

References:

[1] Hunt, Robert G., et al. "Case studies examining LCA streamlining techniques." The International Journal of Life Cycle Assessment 3.1 (1998): 36-42.

[2] Joshi, Satish. "Product environmental life-cycle assessment using input-output techniques." Journal of industrial ecology 3.2-3 (1999): 95-120.

[3] Suh, Sangwon, et al. "System boundary selection in life-cycle inventories using hybrid approaches." Environmental science & technology 38.3 (2004): 657-664.

[4] Bakshi, Bhavik, Guy Ziv, and Michael Lepech. "Techno-ecological synergy: A framework for sustainable engineering." Environmental Science & Technology 49.3 (2015): 1752-1760.

Systematic Analysis and Optimization of Power Generation in Pressure Retarded Osmosis: Effect of Multi-Stage Design

Mingheng Li

November 2, 2017

Outline

Motivation

Mingheng Li | Multi-stage PRO | AIChE 2017 | 3 / 24

Global Grand Challenges in Energy and Water

Humanity's Top Ten Problems for next 50 years

1. **ENERGY**
2. WATER
3. FOOD
4. ENVIRONMENT
5. POVERTY
6. TERRORISM & WAR
7. DISEASE
8. EDUCATION
9. DEMOCRACY
10. POPULATION

2003 6.3 Billion People
2050 8-10 Billion People

[1] Professor Richard Smalley, Nobel Laureate

Mingheng Li | Multi-stage PRO | AIChE 2017 | 4 / 24

Membrane-Based Processes

- Reverse Osmosis (RO)

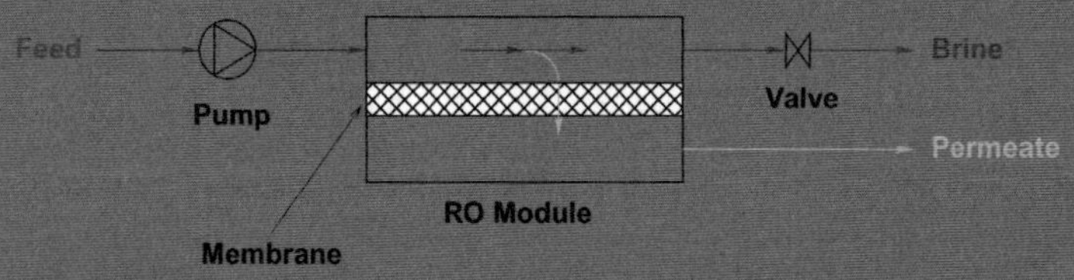

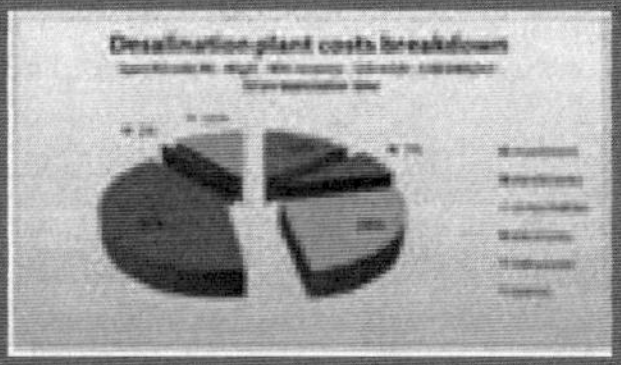

- Pressure Retarded Osmosis (PRO)

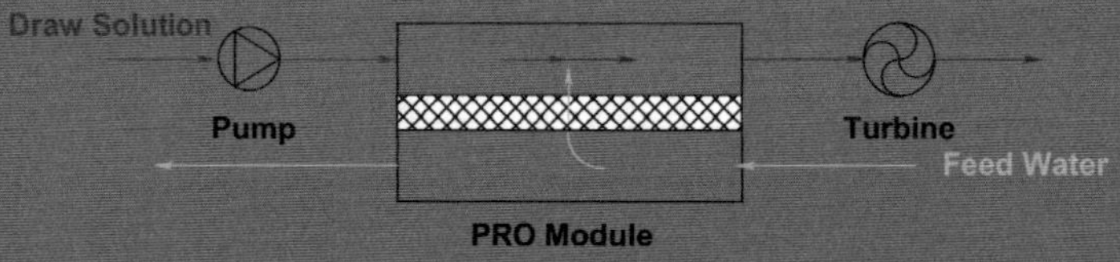

[1] https://www.lenntech.com/processes/desalination/energy/general/desalination-costs.htm

Need for System-Level Understanding of PRO

- There is a lack of computational framework to address scale-up issues in PRO
  - Optimal applied pressure much lower than $\Delta\pi/2$ is observed in some experiments (Xu et al., JMS, 2009; She et al., JMS, 2012; Sharif et al., Membranes, 2014).
  - Power generation in PRO is much lower than theoretical prediction.
- Mathematical models helps elucidate the understanding of energy issues in PRO.

Methodology

Characteristic Equation of RO and PRO

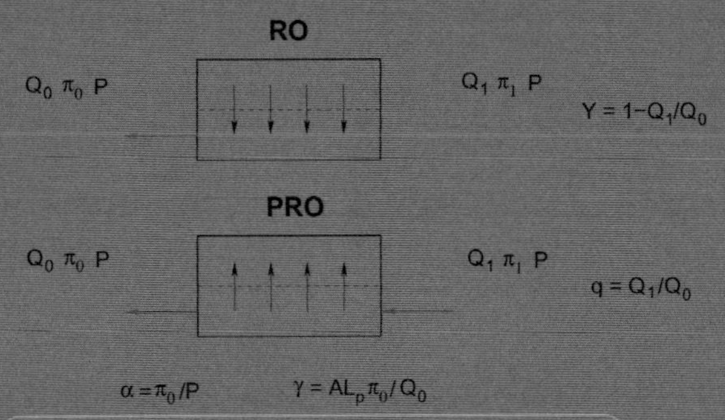

Assumptions:

- Negligible pressure drop
- Negligible concentration polarization
- Perfect transverse mixing of salt

$\Rightarrow$

γ:	Dimensionless membrane area	→ Design parameter
α^{-1}:	Dimensionless applied pressure	→ Operating parameter
q:	Dimensionless flow rate ($Y = 1 - q$)	→ Performance parmater

[1] Li, IECR, 2010; AIChE J., 2015

Model Accounting for PRO Efficiency

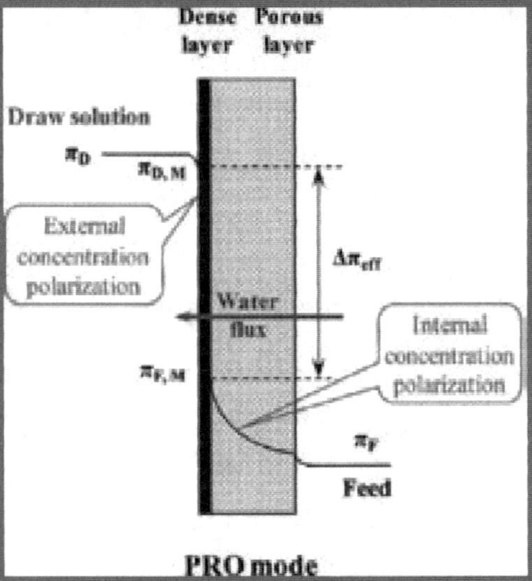

- Detrimental Effects in PRO
  - External Concentration Polarization (ECP)
  - Internal Concentration Polarization (ICP)
  - Reverse Salt Leakage
- Equation accounting for PRO efficiency

[1]Wang, Ong, and Chung, IECR, 2010
[2]Li, AIChE J., 2017

Energy Production in PRO

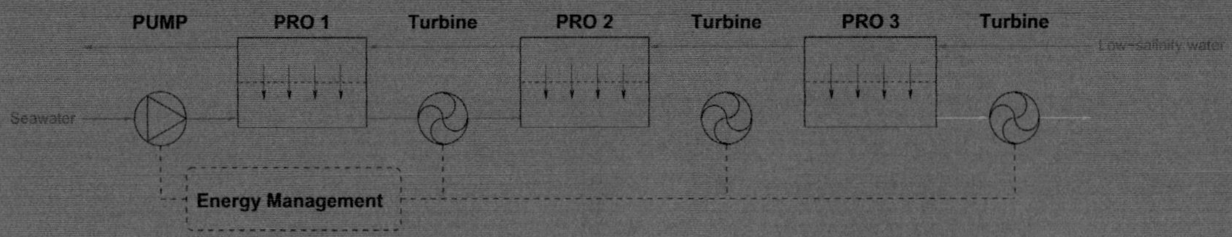

- Normalized Specific Energy Production (NSEP) in PRO
- NSEP in Single- or Multi-stage PRO

[1]Li, AIChE J., 2017

Optimization Model for PRO

[1] Li, AIChE J, 2017

Results & Discussion

Results of NSEP in PRO

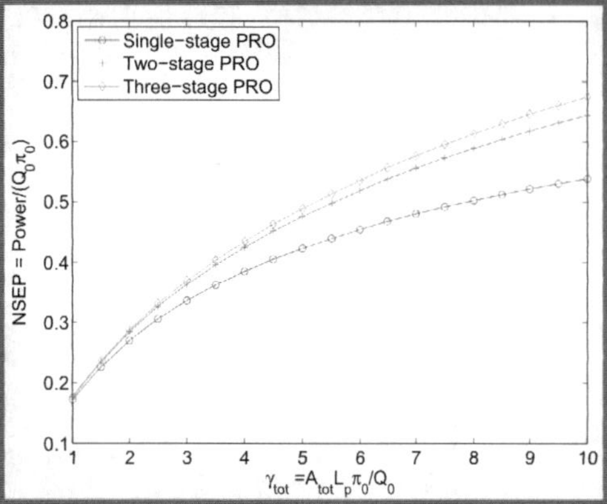

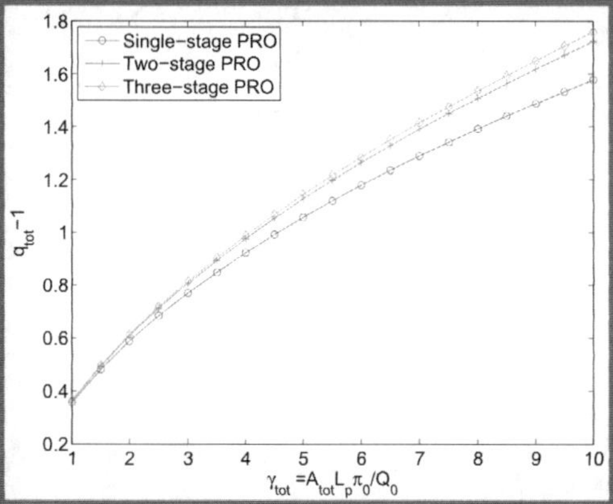

- The increased volume gain is not the sole cause to the boosted power generation in multi-stage design.
- $NSEP = \sum_{j=1}^{k}(q_j - 1)/\alpha_j$ (if ignoring equipment efficiencies), implying pressure profile has a contribution to NSEP as well.

Pressure Profile $\gamma_{tot} = 4$ (top) and 10 (bottom)

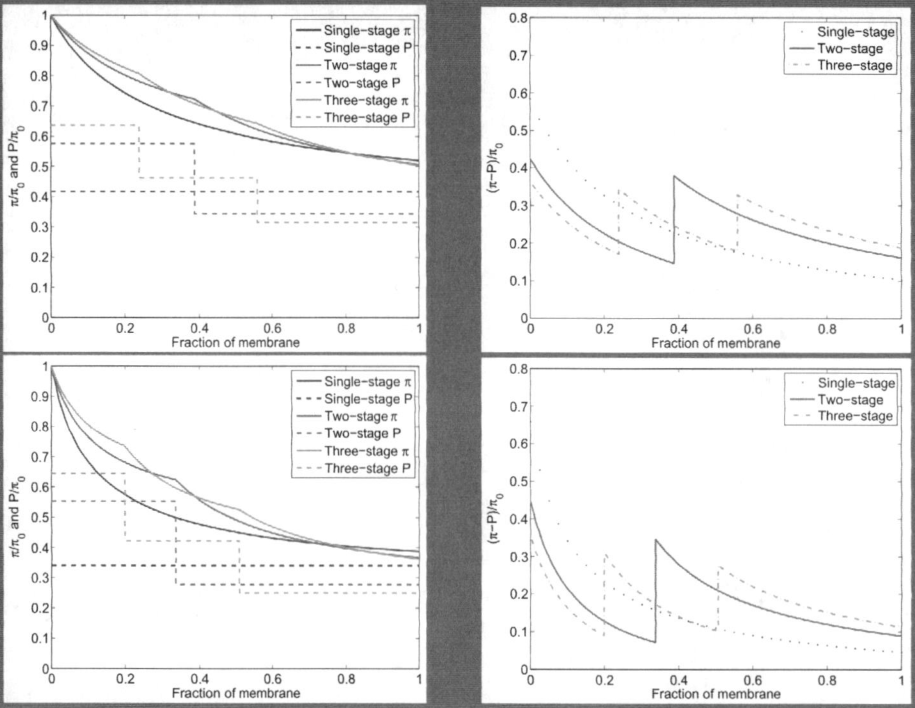

Power Density

- Relationships between NSEP and Power Density (PD)

- *PD* vs L_p — *PD* vs π_0 — *PD* vs Q_0/A

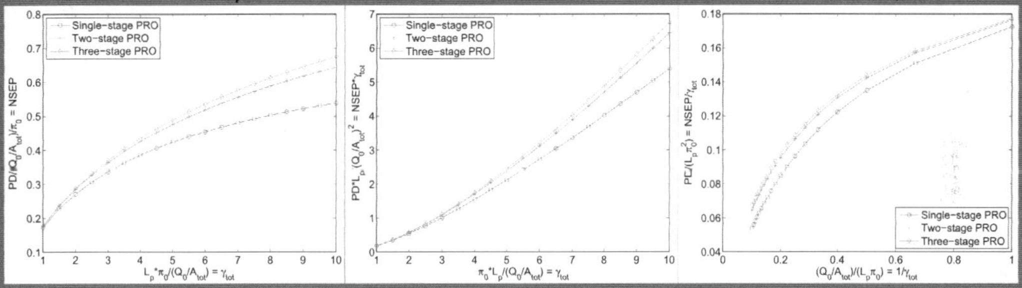

Optimization at the Thermodynamic Limit

- As $\gamma_j \to \infty$, driving force $\to 0$. i.e. $\alpha_j = q_j$
- Optimization model

- Solution

Optimization at the Thermodynamic Limit

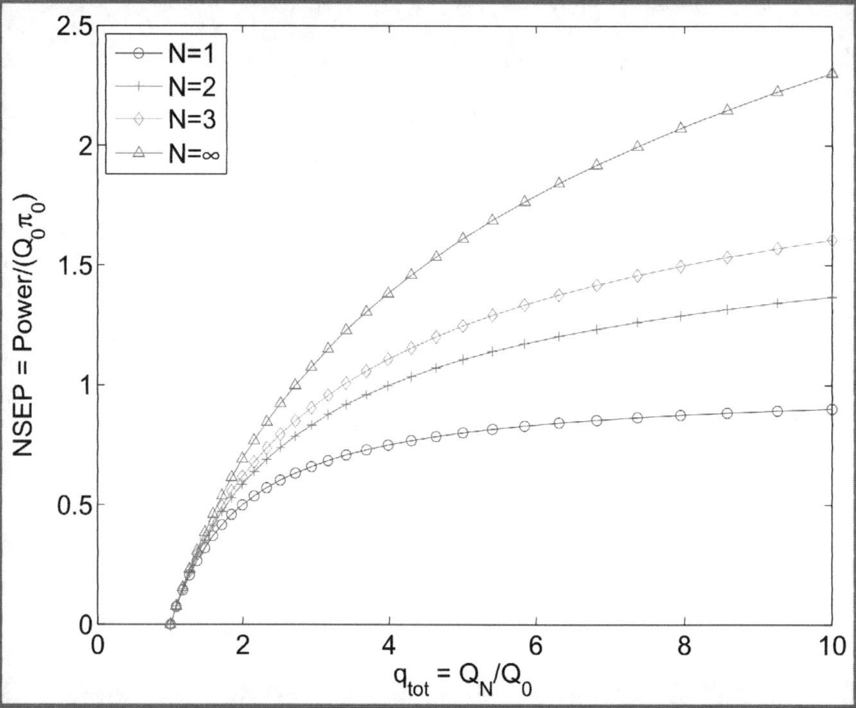

Infinite Stages at the Thermodynamic Limit

- Infinite PRO stages at the thermodynamic limit

- Fully reversible PRO

How Far from the Thermodynamic Limit

γ_{tot}=4	Single-Stage	Two-Stage	Three-Stage
q_{tot}	1.9241	1.9774	1.9895
NSEP	0.3852	0.4251	0.4347
NSEP (eq)	0.4803	0.5777	0.6147
NSEP(γ_{tot}=4)/NSEP(eq)	80%	74%	71%
γ_{tot}=10	Single-Stage	Two-Stage	Three-Stage
q_{tot}	2.5784	2.7234	2.7595
NSEP	0.5388	0.6440	0.6751
NSEP (eq)	0.6122	0.7881	0.8612
NSEP(γ_{tot}=10)/NSEP(eq)	88%	82%	78%

Optimization Accounting for Process Efficiencies

- η_{PRO} = 80%, η_{turb} = 80%, and η_{pump} = 80%

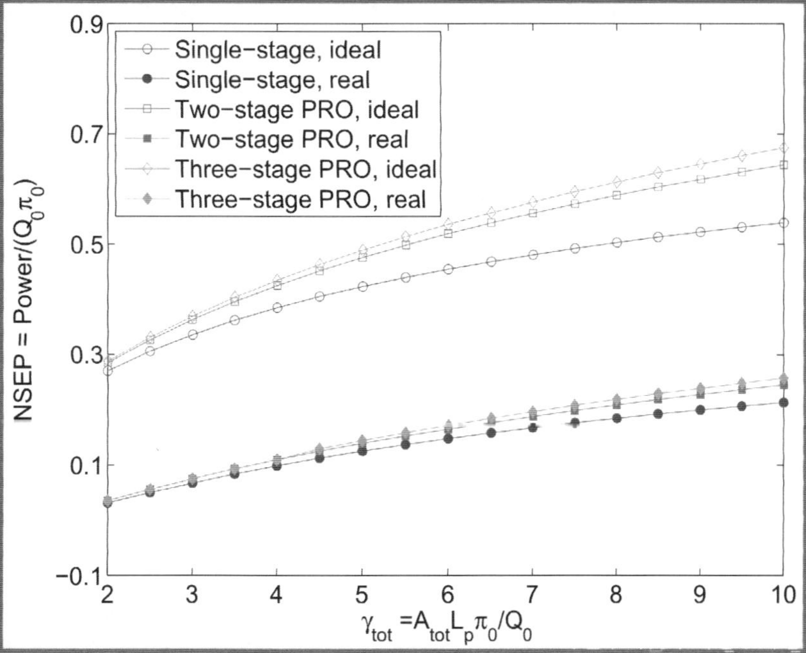

Conclusions

Conclusions

- NSEP increases with number of stages and $\gamma_{tot} = A_{tot} L_p \pi_0 / Q_0$.
- Multistage not only improves volume gain ratio, but also allows a stage-dependent, progressively decreasing pressure, both enhancing power production.
- At the thermodynamic limit where γ_{tot} goes to infinity, the theoretical maximum NSEP by an N-stage PRO system is $N\left(1 - q_{tot}^{-1/N}\right)$. For one stage, NSEP is no more than 1.
- For infinite number of stages, the NSEP at the theoretical limit becomes $\ln q_{tot}$, which is equivalent to that of a fully reversible PRO.
- Process efficiencies have a significant effect on NSEP.

Acknowledgment and References

- Acknowledgement
  - American Chemical Society Petroleum Research Fund (No. 55347-UR9)
- References

  - Li, M. "Systematic Analysis and Optimization of Power Generation in Pressure Retarded Osmosis: Effect of Multi-Stage Design," *AIChE Journal*, in press, 2017.
  - Li, M. "Reducing Energy Consumption of Seawater Desalination: Staged RO or RO-PRO? " *Desalination*, **422**, 124-133, 2017.
  - Li, M. "Analysis and Optimization of Pressure Retarded Osmosis for Power Generation, " *AIChE Journal*, **61**, 1233-1241, 2015.

499267 Removal and Prevention of Biofouling in Forward Osmosis Membrane Bioreactors

Abstract:

With increasing populations and declining clean water supplies, treatment of municipal wastewater for reuse is becoming more viable. Membrane treatment processes will be an integral part in reuse treatment, however membrane fouling results in decreased process efficiencies. Forward osmosis (FO) is a membrane-based separation process that uses the osmotic pressure resulting from a salt gradient rather than a hydraulic pressure as the driving force, resulting in energy savings over traditional membrane treatment processes. The objectives of this project were to 1) design and build a submerged FO membrane bioreactor (MBR) with three independent membrane loops, and 2) to evaluate alternative membrane cleaning mechanisms to improve and maintain flux performance.

As part of the first objective, the hardware and programming logic code for the FO MBR system were developed and integrated to create a system that was able to measure and compare membrane flux over time for three independent FO membrane cassettes where all three were exposed to the same biological fouling conditions. The MBR contained *E. coli* JW3818 Î"*rfaH* grown in LB broth with 50 mg/L kanamyacin, and half strength LB with kanamycin was used to replace volume lost from the MBR through the FO membrane. As part of the second objective, various thermal osmotic backwashing methods were evaluated for flux recovery and propensity for re-fouling. It was found that osmotic backwashing at 70 Â° C for 30 min resulted in 15% greater flux recovery compared to the baseline case of 60 min at ambient temperature. During this process, the hot water removes the adhered biofilm allowing for increased flux recovery. It is hypothesized that the elevated temperature of the thermal backwash results in cell death at the membrane surface, thereby enhancing the osmotic backwashing efficiency. Confocal Scanning Laser Microscopy (CSLM) confirmed that less biofilm was present after thermal backwashing than with ambient backwashing. Based on these preliminary results, hot osmotic backwashing is a promising membrane cleaning technique to remove and decrease reoccurrence rate of biofouling in FO MBR operations.

502721 Understanding the Performance of Forward Osmosis Hollow Fiber Membranes at the Module Scale

Abstract:

This study evaluates an approach to make thin film composite (TFC) hollow fiber membranes for forward osmosis (FO) by simply employing commercial ultrafiltration (UF) membranes as support material. A thin polyamide film with excellent selective properties was synthesized on the inner surface of hollow fibers via interfacial polymerization. Besides demonstrating good performance as TFC FO membrane supports, the commercial hollow fiber UF membranes also provide a systematic platform with consistent properties to study structure-performance relationship. A series of commercial UF membranes were used to evaluate how support layer surface pore size (MWCO) impacted the properties of the selective layer and overall performance of the TFC membranes.

Aside from using commercial hollow fibers as FO membrane supports at bench scale, we also demonstrate that the TFC hollow fiber FO membranes can be made on existing modules at pilot scale. A series of commercial modules with different fiber size were used to make TFC hollow fiber modules. The resultant modules were evaluated under various operating conditions. While we evaluate how performance was impacted by module operating conditions, overall the modules demonstrate impressive FO results.

With the availability of reproducible membranes and modules, we developed a computational fluid dynamics (CFD) model with COMSOL Multiphysics to optimize module design for FO application. The FO process in a hollow fiber membrane was simulated in a 2D axis-symmetry geometry and described by fluid dynamics coupled with mass transfer. We verified the model accuracy by constructing modules with the same dimensions and fibers and testing under pertinent conditions. The agreement between the simulation and experimental results provided insight into how independent variables would impact module performance. To allow our colleagues to take the advantage of this model, we also built it into an application and will release it for the community to use for free.

497043 Understanding the Impact of Membrane Properties and Transport Phenomena on the Energy Efficiency of a Membrane Distillation Desalination System

Abstract:

Direct contact membrane distillation (DCMD) is a promising thermal desalination process that can treat high salinity waters using low-grade heat. DCMD has several advantages from the perspective of the water-energy nexus. These include the utilization of waste heat (below 100ºC), perfect rejection of nonvolatile solutes, low areal footprint, and high scalability. However, currently, the energy efficiency of DCMD is relatively low compared to other work-based and thermal desalination processes. In this study, we aim to quantify how membrane properties and process conditions affect the exergy or second-law efficiency (η_{II}) of a DCMD desalination system with external heat recovery. We begin by developing a fundamental understanding of mass and heat transport through an MD membrane, before modeling how membrane properties and transport phenomena affect entropic losses in an MD system. In particular, we analyze how the membrane permeability coefficient (B) and thermal conduction coefficient (K) impact MD performance. We show that increasing the B value of a membrane by reducing its thickness, initially leads to an increase in η_{II} before conductive heat loss through the membrane causes η_{II} to fall. For a typical MD membrane with a porosity of 0.90, material thermal conductivity of 0.20 W m^{-1} K^{-1}, and a nominal pore diameter of 0.6 μm, we find that the optimal permeability coefficient is 1.59×10^{-6} kg m^{-2} s^{-1} Pa^{-1} (572 kg m^{-2} h^{-1} bar^{-1}). This value corresponds to an optimal membrane thickness of around 95 μm. Our analysis stresses the importance of effective heat recovery in DCMD. We show that an external heat exchanger with a minimum approach temperature of 10ºC reduces energy consumption by 63%. Finally, we demonstrate that increasing the ratio B/K, rather than just the B value, is key to increasing the exergy efficiency of DCMD desalination. For example, increasing membrane porosity from 0.70 to 0.90, which yields a 160% increase in B/K, leads to a 42% increase in η_{II} from 5.3% to 7.6%. The advantages of reducing the bulk pressure (P) in the membrane pores are also explored. For a typical membrane, halving P from 1.0 bar to 0.5 bar, results in a 21% increase in η_{II} from 7.0% to 9.2%. We conclude by identifying that the maximum exergy efficiency achievable as membrane porosity tends to unity is 10% for a bulk membrane pressure of 1.0 bar and 12% for a bulk membrane pressure of 0.5 bar, given perfect heat recovery.

498547 Development of a Multi-Stage Membrane Distillation - Electrocoagulation Process for Treatment of Hydraulic Fracturing Flow Back Waters

Abstract:

There is an urgent need to develop innovative and more effective technologies for treatment of wastewaters. Many of these wastewaters such as hydraulic fracturing flow back waters, contain very high dissolved solids (TDS). Treatment of hydraulic fracturing flowback waters can be very challenging as they often contain TDS in excess of 200,000 ppm as well as dissolved surfactants and small organic compounds. Pressure driven membrane processes such as reverse osmosis are impractical for treating very high salinity wastewaters due to the high osmotic back pressure that must be overcome. Membrane distillation has been proposed as a new unit operation for treatment of very high TDS wastewaters. One of the advantages of membrane distillation is the fact that low grade waste heat may be used.

Here we have developed a multi-staged membrane distillation system for maximizing water recovery from actual hydraulic fracturing flowback waters. The multi-stage system has been designed to different on-site applications for wastewaters with specific TDS levels. We show that ideal membrane properties vary as the TDS of the fee increases above 300,000. At lower TDS values membranes should be selected that maximize flux. However at TDS values above 300,000, membrane surface properties such as roughness becomes important. At theses higher TDS values leakage of the feed through the membrane pores and into the permeate side is a major concern. In addition, appropriate pretreatment of the feed to remove dissolved organic compounds is essential prior to membrane distillation. Here we have developed an efficient electrocoagulation process for pretreatment of the hydraulic fracturing flow back waters.

Here different membranes have been used in the different stages and performance of the system in terms of water recovery potential and specific energy consumption has been evaluated. Significant concentration of the feed has been conducted and over 60% water recovery has been achieved for actual hydraulic fracturing flowback waters from Marcellus shale containing over 100,000 ppm TDS.

503181 Sustainability Metrics for Miscanthus in Building Materials

Abstract:

There are large areas of land that are contaminated from past military, mining, dumping, or other activities. One aspect of sustainable development is to improve soil quality and simultaneously produce a useful product. Miscanthus is a perennial grass that has large biomass yields, has a low nitrogen fertilizer requirement, and moderate water needs. Research is in progress to grow miscanthus on land contaminated by past military operations. One of the uses for miscanthus is to incorporate it into building materials such as composites and fiber board products. Sustainability metrics related to the process of producing miscanthus while improving soil quality and using it in building material products will be the focus of the presentation.

488970 Environmental Life Cycle Assessment of Peracetic Acid Application in the Pulp and Paper Industry

Abstract:

Environmental Life Cycle Assessment of Peracetic Acid Application in the Pulp and Paper Industry

Darlene Echeverria, Yuan Yao, Richard Venditti, Hasan Jameel

Department of Forest Biomaterials, NC State University.

Raleigh, NC 27606, USA

ABSTRACT

The United States is one of the largest producers of paper and paper products in the world. In 2014, the global pulp and paper industry produced 406.5 million tonnes of paper and paperboard, in which 20% were produced in the United States (RISI, 2015). The large-scale production of pulp and paper has traditionally generated a significant amount of wastes and emissions, and now the industry is one of the most environmentally regulated sectors in the United States (Bussing-Burks). Many pulp and paper mills have responded by investing in new technologies and methodologies that maintain profitability while concurrently reducing wastes and pollutants (Smook, 2016). One of the major improvements is the development of new bleaching technology to address the environmental concerns over elemental chlorine and its derivatives that are traditionally used as bleaching agents. Two bleaching technologies have been developed, Elemental Chlorine-Free (ECF) and Total Chlorine-Free (TCF). In 2012, 93% of the worldwide chemical pulp were produced by ECF that utilizes chlorine dioxide as the main bleaching agent, and about 5% of them were produced by TCF that uses oxygen, ozone, hydrogen peroxide and peracetic acids (PAA) as bleaching agents (AET, 2012; Pia Jour, 2015). Compared to other agents, PAA does not produce toxic by-products and has high selectivity in removing lignin and maintaining cellulose intact. Although PAA has been considered as an environmentally benign bleaching agent, few studies have quantified the environmental benefits or total environmental impacts of PAA bleaching process.

In this work, a comparative cradle-to-gate Life Cycle Assessment (LCA) is conducted to quantify the potential environmental benefits of PAA compared to traditional bleaching agents in the pulp and paper production. Different pathways of PAA production are included in this study to understand the variances led by different feedstocks. A techno-economic analysis is performed to define the economics of PAA application and integrated with LCA results to understand the trade-offs between environmental benefits and economic viability. Sensitivity analysis is used to test the robustness of the results and quantify the impacts of system variances and uncertainties on the comparative results of PAA and traditional bleaching agents. The results of this study would be beneficial to inform decision-making towards more sustainable PAA production and pulp bleaching. From an end-user perspective, quantitative understandings of potential environmental benefits brought by PAA could foster partnerships among PAA suppliers and the pulp and paper industry to develop more sustainable production and supply chains. In addition, the results of this work may shed light on the potential and practical applications for PAA bleaching in other industries for reducing their environmental impacts.

References

AET (Alliance for Environmental Technology) (2012). Trends in World Bleached Chemical Pulp Production: 1990-2012. Retrieved from: http://www.aet.org/science_of_ecf/eco_risk/2013_pulp.html

Bussing-Burks, M. The National Bureau of Economic Research. Retrieved from: http://www.nber.org/: http://www.nber.org/digest/feb03/w9125.html

Pia Jour, K. H. (2015). Life cycle assessment of ECF bleaching sequences with focus on carbon footprint. Sweden: TAPPI Journal.

RISI | Pulp and Paper Industry Intelligence (2015). Retrieved from: http://www.risiinfo.com/: http://www.risiinfo.com/press-release/global-production-of-paper-and-board-hit-record-levels-in-2014/

Smook, G. A. (2016). Handbook For Pulp & Paper Technologists. Peachtree Corners: TAPPI Press.

492692 Evaluating Indicators and Life Cycle Inventories for Processes in Early Stages of Technical Readiness

Abstract:

This presentation examines different methods for analyzing manufacturing processes in the early stages of technical readiness. Before developers know much detail about their processes, it is valuable to apply various assessments to evaluate their performance. One type of assessment evaluates performance indicators to describe how closely processes approach desirable objectives. Another type of assessment determines the life cycle inventories (LCI) of inputs and outputs for processes, where for a functional unit of product, the user evaluates the resources used and the releases to the environment. These results can be compared to similar processes or combined with the LCI of other processes to examine up- and down-stream chemicals. The inventory also provides a listing of the up-stream chemicals, which permits study of the whole life cycle.Â

Performance indicators are evaluated in this presentation with the U.S. Environmental Protection Agencyâ€™s GREENSCOPE (Gauging Reaction Effectiveness for ENvironmental Sustainability with a multi-Objective Process Evaluator) methodology, which evaluates processes in four areas: Environment, Energy, Economics, and Efficiency. The method develops relative scores for indicators that allow comparisons across various technologies. In this contribution, two conversion pathways for producing cellulosic ethanol from biomass, via thermochemical and biochemical routes, are studied. The information developed from the indicators and LCI can be used to inform the process design and the potential life cycle effects of up- and down-stream chemicals.

The views expressed in this presentation are those of the authors and do not necessarily represent the views or policies of the U.S. Environmental Protection Agency or U.S. Department of Energy.

497807 Estimating the Life Cycle Impact of Chemicals from Molecular, Thermodynamic and Charge Density Information Via Mixed-Integer Optimization Techniques

Abstract:

The chemical industry is at present striving to decrease its environmental footprint in the transition towards a more sustainable manufacturing sector. In this context, systematic methods are required to assist in the environmental assessment and optimisation of chemicals, identification of critical hotspots across products' supply chains and definition of guidelines to effectively retrofit processes so that they adhere to sustainability principles.

The amount of new chemicals produced annually has exponentially increased in the last decades. The database CAS registry[1] from Chemical Abstract Service, a division of the American Chemical Society devoted to authoritatively collect disclosed chemical substance information, presently contains more than 120 million organic and inorganic substances and 66 million sequences, with approximately 15,000 new substances being added each day. Hence, an enormous amount of chemical species exists, but yet there is a lack of information regarding their chemical hazards and potential environmental impact. This information, however, is critical for the proper evaluation of their wider sustainability impacts and future market feasibility.

Under the above scenario, the recent trend towards the development of more sustainable products has led to a plethora of environmental assessment tools. From these, Life Cycle Assessment (LCA) has gained popularity in recent years, currently being one of the most extended sustainability assessment methods. LCA is based on the evaluation, for all stages of the life cycle of a product, of all possible interactions of the activities carried out and the environment.

Unfortunately, LCA requires large amounts of data across the products' supply chain, some of which might be hard to gather in practice. This constitutes one of its main obstacles, and more so in the chemical industry, which contains many complex networks that exchange mass and energy. In these cases, when a full LCA cannot be applied, a simplified version is used instead. These Streamlined LCA (SLCA) approaches follow the same basis as LCA, but generally either simplify the scope of the analysis and/or reduce the amount of information required in the assessment. The precise simplifications to be done (and the assessment discrepancy with the full LCA) have to be specifically considered for the process or activity assessed.

Under these principles, we present herein a novel SLCA approach for the estimation of the impact of chemicals that make use of a set of specific molecular properties. Previous studies demonstrated the prediction capabilities of molecular and thermodynamic attributes[2]. Here we extend these works by considering the σ-profile of chemicals as attributes. The sigma profile provides the probability distribution of a molecular surface segment having a specific charge density. In this work, we discretize this profile and use it together with molecular and thermodynamic descriptors to estimate eight well known LCA indicators, including the cumulative energy demand (CED), Eco-indicator 99 and global warming potential, among others. The regression models are built using mixed-integer optimization techniques that automatically identify the attributes that best describe the specific impact categories.

We applied our approach to a database consisting of 83 chemicals and considering 15 molecular descriptors, 17 thermodynamic attributes and 8 σ-profile-based descriptors, generating estimates of enough accuracy for the purpose of a standard LCA (i.e. 15-25%). Our framework ultimately leads to simple linear models that can be implemented in computer aided molecular design software, thereby enhancing the application of sustainability principles in the early stages of the development of new chemicals.

References

(1) http://www.cas.org/. Chemical Abstracts Service Home Page http://www.cas.org/ (accessed Oct 18, 2016).

Chemical Inventories using Neural Networks. *Environ. Sci. Technol.* 2008, *42* (17), 6717–6722.

AUTHOR INDEX

AUTHOR INDEX

AUTHOR INDEX

AUTHOR INDEX

AUTHOR INDEX

AUTHOR INDEX

AIChE
120 Wall Street, FL 23
New York, NY 10005-4020

ISBN 978-1-5108-5798-8

CURRAN ASSOCIATES INC.
proceedings.com